WATER FOR FOOD SECURITY AND WELL-BEING IN LATIN AMERICA AND THE CARIBBEAN

This volume provides an analytical and facts-based overview of the progress achieved in water security in Latin America and the Caribbean (LAC) region during the last decade, and its links to regional development, food security and human well-being. Although the book takes a regional approach, covering a vast amount of data pertaining to most of the LAC region, some chapters focus on seven countries (Argentina, Brazil, Chile, Colombia, Costa Rica, Mexico and Peru).

A full understanding of LAC's progress requires framing this region in the global context: an ever more globalized world where LAC has increasing geopolitical power and a growing presence in international food markets. The book's specific objectives are: (1) exploring the improvements and links between water and food security in LAC countries; (2) assessing the role of the socio-economic 'megatrends' in LAC, identifying feedback processes between the region's observed pattern of changes regarding key biophysical, economic and social variables linked to water and food security; and (3) reviewing the critical changes that are taking place in the institutional and governance water spheres, including the role of civil society, which may represent a promising means to advancing towards the goal of improving water security in LAC.

The resulting picture shows a region where recent socioeconomic development has led to important advances in the domains of food and water security. Economic growth in LAC and its increasingly important role in international trade are intense in terms of use of natural resources such as land, water and energy. This poses new and important challenges for sustainable development. The reinforcement of national and global governance schemes and their alignment on the improvement of human well-being is and will remain an inescapable prerequisite to the achievement of long-lasting security. Supporting this bold idea with facts and science-based conclusions is the ultimate goal of the book.

Bárbara A. Willaarts is a Researcher at the Water Observatory – Botín Foundation and a Postdoctoral Researcher at the Research Centre for the Management of Agricultural and Environmental Risks (CEIGRAM), Technical University of Madrid, Spain.

Alberto Garrido is a Professor in Agricultural Economics, Technical University of Madrid, Director of the Research Centre for the Management of Agricultural and Environmental Risks (CEIGRAM) and Deputy Director of the Water Observatory – Botín Foundation, Madrid, Spain.

M. Ramón Llamas is the Director of the Water Observatory – Botín Foundation and Emeritus Professor at Complutense University, Madrid, Spain.

Authors of the Water Observatory – Botín Foundation:

Bárbara A. Willaarts
Alberto Garrido
Maite M. Aldaya
Lucia De Stefano
Elena López-Gunn
Pedro Martínez Santos
Emilio Custodio
Enrique Cabrera
Fermin Villarroya
Daniel Chico
Aurélien Dumont
Insa Flachsbarth
Marta Rica
Gloria Salmoral

Latin American team leaders and authors:

Pedro R. Jacobi and Vanessa Empinotti (PROCAM /IEE Universidade de São Paulo, Brazil)
Rosario Pérez Espejo (Universidad Nacional Autónoma de México)
Guillermo Donoso (Pontificia Universidad Católica de Chile)
Diego Arévalo (Centro de Ciencia y Tecnología de Antioquia, Colombia)
Julio Kuroiwa (Universidad Nacional de Ingeniería, Peru)
Patricia Phumpiu (Centro del Agua para América Latina y el Caribe, Mexico)
María Josefa Fioriti (Subsecretaría de Recursos Hídricos – Ministerio de Planificación, Argentina)
Andrea Suárez (Universidad Nacional de Costa Rica)

EARTSHCAN – Fundación Botín, Santander (Spain)
London-Sterling, VA

Language editor:

Ruth Cunningham

Editorial Assistants:

Daniel del Olmo Rovidarcht
Olga Fedorova
Desireé Torrente

Designer:

María Carmona www.cedecarmona.com

WATER FOR FOOD SECURITY AND WELL-BEING IN LATIN AMERICA AND THE CARIBBEAN
Social and Environmental Implications for a Globalized Economy

Edited by Bárbara A. Willaarts, Alberto Garrido and M. Ramón Llamas

LONDON AND NEW YORK

First published 2014
by Routledge
2 Park Square, Milton Park, Abingdon, Oxon OX14 4RN

And by Routledge
711 Third Avenue, New York, NY 10017

First issued in paperback 2018

Routledge is an imprint of the Taylor and Francis Group, an informa business

British Library Cataloguing-in-Publication Data
A catalogue record for this book is available from the British Library

Library of Congress Cataloging-in-Publication Data
Water for food security and well-being in Latin America and the Caribbean:
social and environmental implications for a globalized economy / edited
by Bárbara A. Willaarts, Alberto Garrido, M. Ramón Llamas.
pages cm
Includes bibliographical references and index.
TD227.5.W38 2014
338.1'98–dc23
2013043668

ISBN 13: 978-1-138-61823-7 (pbk)
ISBN 13: 978-0-415-71368-9 (hbk)

Typeset in Futura Std light
by M. Carmona

Publisher's note
This book has been prepared from camera-ready copy provided by the authors.

Contents

List of figures   VIII
List of tables   XII
List of contributors   XIV
Foreword   XVI
Preface   XVII
Acknowledgements   XX
List of abbreviations   XXI

Part 1 Introduction

Chapter 1. Water and food security in Latin America and the Caribbean: 3
            regional opportunities to cope with global challenges

Part 2 Setting the scene

Chapter 2. Water resources assessment   27
Chapter 3. Trends in land use and ecosystem services in Latin America   55
Chapter 4. Socio-economic megatrends for water and food security in Latin America   81
Chapter 5. Globalization and trade   119
Chapter 6. Tracking progress and links between water and food security in Latin   143
            America and the Caribbean

Part 3 Water for food and non-food

Chapter 7. Water and agriculture   177
Chapter 8. Water security and cities   213
Chapter 9. Water, energy, bioenergy, industry and mining   239

Part 4 Economic, legal and institutional factors for achieving water and food security

Chapter 10. Water efficiency: status and trends   261
Chapter 11. Reforming water governance structures   285
Chapter 12. The role of stakeholders in water management   317
Chapter 13. Economic instruments for allocating water and financing services   343
Chapter 14. Legal framework and economic incentives for managing ecosystem   365
            services
Chapter 15. Rethinking integrated water resources management: towards water   385
            and food security through adaptive management

Index   419

Figures

1.1	Biophysical dimensions of human well-being – water and food security – in LAC and in the rest of the world	8
1.2	The book's framework: topics, inter-dependencies, drivers and focus	9
2.1	Long-term annual rainfall in selected Latin American countries	31
2.2	Renewable resources per capita over the last twenty years in selected countries	32
2.3	Regional rainfall variability in Chile	33
2.4	(A) Total rainfall (1961–1990) and (B) Water use across the world	37
2.5	Water withdrawals per sector in the Latin American region	38
2.6	Water Pollution Level for nitrogen (N) per river basin in Latin America (year 2000)	44
2.7	Water Pollution Level for phosphorus (P) per river basin in Latin America (year 2000)	44
2.8	Observed (left) and expected (right) impacts linked to Climate Change in Latin America	49
3.1	Land use in Latin America and the Caribbean (LAC) in 1990 and 2010 (in million hectares	58
3.2	Land use and land cover changes occurred in Latin America and the Caribbean between 1993 and 2009	61
3.3	Evidence of forest transition in São Paulo State (Brazil) according to four different data sources	63
3.4	Factor analysis explaining drivers of forest area change in Latin America and the Caribbean between 1990 and 2010	66
3.5	Trends in Ecosystem Service provision in Latin America and the Caribbean between 1990 and 2010	69
3.6	Greenhouse Gas Emission (GHG) inventory by sector in LAC countries	70
3.7	Annual growth rates of agricultural land, yields and net production value	75
4.1	Trends in urban population between 1950 and 2000	86
4.2	Population living in slums and population with access to piped water	87
4.3	Annual rate of urban–rural population change (%)	89
4.4	Comparative Evolution of GDP per Capita (GDP, logarithmic scale). (Historic and Projections: 1800–2030)	91
4.5	Comparative Evolution of GDP per Capita (GDP, logarithmic icab). (Historic and Projections: 1800–2030)	91
4.6	Population growth in LAC (1990–2000), Water consumption in LAC (1990–2000) and Evolution of GDP (1990–2000)	92
4.7	Annual freshwater withdrawals per capita vs GDP per capita (1977–2011)	93
4.8	Percentage of population below poverty line	96
4.9	Percentage of population below indigence line	96
4.10	Water Poverty Index in LAC countries	98
4.11	Annual GDP per capita growth (expressed in current US$) for the time period 1980–2010	98
4.12a	Inequality in income distribution. Percentage of the income share held by highest 20% subgroup of population	99
4.12b	Inequality in income distribution. Percentage of the income share held by lowest 20% subgroup of population	99
4.13	Informal employment and the informal economy as part of GDP in LAC	100
4.14	Mexican food consumption pattern. Quantity, energy, water footprint of main food products. 1992 and 2010	103

4.15	Development scenarios for Latin America in 2030	107
4.16	Trends in entrepreneurship and access to information and ICT	107
4.17	Net enrolment rate in first-level education	109
4.18	Net enrolment rate in second-level education	109
4.19	Gross enrolment rate in third-level education	109
4.20	Type of natural hazards and population affected in selected countries in Latin America	112
5.1	Commodities price indices (1960–2011)	123
5.2	Trade as a share of gross domestic product (GDP) (1961–2011)	123
5.3	Inward foreign direct investment flows, annual, 1970–2011(in million US$)	124
5.4	Official development assistance in agriculture & infrastructure by area, in 1995, 2002 and 2009	124
5.5	Private participation in infrastructure by area in 1995, 2002 and 2009	125
5.6	Value of imports and exports of LAC between 1992 and 2011 expressed in nominal US dollars	127
5.7	Breakup of exports from Latin America and the Caribbean to different world regions in 2000, 2005 and 2011 (%)	128
5.8	Trade agreements in the LA region	130
5.9	Trade partners in the LA region	130
5.10	Changes in extensive and intensive trade margins in the LA region	131
5.11	Trade and agricultural growth nexus in five LAC countries (1995–2010)	136
5.12	Trade and income of the poorest decile in five LAC countries (1996–2010)	137
5.13	Trade and poverty rates in five LAC countries (1996–2010)	137
6.1	Indicators and operational frameworks for measuring water security	149
6.2	Existing food and nutrition indicators	152
6.3	Blue water scarcity and population distribution estimates for 2010 in Latin American	153
6.4	Percentage of population with access to drinking water and sanitation coverage in urnban (left) and rural (right) areas in LAC	155
6.5	Irrigation efficiency (measured in terms of water requirement ratios) for Latin American countries, average for the period 1990–2012	155
6.6	Allocation of public investments in water supply and sanitation in LAC, 2000–2010	157
6.7	Economic losses (expressed in % of annual GDP, bars) attributed to water-related hazards (storms, floods and droughts) and GDP evolution (in USD, line) in Latin America and the Caribbean, 1980–2012	157
6.8	Water security performance in LAC countries	158
6.9	Percentage of undernourished and overweight children under five years old (2000–2009)	164
6.10	Food consumption pyramids (in consumed kg per capita per year) for Brazilians during the last two decades	166
6.11	Three pairs of water and food security indicators measured in 2000 and 2010 (countries of the first quartile of per capita income in 2010)	167
6.12	Three pairs of water and food security indicators measured in 2000 and 2010 (countries of the second quartile of per capita income in 2010)	167
6.13	Three pairs of water and food security indicators measured in 2000 and 2010 (countries of the third quartile of per capita income in 2010)	168
6.14	Three pairs of water and food security indicators measured in 2000 and 2010 (countries of the fourth quartile of per capita income in 2010)	168

7.1   Green and blue water footprint (in cubic gigametres per year) of agricultural   183
        production for the LAC region (average 1996–2005)
7.2   Distribution of the agricultural green and blue water footprint (in cubic hectometres   185
        per year) of Mexico, Brazil, Argentina, Peru and Chile (average for the years
        1996–2005)
7.3   Water footprint (in cubic metres per inhabitant per year) of the consumption of   187
        agricultural products (green and blue) in the LAC region (average 1996–2005)
7.4   Composition of the agricultural grey water footprint (in cubic hectometres per year)   190
        by crops in Brazil, Mexico, Argentina, Chile, Colombia and Peru
7.5   Largest total (green and blue) net virtual water importers and blue net virtual water   192
        importers (in cubic Gigametres per year) of agricultural products in the LAC region
        (average 1996–2005)
7.6   Green (above) and blue (below) virtual water exports (in million cubic metres) per   194
        country and main products (1996–2009)
7.7   Green (above) and blue (below) virtual water imports (in million cubic metres) per   196
        country and main products (1996–2009)
7.8   Blue and green virtual water exports and imports (in million cubic metres) between   199
        1996 and 2010 in LAC
7.9   Compound growth rate (%) of land physical (t/ha) and economic productivity ($/   204
        ha) between av. 1991–1993 and av. 2008–2010 for selected countries and
        crops
7.10   Average cultivated area (1,000ha/yr), economic water productivity (US$/m³)   206
        and share of blue WF in crop WF for selected countries and crops. The data
        shown corresponds to an average of the years 2007–2010
8.1   Median age of population and of water pipes in the USA   225
8.2   Access to water and sanitation (% population) and child mortality (deaths per   226
        1,000 born) in different American Countries
8.3   Water network losses in representative urban areas in LA   228
10.1   The relation between the blue water footprint of production (upper) and consumption   263
        (lower) and the level of economic development
10.2   Global irrigation efficiencies, year 2000   271
10.3   The water footprint of national production in LAC (Mm³/yr). Period 1996–2005   272
10.4   Water footprint of domestic water supply by national production (Mm³/yr). Period   273
        1996–2005
10.5   Water footprint of industrial production (Mm³/yr). Period 1996–2005   273
10.6   Total water footprint of agricultural crop production for the LAC region (average   274
        1996-2005)
10.7   Water footprint of livestock production (Mm³/yr). Period 1996–2005   275
10.8   Economic water productivity (US$/m³) in agriculture and industry in LAC countries   276
        (2011)
10.9   Economic water efficiency of industrial production for the LAC region (average   277
        1996–2005) (US$/m³)
10.10   Economic water efficiency of agricultural production for the LAC region (average   277
        1996–2005) (US$/m³)
11.1   Timeline of the approval of the Water Act, domestic supply legislation and specific   295
        groundwater law in selected LAC countries
11.2   Growth of population and water well drilling in Guanajuato State   298
11.3   Timeline: international legal and political recognition of the human right to safe   299
        water and sanitation

11.4	Map on voting for UN General Assembly resolution recognizing the human right to safe drinking water and sanitation	300
11.5	Map on inclusion of Human Right to safe drinking water and sanitation (HRWS) in constitutions	301
11.6	Map with examples of the implementation of the Human Right to Water and Sanitation	304
11.7	Water-related expenditures that need to be funded and sources of incomes in LAC countries	306
11.8	Evolution of international public and private funding to the Latin American water sector over the period 2001–2011	309
11.9	Evolution of international public investment during the period 2001–2011	310
11.10	Global and regional private investment in the water sector	310
11.11	Geographical distribution of investments with private participation in the watersector during the period 2001–2011	311
		323
12.1	Location of mining conflicts in LAC	334
12.2	Timing of approval of information transparency law in LAC	340
14.1	Watershed PES trends in the Latin America region	370
14.2	Constitutional recognition of the right to a clean environment in LA	375
14.3	The percentage of national territory covered by cadastre survey	378
15.1	The 'anthropo-hygeodrogeological' cycle	389
15.2	Population and areas most affected by droughts and floods in the Andean Community	392
15.3	Population and areas most affected by droughts and floods in the Andean Community and Peru	392
15.4	Understanding the nexus. The water, energy and food nexus	395
15.5	Water footprint of electricity production in Latin America	396
15.6	Electricity generation by source and per sub-region (Southern Cone, Mesoamerican, Amazon and Andean) in Latin America	398
15.7	Water consumption and water use for electric generation per sub-region (Southern Cone, Mesoamerican, Amazon and Andean) in Latin America	399
15.8	United Nations Human Development Index versus Carbon Footprint (tons C per capita per year), Water Footprint (cubic metres per capita per year) and Ecological Footprint (global hectares per capita per year)	401
15.9a	Multi-level governance gaps in LAC countries' water policymaking	404
15.9b	Multi-level governance gaps in LAC countries' water policymaking	405
15.10	Preliminary categories of LAC countries	407
15.11	Venn Diagram of dominant, outcast and respected actors in Costa Rica's water management	408
15.12	Social networks of actors in Costa Rica: connections, level of centrality and ease of access	408
15.13	The WRM cycle to achieve water security	413

Tables

1.1	Millennium Development Goals (MDG) progress in Latin America and the Caribbean between 1990 and 2010	17
2.1	Approximate amount of annual precipitation, evaporation and runoff per continent in relation to the water footprint	29
2.2	Renewable water resources and storage capacity in selected countries in Latin America	30
2.3	Water availability in Peru's hydrographic regions	34
2.4	Blue and green water footprint of countries in the Latin America and Caribbean region (those with more than one million inhabitants)	36
2.5	Distribution of water responsibilities in selected countries	47
3.1	Deforestation rates across Latin America between 1990 and 2010	59
3.2	Trends of native and non-native agricultural crops cultivated in Latin America	72
3.3	Changes in ecosystem service supply (expressed in percentage) across Latin America and the Caribbean between 1990 and 2010	73
4.1	Evolution of urban population, percentage living in urban areas by region (1925–2000)	86
4.2	Real per capita income growth 1960–2010	91
4.3	Comparative best and worst cases (or international indexes for the year 2010 using the STEEP (Society–Technology–Economics–Ecology–Politics) approach	94
4.4	Trends of human wellbeing across different regions of LAC in the last two decades	95
4.5	Degree of merchandse trade in LAC	102
4.6	Social vulnerability assessment to climate change in Latin America	111
5.1	Percentage of GDP and population of each region with respect to the world	122
5.2	LAC's busiest ports in thousands of TEUs (twenty-foot equivalent units) in 2011	123
5.3	Participation of LAC in world agricultural trade in dollar terms	125
6.1	Human well- being dimensions considered under different approaches to water security	147
6.2	Evolving definition and scope of the food security concept	150
6.3	Water security progress between 2000 and 2010 in LAC	156
6.4	Food and nutritional security indicators selected to assess Food and Nutritional Security (FNS) performance in Latin America and the Caribbean (LAC)	159
6.5	Food security progress between 2000 and 2010 in LAC	161
6.6	Percentage of people suffering from hunger	162
6.7	External dependencies of wheat and maize in LAC, (average 2007/2008 and 2011/2012)	163
7.1	Irrigation techniques in the LAC region	183
7.2	Evolution of the arable land in Latin American and Caribbean countries, for the years 1995, 2002 and 2011	201
7.3	Yield compound annual growth rate by crop and country	202
7.4	Impact of irrigation by type of system	207
7.5	The green, blue and grey water footprint in the Porce River Basin	209
8.1	Data on some of the largest cities in LA.	215
9.1	General data for the first decade of the 21st century (values rounded up)	243
9.2	Economic productivity of used water in Chile	243
9.3	Energy and water in Latin America and the Caribbean (LAC) in 2005	244
9.4	Average water consumption rates for thermoelectric plants with closed cooling	246

9.5	Geothermal energy in several LA countries	247
9.6	Sugar–cane production and crop area	248
9.7	Approximate costs of producing bio-alcohol and comparative cost of oil	248
9.8	Water needs for fuel production, including processing	250
9.9	Industrial water use in Mexico for the main water-intensive sectors	250
9.10	Current water consumption in mining (values rounded up)	252
11.1	Allocation of responsibilities in water governance at sub-national level and the role of the central government in selected LAC countries	290
11.2	Main challenges in water policy making and their relative importance in selected LAC countries	293
11.3	Ownership of water in selected LAC countries	296
11.4	State recognition of the human right to safe drinking water and sanitation (HRWS) in national constitutions, laws and policies in selected LAC countries	302
12.1	Features of main water conflicts in LAC	321
12.2	Comparative overview of participatory levels in selected LAC countries	326
12.3	Latin American companies involved in water networks and initiatives on water accounting tools	332
12.4	Strengths and weaknesses of the existing transparency laws in several LAC countries	335
12.5	Examples of benchmarking initiatives of water and sanitation utilities companies	336
12.6	Online availability of information about selected issues in five LAC countries	337
13.1	Levies for water use for different zones in Mexico, 2010 (US$ cents per m³, exchange rate Mexican peso /US$ of 2010)	346
13.2	Average monthly bill and average price in the main fourteen water utilities in LA	351
13.3	Payment schemes for watershed protection and water-related ecosystem services in LAC	356
13.4	Main characteristics of water-related PES programs in LAC	357
13.5	WR transactions and prices for the period 2005–2008	359
14.1	Overview of payments for ecosystem services (PES) initiatives found across Latin America and the Caribbean	369
14.2	Main characteristics of water- related payments for ecosystem services programmes	371
14.3	Legal frameworks supporting ecosystem services directly or indirectly	376
14.4	Summary of the advantages and disadvantages of having legal regulation for ecosystem services payment schemes	377
15.1	Comparative features of different components of water resources portfolios	390
15.2	Population prone to suffering droughts and floods in the Andean community countries	392
15.3	Total water footprint and total virtual water flows in Latin American countries	394
15.4	United Nations Human Development Index versus (per capita) Carbon Footprint (CF), Water Footprint (WF) and Ecological Footprint (EF)	402
15.5	Ministries and institutions responsible for the management of water, energy and food resources in different Latin American countries	406

Contributors

Ramón	AGUIRRE	Sistema de Aguas de la Ciudad de México, México
Aziza	AKHMOUCH	Organisation for Economic Co-operation and Development (OECD), Paris, France
Maite M.	ALDAYA	Water Observatory – Botín Foundation, and Complutense University of Madrid, Spain
Virginia	ALONSO DE LINAJE	Universidad Complutense de Madrid, Spain
Diego	ARÉVALO URIBE	Water Management and Footprint. CTA – Centro de Ciencia y Tecnología de Antioquia, Colombia
Pedro	ARROJO AGUDO	Universidad de Zaragoza, and Fundación Nueva Cultura del Agua, Spain
Maureen	BALLESTERO	Global Water Partnership, Costa Rica
Manuel	BEA	Geosys S.L.,Spain
Elisa	BLANCO	Pontificia Universidad Católica de Chile, Santiago, Chile
Emilia	BOCANEGRA	Universidad Nacional de Mar del Plata, Argentina
Wilson	CABRAL DE SOUSA Jr	Aeronautics Technology Institute, São José dos Campos, Brazil
Enrique	CABRERA	ITA, Universitat Politècnica de València (UPV), Spain
Xueliang	CAI	International Water Management Institute (IWMI), Pretoria, South Africa
Claudia	CAMPUZANO	Centro de Ciencia y Tecnología de Antioquia, Colombia
Gerson	CARDOSO DA SILVA Jr	Universidade Federal do Rio de Janeiro, Brazil
Luis F.	CASTRO	School of Civil Engineering, Universidad Nacional de Ingenieria, Lima, Peru
Manuel	CERMERÓN	Aqualogy, Barcelona, Spain
Daniel	CHICO	Water Observatory – Botín Foundation, and CEIGRAM, Technical University of Madrid, Spain
Vanessa	CORDERO	CEIGRAM, Technical University of Madrid, Spain
Emilio	CUSTODIO	Dept. Geo-Engineering, Universitat Politècnica de Catalunya (UPC), Barcelona, Spain
Javier	DÁVARA	Aqualogy - SEDAPAL, Peru
Gabriela	DE LA MORA	Instituto de Investigaciones Sociales, Mexico
Angel	DE MIGUEL	IMDEA Agua – Madrid Institute for Advanced Studies, Madrid, Spain
Lucia	DE STEFANO	Water Observatory – Botín Foundation, and Universidad Complutense de Madrid, Spain
Gonzalo	DELACÁMARA	IMDEA Agua – Madrid Institute for Advanced Studies, Madrid, Spain
Guillermo	DONOSO	Pontificia Universidad Católica de Chile, Santiago, Chile
Aurélien	DUMONT	Complutense University of Madrid, Water Observatory-Botín Foundation, Spain
Marta	ECHAVARRÍA	Ecodecisión, Colombia
Antonio	EMBID IRUJO	Universidad de Zaragoza, Spain
Vanessa	EMPINOTTI	PROCAM /IEE Universidade de São Paulo, Brazil
Juliana S.	FARINACI	Environmental Studies Center (NEPAM) - State University of Campinas, (UNICAMP), Brazil
Olga	FEDOROVA	CEIGRAM, Technical University of Madrid, Spain
Maria Josefa	FIORITI	Subsecretaría de Recursos Hídricos, Ministerio de Planificación, Argentina
Insa	FLACHSBARTH	Water Observatory – Botín Foundation, and CEIGRAM, Technical University of Madrid, Spain
Gabriela	FRANCO	Pontificia Universidad Católica de Chile, Santiago, Chile
Alberto	GARRIDO	Water Observatory – Botín Foundation, and CEIGRAM, Technical University of Madrid, Spain
Luis	GUROVICH	Pontificia Universidad Católica de Chile, Santiago, Chile
Anne M.	HANSEN	Instituto Mexicano de Tecnología del Agua, México
Thalia	HERÁNDEZ- AMEZCUA	Universidad Nacional Autónoma de México, México

Ricardo	HIRATA	Universidad de Sao Paulo, Brazil
Pedro Roberto	JACOBI	PROCAM /IEE Universidade de São Paulo, Brazil
Alejandro	JIMÉNEZ	Stockholm International Water Institute, Sweden
Julio M.	KUROIWA	Laboratorio Nacional de Hidráulica- Universidad Nacional de Ingeniería, Lima, Peru
Jonathan	LAUTZE	International Water Management Institute (IWMI), South Africa
Ramón	LLAMAS	Water Observatory – Botín Foundation, and Complutense University of Madrid, Spain
Elena	LÓPEZ-GUNN	I-Catalist, Complutense University of Madrid, and Water Observatory – Botín Foundation, Spain
Marielena N.	LUCEN	Ministry of Energy and Mines, Peru
Gonzalo	MARÍN	Fundación Canal de Isabel II, Madrid, Spain
Joaquim	MARTÍ	Aguas Andinas, Chile
Pedro	MARTINEZ SANTOS	Universidad Complutense de Madrid, and Water Observatory – Botín Foundation, Spain
Ariosto	MATUS PEREZ	Universidad Iberoamericana, México DF, Mexico
Beatriz	MAYOR	Universidad Complutense de Madrid, Spain
Mesfin	MEKONNEN	University of Twente, The Netherlands
Oscar	MELO	Pontificia Universidad Católica de Chile, Santiago, Chile
Marcela	MOLANO	CEIGRAM, Technical University of Madrid, Spain
Julio I.	MONTENEGRO	School of Civil Engineering,Universidad Nacional de Ingenieria,Lima, Peru
Markus	PAHLOW	University of Twente, The Netherlands
Ignacio	PARDO	Universidad de la Republica, Montevideo, Uruguay
Julio Cesar	PASCALE PALHARES	Embrapa Cattle Southeast, São Carlos, Brazil
Lorena	PEREZ	I-CATALIST, Uiversidad Complutense de Madrid, Spain
Rosario	PÉREZ-ESPEJO	Universidad Nacional Autónoma de México, México
Patricia	PHUMPIU CHANG	Centro del Agua para América Latina y el Caribe – ITESM, Monterrey, Mexico
Marta	RICA	Water Observatory-Botín Foundation, and Universidad Complutense de Madrid, Spain
Gloria	SALMORAL	Water Observatory-Botín Foundation, and CEIGRAM, Technical University of Madrid, Spain
Andrea	SANTOS	Universidade Federal Fluminense, Rio de Janeiro, Brazil
Maria José	SANZ-SÁNCHEZ	Food and Agriculture Organization (FAO), Rome, Italy
Christopher	SCOTT	University of Arizona, Tucson, USA
Miguel	SOLANES	IMDEA Agua- Madrid Institute for Advanced Studies, Spain
Bárbara	SORIANO	CEIGRAM, Technical University of Madrid, Spain
Ursula Oswald	SPRING	Centro Regional de Investigaciones Multidisciplinarias, Universidad Nacional Autónoma México
Laurens	THUY	Utrecht University, Utrecht, The Netherlands
Desiree	TORRENTE	CEIGRAM, Technical University of Madrid, Spain
Roberto C.	TOTO	Universidad Nacional Autónoma de México, México
Natalia	URIBE	WaterLex, Switzerland
Fermín	VILLARROYA	Universidad Complutense de Madrid, Spain
Bárbara	WILLAARTS	Water Observatory-Botín Foundation, and CEIGRAM, Technical University of Madrid, Spain
Erika	ZARATE	Good Stuff International, Switzerland
Guoping	ZHANG	Water Footprint Network, The Netherlands
Pedro	ZORRILLA-MIRAS	Cooperativa Terrativa, Madrid, Spain
Ibon	ZUGASTI	Prospektiker, Spain

Foreword

In 1964 Marcelino Botín Sanz de Sautuola and his wife Carmen Yllera, founded the Marcelino Botín Foundation to promote social development in Cantabria, in the north of Spain. Today the Foundation, faithful to the spirit of its founders and after nearly 50 years of work, is Spain's number one private foundation both in terms of the investment capacity and social impact of its programmes.

The Botín Foundation's objective is to stimulate the economic, social and cultural development of society. To achieve this, it acts in the fields of art and culture, education, science and rural development, supporting creative, progress-making talent and exploring new ways of generating wealth. Its sphere of action focuses primarily on Spain and especially on the region of Cantabria, but also on Latin America. The main office is located in the city of Santander, the capital of Cantabria, in what used to be the Sanz de Sautuola family's house. Its exhibition room is located nearby. Two of the city's emblematic buildings, El Promontorio and Villa Iris, are used for official ceremonies, exhibitions and workshops. The Casa Rectoral in Puente Pumar is the Foundation's centre of operations in the Nansa Valley (Cantabria).

Since the end of 2012 a refurbished, former industrial building in the centre of Madrid houses the Foundation's offices in the capital. In 2014, coinciding with its 50th anniversary, the new Botín Centre will be opened in Santander.

In 2008 the Board of the Botín Foundation decided to create a section devoted to water resources within its Trends Observatory, under the title of Water Observatory of the Botín Foundation. The overarching theme of the Water Observatory is improving water management, using innovative approaches, independent thinking and debates. During the last six years, the water programmes of the Foundation have looked at, among others, groundwater issues, water governance, the role of trade in water resources management, water footprint evaluations and water policy. Carrying out independent research and studies, disseminating the findings and engaging in honest debates with stakeholders, politicians and scientists from all over the world have been the main priorities of the Water Observatory. All our publications and seminar materials can be freely accessed from the Foundation's web page.

The Botín Foundation seeks to make a different in the way water resources are managed and governed in Spain and around the world. This book on water and food security in Latin American is the result of two years of collaborative work with dozens of scientists from both sides of the Atlantic and seven prestigious institutions of Argentina, Brazil, Chile, Colombia, Cost Rica, Mexico and Peru. We hope scientists, politicians and stakeholders from all over the world and, especially from Latin America, find in this book useful ideas and inspiration to lead their work in water issues and contribute to a more equitable and sustainable use of this vital resource.

Íñigo Sáenz de Miera y Cárdenas
General Director of the Botín Foundation

Preface

The Botín Foundation was created in 1964, but began its activity in the field of water resources in 1998 when it launched the Groundwater Project (Proyecto Aguas Subterraneas or PAS), one of the first interdisciplinary assessments of groundwater governance. The project, which I was honoured to coordinate, showcased Spain as an example of many of the ethical dilemmas faced by countries across the world, such as the intensive use of groundwater resources for development. This project has been followed by various workshops organized mainly in Santander since 2003, and then published as part of a series of essays under the following titles: 'Water Crisis: myth or reality?' in 2004; 'Water and Ethics' in 2007; 'Water and Food Security in a Globalised World: ethical issues' in 2010; 'Water, Food and Agriculture in Spain: can we square the circle?' in 2013. The latest workshop on 'Integrated Water Resources Management in the 21st Century: revisiting the paradigm' took place in Madrid in November 2013, and the book of the proceedings will be published in 2014.

In 2008 the Board of the Botín Foundation decided to create a section devoted to water within its Trends Observatory, under the title of *Observatorio del Agua de la Fundación Botín* (Water Observatory of the Botín Foundation). In the last few years, all the water programmes of the Water Observatory (WO) have focused on three goals: to develop independent research and studies, to disseminate the findings and to create a venue for debate and discussions. A team of twelve researchers, assisted by an active advisory board of three members, has devoted its energies, skills and talent to formulating relevant research questions, obtaining rigorous answers and communicating these findings where the ultimate goal is to enhance the quality and relevance of political decision making with regard to water issues in Spain and around the world.

In virtually all its programmes and activities, the WO has sought to team up and create partnerships with the most respected scholars, public officers, business managers and representatives of international organizations from all over the world. Openness and independence have always been the foundations of the Observatory.

Between 2008 and 2012 most of the publications, seminars, workshops and activities have focused on Spain's water problems. Much of the substantive judgement and most of the recommendations drawn from this line of work were published in 2013 in the essay *'Agriculture and the Environment in Spain: can we square the circle?'* and then in Spanish in a shorter volume called *'El agua en España: bases para un pacto de futuro'* ('Water in Spain: the basis for a future pact'). A single sentence synthesizes this line of work by the Water Observatory: water problems in Spain are not related to physical scarcity, but to poor governance. As the book outlined, recent technological and social advances can help to achieve better governance in a way that is socially and economically acceptable. One of the main efforts of the WO is to create a general awareness of this to the society at large.

A fundamental transformation of Spain's water problems and a change in paradigm, focus and thinking came about when the role that international trade with agricultural commodities was outlined in the WO work *'Water Footprint and Virtual Water Trade in Spain: policy implications'*, published in 2010. By looking thoroughly at water uses and traded commodities since 1996, it emerged that Spain was using significantly more 'virtual water' than real physical water in 2004 and thereafter. Furthermore, most of the virtual water was increasingly and massively imported in the form of low-value products (in economic terms) with high virtual water content (cereals, grains and feeds). Spain is also a big exporter of agricultural products, but principally the exports are in the form of more valuable products in terms of economic productivity (livestock products, wine, olive oil, fruits, vegetables and nuts). Another finding of that work was that Spain increasingly relies, as does the rest of the European Union, on imports from the rain-fed based agriculture in South America.

The results and conclusions of several WO's analyses of the water policy in Spain were innovative and in some respects were against what was generally considered 'politically correct'. Therefore, the WO team deemed it appropriate to test its ideas and methods in other countries with different hydrological and socio-economic conditions.

In view of this, and Latin America being a region of special interest in today's globalized world and an area where the Botín Foundation had already a number of activities, in 2011 the WO decided to launch a new project focusing on Water and Food Security in Latin America. The main, though neither the only nor the last, output of this project is this book.

In order to carry out this project in Latin America, the WO created a partnership with seven other institutions. In Argentina, the work was led by María Josefa Fioriti of the Water Resources Office of the Planning Ministry (Subsecretaria de Recursos Hídricos del Ministerio de Planificación Federal); in Brazil by Prof Pedro R. Jacobi and Dr Vanessa Empinotti from Sao Paulo University (Universidade de São Paulo); in Chile by Prof Guillermo Donoso of the Catholic University of Chile (Universidad Pontifica Católica de Chile); in Colombia by Diego Arévalo, from the Technological Water Institute of Antioquia (Centro Tecnológico del Agua de Antioquia); in Costa Rica by Andrea Suárez and Dr Patricia Phumpiu from the National University (Universidad Nacional Costa Rica) and the Technological Centre of Monterrey (Centro Tecnológico de Monterrey) respectively; in Mexico by Dr Rosario Pérez Espejo of the Autonomous University of Mexico (Universidad Nacional Autónoma de Méjico); and in Peru, by Prof Julio Kuroiwa, of the Hydraulic Laboratory of Engineering University (Labotario de Hidraúlica, Universidad Nacional de Ingeniería). I am very grateful to our Latin American partners and co-authors for the useful and valuable input provided to this book.

Immediately after this partnership was formed, it was clear that the project was a challenging one. Much like playing with matryoshka dolls (Russian dolls), when an issue or topic was addressed, it soon became clear there were others underlying or behind it. The scope of the project continued to widen as we worked and as other key organizations such as the Inter-American Network of Academies of Science (IANAS), the Food and

Agriculture Organization (FAO), the Organization for Economic Cooperation Development (OECD), the World Water Council (WWC), and the Global Water Partnership (GWP) published their reports on the same topics.

On 28 May 2013, in the course of an intense meeting held in Madrid, we were lucky enough to receive comments, criticisms and suggestions from a team of four deeply knowledgeable and world-renowned reviewers, who had read the first manuscript of the book. Our reviewers were Prof Anthony Allan (King's College London, United Kingdom), Prof Ignacio Rodríguez-Iturbe (University of Princeton, USA), Prof Blanca Jiménez Cisneros (UNAM, Mexico, and presently UNESCO) and Maureen Ballestero (Global Water Partnership, Costa Rica). We are indebted to them for their honest and acute criticisms. Their comments guided our work during the summer of 2013 when we thoroughly revised the manuscript. Whatever was left unaddressed will be taken on in our following projects. In any case, the responsibility on the result of that revision is entirely with this book's authors.

The Universidad Politécnica de Madrid (Technical University of Madrid, UPM) also contributed to this project, under the project 'Red Temática UPM-USP-PUC-Análisis de Riesgos Agrarios y Medioambientales: estrategias para mejorar la adaptación y la mitigación al cambio climático (AL12-RT-13)', a joint programme the UPM shares with the Pontificia Universidad Católica de Chile and with the Universidade de São Paulo.

Regarding the project, we had difficulties in deciding the geographical scope of the study. The project began with a special focus on the partner countries (Argentina, Brazil, Chile, Colombia, Costa Rica, Mexico and Peru), but then as we gathered larger databases we widened the scope to most of the countries South of the Rio Grande/Rio Bravo. The focus on Latin America was misleading because this is more of a cultural denomination rather than a classification used by the main regional organizations (IDB-Inter-American Development Bank, ECLAC-Economic Commission for Latin America and the Caribbean, or the FAO-Food and Agriculture Organization). In the end we decided to associate the book to the largest and commonest denomination, Latin America and the Caribbean, although we were unable to gather large amounts of data on some of the smallest island states in the Caribbean. Neither the volume nor any book chapter has the ambition to cover the LAC region entirely, but many chapters provide data and draw conclusions covering a large percentage of LAC countries.

Last but not least, I would like to convey my sincere gratitude to Emilio Botín O'Shea, member of the Board of the Botín Foundation. Emilio has always followed and supported our activities since the beginning in 1998.

Prof M. Ramón Llamas
Director of the Water Observatory of the Botín Foundation

(All cited books can be downloadable for free from the Botín Foundation's website: www.fundacionbotin.org/water-observatory_trend-observatory.htm)

Acknowledgments

This book is the product of the project 'Water and Food Security in Latin America' (2010–2013) led by the Water Observatory of the Botín Foundation. This project has been carried out by a consortium of eight partners, including the Water Observatory. The editors would like to express their gratitude to all authors and contributors and their host institutions of Latin America for the exceptional cooperation in the past months.

The book has also received funding from the Universidad Politécnica de Madrid (Technical University of Madrid), under the project "Red Temática UPM-USP-PUC-Análisis de Riesgos Agrarios y Medioambientales: Estrategias para mejorar la adaptación y la mitigación al cambio climático (AL12-RT-13)", which the UPM shares with the Pontificia Universidad Católica de Chile and with the Universidade de São Paulo.

We are deeply grateful to the book reviewers Prof Anthony Allan (King's College London, United Kingdom), Prof Ignacio Rodríguez-Iturbe (University of Princeton, USA), Prof Blanca Jiménez Cisneros (UNAM, Mexico, and presently UNESCO) and Maureen Ballestero (Global Water Partnership, Costa Rica), for the useful and wise advices.

A special mention should be made to our team of editorial assistants. Olga Fedorova, Daniel del Olmo Rovidarcht and Desireé Torrente checked formats, data and consistency tirelessly, enduring large days of work and providing very valuable contributions. Bárbara Soriano managed the data gathering work, assisting many chapters' authors, and becoming the author of some on her own right. Our designer, María Carmona, did a superb work and understood the complexities of the project. Special thanks go to Ruth Cunningham for providing the editorial assistance and correcting the style and language. Also, the editors would like to thank the staff members of CEIGRAM, Esperanza Luque, Katerina Kucerova, Begoña Cadiñanos and Elena Vivas for their constant and valuable support.

We would also like to express our sincere gratitude to Ashley Wright, Alanna Donaldson and Tim Hardwick from Routledge, who have provided us with useful and timely guidance at all times during the book production process.

As always, all members of the Water Observatory are deeply grateful to the Botín Foundation for confiding in our judgement to carry out this and many other projects.

Abbreviations

ACUMAR:Autoridad de Cuenca Matanza-Riachuelo
ADB:Asian Development Bank
ADERASA:Asociación de Entes Reguladores de Agua Potable y Saneamiento de las Américas
ALADI: Asociación Latinoamericana de Integración
ARESEP:Autoridad Reguladora de los Servicios Públicos de Costa Rica
AWWA:American Water-Works Association
AySA:Agua y Saneamientos Argentinos S.A.
CADER:Cámara Argentina de Energías Renovables
CAF:Cooperativa Andina de Fomento
CAN:Comunidad Andina
CATI:Coordenadoria de Assistência Técnica Integral-Brazil
CAWT:Central America Water Tribunal
CDM:Clear Development Mechanism
CDP:Carbon Disclosure Project
CF: Carbon Footprint
CFS:Committee on the World Food Security
CONAGUA:Comisión Nacional del Agua de Mexico
ECLAC/CEPAL: United Nations Economic Commission for Latin America and the Caribbean
ECP:Emissions Compensation Programmes
EF: Ecological footprint
ENSO: El Niño Southern Oscillation
ERS-USDA: Economic Research Service of the United States Department of Agriculture
EPA:Environmental Protection Agency
ES: Ecosytem Services
FAO: United Nations Food and Agriculture Organization
FCCyT: Foro Consultivo Científico y Tecnológico
FCPF:Forest Carbon Partnership Facilities
FNS:Food Nutritional Security
FONAG:Fondo para la protección del Agua
FONAFIFO:Fondo Nacional de Financiamiento Forestal-Costa Rica
 FTA:Free Trade Agreements
GDP:Gross Domestic Product
GEF:Global Environment Fund
GFSI:Global Food Security Index
GHI:Global Hunger Index
GMOs:Genetically Modified Organisms
GSM/EDGE:Global System for Mobile communications/Enhanced Data for GSM Evolution
GWP:Global Water Partnership
HDI: Human Development Index
HLTF:High Level Task Force
HRC:Human Rights Council
HRCI:Hunger Reduction Commitment Index
HRWS:Human Right to Water and Sanitation
IANAS:Inter-American Network of Academies of Science
IBGE:Instituto Brasileño de Geografía y Estadística
ICESCR:International Convention on Economic, Social and Cultural Rights
ICT:Information and Communication Technology

IDB:Interamerican Development Bank
IFAD:International Fund for Agricultural Development
IICA:Instituto Interamericano de Cooperación para la Agricultura
IISD:International Institute for Sustainable Development
IMECHE:Institution of Mechanical Engineers
IMF:International Monetary Fund
INPE:National Institute for Space Research
IOM: International Organization for Migration
IPCC: Intergovernmental Panel on Climate Change
IT: Information Technology
IUCN:International Union for Nature Conservation
IWA:International Water Association
IWRM:Integrated Water Resources Management
LAC: Latin American and Caribbean
LAWT:Latin America Water Tribunal
MDG:Millenium Development Goals
MERCOSUR: Common Market of the South
NIC:National Intelligence Council
OAS:Organization of American States
OECD:Organization for Economic Cooperation Development
OHCHR:Office of the United Nations High Commission for Human Rights
OMS/WHO:Organización Mundial de la Salud/World Health Organization
OPEC:Organization of the Petroleum Exporting Countries
PES:Payments for Ecosystem Services
PHI:Poverty and Hunger Index
PTA:Preferential Trade Agreements
RBO:River Basin Organizations
REDD+:Reducing Emissions from Deforestation and Forest Degradation
STEEP: Society-Technology-Economics-Ecology-Politics
TARWR:Total Actual Renewable Water Resources
TL: Trade liberalization
TLC:Trans Latina Companies
UNASUR: South American Nations Union
UNDP: United Nations Development Program
UNEP: United Nations Environment Programme
UNESCO:United Nations Educational, Scientific and Cultural Organization
UNFCCC:United Nations Framework Convention on Climate Change
UNGA:United Nations General Assembly
UNICEF: United Nations Children's Fund
USDA:United States Department of Agriculture
US-EPA:United States Environmental Protection Agency
USGS:United States Geological Survey
VWT:Virtual Water Trade
WBCSD:World Business Council for Sustainable Development
WEC:World Energy Council
WF:Water Footprint
WS:Water Security
WSSD: World Summit on Sustainable Development
WTO: World Trade Organization

Part 1

Introduction

1

WATER AND FOOD SECURITY IN LATIN AMERICA AND THE CARIBBEAN: REGIONAL OPPORTUNITIES TO COPE WITH GLOBAL CHALLENGES

Authors:
Bárbara Willaarts, Water Observatory – Botín Foundation, and CEIGRAM, Technical University of Madrid, Spain
Lucia De Stefano, Water Observatory – Botín Foundation, and Complutense University of Madrid, Spain
Alberto Garrido, Water Observatory – Botín Foundation, and CEIGRAM, Technical University of Madrid, Spain

Contributors:
Ramón Llamas, Water Observatory – Botín Foundation, and Complutense University of Madrid, Spain
Emilio Custodio, Dept. Geo-Engineering, Universitat Politècnica de Catalunya, Spain
Fermín Villarroya, Complutense University of Madrid, Spain
Pedro Martínez-Santos, Complutense University of Madrid, Spain
Maite M. Aldaya, Water Observatory – Botín Foundation, and Complutense University of Madrid, Spain

1.1 Setting the scene

1.1.1 Placing Latin America and the Caribbean in the global context

The world has never been so globalized and interconnected as today. Advances in transportation, logistics, telecommunications and global production systems have attained unprecedented levels of economic integration. Agricultural commodities are transported across hemispheres and trade makes consumers believe that food production no longer respects the traditional seasons. Thanks to technological progress, increasing production specialization, and the wide dissemination of scientific knowledge, world food systems have become more integrated and developed than ever before (Prakash, 2011).

Despite these achievements, important questions still exist as to whether the world's agriculture has the potential to feed a growing population, expected to reach 9 billion by 2050, unless significant improvements are made in production efficiency alongside the promotion of healthier consumption habits. In 2012, 870 million people were still suffering from hunger and malnutrition, equivalent to nearly 12.5% of the global population (FAO, WFP and IFAD, 2012). Furthermore, somewhat ironically, today there are more people overweight than people suffering from hunger globally. According to WHO (2013), in 2008 1.4 billion people were overweight, of which nearly 500 million were obese.

Bridging the hunger gap and addressing the high calorie intake of a growing and wealthier population, demand vast amounts of inputs: water, land, minerals, and energy. The challenge of feeding the world thus becomes particularly acute if it is to be accomplished without adding further pressure on natural resources and surpassing critical environmental tipping points. The National Intelligence Council (NIC) has identified the water–food–energy nexus as one of the four 'megatrends' which is likely to have major impacts on the world's future up to 2030, as an increasing, wealthier and more urbanized population will pose a higher demand on these inextricably linked resources (NIC, 2012).

The NIC report also predicts that the diffusion of power and geopolitical gravity shifts are ongoing megatrends that are likely to influence the world's future in the short term. As Naím (2013) claims, power[1] in the world is decaying as a result of a so-called 'triple-M revolution': the *more* revolution, the *mobility* revolution and the *mentality* revolution. Among the 'more revolution' facts that Naím mentions, a few are worth bearing in mind: the world's economic output has increased fivefold since 1950 and income per capita became 3.5 times greater; between 1990 and 2010, the number of people living on less than US$1.25 a day decreased to 700 million, thus meeting the Millennium Development Goal on halving extreme poverty five years earlier than planned; child mortality has dropped by 17% since 2000; undernourishment decreased from 34% in

---

1 Naím defines power as the 'ability to direct or prevent the current or future actions of other groups and individuals' (p.16).

1980 to 17% in 2008; the middle class increased from 1 billion in 1980 to 2 billion in 2012, and will likely reach 3 billion in 2020; 84% of the population is literate, up from 75% in 1990; and, last but not least, between 2000 and 2010 the human development index – an overall measure of global human well-being and living standards – has risen everywhere in the world with just a handful of exceptions. This promising picture of countries and citizens progressing, living longer, with healthier lives and improved basic needs, is crucial to understanding today's shifts and redistributions of power, and why it is becoming harder to obtain power and easier to lose it.

Much of these socio-economic transitions have occurred in Latin America and the Caribbean (LAC), a region that over the course of the last decade has shown great progress in social, institutional, political and economic spheres. Part of the economic success is due to the region's 'natural dividend', related to the relative and absolute abundance of natural resources, ranging from minerals and energy sources to land and water. As Naím (2013) argues, demand and access to abundant resources are in fact one of the main world drivers of power decay for countries that lack them and of power conquest for those that are well endowed. This partly explains why LAC countries with very little global power until recently are now influential members in the G20 (Argentina, Brazil and Mexico), major world energy providers (Bolivia, Colombia, Ecuador, Venezuela), crucial countries for LAC's overall security (Mexico, Colombia), key EU trading partners (Chile, Peru and Colombia, and the Central American states of Costa Rica, Guatemala, El Salvador, Honduras, Nicaragua and Panama), and leaders of the transpacific cooperation, as four countries (Chile, Peru, Mexico, Colombia) have created the Pacific Alliance to enhance cooperation within the region and across the Pacific with Asia. By all accounts, the LAC region has become a key player in global geopolitics. Exploring how these changes play out in the domain of water and food security contributes to understanding what paths of development this region is following and what are the implications regionally and globally.

1.1.2 Water for regional and global food security

Globally, the largest share of consumptive water use is associated to agricultural production, and just a minor fraction (less than 10% on average) is for cities and industries. Because of the prevalence of rain-fed agriculture over irrigation, the largest share of water consumed in agriculture is *green* water, soil moisture. *Blue* water – water taken from rivers and aquifers – represents a smaller fraction of the agricultural water footprint, although it varies amongst countries. The importance of water for agricultural production and the fact that agriculture is the lion's share of water consumption, renders it relevant and necessary to look at water and food security through the double lens of what Allan (2013) defines as 'food-water' water needed to secure agricultural production, either green or blue- and 'non-food water', which refers to the fraction of blue water providing all other water-related services, beyond food, which are important for human development and well-being.

LAC's agriculture is a strategic sector for rural development and poverty alleviation and it plays a key role in overcoming local and global food insecurities. During the last fifteen years, LAC's agricultural sector has grown considerably, to a large extent driven by trade liberalization policies, which have contributed to turning LAC into an increasingly important competitor in the global agricultural market (for both food and biofuel production). Its weight is not so much in terms of economic value, but in calories and vegetal and animal protein supply, making both developed and emerging economies increasingly more dependent on LAC's output. In recent years, this region has captured an increasing share of the global market of agricultural products, and LAC now controls over 18.4% of the world agricultural trade compared to the 11.4% in 1990 (World Bank, 2013a). Oilseeds, soybean, cereal grain and to a lesser extent livestock products accounted for more than half of this export growth, with a few countries such as Brazil, Argentina and Chile generating over 65% of total LAC exports (*ibid*.).

The expansion of agricultural production and exports has been partially stimulated by the peaks in commodity prices seen in 2007, 2008 and 2012. However, increased price volatility has a lingering effect in the minds of those responsible for managing and governing food systems at international and national levels, even after the price crises subsided. Many governments concluded that relying too much on food imports entailed serious economic and social risks. The notion of food security was thus redefined after the price crises, and food sovereignty is now gaining more prominence to the extent that increasing national food production is becoming an overarching objective in all domains of world and national governance. Nevertheless, under the likely scenario of reaching 9 billion people by 2050, the ongoing process of global urbanization and dietary shifts, the reliance on food imports will remain an indispensable strategy in order to overcome global water and land shortages and cope with future food demand. In this context, it is very likely that LAC will be a major supplier in this long-term scenario as it has already demonstrated over the last decade.

1.1.3 Water for economic development and human well-being

If food-water is essential for achieving food security, non-food water is an equally strategic element for human well-being and social progress. Population growth and the aspiration for higher incomes, greater services and job opportunities, have favoured a rapid and sustained migration flow from rural to urban areas over the last decades. Today, LAC is more urbanized than the average 'high-income' country, with almost 80% of the population living in cities in 2012 (World Bank, 2013b). The region holds four of the largest and most populated cities in the world (the megacities of Mexico D.F., Sao Paulo, Buenos Aires and Rio de Janeiro) and a fast-growing number of middle-size urban areas. This booming process of urbanization, often poorly planned, and the resulting high urban density, pose major challenges for managing water and the delivery of key services to citizens. These include securing access to safe water and sanitation, protection against water hazards such as floods, guaranteeing water provisioning services during drought periods or addressing the growing water pollution problem and environmental degradation of freshwater ecosystems resulting from poor wastewater management policies, amongst other factors.

Non-food water is also a critical input for the industrial sector, including mining, energy production and navigation. Hydropower is the main energy source in the LAC region and still has a large growth potential. Yet its development faces growing physical and socio-economic constraints, including the rights of native and local inhabitants and environmental concerns. Similarly, the growth of the mining sector in LAC, particularly in South America and Mexico, is also generating a growing number of water conflicts. On one hand, because it competes with other economic sectors for sometimes scarce water resources. On the other hand, because of the large pollution problems this sector generates for downstream water users and ecosystems.

1.1.4 Development and sustainability goals: confrontation or alignment?

The strategic value of LAC's natural dividend offers a triple-sided topic of research and inquiry. On the one hand, the role of LAC in the world's current food system and its contribution to global food security cannot be emphasized enough. Interestingly, this crucial role has become a reality in just one decade, and the consequences are now beginning to emerge, in both the political and the scientific spheres. On the other hand, the local, national and regional impacts of this plethora of economic and business opportunities pose enormous challenges for LAC governments. In a time of rapid reconfigurations of power, civil society, NGOs and grassroots organizations have advocated bold reforms at the highest political level (reaching the constitutional one) that enshrine basic rights such as those regarding access to food and water. Last but not least, a fundamental question for the region is whether existing development opportunities and sustainability goals should be framed in terms of trade-offs, or they could also be thought of as win–win opportunities. This dilemma is pertinent worldwide, since decisions concerning to LAC's development and natural resource use will have global consequences for biodiversity, the earth energy balance and the world's climate (Rockström et al., 2009; Gloor et al., 2012).

1.1.5 This book's conceptual approach: linking food and water security

Over the last few years numerous authors and organizations have been looking at the consequences of LAC agricultural growth and globalization. Questions like *What are the socio-economic and environmental implications of this trend for regional development? How does it contribute to local water and food security?* and *What is the role of LAC in global water and food security?* are of critical importance to the region, but knowledge remains sparse and the overall picture is unclear. Behind all these key questions there are numerous interrelated phenomena and processes at the global, national and local levels that must be jointly analysed in order to provide convincing explanations that allow valid conclusions to be drawn.

The answers to these questions have to be sought in the linkages between regional development, economic globalization, well-being, water resource use (food-water and non-food-water), and the global dimension of water and food systems in LAC. To tackle this complex phenomenon a first and fundamental concern is the biophysical sphere, the realization that no social and economic progress of human beings exists without an adequate *material stratus*. This link is sketched in Figure 1.1. A crucial feature that distinguishes LAC from other regions is that most of its vast agricultural production is obtained in rain-fed systems, relying thus primarily on green water. This green water embedded in agricultural exports are of critical importance for global food and water security. Likewise, LAC's food-water and non-food water are also crucial for regional development and for meeting its growing domestic consumption needs. In the particular case of LAC, with its booming economy and a heavy reliance on natural resources, one can imagine scenarios where the rest of the world's craving for food and natural resources compromises the livelihoods of future LAC's generations and scenarios where the two positively reinforce each other. The latter implies that the booming economy and social progress run along more sustainable paths. This book is an inquiry into the type of path LAC countries seem to be following.

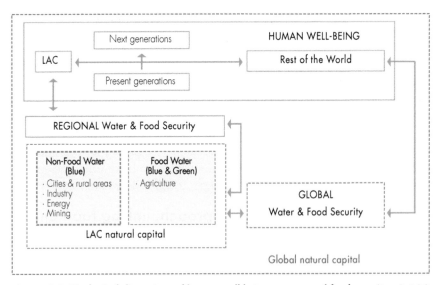

Figure 1.1 Biophysical dimensions of human well-being – water and food security – in LAC and in the rest of the world. *Source: own elaboration.*

A second and equally fundamental concern is the governancel system. If the biosphere represents the material stratus needed for the realization of any kind of security, governance systems represent the *intangible stratus* (Figure 1.2). An underlying theme of this book is that LAC's future depends dramatically on strong governance and institutional frameworks, both within countries and at regional and global levels. Countries' governance systems are where rules for land and water uses are developed and where the bases for water and food security are laid out, as they intervene on how humans interact with the biophysical sphere. Furthermore, the global governance system – e.g. international trade policy, free trade agreements, food safety and sanitary measures – has also become pivotal for food security in LAC and globally. Considering the relatively weak global governance structures of present times, the engagement of national governments in far-seeing and inclusive policies and the demand of citizens of being lead equitably and responsibly are prerequisites for thinking optimistically about the future. This book does not attempt to revise all governance forces operating inside and out of LAC and summarized in Figure 1.2, but to specifically focus on those that have a direct impact on water governance in LAC.

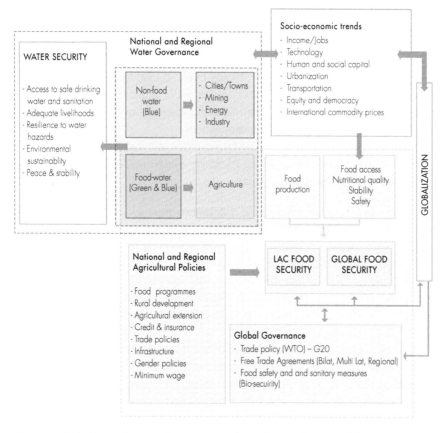

Figure 1.2 The book's framework: topics, inter-dependencies, drivers and focus.
Source: own elaboration.

1.2 The objectives of this book

This book's main goal is to provide an analytical and facts-based view of the progress of LAC's regional water and food security, its contribution to global water and food security and the challenges ahead. A full understanding of these regional changes requires framing LAC in the global picture: a region with increasing geopolitical power in an ever globalized world and a growing presence in global food markets. This overview ultimately aims at facilitating policy debates at national and global levels about these compelling issues. Within this overarching goal, the book has the following specific objectives:

- To diagnose water and food security issues in LAC, using prospective analysis and up-to-date literature. The book pays particular attention to food-water, how it is being used and the links to regional and global food security, without neglecting the importance of non-food water, as it also represents a key asset for development and progress.
- To investigate the role of the socio-economic 'megatrends' in LAC, identifying feedback processes between the region's observed pattern of changes of key biophysical, economic and social variables linked to water and food security.
- To document and analyse the environmental implications linked to the growth of a natural resources-intensive economic model over the last decade, i.e. LAC becoming the world's *food basket* and a key economic actor in domains such as mining and some key industrial products, whilst reviewing the policies in place that have been pursued to mitigate their negative consequences.
- To review the critical changes that are taking place in the institutional and governance water spheres, including the role of civil society, which may represent promising means to advance towards the goal of improving water security in LAC.

Covering a wide array of spheres and databases ranging from biophysical, social and economic variables to detailed records of legal and institutional reform in LAC countries, the book's unique approach offers a complementary view of previous works, including Jiménez-Cisneros and Galizia-Tundisi (2012), Regional Process of the Americas (2012), FAO (2012) and OECD (2013). The first two publications provide considerable updated data on water-related aspects and formulate extremely relevant policy conclusions, while FAO (2012) offers a valuable review of the food security challenges in the LAC region, and OECD (2013) in the world. While this book has a central focus in LAC's water and food challenges, compared to other publications it makes two main contributions: 1) it focuses primarily on the synergies and relationships that both food and water security goals represent for LAC, and 2) it seeks to cover a much vaster domain linking trade and globalization, with water economic uses, pressures on environment and ecosystem services, and water policies, with an overarching view of water and food security for the people and the productive economy in LAC. It does so by first considering international food trade flows and using water accounting techniques to quantify its significance in terms of virtual water movements.

This book will provide an overall picture of LAC's current status and the challenges regarding these compelling issues. But problems and challenges greatly differ across and within countries. LAC is a highly heterogeneous physical territory, even though culturally it is more homogeneous. Whilst this cultural convergence helps in terms of human relations, the different national identities do have an influence on how countries share resources and address common problems, including the widely different standpoint each one has about globalization and the major megatrend. A similar phenomenon can be observed within regions (provinces, states) of the same nation. Thus, although the continental view provides an overall picture, it may also greatly differ from the local vision. It would be impossible to include such a degree of detail within the scope of this book, but in the different chapters some of the striking differences are identified as examples.

1.3 The structure, scope and contents of this volume

This book contains a collection of fifteen essays (including this one) that look at fundamental issues surrounding water for food and human well-being in an increasingly globalized LAC. Most chapters take a regional approach, covering a broad range of data and variables pertaining to most of countries in the region, although a sharper focus is placed in some chapters on seven countries (Argentina, Brazil, Chile, Colombia, Costa Rica, Mexico and Peru), as these are the countries represented in the partners' consortium of the project which lead to the present book (see the Foreword). To cover this vast number of issues, the book generates new data, delves into the vast array of already existing literature and datasets about the region and explores linkages among phenomena and trends.

The book is structured in four parts. Part 1 is this introductory chapter. Part 2 sets the scene for the book looking at the biophysical and socio-economic context of LAC. Part 3 describes the main drivers for land and water uses in the region and for the particular case of the seven aforementioned countries. Part 4 presents the economic, legal and institutional context where those uses occur and where water and food security is to be achieved. In the following sections, the topics, data and approaches of each volume's part and chapter are outlined.

1.3.1 Part 2 on the biophysical and socio-economic context

Chapter 2 provides a general overview of the status and trends of water resources in LAC: its spatial and temporal distribution, its uses and the main challenges that those uses pose to the conservation of water and its associated ecosystems.

Chapter 3 describes the status and main pressures on land and ecosystem services and shows that, as a result of the expansion of agricultural and livestock production, LAC has undergone some of the most noteworthy land use changes in modern history. Associated with these changes significant reductions in the provision of ecosystem services have occurred. The chapter discusses the available options to minimize competition between agricultural land and forests.

Chapters 4 and 5 seek to identify the major socio-economic drivers of change within LAC, looking both at endogenous and exogenous global aspects. Chapter 4 describes and documents a wide selection of socio-economic megatrends of LAC including demographic dynamics, economic growth, migration, income growth, disparity and poverty, human development, education, trade and liberalization, food-consumption patterns, technological change, and climate change. Chapter 5 provides an overview of the trends of direct investment, trade flows and policy, and adds further data about the region's connectivity with the rest of the world. The predominance of trading agricultural and mining commodities stands out for its amount, growth and continuity. The chapter also reviews the literature on the impacts of virtual water trade and some of the most serious concerns, anticipating a more complete presentation of data and discussion in Chapter 7.

Chapter 6 explores the meaning of water and food security in the context of LAC countries, taking a wide perspective and trying to account for all those aspects concerning water and food which are important for human well-being beyond its physical availability. It provides a quantitative analysis on the performance of water and food security indicators between 2000 and 2010 with a view to assessing progress and the links between them. This chapter concludes with a final section assessing the influence of socio-economic factors on water and food security advances.

1.3.2 Part 3 on water demand and drivers

Chapter 7 analyses the challenges and opportunities of water management in the region from the perspective of the agricultural sector. The chapter provides detailed data pertaining to water quantity and quality obtained under the framework of the water footprint indicator. Connecting the data on trade presented in Chapter 5, virtual water trade in the LAC region is also analysed with reference to both countries and time. In the final section, the chapter includes a productivity analysis taking into account social and economic aspects.

Chapter 8 focuses on the urban sector. First, it reviews the major challenges associated with the objective of expanding coverage and sanitation to hundreds of large and middle-size cities which are constantly undergoing processes of expansion and economic growth. It further goes on to analyse the challenge of maintaining the existing infrastructure to provide safe water to hundreds of millions of LAC people. This is illustrated in a number of case studies including Sao Paulo, Rio de Janeiro, Mar del Plata, Mexico D.F., Santiago de Chile, Buenos Aires and Lima. In many cases groundwater has a very significant r ole, even if is not a dominant one. It is also remarked upon that some poor natural water quality problems are a concern, especially in small towns and rural areas.

Chapter 9 focuses on mining, energy and industrial sectors. Each of these is reviewed covering the major challenges each faces as water users and potential pollutants. The mining and industrial sectors stand out for having large impacts on the environment, in addition to wastewater discharges from large cities. The mining sector is potentially subject to water shortages since many mines are in desert areas and they compete for scarce water resources with the urban users and the environment.

1.3.3 Part 4 on the economic, legal and institutional context for achieving water and food security

Chapter 10 reviews the efficiency of water resource use in LAC. To this end, it provides the concepts and definitions together with the drivers for water efficiency. Then, it analyses the efficiency of water resources use in Latin America, looking at the different water users: urban and industrial, mining, agriculture, energy and the environment.

Chapter 11 describes fundamental aspects of water governance, including the constitutional provisions in relation to water, water laws, and the recognition of the human right to water and sanitation. The chapter also analyses financial aspects, funding schemes and investments made and needed in order to ensure the enforcement of constitutional and legal mandates on water.

Chapter 12 focuses on different strategies that stakeholders apply in order to influence water governance in LAC. After reviewing the main sources of tensions regarding water in the region, the chapter looks at practices of activism and advocacy often triggered by disputes that represent informal but important spaces for the participation of civil society. Then the chapter discusses means to achieve transparency, accountability and more robust governance, including, the creation of formal venues of participation as a space for negotiation, the role of the private sector, water certification approaches and legal provisions to ensure access to information.

Chapter 13 explores the role of economic instruments in coping with the most pressing challenges of LAC's water problems. The chapter covers pricing policies, as applied to users of natural resources or mere abstraction activities, and to final users in the urban sector or agricultural sectors. It also reports on a few initiatives with pollution charges and the use of payments for ecosystem services. Since Chile is the only country in the region with experience with water markets, the chapter also offers a brief assessment of how they function and mentions the most recent reforms. The chapter concludes with the potential for improving water and food security indicators by using economic instruments.

Chapter 14 explains how LAC countries are confronting the environmental downside of an economic model based on the intensive use of natural resources and the process of urbanization. It reviews the constitutional and legal approaches and economic initiatives meant to address environmental protection that have been implemented in a large number of LAC countries. It then looks at the impediments and the potential effects private rights and ownership could have. It ends with a technical and detailed discussion of the role of payments for ecosystem services, complementing the brief introduction in Chapter 13.

Chapter 15, the last chapter, relates most of the topics and aspects that have been covered in the book with the changing and ambiguous concept of integrated water resources management (IWRM). The reasons for rethinking the concept of IWRM include a number of innovations and recent findings in fields traditionally not placed at the core of water resource sciences, such as non-conventional water resources, climate science and water globalization.

1.4 Main book's highlights

While it is not prudent to make generalizations for the entire LAC region, as it is obvious that LAC challenges might differ across and within countries, the following section summarizes the main highlights emerging from this volume, grouping them under six main headings: (1) globalization, trade and the role of LAC in international food and water security; (2) implications for LAC's role in the social and environmental spheres; (3) the performance of LAC's indicators of water and food security; (4) the challenges of urbanization, large cities' water, intensive industrial and mining sectors; (5) progress in water governance; (6) democracy, education and good governance as a basis for LAC's natural resources and social dividends.

1.4.1 Globalization and international trade have changed the way of coping with food and water security challenges and LAC is a key player in this new setting

In 2011 the value of traded goods globally was equivalent to 59% of the world's GDP, up from 49% in 2000, and 39% in 1990. With US$1.356 trillion traded in 2011, agricultural products represent the world's third largest sector in traded value, after fuels and non-pharmaceutical chemicals. LAC's agricultural exports now account for 18.4% of global agricultural exports and in value terms they grew by 21% in 2011, mostly because of the increase in commodity prices. In total, LAC's exports of mining and agricultural products represent between 38% and 40% of all goods exports.

The growth in exports of agricultural and mining products has been a major source of income for the wealthiest nations of LAC. But the region's exports have not been sufficiently diversified and hence un-manufactured and less processed products still account for the largest share of LAC agricultural exports. Within the group of the eleven largest economies of LAC, only Argentina (with automobile exports in the third place) and Mexico (with exports of sound and telecommunication equipment in the third place) had, in 2008, a non-agricultural or non-mining sector amongst the three largest exporting sectors (Dingemans and Ross, 2012).

Because of the abundance of agricultural land and the favourable climate, agricultural production in LAC is primarily rain-fed. International demand for agricultural products is mostly satisfied with green water and thus through the use of vast amounts of land. Over 95% of the production water footprint in LAC ($\approx 1060 km^3$/year, an average for the time period of 1996–2005) is for food production and nearly 20% of this 'food water' ($\approx 203 km^3$/year) is exported from LAC, mostly to Asia and Europe.[2] The growing

---

2 According to Dalin et al. (2012), South America exported in 2007 approximately 178 km^3 of virtual water outside the LAC region, i.e. to Asian and European countries. This would imply that roughly over 87% of the 'food water' exported annually by LAC countries is meant to meet the demand from other regions, and only 13% is traded regionally.

international demand for protein crops, oilseed, cereal grains, and meat products has contributed to increasing virtual water exports of 37.5% between 2000 and 2010. The remaining 80% of food water consumption is used for to satisfy the internal demand of a growing and wealthier population.

South America's main trading partners are now in Asia, especially China and India, while Central America and the Caribbean still export primarily to North America. Exports from South America to Asia contributed to 30% of the virtual water trade increase between 1986 and 2007, 95% of which is green water. In this context, Brazil and Argentina are now major players in the global markets of agricultural commodities, providing up to 13% of the global annual green water exports. The expansion of transportation infrastructure connecting ports with vast inland regions will probably enhance the effects of globalization in the more remote areas of the region.

Falkenmark and Rockström (2011), Dalin et al., (2012) and OECD (2013) amongst many others conclude that international trade is a basic element for achieving global food and water security, particularly taking into consideration the future global population and the shifting dietary habits. This points to the key role that global governance architecture, including the World Trade Organization as part of its founding elements, should play in ensuring a fair food trade as a necessary premise for global security. It also suggests that, despite the growing importance the food sovereignty discourse is gaining across many countries, agricultural trade will be still necessary, and LAC is likely to remain a key food provider globally.

1.4.2 Pursuing global water and food security intensively taps into LAC's natural capital and has social and environmental trade-offs

The growth of the agricultural sector in LAC is a result of rapid modernization and competitiveness gains, pushed by technology adoption and innovation, infrastructure development and increasing production efficiency, in both physical and economic terms. LAC still has much potential for scaling up its agricultural output owing largely to its rich natural endowment, especially in terms of land and water. Currently, the appropriation of land for agriculture represents 27% of the total LAC area, a figure comparatively lower than the 38% global average. With less than 13% of this land equipped for irrigation (FAO, 2012), the green water dependency of LAC's agriculture is considered a comparative advantage compared to blue water intensive agricultural production systems. However, relying largely on rain-fed agriculture for food security is not exempt from trade-offs since its expansion implies important environmental impacts and the loss of valuable ecosystem services (e.g. deforestation, widespread pollution, carbon emissions, biodiversity loss).

The growth of rain-fed agriculture in LAC has significantly changed land use patterns. Yet, LAC is the second largest deforestation hotspot in the world, only preceded by Southeast Asia. Between 2000 and 2010, close to 1 million km² of forest have been transformed into agricultural land, an area equivalent to the size of Venezuela, with large consequences for biodiversity and ecosystem services. Deforestation has been particularly intense in South America, with Brazil accounting for 60% of LAC's forest clearing during the last decade. The great majority of the ongoing deforestation in South America is related to the growing international demand for oilseeds grains. In Mesoamerica, deforestation has advanced at a slower pace, and the drivers seem to be related mainly to the low agricultural productivity, which keeps pushing at the agricultural frontier in order to overcome local food insecurity gaps. Annual deforestation rates peaked between 2000 and 2005 and declined slightly in 2005–2010, but are still higher than in 1995–2000.

The sustainable intensification argument was brought up with enthusiasm, as a 'win–win' solution, which may allow the achievement of the triple goal of ensuring food–water–environmental security. However, gains from this sustainable intensification will be slow and require large investments in research and field trials to avoid falling in the 'intensification trap', since as agriculture intensifies, input demands (e.g. energy, fertilizers, water) also rise, and this has additional environmental consequences (Titonell, 2013). Non-point source pollution of water and soil is, jointly with biodiversity loss and built-in resistance to pests and weeds, the main unwanted consequence of agricultural expansion in LAC. Important causes are the extensive application of pesticides and fertilizers, irrigation-induced salinity and the reuse of insufficiently treated wastewater for irrigation. Improvements in agricultural productivity across many countries in LAC will surely help to spare land and reduce the impacts of deforestation, but important challenges remain in order to mitigate the resource-use dependency of agriculture.

1.4.3 LAC's water and food security indicators have improved, but important goals remain and new challenges are emerging

The buoyant global tailwinds that enabled the remarkable economic development of LAC over the last decade have undoubtedly contributed to social progress in the region. Social advances are obvious in the achievements of LAC countries to meet many of the Millennium Development Goals (MDGs) (see Table 1.1). At the continental level, LAC has made notable advances in alleviating extreme poverty (MGD1a), undernourishment (MDG1c) and improving access to drinking water and sanitation (MDG7). Yet progress achieved upon the rest of MDGs, albeit notable, is still not sufficient to meet the 2015 objectives.

When analysing the achievements made by countries separately, the wide divergences in accomplishing the different MDGs become evident. Overall, high and medium-high income countries (e.g. Southern Cone countries, as well as Brazil, Mexico, Costa Rica, Peru, Panama or Ecuador) are on good track to meet at least those MDGs related to basic indicators of water and food security (MDG1 and MDG7). Goals related to improved education, health, equity and female empowerment are progressing but at a slower pace, and there is a risk that they will not be accomplished by 2015 if the prevailing trend continues. In the Caribbean islands there is a large knowledge gap and the information that is available shows slow progress for the most part. Low-income countries also run the risk of not meeting most of the 2015 goals, except in Bolivia, Nicaragua and Honduras.

Table 1.1 Millennium Development Goals (MDG) progress in Latin America and the Caribbean (LAC) between 1990 and 2010.

Target goals on track to be accomplished by 2015 or earlier are represented in green, observed progress but off track if prevailing trends persist are presented in yellow, and off-target ones in red.

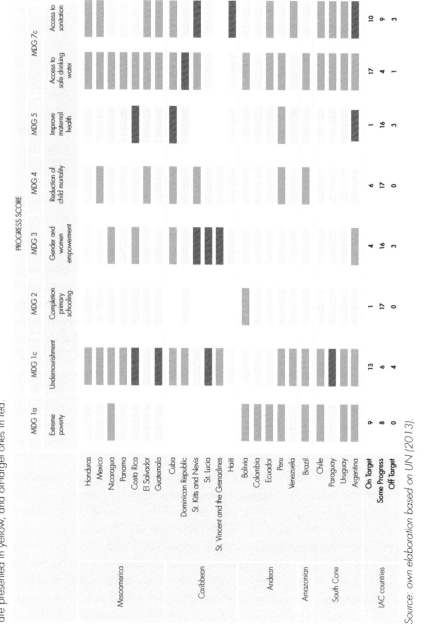

These improvements imply that between 1990 and 2012 the percentage of the population living in poverty in LAC has decreased from 49.4% to 28.8%. Still there are 168 million people living in poverty, the majority living in urban areas (ECLAC, 2013). Income distribution inequality is still the Achilles heel of LAC, but a clear downtrend has been evident since the early 2000s. Nevertheless, by 2011, 30% of the population still received over 60% of the total income (*ibid.*).

With regard to food security improvements, the number of people undernourished has decreased from over 65 million in 1990–1992 to 49 million in 2010–2012 (FAO, 2012). The prevalence of stunting in children under five years has also decreased from 19% in 2000 to 14% in 2010, but the problem remains that one in every seven children born in LAC will have stunted growth. In addition to sub-nutrition, LAC is also facing a growing problem of malnutrition. Obesity now affects nearly 18% of the Latin American population (> 110 million people) and overweight up to 33% (> 200 million people) (Finucane et al., 2011). Malnutrition is particularly affecting middle and high income countries like Chile, Brazil, Uruguay, Argentina and Mexico. Current rates of overweight and obesity in LAC are at least double those of other developing regions and comparable to the ones found in Europe.

Regarding water security, LAC boasts the highest renewable water resources per person among the world's regions, but climatic variability, together with urbanization patterns, generates asymmetries between water demands and water availability across the region, and results in water stress in some of the most economically dynamic areas of the region. Over 100 million LAC citizens currently live in basins which face physical water scarcity.

The number of people without access to an improved sanitation facility has decreased from 146.7 million in 1990 to 103.8 million in 2011 (WHO-UNICEF, 2013). The greatest improvements, however, have been achieved in reducing the number of people without access to safe drinking water from 63.8 million in 1990 to 32.8 million by 2011 (WHO-UNICEF, 2013). These figures mask important differences across countries, between urban and rural areas, as well as within urban areas. Overall, and particularly across the poorest countries, water service deficiencies in rural areas are still very significant.

The vulnerability of countries to growing water hazards stands as another important priority when attempting to increase regional water and food security. The frequency of extreme hydro-meteorological events such as floods has quadrupled between 2000 and 2009, compared to the period of 1970–1979 (EM-DAT, 2013). The social impacts of floods and storms have remained relatively stable (< 3% of the LAC population affected annually) at the LAC scale, but in countries like Belize, Guyana or Cuba, social exposure risk has increased. The economic impacts have grown considerably, and in 2010 they peaked with damages accounting for almost 2% of LAC's GDP. A major reason for this high vulnerability to floods in LAC is related to accelerated urbanization with little or no urban planning, but also to the fact that many cities are located in very flat areas, where large concentrated rain events may produce serious problems such as the 2013 flooding of La Plata. Hydro-climatic variability, in the form of droughts, also represents a major risk for regional food security. Currently, only 13% of the total agricultural area in LAC is equipped with irrigation (FAO, 2012) which makes the agricultural sector in LAC highly vulnerable to drought. Only some parts of Mexico, Chile, Peru, the Northwest of Argentina, and the Northeast of Brazil rely on irrigation water for food production, mostly for the production of value added products such as fruits and vegetables. The potential to expand irrigation is huge but fairly unrealized: FAO (2013) includes Argentina, Brazil, Colombia, Mexico and Peru in the list of twenty world countries with the largest potential, but only Brazil and Mexico stand among the twenty countries with the largest area already equipped with irrigation. However, groundwater salinity, poor drainage in flatlands and droughts make irrigation developments very risky unless large infrastructure investments are made.

1.4.4 The development and operation of the urban water cycle in large cities, intensive industries and the mining sector pose major environmental challenges

Providing a quality drinking water service, improved sanitation and adequate treatment of wastewater is challenging, especially in LAC where the population, particularly in urban areas, has expanded rapidly. The root of the problem is neither economic (scale economies allow for the provision of good quality and sustainable water services at a reasonable cost) nor technical (current engineering can deal with the most complex problems). Barriers to cope with the already diagnosed problems are mostly the lack of governance and institutional leadership, as well as political agendas that often do not include the universal coverage of water supply and sanitation as key priorities.

During the last decade, infrastructure development for domestic supply has been for the most part orientated towards water service provision while sanitation and wastewater management investments have received less attention. In fact, public and private investment on this front has levelled off recently. Another challenge associated with urban water supply is that initial investments for providing access to water have not been followed by stable funding for maintenance and in fact many water services are currently in dire need of replacement and modernization. This and the large population growth, especially in urban areas, are responsible for the deficient quality of supplied services. Regional Process of the Americas (2012) explicitly highlights among these deficiencies: the insufficient water disinfection, the poor surveillance of water abstractions, discontinuous service, insufficient pressure, high leakage percentage (above 40% in many cases) and the limited wastewater treatment. These are big challenges and making use of economies of scale seems to be the most logical and feasible solution to provide a good quality urban water service to the citizen at the lower possible cost, when all involved costs and long-term economic balances are considered (Cabrera et al., 2013). However, this needs a good administrative structure, political support and remarkable leadership amongst decision makers. Although the main focus of water services in large cities is on domestic supply improvements, natural hazards and pollution are also serious concerns in many rural and small urban areas. These are brought to the attention of national authorities through local political and social representatives.

Although water consumption for energy, industry and mining may only be a small percentage of countries' consumption, it can be locally significant, especially in small basins and in the arid and hyper-arid areas of LAC. This consumption may also be economically and socially important, and therefore water quantity and quality should be guaranteed.

Mining and industrial production are emerging sectors in the region and represent an important share of LAC's economy. Furthermore mining is a key source of income and employment. Nonetheless, industrial activities, and in particular mining, contribute to water resources deterioration, threatening water security locally and downstream. This is due to the disposal of water with high salinity, often containing acids and diverse unwanted and noxious solutes. These unwanted constituents are derived from minerals – diverse heavy metals – or from concentration and processing, such as flotation compounds. Quicksilver (mercury) and cyanide can also be found in the case of the many gold mines in LAC, especially the small and artisanal ones. Pollution management is hindered by financial constraints, as well as by insufficient monitoring programmes and wastewater treatment investments. Yet pressures to maintain and expand mining activities will grow because of the world's demand for metals and non-ferrous products. LAC countries currently supply 51% of the world's silver, 45% of its copper and overall 25% of the world's metal market. The production of lithium, a series of secondary metals and coal are also important, as well as gems. Water productivity in the mining and industrial sector is at least one order of magnitude higher than in the agricultural sector.

1.4.5 In LAC water governance is evolving to address the challenges posed by rapid socio-economic changes, however, as is often the case, the implementation of reforms lags behind

Large unexploited natural resources, coupled with the sustained growth pattern of many LAC countries, contribute to create situations where different needs, interests and understanding of the concept of socio-economic development lead to tensions. Poor legal compliance, insufficient legal instruments and lack of funds are often at the root of significant environmental damages and conflicts. Disputes are mainly related to the construction and operation of water works, water diversion, industrial and mining pollution and the privatization of water supply and sanitation coverage in urban areas. This means that most tensions spin around 'non-food water', i.e. a small fraction of the water actually consumed in the region, as high potential of pollution, new risks of flood and fear to lose the precarious water supply in marginal urban areas act as powerful catalysts for stakeholders concern.

Advocacy networks play a key role in empowering and giving national and international visibility to local populations directly affected by environmental degradation or social unfairness. During the past two decades, the demands from civil society for more inclusive, sustainable, efficient and effective governance, as well as the influence of international organizations and supranational agencies, have triggered significant institutional reforms in the region in the form of much legislative activity.

Common elements in those reforms include: a shift towards decentralization, often complemented with the creation of coordination and supervising bodies at a higher level; the formulation of new water laws and policies that include IWRM principles (environmental sustainability, integration, participation, accountability, transparency, cost recovery); and the creation of water use taxes and tariffs for cost recovery. Additionally, in its search for improved water security, LAC has pioneered the recognition of the right to water and sanitation as a human right.

In most of the countries the focus is now on implementing institutional reforms, where the main challenges are related to the lack of integrated planning of water use, the poor coordination of the main stakeholders (both governmental and non-governmental), insufficient local capacity and the need for management instruments that best fit the specific regional differences.

In the spaces for dialogue and participatory decision making created by reforms (e.g. watershed committees, water councils or customary tribunals), formal participation is mainly limited to water users, usually those representing large-scale economic activities. Some accomplishments in participation deserve to be acknowledged, and there are efforts for refining those formal instruments to make them more inclusive and representative of civil society. Nonetheless, the credibility of participation is often questioned due to stakeholders' unequal capacity to participate and the direct access of strong economic lobbies to decision-makers. Other interests not associated to water rights or the perspectives of indigenous population are often underrepresented and social activism still prevails as the main instrument to voice their demands.

Governance failures at different levels have spurred civil society's claims for higher accountability of elected representatives and public authorities. As a reaction, most LAC countries passed, during the last decade, information transparency laws, which apply also to environmental and water-related public information. The actual implementation of the legal obligations to disclose information, however, is still deficient, thus hindering the process of accountability of public authorities before their constituents.

The progressive deterioration of water resources and the need to finance water services provision have fostered the establishment of economic instruments to implement the 'polluter-pays-principle' and increase cost recovery rates. Environmental taxation has been implemented in some LAC countries, but enforcement and collected revenue are still low and do not act as a true deterrent to polluters. After decades of little or no cost recovery rates in irrigating schemes, some countries, such as Argentina, Mexico, Peru and Brazil, have taken steps to make farmers pay for operation and maintenance costs of the infrastructure supplying their water. This may be a tax on exports to compensate for government investments in infrastructures when the product is sold to other country.

Incentives for environmental conservation like payments for ecosystem services (PES) and PES-like schemes have been developed in LAC over the last few years as a complementary instrument to conventional command-and-control and financial instruments. Yet the most successful initiatives have been orientated towards securing availability and quality of water for urban areas (e.g. Produtor Agua in Brazil or Fondo para la Protección del Agua (FONAG) in Ecuador), and thus are geared towards protecting non-food water for cities. The dependency of many PES schemes on international funds, their often weak financial sustainability and the lack of secure land tenure and property rights, amongst other factors, hinder the implementation and long-term sustainability of many other PES initiatives.

1.4.6 Democracy, education and good governance are the basis for using LAC's large natural and human capital for the achievement of human well-being

At present, in LAC, fertility and birth–death rates have decreased, and the population structure is fairly young, with over 50% of working-age. Such a 'demographic dividend', if maintained and accompanied with the corresponding investments and policies, represents a key asset for assuring LAC's socio-economic development in the decades to come. A deeper democratization, the emergence of a powerful civil society, the rise of a middle class, economic openness, and macro-economic stability are also key elements explaining the recent evolution of LAC societies (World Bank, 2013c).

Economic development and the rapid urbanization process have changed societies in LAC, their needs and the way the population use their natural resources. Economic growth and international trade are contributing to changing the dietary habits of LAC citizens, thus affecting the use of water and land. During the past few decades, globalization and the global trade of goods has opened up new development paths and has triggered dynamics whose implications in terms of water and food security in LAC are still difficult to grasp in full.

The opportunities for LAC to achieve a more sustainable and efficient use of their resources, and facilitate a transition towards a green economy are numerous. In fact, there are already a number of successful cases of application and a window of opportunity for the evaluation of trade-offs whilst identifying the potential for significant improvement. The extraordinary natural endowment coupled with the population dividend represents a unique opportunity to foster LAC's socio-economic development.

Nevertheless many challenges still need to be faced, as in several cases economic growth in LAC has been achieved at the expense of land use, energy and water resources intensification, combined with an increase in the levels of pollution and the loss of ecosystems and biodiversity. The reinforcement of national and global governance schemes and their alignment on the achievement of true and universal human well-being, under ethical and moral principles, and will remain an inescapable prerequisite to facing these challenges.

References

Allan, J.A. (2013). Food-water security: beyond water resources and the water sector. In: Lankford B., Bakker, K., Zeitoun, M. and Conway, D. (eds). *Water security: principles, perspectives and practices*, London, Earthscan. In press.

Cabrera, E., Cobacho, R., Espert, V. & Soriano J. (2013). Assessing the full economic impact of domestic water tanks. International Conference on *Asset Management for Enhancing Energy Efficiency in Water and Wastewater Systems*. Marbella, Spain, IWA.

Dalin, C., Konar, M., Hanasaki, N., Rinaldo, A. & Rodriguez-Iturbe, I. (2012) Evolution of the global virtual water trade network. *Proceedings of the National Academy of Sciences of the United States of America* 109 (16): 5989–5994.

Dingemans, A. & Ross, C. (2012). Free trade agreements in Latin America since 1990: an evaluation of export diversification.CEPAL *Review*, 108: 27–48.

ECLAC (2013). Economic Comission for Latin America and the Caribbean. *Social Panorama of Latin America 2012*. Technical Report. 238 pp.

EM-DAT (2013). Emergency Events Database on Disasters. World Health Organization and Centre for Research on the Epidemiology of Disasters Database. [Online] Available from: www.emdat. be/ [Accessed May, 2013].

Falkenmark, M. & Rockström, J. (2011). Back to basics on water as constraint for global food production: opportunities and limitations. In: Garrido, A. and H. Ingram (eds). *Water for Food in a Changing World*. 2nd –Rosenberg Volume Series. London, Routledge Publishers. pp.103–116.

FAO (2012). Food and Agriculture Organization of the United Nations. *Panorama de la Seguridad Alimentaria y Nutricional 2012 en América Latina y el Caribe*. Rome, FAO. 150 pp.

FAO, WFP and IFAD (2012). *The State of Food Insecurity in the World 2012. Economic growth is necessary but not sufficient to accelerate reduction of hunger and malnutrition*. Rome, FAO. 65 pp.

FAO (2013). Food and Agriculture Organization of the United Nations. FAOSTAT *Database*. [Online] Available at: www.faostat.fao.org/ [Accessed in September, 2013].

Finucane M.M., Stevens, G.A., Cowan, M.J., Danaei, G., Lin, J.K., Paciorek, C.J., Singh, G.M., Gutierrez, H.R., Lu, Y., Bahalim, A.N., Farzadfar, F., Riley, L.M. & Ezzati, M. (2011). Global Burden of Metabolic Risk Factors of Chronic Diseases Collaborating Group (Body Mass Index) national, regional, and global trends in body-mass index since 1980: systematic analysis of health examination surveys and epidemiological studies with 960 country-years and 9·1 million participants. *Lancet 377*: 557–567.

Gloor, M., Gatti, L., Brienen, R.J.W., Feldpausch, T., Phillips, O., Miller, J., Ometto, J.P., Ribeiro da Rocha, H., Baker, T., Houghton, R., Malhi, Y., Aragao, L., Guyot, J.L., Zhao, K., Jackson, R., Peylin, P., Sitch, S., Poulter, B., Lomas, M., Zaehle, S., Huntingford, C. & Lloyd, J. (2012). The carbon balance of South America: status, decadal trends and main determinants. *Biogeosciences Discussions*, 9(1): 627–671.

Jiménez-Cisneros, B. & Galizia-Tundisi, J (Coord.) (2012). *Diagnóstico del Agua en las Américas*. Red Interamericana de Academias de Ciencias (IANAS) and Foro Consultivo Científico y Tecnológico (FCCyT). [Online] Available from: www.foroconsultivo.org.mx/home/index.php/libros-publicados/diagnosticos-y-analisis-de-cti/991-diagnostico-del-aguaen-las-americas. [Accessed February, 2013].

Naím, M. (2013). *The end of power: from boardrooms to battlefields and churches to states, why being in charge isn't what it used to be.* New York, Basic Books.

NIC (2012). National Intelligence Council of US. *Global Trends 2030: Alternative Worlds* [Online] Available from: www.dni.gov/files/documents/GlobalTrends_2030.pdf [Accessed September, 2013].

OECD (2013). Organisation for Economic Co-operation and Development. *Global food security: challenges for the food and agricultural system.* Paris, OECD Publishing. 160 pp.

Prakash, A. (2011). *Safeguarding food security in volatile global markets.* Rome, FAO. 619 pp.

Regional Process of the America (2012). *Americas' water agenda: targets, solutions and the paths to improving water resources management.* February 2012. Available at: www.gwp.org/Global/GWP-CAm_Files/Americas'%20Water%20Agenda.pdf

Rockström, J., Steffen, W., Noone, K., Persson, Å., Chapin, F.S., Lambin, E., Lenton, T.M., Scheffer, M., Folke, C., Schellnhuber, H. J., Nykvist, B., Wit, C.A. de, Hughes, T., Leeuw, S. van der, Rodhe, H., Sörlin, S., Snyder, P.K., Costanza, R., Svedin, U., Falkenmark, M., Karlberg, L., Corell, R.W., Fabry, V. J., Hansen, J., Walker, B. & Liverman, D. (2009). Planetary boundaries: exploring the safe operating space for humanity. *Ecology and Society,* 14 (2).

Tittonell, P.A. (2013). *Farming systems ecology towards ecological intensification of world agriculture.* Inaugural lecture upon taking up the position of Chair in Farming Systems Ecology. Wageningen University. Available at: www.wageningenur.nl/upload_mm/8/3/e/8b4f46f7-4656-4f68-bb11-905534c6946c_Inaugural%20lecture%20Pablo%20Tittonell.pdf [Accessed June, 2013].

UN (2013). United Nations. *Sustainable development in Latin America and the Caribbean: follow-up to the United Nations development agenda beyond 2015 and to Rio+20.* Preliminary version. Available at: www.eclac.org/rio20/noticias/paginas/8/43798/2013-273_Rev.1_Sustainable_Development_in_Latin_America_and_the_Caribbean_WEB.pdf [Accessed August, 2013].

World Bank (2013a). *Agricultural exports from Latin America and the Caribbean: harnessing trade to feed the world and promote development.* Washington DC, World Bank.

World Bank (2013b). *World Development Indicator Database.* [Online] Available at: databank.worldbank.org/data/views/variableSelection/selectvariables.aspx?source=world-development-indicators [Accessed March, 2013].

World Bank (2013c). *Shifting gears to accelerate shared prosperity in Latin America and the Caribbean.* Washington DC, World Bank. [Online] Available at: www.worldbank.org/content/dam/Worldbank/document/LAC/PLB%20Shared%20Prosperity%20FINAL.pdf [Accessed September, 2013].

WHO (2013).World Health Organization. *Obesity and Overweight* Fact sheet No. 311. [Online] Available at: www.who.int/mediacentre/factsheets/fs311/en/ [Accessed October, 2013].

WHO-UNICEF (2013). Joint Monitoring Programme (JMP) for Water Supply and Sanitation data. [Online] Available at: www.wssinfo.org/data-estimates/table/ [Accessed July 2013].

Part 2

Setting the scene

2

WATER RESOURCES ASSESSMENT

Coordinator:
Pedro Martínez-Santos, Universidad Complutense de Madrid, and Water Observatory – Botín Foundation, Spain

Authors:
Claudia Campuzano, Centro de Ciencia y Tecnología de Antioquia, Colombia
Anne M. Hansen, Instituto Mexicano de Tecnología del Agua, México
Lucia De Stefano, Water Observatory – Botín Foundation, and Universidad Complutense de Madrid, Spain
Pedro Martínez-Santos, Universidad Complutense de Madrid, and Water Observatory-Botin Foundation, Spain
Desiree Torrente, CEIGRAM, Technical University of Madrid, Spain
Bárbara A. Willaarts, Water Observatory – Botín Foundation, and CEIGRAM, Technical University of Madrid, Spain

Contributors:
Elisa Blanco, Departamento de Economía Agraria – Pontificia Universidad Católica, Santiago, Chile
Luis F. Castro, School of Civil Engineering, Universidad Nacional de Ingenieria, Lima, Peru
Guillermo Donoso, Pontificia Universidad Católica de Chile, Santiago, Chile
Gabriela Franco, Departamento de Economía Agraria Pontificia Universidad Católica, Santiago, Chile
Julio Kuroiwa, Laboratorio Nacional de Hidráulica – Universidad Nacional de Ingeniería, Lima, Peru
Marielena N. Lucen, Ministry of Energy and Mines, Lima, Peru
Julio I. Montenegro, School of Civil Engineering, Universidad Nacional de Ingenieria, Lima, Peru
Markus Pahlow, Department of Water Engineering & Management, University of Twente, The Netherlands
Guoping Zhang, Water Footprint Network, The Netherlands

Highlights

- Latin America boasts some of the world's largest rivers, lakes and aquifers. Overall, these store and yield more water per person than any other region in the planet.

- Climatic variability, together with urbanization patterns, generates strong asymmetries between water demands and water availability across the region. This results in severe water stress in some of the most economically dynamic areas.

- Despite the abundance of surface water resources, groundwater use is gradually increasing. This is partly because of the growing costs associated with surface water storage and treatment, and partly because the advantages of groundwater use are becoming better known and accepted.

- Water quality poses a major cause for concern across the region. Pollution management is complicated by financial constraints, as well as by the absence of adequate monitoring programmes and wastewater treatment facilities.

- A much needed step towards protecting the environment and the health of water-related ecosystems is to implement integral management systems that cater for the maintenance of forests, wetlands, lagoon systems and coastal estuaries.

- Latin American water resources face threats derived from population growth, urbanization, changes in land use patterns and climate change.

2.1 Introduction

The Latin America and Caribbean region is water-abundant. It boasts some of the world's largest rivers, lakes and aquifers, which yield more water per person than any other region in the planet. However, water is irregularly distributed in time and space due to climatic variability. While heavy rainfall takes place across the year in the Amazon rain-forests, it barely ever rains in the Atacama Desert. Besides, the majority of the population is concentrated in cities. This generates strong asymmetries between water demands and water availability. Largely as a result, many freshwater ecosystems are endangered by a wide array of different pressures. Adaptation to climate change, universal access to water and sanitation services, pollution control and an integrated approach to transboundary water resources management are the main challenges ahead.

2.2 Water availability

Latin America only accounts for 13% of the total emerged lands and 6% of the global population, but it produces over one-third of the world's total runoff (Table 2.1). This region is home to some of the world's most important rivers, including the Amazon, Parana, Orinoco, and Magdalena, as well as some of the largest lakes. Take for instance the Titicaca Lake in Bolivia and Peru, the Nicaragua Lake, and Lake Chapala in Mexico. Surface water accounts for over 80% of Latin America's renewable resources, but the region is also endowed with abundant groundwater (Table 2.2). This includes the Guarani aquifers which are shared by Argentina, Brazil, Paraguay and Uruguay. Groundwater also represents a strong environmental element, discharging an estimated 3,700km^3/year into Latin America's rivers. From an economic viewpoint, groundwater storage is particularly important because it remains relatively stable over time and is comparatively better protected from domestic, agricultural and industrial pollution sources (Rebouças, 1999).

Looking at these facts one would think that water scarcity is hardly a matter of concern in Latin America. Overall figures are, however, misleading, as Latin America is diverse within itself. The irregular distribution of water, in both time and space, natural quality problems and an asymmetric occupation of the land imply that the above situation is not representative of all basins across the region. As a result, some are subject to mounting pressures, if not already confronted with water scarcity. For instance, the basins of the Gulf of Mexico, the South Atlantic and the Río de la Plata cover some 25% of Latin America's territory and are home to more than 40% of the population, but contain just 10% of the available water resources (WWC, 2000). Meanwhile, about 53% of the region's total renewable water supply comes from just the one river, the Amazon.

Table 2.1 Approximate amount of annual precipitation, evaporation and runoff per continent in relation to the water footprint

REGION	SURFACE (1000 km²)	POPULATION (million)	AVERAGE RAINFALL (mm)	TOTAL RAINFALL (km³)	AVG. EVAP. (mm)	TOTAL EVAP. (km³)	RUNOFF (km³)	WATER FOOTPRINT (km³)	WATER FOOTPRINT (% of rainfall)
Asia	43,820	4,216	650	28,500	410	18,000	10,500	4,850	17.0
Africa	30,370	1,072	740	22,500	630	19,000	3,500	1,400	6.2
North America	24,490	346	800	19,500	470	11,500	8,000	970	5.0
South America	17,840	596	1,600	28,500	900	16,000	12,500	1,130	4.0
Europe	10,180	740	820	8,400	590	6,000	2,400	1,250	15.0
Oceania	9,010	37	440	4,000	400	3,500	500	45	1.1

Source: Martínez-Santos et al. (2014)

Table 2.2 Renewable water resources and storage capacity in selected countries in Latin America

COUNTRY	RENEWABLE SURFACE WATER (km³/yr)	RENEWABLE GROUNDWATER (km³/yr)	RESERVOIR STORAGE CAPACITY (km³)
Belize	19	8	–
Mexico	409	139	180
Costa Rica	75	37	–
El Salvador	25	6	–
Guatemala	103	3	–
Honduras	87	39	9
Nicaragua	193	59	0.5
Panama	145	21	–
Guyana	241	103	–
Suriname	122	80	22.7
Bolivia	596	130	0.3
Colombia	2,132	510	–
Ecuador	424	134	–
Peru	1,913	303	3.9
Venezuela	1,211	227	164.1
Brazil	8,233	1,874	513.1
Argentina	814	128	–
Chile	922	140	4.7
Paraguay	336	41	37.7
Uruguay	139	23	18.8
Total	18,139	4,005	955

Source: FAO (2013)

Brazil alone generates 37% of Latin America's surface runoff, while no other country reaches 10%. In contrast, arid zones have no surface runoff, except during rare and extreme rainfall events. Rainfall averages 1,600mm/year across the region (Figure 2.1), but ranges from 20mm/yr in the Atacama Desert to over 2,000mm/yr in the mountains of southern Chile (Box 2.1). Rainfall is also characterized by its strong seasonal component. Take for instance Central America, where about half of the precipitation occurs from August to October, and only 7% between February and April. In South America, 35% of stream flows take place between May and July, whilst only 17% corresponds to the November–January period (Shiklomanov, 1999).

Seasonal variability is influenced by cyclic atmospheric phenomena known as El Niño and La Niña. Both are associated with major temperature fluctuations in the tropical Pacific Ocean. El Niño is an abnormal warming of the sea surface temperature, whereas La Niña is a cool ocean phase. During El Niño, droughts take place along the Pacific coast of Central America. Conversely, higher precipitations occur in the Caribbean coasts. In South America, El Niño generates the opposite response. In the Pacific and Southern Atlantic coasts, rains become more intense, whereas in the tropical Atlantic coast drought frequency increases. The cold phase La Niña causes different patterns. Rain events increase along the Pacific coast of Central America. The same occurs with the frequency of hurricanes in the Caribbean Sea. Along the Southern Pacific coast of South America, droughts tend to become more frequent and temperatures drop. Overall, El Niño events are more frequent than La Niña. In the last decades a higher frequency of El Niño events has been recorded, leading some experts to contend that this might be related to climate change (Magrin et al., 2007).

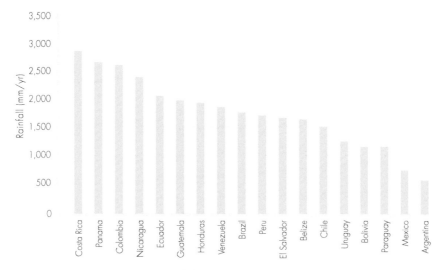

Figure 2.1 Long-term annual rainfall in selected Latin American countries. *Source: FAO (2013)*

Overall, water availability per capita has steadily decreased over the last decades, mostly due to the fact that the population has grown from 420 to 550 million inhabitants between 1992 and 2011. Currently, water availability ranges from Mexico's 3,500m³/person/yr to Peru's 55,000m³/person/yr (Figure 2.2). In other words, all of Latin America's countries are safely located above Falkenmark's 1,700m³/person/yr threshold for water scarcity. The regional average is around 25,000m³/person/yr, well above Europe's 8,500m³/person/yr or Asia's 3,600m³/person/yr. However, while most standard indicators underline Latin America's privileged position in terms of water resources, water scarcity does occur at the regional scale. This is because water resources are mostly located in the inland, while urbanization and land development followed the path of decisions made in colonial times. Thus, cities and economic activities were concentrated either near the coast to facilitate exports to Spain and Portugal, or close to the main cities of the Aztec and Inca empires to take advantage of the abundant labour (Mejía, 2010). In practice, this means that large countries such as Venezuela, Mexico and Peru show strong asymmetries between water availability and population density (Box 2.2).

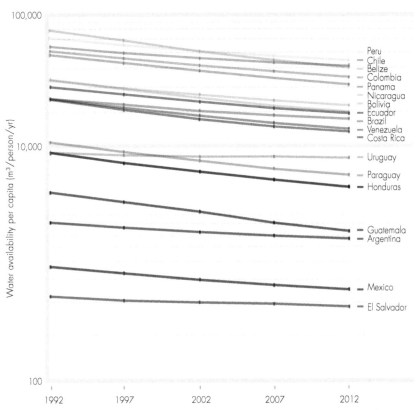

Figure 2.2 Renewable resources per capita over the last twenty years in selected countries.
Source: FAO (2013)

Box 2.1 Large countries are naturally diverse: rainfall variability in Chile

Although Latin America is best described as a water-rich region, average water availability is decreasing. Chile is an excellent example since the country's unique geography, including a number of short river valleys running from the Andes to the Pacific Ocean, provides a variety of climatic conditions. Two primary mountain ranges, the Andes and the Coastal Mountains, span the length of Chile and provide the limits between the coastal plain and the central valley. Average precipitation ranges from near zero in the north to about 2,000mm/yr in the south (Figure 2.3).

The rainy season is in winter, from June to September, and much of the precipitation is stored in the snowpack of the Andes. Water flows in most river basins have a mixed origin, since waters come from winter precipitations and summer snow melt, presenting highest flows in summer (November–February) due to said snow melt, and pronounced reductions in winter (from April to June). Additionally, rainfall fluctuations show greater variability in the arid and semi-arid north (between the Arica-Parinacota Region and the Coquimbo Region). South of 37°S latitude, rainfall becomes more uniform. Therefore, the hydrological regime of Chile is rather irregular.

Within the global context, Chile as a whole may be considered privileged in terms of water resources. The total runoff is on average equivalent to 53,000m^3/person/yr (World Bank, 2011a), a value considerably higher than the world average (6,600m^3/person/yr). However, there exist significant regional differences: north of the city of Santiago, arid conditions prevail with average water availability below 800m^3/person/yr, while south of Santiago the water availability is significantly higher, reaching over 10,000m^3/capita/yr.

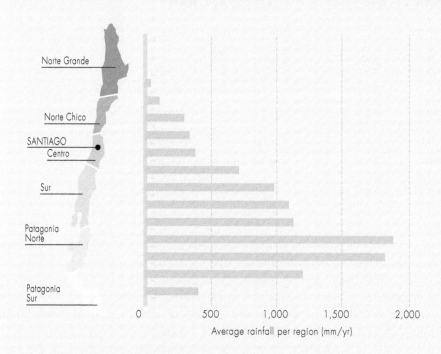

Figure 2.3 Regional rainfall variability in Chile. *Source: modified from Donoso (2014)*

Box 2.2 Water and population asymmetries: Mexico, Peru and Venezuela

Many Latin American countries show a significant disparity between water resources abailavility and the population distribution. Take for instance Mexico. In this country, 77% of the population, 84% of the economic activity and 82% of the irrigated land is located in the central and northern plateaus, some 1,000 metres above sea level. In contrast, 72% of water availability occurs in the south and below that altitude. Another example is Venezuela, where 90% of population and economic activity is located in the north of the country with less than 10% of water availability. In contrast, most of the water availability is found south of the Orinoco River away from the northern coast. But perhaps the most startling case is Peru. Rainfall in the Peruvian part of the Amazon basin, which is home to 30% of the country's population, accounts for 97.5% of the country's surface water. Conversely, the Pacific basin hosts 65% of the population and produces only 1.8% of the water resources of the nation. Rainfall in the capital, Lima, is 10mm/yr or lower. This asymmetry makes the most economically dynamic regions of Peru severely water stressed.

Table 2.3 Water availability in Peru's hydrographic regions

BASIN	AREA (1,000km²)	RAINFALL (mm/yr)	WATER AVAILABILITY (hm³/yr)	WATER AVAILABILITY (% of total)	POPULATION (million)	POPULATION (% of total)	WATER AVAILABILITY (m³/person/yr)
Pacific	279.7	274	37,363	1.8	18,315,276	65	2,040
Amazon	958.5	2,061	1,998,752	97.7	8,579,112	30	232,979
Titicaca	47.2	814	10,172	0.5	1,1326,376	5	7,669
Total	**1,285.20**		**2,046,268**	**100**	**28,220,764**	**100**	**72,510**

Source: Kuroiwa et al. (2014)

Development of non-conventional water resources remains relatively uncommon. Take for instance desalination. Peru and Chile are Latin America's premier users of desalinated seawater, on which they rely for specific developments. Most of the investments in Chilean desalination projects are located in the dry north of the country. These have been designed to underpin mining activities, as well as urban supply. In coastal Peru, desalination provides water for the industrial sector, households and agriculture.

Drinking water and sanitation services reach a relatively large share of the Latin American population. Total coverage amounts to 87% in the case of water supply and 78% in the case of sanitation. However, these figures hide an uneven distribution. For instance, important variations in drinking water coverage are observed across countries.

In Brazil, Mexico, Costa Rica, and Colombia, this figure exceeds 90%, whereas in Peru it is lower than 75%. In terms of rural areas, only Mexico and Costa Rica exceed 85%. Few other countries reach 60%. Sanitation systems are largely insufficient to meet demands. Coverage is similar to that of water supply, exceeding 80% in some urban agglomerations, but rural areas rarely ever reach 50%. Chile poses a remarkable exception, having increased its water services dramatically over the last decade (World Bank, 2011a). Currently, it exceeds 95% in terms of water supply and sanitation coverage in urban areas and 60% in rural regions. The vast majority of sewage goes untreated, thus generating downstream pressures. Less than 40% of sewage is treated in countries such as Argentina, Brazil or Colombia. All these issues will be discussed in more depth in Chapters 6 and 8.

2.3 Water uses

The available water data mostly refer to water withdrawals within each country. In other words, it does not distinguish between water use for production, for domestic consumption or for producing goods for exportation, and exclude virtual water. Moreover the lack of sufficient data on climate, soils and growing seasons in most countries is often the factor limiting the ability to produce meaningful information on consumptive uses. This is most often due to inadequate databases or to the absence of data. In this sense, it is important to distinguish between consumptive uses and withdrawals. Not all water withdrawals result in consumptive water use. This is due to the fact that a large share of withdrawn waters goes back into the hydrological cycle in the form of pipeline losses, wastewater or irrigation returns. On the other hand, not all consumptive uses stem directly from withdrawals. Rain-fed agriculture, for instance, represents a significant fraction of the total water use without being responsible for any direct extraction from the water cycle.

Despite these clarifications, which apply to water figures across the world, Latin America is known to be less water-stressed than other regions (Figure 2.4). Unlike Asia, where a significant part of the water resources are already in use, a large share of Latin America's waters remains untapped. Figure 2.5 shows the distribution of water withdrawals per sector and sub region in Latin American and Caribbean regions. Agriculture comprises irrigation and livestock. Consumption due to water uses such as hydropower, navigation, fishing or recreation is considered negligible for practical purposes.

Rainwater can be split into 'green' and 'blue' water. Green water refers to the share of rainwater that stays in the soil rather than running off or recharging groundwater. In other words, green water is that which underpins rain-fed agriculture. On the other hand, blue water refers to the water in rivers, lakes, reservoirs, ponds and aquifers. Irrigated agriculture uses blue water as a supplement to rainfall. As will be discussed in Chapters 6 and 7, green and blue water have different implications for the purpose of water and food security.

The water footprint provides a useful indicator of water use. As shown in Table 2.4, green water agriculture accounts for the largest share of the region's water footprint (Mekonnen and Hoekstra, 2011). Blue water follows in magnitude, well ahead of industrial and domestic use. Irrigation efficiency is however low. Efficiency is measured by taking into account the difference between the volume of water captured and the actual delivery to the farms, and is mostly dependent on the type of irrigation system. In many Latin American countries, irrigation efficiency ranges between 30% and 40% (San Martín, 2002). Inefficient irrigation technologies do not necessarily imply a wasteful water use, as the losses return to the hydrological cycle. However, these rank among the main causes behind the loss of fertile soils and are largely a consequence of policies that promote production. This represents one of the main threats to agricultural sustainability across the region.

Table 2.4 Blue and green water footprint of countries in the Latin America and Caribbean region (those with more than one million inhabitants). All consumption figures are rounded to the nearest decimal.

COUNTRY	POPULATION (million)	CONSUMPTION OF AGRICULTURAL PRODUCTS				CONSUMPTION OF INDUSTRIAL PRODUCTS		DOMESTIC WATER CONSUMPTION
		INTERNAL (hm³/yr)		EXTERNAL (hm³/yr)		INTERNAL (hm³/yr)	EXTERNAL (hm³/yr)	(hm³/yr)
		GREEN	BLUE	GREEN	BLUE	BLUE	BLUE	BLUE
Argentina	37.1	47,746	3,258	1,298	146	116	61	491
Bolivia	8.4	25,764	377	2,489	124	4	4	18
Brazil	175.3	288,345	8,498	27,981	2,075	420	147	1,202
Chile	15.5	6,994	2,101	5,071	278	93	32	142
Colombia	40.1	35,863	1,386	9,101	716	16	33	539
Costa Rica	4	2,725	148	1,381	187	12	11	79
Cuba	11.1	13,194	831	1,944	130	47	9	156
Dominican Rep.	8.9	6,590	826	3,263	211	2	13	109
Ecuador	12.4	17,175	1,440	2,464	134	33	12	212
El Salvador	5.9	3,441	42	1,482	215	7	7	32
Guatemala	11.4	8,137	149	1,553	202	10	13	13
Haiti	8.7	6,809	225	1,230	434	0	2	5
Honduras	6.3	5,754	113	777	172	2	4	7
Jamaica	2.6	2,162	45	1,510	164	3	6	14
Mexico	99.8	83,841	8,654	65,986	8,475	135	358	1,359
Nicaragua	5.1	3,498	134	536	99	1	4	19
Panama	3	2,226	54	928	95	1	8	55
Paraguay	5.4	9,673	214	14,	59	2	6	10
Peru	26.2	13,142	3,299	8,050	404	74	18	168
Trinidad & Tobago	1.3	0	1,588	115	2	4	21	1,590
Uruguay	3.3	177	1,286	13	2	8	8	1,268
Venezuela	24.6	21,551	1,194	12,985	547	16	21	381

Source: Mekonnen and Hoekstra (2011)

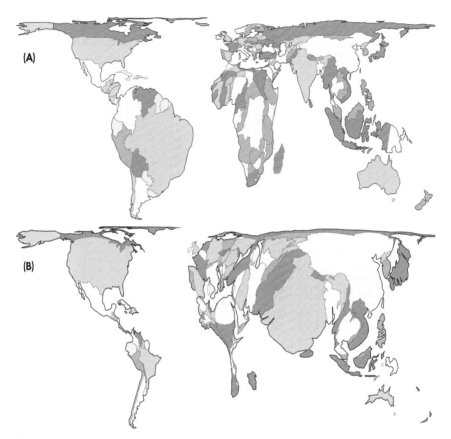

Figure 2.4 (A) Total annual rainfall (av. period 1961–1990) and (B) Water use across the world. Country areas are deformed as a function of total rainfall and water use, i.e. the larger the country is represented, the larger is its proportional share relative to other countries. Source: modified from Sasi Group (University of Sheffield) and Mark Newman (University of Michigan)

Industry uses relatively little water in comparison with other sectors. Water-intensive industries include food processing, pulp and paper, petro-chemical and textile sectors. These demand raw materials that are abundant in the region, creating significant multiplier effects in the local and national economies (San Martín, 2002). However, industries are responsible for environmental degradation by dumping untreated sewage into rivers and aquifers. This is particularly true of the mining industry, whose water use is relatively low, but which is considered one of the main water polluters across the region (Chapter 9).

While surface water is the preferred source of water in the region, groundwater use has increased in recent decades (Box 2.3). This is partly because of the growing costs associated with surface water storage and treatment and partly because the advantages of groundwater use are becoming more accepted (Llamas and Martínez-Santos, 2005). Most of the existing groundwater-based developments are concentrated in areas of economic or political interest, or where surface water is under stress. In contrast, fundamental hydrogeological knowledge is still under development in many parts of Latin America and there are vast regions where groundwater data are scarce or non-existent (Ballestero et al., 2007). Besides, natural water quality problems, such as elevated concentrations of arsenic, are yet to be fully assessed in countries such as Argentina, Bolivia, Chile, Ecuador, El Salvador, Mexico, Nicaragua and Peru.

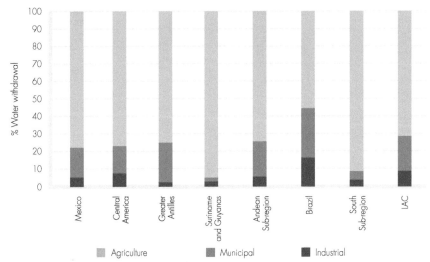

Figure 2.5 Water withdrawals per sector in the Latin American and Caribbean region. Central America comprises Costa Rica, Honduras, Guatemala, Belize, El Salvador, Panama and Nicaragua; the Greater Antilles include Cuba, Haiti, the Dominican Republic and Jamaica; the Andean sub-region refers to Colombia, Ecuador, Venezuela, Bolivia and Peru; and the South sub-region includes Paraguay, Uruguay, Argentina and Chile. *Source: FAO (2013)*

Groundwater use is especially relevant in Argentina, where it accounts for 30% of the total water withdrawals. Likewise in Chile, where it is of particular importance in the mining sector. In this country, 63% of the water used in mining and 46% of domestic water supply comes from aquifer sources. Groundwater is also the primary source for human consumption in Costa Rica and in Mexico, where groundwater accounts for 50% of industrial demands, 70% of domestic supply in cities and practically all domestic supply in rural areas.

The wealth of water resources in Latin America is reflected in the region's natural resources and the environmental services that these provide (UNEP, 2003). Natural forests cover 47% of the total surface area of the region, the northern part of the Amazon and the Guyana area being home to the largest expansion of virgin forest in the world. About 95% of the green surface corresponds to the tropical rainforests of Central America, the Caribbean and the South American sub-tropics. The remainder is located in temperate South America, primarily Argentina, Chile and Uruguay.

Box 2.3   Groundwater mining in Latin America

[By Prof Emilio Custodio (Universitat Politècnica de Catalunya, Spain)]

As is the case of diverse regions of the world, and especially in arid and semi-arid areas (Custodio, 2010, 2011), in some of the driest areas of Latin America groundwater reserves are being depleted due to intensive exploitation, at a rate much higher than they are being replenished. Groundwater mining is mostly produced in two areas. One corresponds to the hyper-arid areas of the Andean Region, comprising coastal Peru, northern Chile, southwestern Bolivia and northwestern Argentina, where groundwater renewal is scarce to nil. Groundwater abstraction takes place primarily to supply the mining of metal ores and also for brine extraction in terminal salt lakes ('salares') in order to exploit some solutes such as lithium, potassium and nitrate. The sustainability of small springs and groundwater discharges that are important for some human settlements, tourism in the area and have a significant ecological value, such as the high altitude wetlands ('bofedales'), is of special concern. Rainfall in the intermediate depressions is a few mm/yr on average and the limited replenishment is occasionally produced by some sporadic floods in gullies whose headwaters are in the highlands ('Altiplano'). Albeit rainfall in the Altiplano is scarce, a combination of almost bare soil with low humidity retention (mostly acidic ignimbrites) and rainfall retention in the seasonal snow cover favour some recharge. This manages to sustain some springs that yield water with a very long turnover time. Although mining may deplete groundwater reserves and their reco-very may take centuries, there are no specific studies on groundwater reserve depletion.

Other groundwater mining areas can be found in the dry areas of Mexico, where reserves are being depleted at a rate greater than recharge, even if recharge is still significant. In this case groundwater is mostly used for irrigation, but also for mining and industrial activities. In some coastal areas freshwater in the aquifer is being replaced by laterally or vertically intruding saline water, as in Sonora's coastal aquifers. In Mexico, 104 of the existing aquifer systems are considered over-exploited by the Federal Water Authority (Comisión Nacional del Agua). Even though this is a small fraction of the existing aquifers receiving a groundwater recharge of $2{,}500 m^3/s$, these aquifers yield $800 m^3/s$. This amounts to 80% of used groundwater (Jiménez-Cisneros and Galizia-Tundisi (2012). About 20–25% of groundwater reserves, equivalent to $171 m^3/s$, have been ruined, which is equivalent to about half the water used for public supply.

Some cases of groundwater resources depletion are located in the agricultural valleys of western Peru, such as Ica–Villacurí. The main problem here is the integrated management of water resources and the adequate use of the aquifer as a storage reser-voir. A key issue is the mixing of freshwater and old saline water. Similar problems are found in the dry northeast of Brazil and also in the dry areas of the Argentinean Pampas (the Chaco-Pampean region), where arsenic and fluoride groundwater quality problems and deep-seated relict saline water upcoming add to periodic, non-permanent depletion of water reserves.

Other well-known groundwater problems, such as seawater intrusion in Mar del Plata and Recife, or land subsidence around Mexico City or Queretaro, are better described as they hydrodynamic results of intensive groundwater development, rather than as groundwater mining problems.

Computing environmental requirements for ecosystems is a recurrent stumbling block for academics and managers across the world. Latin America's ecosystems are no exception. Although the importance of marine and coastal flora and fauna is widely acknowledged (WSSD, 2003), there is less recognition of water needs to support ecosystems, which are themselves legitimate water users (UNEP, 2012). Ecosystems, which provide life-supporting goods and services, need water of adequate quantity and quality. Appropriate timing is also crucial in many cases. A much needed step towards protecting the environment and health of water-related ecosystems is to implement integral management systems that cater for the maintenance of forests, wetlands, lagoon systems and coastal estuaries.

Hydropower is the main non-consumptive use across the region. Take for instance Chile, whose hydropower sector has grown to account for 38% of its total energy production, and whose current flow rate is in the order of 4,190m^3/s (Ayala, 2010). This is largely explained by sustained economic growth over the last three decades, which has led to a significant increase in energy demands. About 82% of Colombia's reservoirs are devoted to energy generation, but even so, it is estimated that the country is only taking advantage of 10% of its potential for producing energy, since many of its rivers are still unregulated. Mid- and long-term developments are therefore expected, the main challenge being the need to balance environmental, social and economic constraints. Hydropower is an important water user in several other countries, including Brazil. In Costa Rica, this industry holds 82% of the water licenses.

Due to the size of many Latin American rivers, navigation is another relevant user. It is particularly important in countries such as Argentina, which operates along the Paraguay-Paraná and Alto Paraná waterways. Buenos Aires boasts South America's most important harbour. Maintaining adequate navigation conditions implies continuous work on the Río de la Plata. The Patagonian harbours, to the south of the country, have experienced notable development over the last decades in order to favour tourism.

Fishing is an important activity in many regions, allowing for the economic subsistence of local communities. It is also an established industry and features highly among tourist destinations. However, this sector has experienced setbacks in recent years due to the construction of reservoirs, over-fishing, the introduction of exotic species and contamination. Water pollution has proved particularly detrimental to recreational uses. Indeed, poor or non-existent treatment of wastewater effluents has endangered tourism and ecosystems in freshwater bodies across the region. Many Latin American lakes, including Lake Chapala in Mexico and Lake Titicaca in Bolivia and Peru, are at present severely polluted.

Water demands are leading to increasingly important conflicts between users (Chapters 11 and 15). In terms of consumptive uses, agriculture is usually displaced by the domestic and industrial sectors. In most cases, however, the environment is the net loser. There is a general consensus that contamination due to untreated wastewater, industrial and mining effluents, and widely dispersed agricultural pollutants are serious problems in many areas across the region.

2.4 Water quality

Water pollution in Latin America is caused by human activities and refers in general to the presence of pollutants from anthropogenic sources. In addition, natural phenomena such as volcanic activities, storms or earthquakes cause changes in water quality. Pollutants may cause water to be unfit for human consumption or to sustain aquatic life.

Water quality is associated with the use it is given. García (2006) explains that a water body is polluted when it contains substances that make it inadequate for certain uses, and contaminated when it contains substances that endanger human health. Therefore, a water body may be polluted and not contaminated. Conversely, if it is contaminated, it is polluted. Due to its capability to dissolve chemicals, natural and residual waters, as well as water for human consumption, always contain dissolved substances. Depending on their concentrations, all pollutants have the potential to become contaminants.

Water pollutants include both organic and inorganic chemical substances as well as pathogens. These substances can be man-made or of natural origin, such as plant residues. Some are found naturally in Latin American water bodies and their concentrations may assist in defining their natural origin or classification as contaminants. Many chemicals are toxic and some of them are biodegradable, thus consuming oxygen dissolved in water.

In Latin America pollutants are frequently discharged into water bodies from both point and non-point sources, producing physical, chemical, and biological changes that cause adverse effects in humans and in ecosystems. Point source pollutants are those that enter water bodies through discharge pipes or channels. They include municipal and industrial wastewater discharges, with or without previous treatment, and urban runoff drains. Diffuse source pollution does not come from a single source but is the accumulation of pollutants after runoff from areas with diverse land uses. This type of pollution is the main cause of water eutrophication, which refers to the increase in concentration of nutrients. This, in turn, may increase the primary production in water bodies, causing anoxia and decreased water quality, affecting ecosystems and other water uses.

Land uses in Latin American watersheds and water uses for human purposes introduce changes in the natural cycle of precipitation, absorption, water flow, infiltration, and evapotranspiration. While part of the used water is consumed, part is returned to the water bodies but most often with different quality. While agricultural use return flows contain salts, nutrients, pesticides, and organic matter, industrial discharges contain organic matter, metal ions, chemical residues and salts and what's more, at higher temperatures. Domestic discharges carry grease, detergents, dissolved solids, bacteria, and viruses (García, 2006).

Agricultural effects on water quality are mostly due to chemical contamination of fertilizers and pesticides that accumulate in some aquifers, and reuse of sewage effluents for irrigation that can transmit a number of pathogens, even after secondary water treatments (World Bank, 2011b). Significant water pollution due to irrigation has been reported in Barbados, Mexico, Nicaragua, Panama, Peru, Dominican Republic and Venezuela (Biswas and Tortajada, 2006; FAO 2004; LA-Mexico, 2012).

Salinity due to irrigation has been a serious constraint in countries such as Argentina, Cuba, Mexico, and Peru, and, to a lesser extent, in the arid regions of northeastern Brazil, north and central Chile and some small areas of Central America (ibid). The reuse of domestic wastewater for irrigation has been established as a common practice on the outskirts of the cities located in arid and semi-arid areas, where intense competition for water for agriculture and urban uses often occurs.

Arsenic and fluoride pose groundwater quality concerns in several parts of the region. Arsenic content in groundwater is sometimes natural, but can also be attributed to economic activities such as gold or lead mining or to industrial effluents. High arsenic concentrations are known to be a problem in parts of Mexico, the Andean range and Argentina. High fluoride concentrations are often associated to sodium-bicarbonate waters found in weathered alkaline and metamorphic rocks, coastal aquifers affected by cation exchange or aquifers affected by evaporation. Thus, high fluoride concentrations have been observed in parts of Brazil and the Andes.

As indicated by Biswas and Tortajada (2006), water is becoming increasingly polluted in Latin America. Such pressures vary in the different sub-regions, and some sectors, such as mining and agriculture as well as large cities, are quite conspicuous, representing specific local water quality concerns for both surface and groundwater (Box 2.4). While large mining companies recycle and treat discharges, most small and artisanal mining companies do not have control and measures of their water pollution, and constitute important sources of contaminants to adjacent water bodies (World Bank, 2011b).

Box 2.4 Pollution by metals, metalloids and other contaminants in Chile

Rivers in the north of Chile have relatively high concentrations of metals from both natural sources and mining activities. Recent studies address the variation in concentration of heavy metals and sulphates, which is also a by-product of mining, in eleven rivers in the north of Chile. These show high concentrations of heavy metals and sulphates that in many cases exceed Chilean regulations. Arsenic is an important contaminant related to natural pollution in Chile. High evaporation and increased extraction of water have caused higher contents of salts in water. Aluminium is also an important pollutant in the central zone of the country. In order to control changes in waters quality, regulations of discharges must be fulfilled and norms are being developed that specify water quality limits for these releases to aquatic systems, considering the specific conditions of the receiving water bodies (Jiménez-Cisneros and Galizia-Tundisi, 2012).

Barrios (2006) points out that water quality management is not a substitute for efficient water management but a strategic issue that requires the integration of water quantity, pollution control, efficient use of water, environmental considerations and human health implications. Since Latin American countries are heterogeneous in terms of physical, climatic, economic, social, institutional, and environmental conditions (Biswas and Tortajada, 2006), water quality management should be specifically planned and developed and be an integral component of water management policies. The region's water quality management is complicated by the lack of wastewater treatment, financial constraints, difficulties in complying with standards and criteria of receiving waters, and the lack of monitoring programmes (García, 2006).

Not all of Latin America faces the same water quality problems, since these vary according to development and types of economic activity. While standards have been established to control point source pollution, García (2006) affirms that the resulting water quality is still not adequate. Among the problems are the disposal of sewage and lack of wastewater treatment. Besides, the attention has mostly been towards industrial discharges, ignoring municipal and non-point source pollution. The lack of monitoring and assessment has prevented the development and application of receiving waters criteria for more efficient basin-wide approaches to cope with such problems.

Given the magnitude of non-point source pollution's contribution to water quality losses, there is widespread agreement that many water quality goals cannot be reached without reducing this type of pollution. The cost-effectiveness of controlling non-point source pollution is generally recognized as opposed to narrowing regulations so that tertiary treatments of point source discharges are required (Russell and Clark, 2006).

Hoekstra et al. (2011) developed the Water Pollution Level (WPL) as an indicator of the level of water pollution. WPL is defined as the ratio between the total grey water footprint in an area or a watershed to the actual runoff. In Latin America, the overall WPL related to nitrogen (N) that are close to or higher than 1.0 are widespread over the entire region (Figure 2.6), while those related to phosphorus (P) that are close to or higher than 1.0 are mostly in Mexico and to the south and east of the region (Figure 2.7).

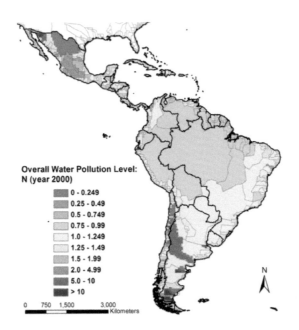

Figure 2.6 Water Pollution Level for nitrogen (N) per river basin in Latin America (year 2000).
Source: Liu et al. (2012)

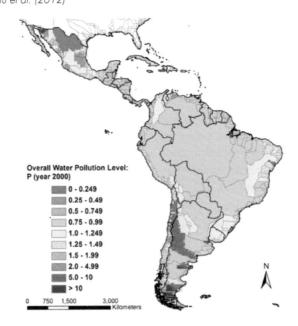

Figure 2.7 Water Pollution Level for phosphorus (P) per river basin in Latin America (year 2000). *Source: Liu et al. (2012)*

Water quality is acquiring great relevance because of the role of water in transporting contaminants and the growing concern over emergent forms of pollution such as endocrine-disrupting substances, in addition to persistent organic pollutants and other toxic compounds. Very few developing countries are prepared to face these concerns (Barrios, 2006) and, although there are specific case studies in Latin America that relate to the presence of these pollutants, to date there is no general overview of their presence in water bodies of the region. No permanent programmes exist for the monitoring of persistent organic pollutants, emerging pollutants, and other toxic compounds, and there are therefore no inventories or formal valuations of the exposure and risks associated with these substances (Box 2.5).

Box 2.5 Water quality policies in Mexico

The priorities of the Mexican water policy are to assure enough water of appropriate quality, recognize the strategic value of water, efficient use of water, protect water bodies, and to ensure the sustainable development and environmental conservation (CONAGUA, 2008).

The National Water Law (DOF, 2012) establishes the water quality requirements depending on its use, with the priority on human consumption relative to other uses of water. The norm NOM-127-SSA1-1994 (permissible limits of water quality and treatments for water purification) and NOM-179-SSA1-1998 (monitoring and evaluation of water quality control for human use and consumption, of water distributed by public supply systems), establish limits for human use and consumption. On the other hand, NOM-001-SEMARNAT-1996 establishes the limits for discharges to waters and national properties and NOM-002-SEMARNAT-1996 establishes the limits for discharges to municipal and urban sewage systems. The ecological criteria for water quality, CE-CCA-001/89, include limits for urban public use, recreation with direct contact, irrigation, livestock and aquatic life.

Currently, the evaluation of water quality in Mexico is based on three basic indicators: biochemical oxygen demand (BOD), chemical oxygen demand (COD), and total suspended solids (TSS). In 2009, twenty-one of 1,471 river basins were classified as heavily contaminated according at least one of these indicators. Nearly 13% of the Mexican surface water was polluted owing to BOD, 31%, to COD, and 7.5%, to TSS.

Hansen and Corzo-Juárez (2011) highlighted the priorities and requirements for the evaluation of pollution of watersheds, referring to the policy of water management in Mexico, and the above-mentioned regulations. They remark that the national programme for monitoring and evaluation of toxic persistent and bioaccumulable substances (STPB) is recently being implemented and up until now there had been no formal valuations or cataloguing of the substances and the associated risks. A proposed list of substances to be included in a monitoring program of STPB in watersheds and aquifers has been presented by Hansen (2012).

Water quality problems are not only solved by constructing and operating wastewater treatment plants. Water quality management should include the formulation and implementation of water policies, monitoring and evaluation of water quality, installation of appropriate legal and institutional frameworks, capacity building, and evaluation and control of non-point sources pollutants.

2.5 Transboundary resources

A significant number of the region's basins are shared by two or more countries. These transboundary basins cover an area where a relatively large fraction of the population is concentrated. Take for instance South America, where there are thirty-eight international water basins that cover almost 60% of the continent area and that are home to more than one hundred million people (nearly 30% the population) (UNEP, 2007). Despite this, only four transboundary basins in South America have transboundary agreements in place (La Plata, Titicaca, Amazon and Lagoon Mirim). Remarkably, the Orinoco and Essequibo basins, i.e. the third and fourth largest of the continent, are not governed by international treaties (De Stefano et al., 2012).

Another important factor influencing the territorial structure of water management is that four of the largest countries in the continent are federal (Brazil, Mexico, Argentina and Venezuela). This means that most transboundary basins are directly or indirectly influenced by federalism. In those cases, strong state-level authorities will determine land and water use based on social, economic and political interests that may not take into account the interests of upstream or downstream users.

The distribution of water management responsibilities in Latin American countries is diverse (Table 2.5). Water resources commissions and river basin organizations have often demonstrated themselves to be useful bodies to coordinate inputs from sectors and stakeholders acting at the chosen management scale. This can be seen in the institutional evolution of several countries in Latin America. In Mexico, for example, management units include basins and sub-basins, and basin organizations at both scales. Mexico together with Brazil and Argentina have a tradition of river basin organizations, whereas in other countries, e.g. Peru, such entities are still being set up. River basin organizations have had deficiencies since their creation, partly due to weak institutional and policy frameworks, weak investment or financing methods (Dourojeanni, 2011). Take for instance Argentina, where the lack of financial autonomy of the river basin committees makes them highly dependent on provincial and local governments (OECD, 2012). In some cases, decentralized watershed management exists but is isolated and not formally recognized, stemming from local initiatives or pursued by sub-national authorities through informal processes and without the support of national political elites (see for instance Ecuador, Kauffman, 2011). Dourojeanni (2001) identified several challenges for river basin organizations, including the clarification of their role (and the potential competition with other authorities), economic viability and funding.

Table 2.5 Distribution of water responsibilities in selected countries

COUNTRY	WATER RESOURCES	DOMESTIC SUPPLY	RIVER BASIN ORGANIZATIONS	WATER USER ASSOCIATIONS
Argentina	Provinces	Provinces, municipalities	Yes	Yes
Brazil	Central Government, Water-specific bodies, RBO	Municipalities	Yes	Yes
Chile	Central Government	Central Government, water-specific bodies, local rural, committees	No	Yes
Costa Rica	Central Government	Municipalities	No	No
El Salvador	None	Municipalities, inter-municipal bodies, water-specific bodies, RBOs	n/a	No
Guatemala	RBOs	Municipalities	Yes	Yes
Honduras	Municipalities, inter-municipal bodies, water-specific bodies	Municipalities, inter-municipal bodies, water-specific bodies	n/a	No
Mexico	Regions, municipalities, inter-municipal bodies, RBOs	Regions, municipalities, inter-municipal bodies, RBOs	Yes	Yes
Nicaragua	Regions, municipalities, inter-municipal bodies, water-specific bodies, RBOs	Regions, municipalities, RBOs	Yes	Yes
Panama	None	Municipalities, others	n/a	No
Peru	Regions, municipalities, water-specific bodies, RBOs	Regions, municipalities, water-specific bodies, RBOs	Yes	Yes

Source: OECD (2012); OECD (2011); LA–Chile (2012); LA–Costa Rica (2012)

2.6 Climate change and water resources

Many countries of LAC have reported multiple evidences of a changing climate.[1] The most frequent impacts reported include an increase in average temperature, higher frequency of extreme rainfalls, sea level rise and coastal retreat, droughts, hurricanes and strong winds, and glacier melting. The magnitude and importance of each impact differs across regions and within countries (see Figure 2.8).

Based on the number and frequency of recorded impacts, the Andean region is the most vulnerable zone to climate change. Mean temperature has increased in all countries,

---

1 Most of the information under this heading stems from National Communication reports (NCs) in compliance with the United Framework Convention on Climate Chagne (UNFCC) by twenty LAC countries (Non-annex I parties), including: Mexico, Belize, Guatemala, Honduras, El Salvador, Costa Rica, Panama, Nicaragua, Peru, Bolivia, Colombia, Ecuador, Venezuela, Brazil, Suriname, Guyana, Chile, Argentina, Paraguay and Uruguay.

most importantly in the higher altitudes, e.g. in the Bolivian Andean altiplano (between 1.1 to 1.7°C) and in the high Colombian plains (up to 1°C). In these parts of the Andes rainfall has decreased and droughts are becoming more frequent. Such trends are probably behind observed glacier melts, particularly in Peru and Colombia. In contrast, the Andean lowlands are becoming wetter and more prone to extreme rainfall. These changes have been linked to the intensification of El Niño events.

The South Cone also appears to be suffering important changes. The most frequently recorded phenomenon is an increase in extreme rainfall events, particularly in the northern part of the region. In central and northern Argentina the number of extreme rainfalls has increased fourfold since the 1960s. Also, a sea level rise of up to 4mm/year has been recorded on the coast of Rio de la Plata during the last two decades. The persistence of both trends is worrisome given the population density in this area. Elsewhere, along the Andean mountains of Chile and Argentina, the frequency and length of droughts have increased. Dryness has been associated with the intensification of La Niña. In Chile for instance, the number of dry years has increased substantially over the last century, e.g. during the first quarter of the 20th century the frequency of dry years was 15%, during the last fifty years, the frequency has increased to 50%. In the South Cone, an increase in mean temperature has mostly occurred in the Patagonian region (up to 1°C) and the Andean Mountains (+0.25°C) but not along the coastal areas. Eighty-seven out of one hundred glaciers under study along the Andean region have receded during the last century.

The intensification of El Niño events has been linked to the increasing frequency of extreme rainfall events and hurricanes along the Caribbean Coast. In Belize, for instance, four out of the eight major storms recorded during the 20th century have occurred in the last twenty five years. Likewise, Honduras is the third country in the world with the highest record of extreme event occurrence between 1990 and 2008. Extreme rainfall has caused nearly sixty floods in Costa Rica over the last six decades. While the Caribbean coast is becoming wetter and rain events more extreme, droughts are increasing along much of the Pacific coast of Central America. In the north of Costa Rica the frequency of dry years has increased remarkably between 1960 and 2005, and the average reduction in precipitation during these dry years surpasses 32% of the mean annual precipitation.

These observed trends largely coincide with the climate projections made by the Intergovernmental Panel on Climate Change (IPCC) for the region (Figure 2.8). According to Galindo et al. (2010), 2100 climate projections show an increasing frequency of hurricanes in the Caribbean and Central America, as well as a higher drought frequency and a reduction in annual rainfall. Glacier melting will continue along much of the Andean tropical glaciers of Colombia, Ecuador, Peru, as well as in Chile and Argentina.

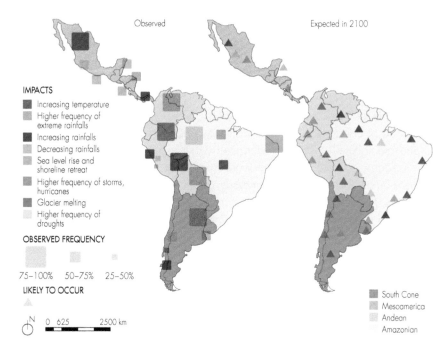

Figure 2.8 Observed (left) and expected (right) impacts linked to Climate Change in Latin America. *Source: own elaboration based on the information on observed impacts recorded from the National Communications (NCs) performed by twenty Latin American countries and summarized by major regions (UNFCCC, 2013); and expected climate change impact projections for the year 2100 in LAC as summarized in Galindo et al. (2010).*

2.7 Future challenges

Latin American water resources face important threats derived from population growth, urbanization, land use patterns and climate change, among others (Jones and Scarpati, 2007). United Nations' estimates suggest that the population will increase significantly in the coming years. By 2030, the population in northwest South America, from Venezuela to Bolivia, is expected to grow by one-third. Countries such as Brazil, Argentina or Chile will experience a demographic growth of about 20%. In addition, Latin America is experiencing other changes, namely, the shift of population from the countryside into the cities. As a result, per capita water consumption is rising dramatically in urban areas (see Chapter 8). This increases the pressure on local resources, such as Mexico City's aquifer, leading to problems of groundwater quality degradation, aquifer depletion and subsidence. Besides, the increase in paved areas, coupled with inadequate drainage, favours devastating floods such as the ones that have occurred in Sao Paulo, Mexico City, Rio de Janeiro or Buenos Aires in the recent past (Regional Process of the Americas, 2012).

Climate change is likely to cause increasing variability in precipitation and runoff, in both time and space, resulting in the excess or scarcity of water, and extreme events.

Inevitably this will cause changes in hydropower generation, agriculture, industry and domestic water supply. Some of the practical effects of climate change include the gradual substitution of Amazon rainforest with savannahs, changes in crop patterns and yields across the region, increased vulnerability to floods and droughts in Central and South America, augmented effects of the El Niño and La Niña oscillation phenomena and glacier melting in the Andes (EuropeAid, 2009).

In a context of unevenly distributed water resources and increasing drought in some regions and precipitation in others, enhanced water efficiency and management poses a major challenge, not only for direct water users and managers, but also for indirect water users such as policy makers, businesses, agricultural commodity trading companies and consumers. In contrast, consistent water accounting systems are yet to be developed. Quantifying and accounting for water flows within the economy (including environmental needs) and related impacts on the appropriate time and spatial scales would allow transparent information to be attained and thus contribute to the development of robust allocation and management systems needed to underpin a green economy (UNEP, 2010).

Deteriorating water quality due to urban and agricultural waste has long threatened public health and ecosystems. Full integration of water quality into the management debate is needed in order to ensure the preservation of water resources for the future. In this regard, systematic water quality monitoring, pollution control and wastewater treatment programmes are perceived as both urgent and essential.

Although some encouraging steps have been taken in the last few years, integrated water resources management is still absent in most countries (Chapter 15). Water governance opportunities are associated with the administration of water resources, the need to broaden and strengthen the capacity of public institutions, the establishment of clear and effective regulations for the provision of efficient services or the formulation and implementation of effective policies, with the subsidiary action of governments and with the participation of all water users including public–private cooperation strategies at local, sub-national and national levels (Regional Process of the Americas, 2012). Amongst all the challenges not least is the need to devise adequate governance frameworks for shared basins.

References

Ayala, L. (2010). *Aspectos Técnicos de la Gestión Integrada de las Aguas (GIRH). Primera etapa: Diagnóstico.* Informe preparado para el diagnóstico de la gestión de los recursos hídricos, Santiago, Chile.

Ballestero, M., Reyes, V. & Astorga, Y. (2007). Groundwater in Central America: its importance, development and uses, with particular reference to its role in irrigated agricultures. In: Giordano, M. & Villholth, K.G. (eds). *The Agricultural Groundwater Revolution: Opportunities and Threats to Development,* London, CAB International. pp.100–128.

Barrios, J.E. (2006). Water quality management: missing concepts for developing countries. In: Biswas, A.K., Tortajada, C., Braga, B. & Rodriguez, D.J. (eds). *Water Quality Management in the Americas. Water Resources Development and Management,* The Netherlands, Springer. pp. 137–146.

Biswas, A.K. & Tortajada, C. (2006). Preface. In: Biswas, A.K., Tortajada, C., Braga, B. & Rodriguez, D.J. (eds). *Water Quality Management in the Americas. Water Resources Development and Management,* The Netherlands, Springer. pp. 5–9.

CONAGUA (2008). Comisión Nacional del Agua. *Programa Nacional Hídrico 2007–2012. Secretaría del Medio Ambiente y Recursos Naturales.* [Online] Available from: www.conagua. gob.mx/CONAGUA07/ Contenido/Documentos/PNH_05-08.pdf [Accessed July, 2013].

Custodio, E. (2010). Intensive groundwater development: a water cycle transformation, a social revolution, a management challenge. In: Martínez-Cortina, L., Garrido, A. & López-Gunn, E. (eds). *Rethinking Water and Food Security,* Boca Raton,FL, CRC Press/Balkema. pp. 259–298.

Custodio, E. (2011). Hidrogeología en regiones áridas y semiáridas. VII Congreso Argentino de Hidrología. Temas Actuales de la Hidrología Subterránea/Hidrogeología Regional y Exploración Hidrogeológica. *Salta*: 1–17.

De Stefano, L., Duncan, J., Dinar, S., Stahl, K., Strzepek, K. & Wolf, A.T. (2012). Climate change and the institutional resilience of international river basins. *Journal of Peace Research*, 49(1): 193–209.

DOF (2012). Diario Oficial de la Federación. *Ley de Aguas Nacionales.* Latest reform published on June 8, 2012. pp. 1–106.

Donoso, G. (2014). Integrated water management in Chile. In: Martínez-Santos, P., Aldaya, M.M. & Llamas, M.R. (eds). *Integrated Water Resources Management in the 21st Century: Revisting the paradigm.* Boca Raton,FL, CRC Press. Forthcoming.

Dourojeanni, A. (2001). Water management at the river basin level: challenges in Latin America. Serie Recursos Naturales e Infraestructura: ECLAC. Report No. 29.

Dourojeanni , A. (2011). El error de crear organizaciones de cuenca sin las atribuciones necesarias para cumplir sus roles. *Revista Virtual REDESMA* [Online] 5(1). Available from: www.cebem. org/cmsfiles/articulos/REDESMA_11_art05.pdf [Accessed September, 2013].

EuropeAid (2009). *Climate change in Latin America.* AGRIFOR consult Group. [Online] Available from: ec.europa.eu/europaid/index_en.htm [Accessed July, 2013].

FAO (2004). Food and Agriculture Organization of the United Nations. AQUASTAT. [Online] Available from: www.fao.org/ag/agl/aglw/aquastat. [Accessed February, 2013].

FAO (2013). Food and Agriculture Organization of the United Nations. AQUASTAT. [Online] Available from: www.fao.org/ag/agl/aglw/aquastat. [Accessed February, 2013].

Galindo, L.M., de Miguel, C. & Ferrer, J. (2010). *Gráficos vitales del cambio climático para América Latina y el Caribe.* Edición especial para la CP16/CP-RP 6, México. Ciudad de Panamá. [Online] Available from: www.cridlac.org/digitalizacion/pdf/spa/doc18216/ doc18216.htm [Accessed February, 2013].

García, L.E. (2006). Water quality issues in Latin America. In: Biswas, A.K., Tortajada, C., Braga, B. & Rodriguez, D.J. (eds). *Water Quality Management in the Americas*, Berlin, Springer. pp. 1–15.

Hansen, A.M. (2012). Programa de monitoreo y evaluación de STPB en cuencas hidrológicas y acuíferos. *Tecnología y Ciencias del Agua*, III (4): 167–195.

Hansen, A.M. & Corzo-Juárez, C. (2011). Evaluation of pollution in hydrological basins: priorities and necessities. In: Oswald-Spring, U. (ed). In: *Water Resources in Mexico. Scarcity, Degradation, Stress, Conflicts, Management, and Policy.* Berlin-Heidelberg, Germany, Springer. pp. 201-230.

Hoekstra, A.Y., Chapagain, A.K., Aldaya, M.M. & Mekonnen, M.M. (2011). *The Water Footprint Assessment Manual: Setting the global standard*. London, UK, Earthscan.

Jiménez-Cisneros, B. & Galizia-Tundisi, J (Coord.) (2012). *Diagnóstico del Agua en las Américas*. Mexico. Mexico, Red Interamericana de Academias de Ciencias (IANAS) and Foro Consultivo Científico y Tecnológico (FCCyT). [Online] Available from: www.foroconsultivo.org.mx/home/index.php/libros-publicados/diagnosticos-y-analisis-de-cti/991-diagnostico-del-agua-en-las-americas [Accessed February, 2013].

Jones, J.A. & Scarpati, O.E. (2007). Water resources issues in South America. *GeoJournal*, 70: 227–231.

Kauffman, C. (2011). *Transnational Actors and the Power of Weak Laws: Decentralizing Watershed Management Without the State*. Paper presented at the 2011 Annual Meeting of the American Political Science Association. Seattle, Washington, September 1–4, 2011.

Kuroiwa, J.M., Castro, L.F., Lucen, M.N. & Montenegro, J.I. (2014). Integrated water resources management in Peru: the long road ahead. In: Martínez-Santos, P., Aldaya, M.M & Llamas, M.R. (eds). *Integrated Water Resources Management in the 21st Century: Revisting the paradigm*. Boca Raton, FL, CRC-Press. Forthcoming.

LA–Chile (2012). *Report of Chile*. Contribution to the Water and Food Security in Latin America Project. Water Observatory Project. Madrid, Spain, Fundación Botín.

LA–Costa Rica (2012). *Report of Costa Rica*. Contribution to the Water and Food Security in Latin America Project. Water Observatory Project. Madrid, Spain, Fundación Botín.

LA–Mexico (2012). *Report of Mexico*. Contribution to the Water and Food Security in Latin America Project. Water Observatory Project. Madrid, Spain, Fundación Botín.

Liu, C., Kroeze. C., Hoekstra. A.Y. & Gerbens-Leenes, W. (2012). Past and future trends in grey water footprints of anthropogenic nitrogen and phosphorus inputs to major world rivers. *Ecological Indicators*, 18: 42–49.

Llamas, M.R. & Martínez-Santos, P. (2005). Intensive groundwater use: silent revolution and potential source of social conflict. *Journal of Water Resources Planning and Management*, 131(5): 337–341.

Magrin, G., Gay García, C., Cruz Choque, D., Giménez, J.C., Moreno, A.R., Nagy, G.J., Nobre, C. & Villamizar, A. (2007). Latin America. In: Parry, M.L., Canziani, O.F., Palutikof, J.P., van der Linden P.J. and Hanson, C.E. (eds) *Climate Change 2007: Impacts, Adaptation and Vulnerability. Contribution of Working Group II to the Fourth Assessment Report of the Intergovernmental Panel on Climate Change*. Cambridge, Cambridge University Press. pp. 581–615.

Martínez-Santos, P., Aldaya, M.M., Llamas, M.R (2014). Integrated water resources management in Peru: state of the art and the way forward. In: Martínez-Santos P., Aldaya M.M. & Llamas M.R (eds). *Integrated Water Resources Management in the 21st Century: Revisting the paradigm*. Boca Raton FL, CRC-Press. Forthcoming.

Mejía, A. (2010). *Water For The Americas: Challenges & Opportunities*. Rosenberg International Forum on Water Policy. Buenos Aires, Argentina.

Mekonnen, M.M. & Hoekstra, A.Y. (2011). *National Water Footprint Accounts: The green, blue and grey water footprint of production and consumption*. UNESCO-IHE, Value of Water Research Report Series No. 50.

OECD (2011). Organisation for Economic and Cooperation Development. *Water Governance in OECD Countries A Multi-level Approach OECD Studies on Water*. Paris, OECD Publishing.

OECD (2012). Organisation for Economic and Cooperation Development. *Water Governance in Latin America and the Caribbean: A Multi-level Approach, OECD Studies on Water.* Paris, OECD Publishing.

Rebouças, A. (1999). Groundwater resources in South America. *Episodes,* 22(3): 232–237.

Regional Process of the Americas (2012). *Americas' Water Agenda. Targets, Solutions and the Paths to Improving Water Resources Management,* Marseille, 6th World Water Forum, [Online] Available from: www.unesco.org.uy/phi/fileadmin/phi/infocus/Americas__Water_Agenda. pdf Accessed July, 2013].

Russell, C.S. & Clark, C.D. (2006). Economic instruments and non-point source water pollution. In: Biswas, A.K., Tortajada, C., Braga, B., Rodriguez , D.J.(eds.) *Water Quality Management in the Americas.* Berlin, Springer. pp. 17–45.

San Martín, O. (2002). *Water Resources in Latin America and the Caribbean: Issues and Options.* Inter-American Development Bank. Technical Report No. 65.

Shiklomanov, I.A. (1999). *World Water Resources at the Beginning of the 21st Century.* UNESCO/ IHP.

UNEP (2003). United Nations Environment Programme. *Water Resources Management in Latin America and the Caribbean.* Technical Report No. 26.

UNEP (2007). United Nations Environment Programme. *Hydropolitical Vulnerability and Resilience along International Waters. Latin America and the Caribbean.* Nairobi, UNEP.

UNEP (2010). United Nations Environment Programme. *Measuring Water Use in a Green Economy. A Report of the Working Group on Water Efficiency to the International Resource Panel.* Technical Report No. 91.

UNEP (2012). United Nations Environment Programme. *Global Environmental Outlook 5* (GEO5). Technical Report No. 591.

UNFCCC (2013). United Nations Framework Convention on Climate Change. *National Communications of non-Annex I parties of United Nation Framework. Convention on Climate Change.* [Online] Available from: unfccc.int/national_reports/non-annex_i_natcom/ items/2979.php [Accessed May, 2013].

World Bank (2011a). *Chile: Diagnóstico de la gestión de los recursos hídricos. Departamento de Medio Ambiente y Desarrollo Sostenible para América Latina y el Caribe.* Santiago, Chile. [Online] Available from: www-wds.worldbank.org/external/default/WDSContentServer/ WDSP/IB/2011/ 07/21/000020953_20110721091658/Rendered/PDF/633920ES W0SPAN0le0GRH0final0DR0REV.0doc.pdf. [Accessed April, 2013].

World Bank (2011b). *Water and Sanitation Program. American Water, a window of information to support water and sanitation sector.* [Online] Available from: www.wsp.org/featuresevents/ features/agua-latina-window-access-knowledge-and-information [Accessed April, 2013].

WSSD (2002). World Summit on Sustainable Development. *Johannesburg Plan of Implementation. World Summit on Sustainable Development. JPOI Response Strategy.* [Online] Available from: www.cooperazioneallosviluppo.esteri.it/pdgcs/documentazione/ AttiConvegni/2003-01-01_JohannesburgPlanImplementation.pdf [Accessed April, 2013].

WWC (2000). World Water Council. *World Water Vision: Making Water Everybody's Business.* London, Earthscan, [Online] Available from: www.worldwatercouncil.org/library/archives/ world-water-vision/vision-report/ [Accessed June, 2013].

TRENDS IN LAND USE AND ECOSYSTEM SERVICES

Authors:
Bárbara A. Willaarts, Water Observatory – Botín Foundation, and CEIGRAM, Technical University of Madrid, Spain
Gloria Salmoral, Water Observatory – Botín Foundation, and CEIGRAM, Technical University of Madrid, Spain
Juliana S. Farinaci, Environmental Studies Center (NEPAM) – State University of Campinas, (UNICAMP), Brazil
Maria José Sanz-Sánchez, FAO, Roma, Italy

Highlights

- The land used for agricultural production in Latin America and the Caribbean (LAC) comprises 26% of its total surface area: 10% for crops and 16% for livestock grazing. This share still remains below the global average land appropriation (38%).

- Between 1990 and 2010 LAC lost approximately 92 million hectares of forests, becoming the second most important deforestation hotspot worldwide, only preceded by Southeast Asia. Some 88% of this forest loss has occurred in South America and 12% in Mesoamerica. Brazil alone accounts for 60% of LAC's deforestation. In the Caribbean forest area has increased.

- Agriculture is the major driver of deforestation in LAC. In South America the cultivation of agricultural commodities, mostly oilseeds and grains, underpin much of the ongoing deforestation together with the sharp expansion of the livestock sector. In Mesoamerica, the low agricultural productivity keeps pushing the agricultural frontier in order to overcome national food in security problems.

- LAC has outstanding natural capital and contributes to the provision of multiple ecosystem services on a wide range of scales. Yet, land use changes are a major driver of ecosystem services loss even above climate change.

- The deep transformations that have occurred in LAC over the last two decades have had important impacts on the provision of key ecosystem services. Regulating services such as carbon sequestration and biodiversity conservation have experienced the largest impacts, with an average loss of 9%. Also, native agro-diversity has shrunk almost 6%. Cultural services like ecotourism has grown over 150% and provisioning services like forestry and water provision have also increased (35% and 6%, respectively).

- Deforestation rates are slowing down. Yet, the growth of agriculture in LAC is increasingly being decoupled from expanding the agricultural frontier and more based on increases in agricultural yields.

- To cope with the increasing world food demand while ensuring the conservation of LAC's natural capital and ecosystem services, it is necessary to develop integrated land use approaches, including agricultural oriented measures (e.g. land sparing and land sharing) and conservation initiatives (e.g. Reducing Emissions from Deforestation and Forest Degradation- REDD+).

3.1   Introduction

Latin America and the Caribbean region (LAC) is currently facing a daunting challenge: producing food, fibre, and fuel to satisfy an increasing internal and international demand and at the same time preserve its outstanding natural capital and related ecosystem services (ES) (Martinelli, 2012). Compared to other regions, LAC has a major advantage to achieve this double goal due to its rich natural endowment in terms of land, water and its low population density.

Ongoing pressure on LAC natural resources is linked to internal development but also to economic globalization, population growth and principally changing diets throughout the world. FAO (2009) estimates that by 2050 agricultural production will need to double in order to satisfy the increasing world food and biofuel demand. This future demand can partly be met by intensifying existing agricultural land and improving resource use efficiency (e.g. bridging the yield gap, the development of genetically modified crops-GMOs, etc.), however, most experts agree that between 50 and 450 million hectares of additional agricultural land will also be required (FAO, 2009; Fisher et al., 2009; Lambin and Meyfroidt, 2011). This additional land demand is most likely to be absorbed by developing countries that have the greatest land availability, primarily sub-Saharan Africa and LAC (Smith et al., 2010).

Food and fibre are key provisioning ES to LAC as they provide important benefits which are contributing to overcome local and global food insecurity gaps and at the same time allow for regional economic development. By 2011 annual gross revenues of LAC's agriculture accounted for over 120,000 million US$ (FAO, 2013), and generated approximately 18% of the employment (World Bank, 2013). In some of the major agricultural producing countries, like Brazil, agro-industry accounted for 22% of the national GDP in 2011 (CEPEA, 2013). A large part of this agricultural market expansion is taking place at the expenses of replacing natural ecosystems, mostly tropical savannahs and forests. The ecosystem productivity of these tropical forests ranks among the highest in the world due to their extension and quality, particularly along the Amazon basin and much of Central America (Pfister et al., 2011). Their replacement entails important trade-offs for the provision of other key non-market ES, like carbon sequestration, pollination, water flow regulation or biodiversity conservation. Balancing these ES trade-offs are key to LAC but also globally since the Amazon tropical forests play a key role in the global carbon and water cycle (Rockström et al., 2009; Gloor et al., 2012).

Despite the pressure, significant improvements in agricultural production have been achieved in many LAC countries, in an attempt to increase efficiency, decouple production from water and land resource consumption and thus minimize existing ES trade-offs. Efforts in this direction are critical since deforestation, as opposed to climate change, causes abrupt changes in ecosystems, limiting and often precluding opportunities for adaptation.

Accordingly, this chapter aims to explore: 1) what major changes in land use have occurred in LAC during the last two decades of significant economic changes; 2) what are the drivers behind these land changes; 3) how are those changes influencing the flow of ES across the region; and 4) what policy options are in place to safeguard LAC's natural capital while contributing to global food security.

3.2 What have been the main land use trends over the last decades?

As Chapters 4 and 5 describe, LAC has experienced significant changes over the last decades as a result of its great economic acceleration and the strong development of its agricultural sector. This growth has been accompanied by the expansion of LAC's agricultural area by almost 57 million hectares (see Figure 3.1). Such increase is related to the expansion of pastures for livestock production and arable land. Likewise, shrublands and secondary forests have also experienced an important area increase (≈ +27 million hectares). Much of these land uses have grown at the expense of replacing natural meadows and even more notably, natural forests, which have shrunk 92 million hectares, an area equivalent to the size of Venezuela. This forest reduction represents 46% of the total forest losses occurred in the southern hemisphere over the last two decades (FAO, 2010; Rademaekers et al., 2010), demonstrating that LAC, and particularly South America, is one of the most important global deforestation hotspots.

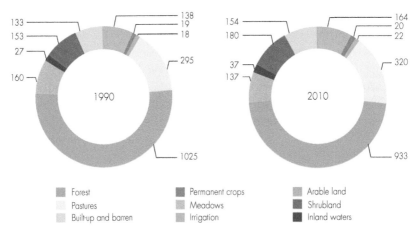

Figure 3.1 Land uses in Latin America and the Caribbean (LAC) in 1990 and 2010 (in million hectares). *Source: own elaboration based on FAO (2013)*

Within LAC, the most important deforestation hotspots are located in Brazil and to a lesser extent in Venezuela, Bolivia and Argentina (Table 3.1). Since 1990, Brazil alone has lost over 55 million hectares, although the rates of deforestation have slowed down significantly over the last years. According to the National Institute for Space Research (INPE) deforestation rates in the Brazilian Legal Amazon have diminished from about 2.9 million hectares per year in 2004 to 0.47 million hectares per year in 2012

(INPE, 2012). Deforestation rates in other Brazilian biomes (e.g. Cerrado, the Brazilian savannah) remain high, but overall it is patent the progressive regression of deforestation on a national level. This slow down in forest cover loss has not been observed yet in Venezuela, Bolivia and Ecuador, where deforestation rates have remained stable or even increased in the last years. In Mesoamerica, the largest forest losses have occurred in Mexico, Honduras, Nicaragua and Guatemala. In the Caribbean region the trend points into a different direction, since forest area has increased over 10,300 hectares between 2000 and 2010 (FAO, 2010).

Table 3.1 Deforestation rates across Latin America between 1990 and 2010. Figures have been rouded to the nearest decimal.

COUNTRY	ANNUAL RATE OF DEFORESTATION (million ha/yr)			TOTAL DEFORESTATION (million ha)
	1990–2000	2000–2005	2005–2010	1990–2010
BRAZIL	2.9	3.1	2.2	55.3
VENEZUELA	0.3	0.3	0.3	5.8
BOLIVIA	0.3	0.3	0.3	5.6
ARGENTINA	0.3	0.3	0.2	5.4
ECUADOR	0.2	0.2	0.2	4.0
PARAGUAY	0.12	0.2	0.2	3.6
PERU	0.1	0.1	0.2	2.2
COLOMBIA	0.1	0.1	0.1	2.0
MEXICO	3.5	1.2	0.8	5.5
HONDURAS	1.7	0.6	0.6	2.9
NICARAGUA	0.7	0.4	0.4	1.4
GUATEMALA	0.5	0.3	0.3	1.1

Source: FAO (2010)

Figure 3.2 shows the prevailing land use trends across LAC's territory since the 90s.[1] Overall, LAC's territory has been very dynamic during the last two decades, with 40% of the territory (over 900 million hectares) experiencing either a change in land use or in land cover. This dynamism is the result of two major trends: (1) a pronounced reduction of the forest cover, either due to large-scale deforestation for cultivation or through small to

---

1 The land use trends have been obtained from the land use transition matrix created by combining the 1993 Global Land cover (USGS 2008) and the 2009 Glob Cover Map (ESA 2010) for LAC. Map sources have different spatial resolutions and legends, therefore figures on land use trends need to be considered as a first gross approximation to the real size of ongoing land use trends in LAC.

medium-scale forest clearing for cattle, mining and subsistence agriculture; and (2) a less pronounced but growing trend of reforestation, which combines processes of secondary natural succession, human-induced afforestation and woody encroachment on previous cultivated areas.

Deforestation and expansion of the agricultural frontier has been the dominant trend in LAC in the last two decades (Figure 3.2). The greatest expansion of pastures and arable land has occurred in South America, mostly in Brazil, Argentina and Paraguay. In Mesoamerica, countries like Nicaragua, Honduras, Panama and Guatemala have also seen an increase of their agricultural area, mostly arable land but also permanent pastures for grazing.

Although less intensive, the progressive trend of forest degradation observed in many parts of the region is still important. This can be seen along the northern part of Mexico, in the region of Los Llanos in Venezuela, northwest of Colombia, the Amazonian belt in Brazil, and along much of the Andean region of Peru, Ecuador and Colombia. This trend of forest degradation comes from the clearing of natural forest and shrubs to be turned into pastures. The underlying reasons of this trend might be diverse but some common causes include the extended practice of slash and burn agriculture, extensive livestock grazing, gold mining, illegal logging and crop plantation.

Despite this reduction in LAC's forest area, symptoms of forest recovery, the so-called 'forest transitions' (Mather, 1992), are emerging in some areas. The clearest example of this forest transition is the emergence of new forests on previously cultivated areas or pastures. These new forests are either naturally regenerated or planted (afforested). Such trend is widespread in the southeast and northeast of Brazil and across various areas of northern Mexico (Figure 3.2). Another important reforestation trend is the development of new shrub areas in previously cultivated or grazed areas. The development of this woody vegetation is a natural ecological response to the abandonment of agriculture or grazing activities. In grasslands the ceasing of agriculture normally ends with the encroachment of shrubs, whereas in forest areas, the appearance of this woody vegetation could represent an early successional stage of forest regeneration. Across LAC, this shrub encroachment has mostly occurred in the central-north region of Brazil and in the Argentinean Pampa. These processes of forest recovery largely overlap with the reforestation hotspots identified by Aide et al. (2012), although the size of the reforestation trends seem to be greater in our study. Differences in methodologies, scales and data sources might explain the divergences found across both studies, highlighting the need for further investigation and the difficulties in providing precise figures. Overall, according to our analysis, reforestation in all its forms i.e. through forest natural succession, afforestation, or woody development represents at least 20% of the current forest area in LAC. The extent to which these new 'secondary' forests have or fulfil the same ecological processes as those of primary forests remains unclear and needs further investigation (Lambin and Meyfroidt, 2011).

Grau and Aide (2008) argue that a main driver underpinning reforestation in LAC is related to the industrialization of agriculture, which has contributed to the concentration of production to the most fertile areas, while marginal agriculture has progressively been abandoned, leading to ecosystem recovery. In addition to the changes in the agricultural production system, the strong rural–urban migration flow together with the implementation of conservation policies in many rural areas (*ibid.*) has also favoured forest regeneration. Such evolution of the land use pattern, in which agricultural areas have become highly intensified on the most fertile or suitable lands, and natural areas tend to stand along the slopes or less accessible zones, resembles the land use path followed by other regions such as Europe. Box 3.1 summarizes the complexity of the factors underlying forest transitions and reforestation processes in southeast Brazil.

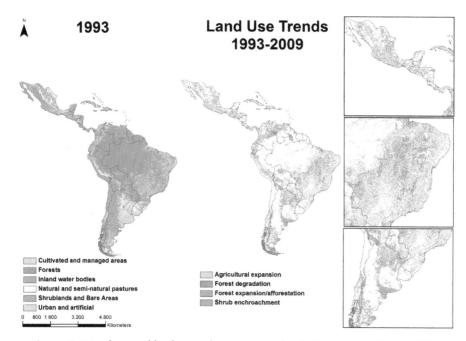

Figure 3.2 Land use and land cover changes occurred in Latin America and the Caribbean between 1993 and 2009. *Source: own elaboration based on 1993 Global Land cover (USGS, 2008) and the 2009 Glob Cover Map (ESA, 2010)*

Overall, agricultural expansion is the predominant land use trend in LAC, although deforestation rates seem to be slowing down and in some cases even reversing. As described in Chapters 1, 5 and 7, the growth of the agricultural sector in LAC is largely related to a growing internal demand for food and energy and ongoing dietary shifts, but is also driven by the rising international demand for oilseeds and cereal grains. To understand past, but foremost, future land use decisions in LAC and develop possible solutions for curbing deforestation and environmental degradation, it is crucial to understand the drivers underpinning the increasing need for agricultural land in this part of the world.

Box 3.1   Drivers of forest transition: theory and practice in São Paulo, southeastern Brazil

Forest transitions – the change in land use characteristics from a period of constant reduction of forest cover to a period of net forest increase – have diverse drivers, including a variety of socio-economic, cultural and political factors. In the last decades some 'pathways' have been proposed to explain the processes and factors behind observed forest recovery across countries (see e.g. Rudel et al., 2005; Lambin and Meyfroidt, 2010). The most common argument is the so-called 'economic development' pathway: economic development associated with industrialization, urbanization, and land use intensification results in agricultural land abandonment and reforestation through secondary succession or tree planting. Also, forest transition would occur when a lack of forest products prompts governments and landowners to plant trees – the 'forest scarcity' pathway (Rudel et al., 2005).

Much of the research conducted in LAC countries like Argentina (Grau and Aide, 2008), Brazil (Perz and Skole, 2003; Baptista, 2008; Walker, 2012), El Salvador (Hecht et al., 2006), and Mexico (Klooster, 2003; Bray and Klepeis, 2005), raised doubts about the broad applicability of forest transition models based on economic development or forest scarcity, emphasizing the importance of a variety of factors linked in a complex network of institutional, social, biological, cultural and physical interactions. In this sense, Lambin and Meyfroidt (2010) proposed the 'globalization', the 'state forest policies' and the 'smallholder, tree-based land use intensification' pathways, which offer more refined explanations of processes involved in forest transitions.

In Brazil, although deforestation rates are greater than forest recovery, forest increase seems to be occurring in some regions. In São Paulo, a southeastern state, evidence suggests that a forest transition took place in the 1990s at the state level, which coincides with a period of overall economic growth in the country (Farinaci and Batistella, 2012) (see Figure 3.3).

Considering only a broad scale, it would be reasonable to explain the forest transition in São Paulo in terms of the 'economic development' pathway, as the state became increasingly urbanized, industrialized and wealthy. However, analysing the processes occurring on a smaller spatial scale, Farinaci (2012) concluded that the transitions observed in municipalities in eastern São Paulo were more influenced by crises and economic stagnation in late 1980s and 1990s – a period in which sustainable development became part of the political discourse in different sectors of society – than by the acceleration of economic growth during the 2000s. Moreover, at the intra-municipality level, forest recovery was not driven by local economic development or agricultural adjustment, but rather by the failure of production systems to ensure the livelihoods of rural population. In São Luiz do Paraitinga, which exemplifies changes occurring in rural areas in eastern São Paulo over the last few decades, the decline of dairy farming was the most important factor influencing recovery of native forest,

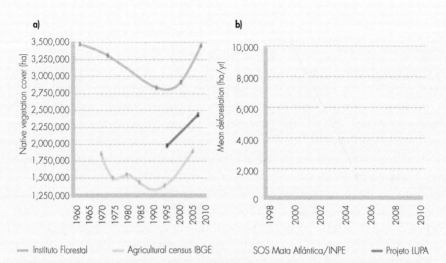

Figure 3.3 **Evidence of forest transition in São Paulo State (Brazil) according to four different data sources.** *(a) Temporal variation on native vegetation cover (b) Deforestation rates between 2000 and 2010 (annual mean values for each period) - (Sources: Kronka et al., 1993, 2005; SIFESP, 2010; Fundação SOS Mata Atlântica and INPE, 2008, 2009, 2010; IBGE (2009); SAA, CATI and IEA (2009).*

predominantly via secondary succession. Modernization of the dairy sector, shortage of rural jobs, lack of public investment on rural infrastructure, and competition with other regions contributed to a decline in dairy farming. Moreover a reduction in soil fertility and rugged relief restricted the possibilities for alternative land uses. Concurrently, an increasing number of people who are willing to purchase land for second residences or tourism activities, often motivated by conservation values, favoured forest recovery. In addition, laws restricting tree cutting and hunting, improvement of fire monitoring systems, and protected areas were important prompters of forest conservation and reforestation. When smaller-scale processes are considered, and put into socio-economic, political and cultural contexts, it is clear that the 'globalization' pathway in association with the 'state forest policies' pathway, as proposed by Lambin and Meyfroidt (2010), provide more comprehensive explanations of the processes leading to forest transitions as observed by Farinaci (2012) in São Paulo.

3.3 What are the drivers of the observed deforestation trends?

Deforestation and land appropriation is an ancient and constant process throughout human history, although driving forces have evolved over time. Around the tropics, deforestation between the 1970s and the early 1990s was largely 'state-driven' to promote rural development (Rudèl, 2007). Government policies varied from region to region, but generally provided incentives for the colonization of remote forests, such as cheap land, and investments in infrastructure (e.g., road building) in order to foster the development process. In the case of LAC, since the 1990s different structural adjustment programmes endorsed by the World Bank, International Monetary Fund (IMF) and other international donors favoured the development of trade liberalization policies. Ever since then, deforestation in LAC has been primarily 'enterprise-driven', particularly by large multinationals (Rudel, 2007). Yet, governments still contribute to these efforts indirectly, e.g. through tax incentives for businesses to settle and also by developing infrastructures, which facilitate and speed up the transportation of goods and natural resources to the nearest harbours (Rudel et al., 2009; DeFries et al., 2010). Tree felling, agricultural industrialization, trade, mining and biofuel are the dominant drivers of current deforestation in many tropical countries (Butler and Laurance, 2008).

Figure 3.4 summarizes some of the main drivers explaining ongoing deforestation trends in LAC.[2] Economic globalization (Factor 1), and particularly the specialization of LAC's economies in the exportation of agricultural commodities (e.g. cereals and oilseeds), explains approximately 21% of the observed forest losses in LAC between 1990 and 2010. This factor is the underlying reason for most of the deforestation in South American countries like Brazil, Bolivia, Argentina, Ecuador and Paraguay. Despite the migration of rural population to the cities, the ongoing efforts to increase the area under protection and the yield improvements, deforestation in these countries has not halted. Whether deforestation is likely to continue in LAC is very much linked to the major drivers underpinning the expansion of agriculture (e.g. international food and biofuel demand, agricultural specialization) and undoubtedly the set of policy instruments and economic incentives (e.g. increases in agricultural productivity, Reduced Emissions from Deforestation and Forest Degradation – REDD+) that may be put in place to reverse deforestation and promote a greener economy. According to FAO (2010), Brazil is responsible for almost 60% of current LAC deforestation, therefore this country is called on to play a key role in this respect, and more recent data suggests that government measures are starting to be effective (Table 3.1).

Nevertheless, the globalization of LAC's economies does not always lead to deforestation. In fact those countries with a high GDP per capita, high agricultural productivity, greater agricultural investments (e.g. in machinery) and with a powerful forestry sector (e.g. Chile or Uruguay) have experienced a net forest area increase despite their strong exporting policies. The extent to which these new secondary forests provide an equivalent flow of ES as the native ones requires further investigation as was mentioned previously.

---

2 To assess the factors underpinning ongoing land use trends in LAC we conducted a multivariate factor analysis (FA) by combining information from twenty-four different socio-economic variables. All variables represent national values for the time period 1990–2010.

Another critical factor of LAC deforestation beyond globalization is the high reliance of many countries on a primary-based economy (see Figure 3.4. Factor 2). High rates of deforestation overlap with countries where agriculture and mining represent a large percentage of their GDP. This factor could explain much of the deforestation observed in Mesoamerican countries like Guatemala, Honduras or Nicaragua, where around 23% of their national GDP is linked to agriculture. These countries have low yields and are mostly land stressed, i.e. they have a low land per capita availability and over 67% of the actual agricultural area is used to produce staples like maize, beans and export crops like coffee. Deforestation in these countries is probably less related to the growth of agricultural exports, and more influenced by the expansion of agriculture to overcome food insecurity problems. The development of the mining industry, mostly in South American countries like Brazil, Peru, Colombia and Ecuador, also appears to be influencing deforestation. Likewise, the development of the livestock sector is an important driver of tropical deforestation. The majority of cattle in LAC is produced extensively in pastures, making the growth of this sector highly dependent on land availability. Since 1990 livestock production has increased 21% in the Caribbean, 44% in South America and 53% in Mesoamerica (FAO, 2012). The value of livestock products in two decades has increased by almost 10,000 million US$ in Mesoamerica and up to 32,000 million US$ in South America (World Bank, 2013). In the Caribbean region, the predominance of a service-oriented economy largely relying on fuel exports and tourism has contributed to preserve and even augment the forest area.

Nevertheless, and despite the importance of the two drivers mentioned above, agricultural expansion and forest area change are also influenced by many other socio-political and legal aspects. For instance, in Colombia much of the reforestation observed between 2001 and 2010 (about 1.7 million hectares) is due to the coca crops eradication programmes enforced by the government (Sánchez-Cuervo et al., 2012). Land tenure and undefined property rights may also be a driver on land use change and its influence will depend on site specific socio-economic dimensions. In Mexico, Bonilla-Moheno et al. (2013) show that the private-common-pool dichotomy was not the dominant explanatory dimension for deforestation; since the greatest differences occurred between types of common-pool systems. Physical variables like altitudinal differences, usually not included in most models of deforestation, can also play an important role in identifying intraregional drivers. One example can be seen in the differences between lowland and montane forest cover changes in Colombia, due in part to the accessibility of forests and differences in wealth and economic activities (Armenteras et al., 2010). The energy sector (e.g. dam construction) is most likely to be an important driver of actual deforestation but no data was found to include this variable in the assessment. All these factors need to be jointly considered in order to identify sustainable land use options at the local level and hence providing opportunities for development and the minimization of environmental trade-offs.

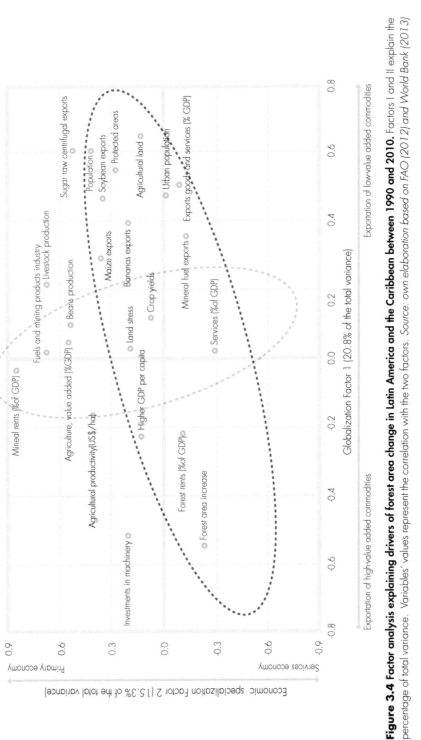

Figure 3.4 Factor analysis explaining drivers of forest area change in Latin America and the Caribbean between 1990 and 2010. Factors I and II explain the percentage of total variance. Variables' values represent the correlation with the two factors. *Source: own elaboration based on FAO (2012) and World Bank (2013)*

3.4 Impacts of land use changes on ecosystem services

The observed changes in land use in LAC have deep implications for the provision of ES. Yet knowledge of the performance of ES in LAC is sparse across countries but overall significant (Balvanera et al., 2012). Much of the existing knowledge on ES is primarily focused on provisioning services, e.g. timber production and freshwater provisioning and regulating services such as water flow regulation or carbon sequestration (*ibid.*). However, less knowledge is available on other key ES, e.g. pollination and pest regulation. Figure 3.5 summarizes the quantification of six ES at the national scale and their trends between 1990 and 2010.

3.4.1 Carbon sequestration

Carbon (C) stocks vary depending on the type of biome and the management practices. Across LAC, the largest aboveground C pools are found in the native tropical forests of Brazil, Peru, Colombia, Venezuela and Bolivia (FAO, 2010). Together these countries store 87,280MtC (million tons of carbon); around 84% of the total aboveground C stock of LAC. The importance of these stocks is related to the extension of their tropical forests but also to the average C content per hectare (>105t/ha), which is above the LAC average.

Between 1990 and 2010 approximately 8,600t C have been lost which is equivalent to 10% of LAC's total C stock. Some 80% of these C emissions have occurred in the aforementioned countries (Brazil, Peru, Colombia, Venezuela and Bolivia). Nowadays land use changes, and particularly deforestation, is the most important source of green house gas emissions (GHG) across most LAC countries, and therefore represents a major driver of climate change (see Figure 3.6). Among some of the most important initiatives currently under negotiation to halt deforestation and mitigate climate change in LAC is through the Reducing Emissions from Deforestation and Forest Degradation (REDD+) (see Box 3.2).

Box 3.2 Enhancing forest conservation through Reducing Emissions from Deforestation and Forest Degradation (REDD+)

Since the end of 2006 negotiations have been held under the United Nations Framework Convention on Climate Change (UNFCCC) to support developing countries in reducing greenhouse gas emissions (GHG) and enhancing forest carbon sinks as a key mitigation strategy. Initially only emission reductions from deforestation and forest degradation were considered, the so-called REDD strategy. But soon given the different national circumstances and the position on the forest transition curve (Perz, 2007a and b) of tropical developing countries, in addition to reducing emissions from deforestation and degradation, the negotiations expanded to further include the conservation of forest carbon stocks, sustainable management of forests and enhancement of forest carbon stocks. This wider scope was agreed upon to allow broad non-Annex I parties (mostly developing countries), based on differing national circumstances, and was renamed REDD+. This climate change solution for developing countries has been endorsed by different initiatives (e.g. the UN-REDD programme, the Forest Carbon Partnership Facility (FCPF) and the Forest Investment Program (FIP), hosted by the World Bank). Currently the UN-REDD programme supports different activities in forty-six countries, including Bolivia, Panama and Ecuador.

Negotiations relating to REDD+ can be traced back to the 11th session of the UNFCCC Conference of Parties (COP) in Montreal (2005), where it was raised as an agenda item that later initiated a two-year process under the UNFCCC's Subsidiary Body for Scientific and Technological Advice (SBSTA), including several technical workshops on the issue. This lead to the introduction of REDD+ as part of the Bali Action Plan at COP13 in 2007, as Decision 2/CP.13, that also provided some early methodological guidance. At COP 15 (Copenhagen in 2009), several principles and methodological guidelines were defined further (Decision 4/CP.15). Parties at COP16 (held in Cancun, 2010), adopted Decision 1/CP.16, section C, defined guidance and safeguards, the need of a phase approach and the five activities under REDD+ in its paragraph 70 by saying: 'Encourages developing country Parties to contribute to mitigation actions in the forest sector by undertaking the following activities, as deemed appropriate by each Party and in accordance with their respective capabilities and national circumstances: Reducing emissions from deforestation; Reducing emissions from forest degradation; Conservation of forest carbon stocks; Sustainable management of forests; Enhancement of forest carbon stocks.'

Since the Bali Action Plan (2007) put forest in the UNFCCC agenda, there is not one single understanding of REDD+ and even greater diversity of views on how best to slow or halt deforestation, but there is a wide recognition of the complexity and that progress is being made in understanding diversity and the importance of national circumstances and drivers of the deforestation and forest degradation. For example, some view REDD+ strictly as a mechanism that provides financial payments for verified emission reductions while for others it is a broader suite of actions and incentives that, when combined, reduce emissions from deforestation and forest degradation.

In light of the new challenges, the lessons learnt during the past three years and the recent discussion at COP18 in Doha, it seems several pathways may be considered for the financing of REDD+ activities and allow countries to adopt alternative development pathways in which deforestation is reduced by tailoring the measures to their needs and national circumstances. However, when creating a forest protection climate agreement, which includes international incentives, it is important to note that if markets have to be considered, deeper commitments from major emitters, with their large mitigation potential, would be required if they need to be environmentally acceptable or politically palatable.

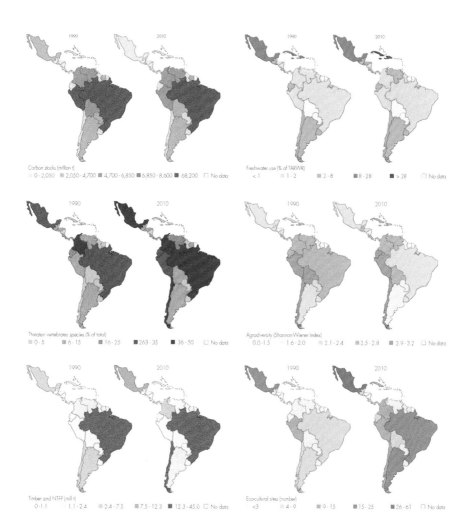

Figure 3.5 **Trends in Ecosystem Service provision in Latin America and the Caribbean between 1990 and 2010.** *Data and indicators to measure the ES performance are as follows. Carbon sequestration was measured using data on aerial carbon pools obtained from the Global Forest Resource Assessment (FRA) performed by FAO (2010) and the indicator used accounts for the total amount of carbon stored aboveground. Soil carbon stocks are not considered here. Freshwater use data was obtained from FAO (2013) and refers to the % of total actual renewable water resources (TARWR) withdrawals for human uses. Biodiversity data was obtained from the Red-list database of the International Union for Nature Conservation (IUCN, 2013). In order to account for the LAC's agro-diversity, we used the Shannon-Wiener index to measure the variety of crops grown in each country and the relative importance of each one (in terms of area dedicated to its cultivation) during two time periods (1990–2000 and 2000–2010). Timber and non-timber forest products (NTFP) data was obtained from FAO (2010) and the number of ecosites represents the sum of World Heritage Sites (WHS) and Biosphere Reserves (BR) by country and was obtained from UNESCO (2013).*

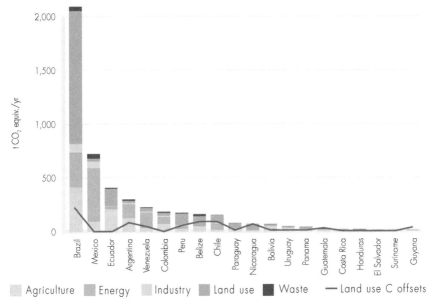

Figure 3.6 Greenhouse Gas Emissions (GHG) by sector in LAC countries. *Source: own elaboration UNFCCC (2013)*

3.4.2 Freshwater use

LAC is an extremely well-endowed region in terms of blue water availability. As described in Chapters 2 and 6, this region holds one-third of the global renewable blue water resources and the average blue water availability per capita for the whole region exceeds the 30,000m³/cap/yr (FAO, 2013). Over the last few decades water withdrawals have increased, both as a result of endogenous factors such as irrigation development, population growth and urbanization and as a result of exogenous factors such as the globalization of LAC's economies and the increase in exports of agricultural virtual water trade (see Chapter 7).

Freshwater abstractions in LAC have increased nearly 5% between 1990 and 2010, from 277 million cubic metres in 1990 up to 290 in 2010 (FAO, 2013). Such increase implies that 5% of the total actual renewable water resources (TARWR)[3] of LAC is extracted for human uses (Figure 3.5). Only in Mexico, Cuba or Dominican Republic water extractions surpass 15% of the national TARWR. Despite these positive figures, regional water scarcity problems exist in countries like Mexico, Chile, Argentina or Brazil where at least 13% of the population lives in water-scarce basins (see Table 6.3, Chapter 6). Also, as Chapter 6 also outlines, in the majority of countries, pollution rather than over-abstractions represents a greater threat for maintaining this provisioning ES in the medium and long run.

---

3 TARWR stands for total annual renewable water resources

3.4.3 Biodiversity conservation

LAC is home to seven out of twenty-five world biodiversity hotspots for Conservation Priority (IUCN, 2013). Mega-diverse countries such as Brazil, Colombia, Ecuador, Mexico, Venezuela and Peru alone cover less than 10% of the world's terrestrial surface but contain approximately 70% of the world's mammals, birds, reptiles, amphibians, plants and insects (*ibid.*). Yet 11% of the total number of vertebrate species identified in LAC are threatened (IUCN, 2013) as shown in Figure 3.5. Yet the countries with the largest ratio of threatened species are: Chile (50%), Brazil (43%), Colombia (42%) and Mexico (41%). Countries with ranges of threatened species varying between 20 and 40% are: Ecuador (32%), Peru (28%), Argentina (25%) and Venezuela (20%). The underlying drivers of this decline in order of importance are (IUCN, 2013): agricultural expansion and habitat change (in 25% of the cases); tree felling and wood harvest (22%); urbanization (13%); agricultural and forestry pollution (12%); and alien and invasive species (10%). In less than 10% of the cases climate change was the underlying driver of species pressure, which highlights a key fact: among global drivers, land use changes by far exert the largest pressure on biodiversity, even above climate change.

3.4.4 Agro-diversity

LAC is the home to some key food components of our diets. The highest agro-diversity within LAC is found in the Andean region and Brazil, although in the last two decades, this agro-diversity has decreased sharply (see Figure 3.5). This loss of agricultural diversity is very much related to the progressive trend of agricultural specialization into oilseed and cereal grain production (mostly soybeans, maize, wheat, barley) and also into bio-fuels such as sugar cane. Among all the crops grown in LAC, over sixteen were originally domesticated in this part of the world (see Table 3.2). Cotton, beans and sunflower are the native crops that have experienced the greatest reduction in area cultivated since the 1990s. Maize on the other hand has experienced a sharp increase, particularly in Argentina, Paraguay, Brazil and also Nicaragua and Venezuela. Much of the loss in agricultural area of native species has been due to the expansion of non-native crops like soybean, which has increased its area 2.5 times since 1990. Sugar cane area has also increased substantially. Soybean expansion in Brazil is mostly related with the increasing demand of animal feed by the EU27 and more recently China, whereas sugar cane production has mostly increased as a result of internal biofuel demand.

3.4.5 Forest products

Commercial forestry in LAC is mostly oriented towards the production of non-timber forest products (NTFP) such as pulp. This pulp comes predominantly from softwood tree plantations of Eucalyptus spp. and *Pinus radiata* and it is used to produce paper. The development of the paper industry in LAC is relatively new compared to other parts of the world. To a large extent this has been driven by government policies that have boosted forestation based on high-yielding species to promote the paper industry. Brazil, Chile and Uruguay are currently the three leading countries in the paper industry within

Table 3.2 Trends of native and non-native agricultural crops cultivated in Latin America

	COMMON NAME / SCIENTIFIC NAME	ORIGINALLY FROM	DOMESTICATION DATE	AREA 1990 (ha)	AREA 2010 (ha)
NATIVE	Beans *Pachyrhizus ahipa* *Pachyrhizus tuberosus* *Phaseolus vulgaris*	Andean Region	<1000 BC	8,178,705	6,788,716
	Squash and pumkins *Cucurbita pepo*	Mesoamerica	7000 BC	152,556	6,788,716
	Maize *Zea mays*	Mesoamerica	6000 BC	24,893,987	28,735,226
	Manioc/cassava *Manihot esculenta*	Lowland South America	6000 BC	2,744,838	2,697,564
	Avocado *Persea americana*	Mesoamerica	2000 BC	160,276	272,564
	Chilli peppers *Capsicum annuum*	Mesoamerica	5000 BC	139,843	237,227
	Chilli peppers *Capsicum baccatum*	Andean Region	4000 BC		
	Cotton *Gossypium hirsutum*	Mesoamerica	5000 BC	3,723,923	1,617,139
	Sunflower *Helianthus annuus*	Eastern North America	2000 BC	2,948,417	2,054,437
	Sweet potato *Ipomoea batatas*	Andean Region	4000 BC	252,571	273,136
	Tobacco *Nicotiana tabacum*	Andean Region	1000 BC	473,209	609,169
	Pinaple *Ananas comosus*	Lowland South America	<1000 BC	96,227	222,481
	Cocoa *Theobroma sp*	Mesoamerica	2000 BC	1,490,618	1,529,507
	Quinoa *Chenopodium quinoa*	Andean Region	4000 BC	47,585	99,499
NON-NATIVE	Soybean *Glycine max*	East Asia		18,035,280	46,181,492
	Sugar cane *Saccharum ssp*	South Asia		7,932,457	12,014,797
	Wheat *Triticum spp*	Near East		10,673,991	8,819,368

Source: own elaboration based on Pickersgill (2007) and FAO (2013)

LAC. The availability of space for cultivation together with the advantageous climatic conditions are two important factors explaining its comparative advantage and much of the growth of this sector, particularly since the mid-20th century (Lima-Toivanen, 2012). In fact Brazilian and Chilean pulp and paper producers are among the most profitable companies producing fast-growing eucalyptus trees and have become cost leaders in the production of market pulp (Gurlit et al., 2007).

Brazil, Chile and Mexico are the largest producers of pulp and accrue over 80% of the continental production. Argentina used to be an important producer in the 1990s, but lately it has lost its market share within LAC (from 11% of total LAC pulp production to less than 2%). According to FAO (2010), since 1990, pulp production has increased sharply among the largest producers and also amongst medium producers such as Colombia and Uruguay (see Figure 3.5).

3.4.6 Eco-tourism

The rich diversity of species and ecosystems found in LAC together with its diverse indigenous cultures, provide a wealth of opportunities for recreation and tourism. On the continental scale it is difficult to measure the performance of this cultural ES, as it is determined by a large set of natural, cultural and economic factors. As a proxy indicator to account for the eco-cultural importance of LAC we used the number of World Heritage Sites (WHS) and Biosphere Reserves (BR) as defined by UNESCO (2013).

Mexico, Brazil and Peru are the countries holding the largest number of WHS and BR, here grouped under the name of 'eco-cultural' sites (see Figure 3.4). These three countries also account for the majority of the new WHS and BR declared since 1990. The Caribbean region, except Cuba, has a very small number of 'eco-cultural' sites. In South America, countries like Argentina and Bolivia have experienced a significant increase. The number and progress of WHS and BR in a way represents the effort that regional and national governments are performing to preserve important natural and cultural features and promote them amongst national and international tourists.

Table 3.3 summarizes the trends in ES performance across LAC regions between 1990 and 2010. The general trend points towards a reduction in performance of regulating and some cultural services, whereas production and other cultural services such as eco-tourism opportunities are increasing. The Caribbean region, however, follows an

Table 3.3 Changes in ecosystem service supply (expressed in percentage) across Latin America and the Caribbean between 1990 and 2010. Green values refer to an increase in service supply, whereas orange values stand for service's reduction. Note: ES classification is based on MA (2005)

REGION	REGULATING		CULTURAL		PROVISIONING	
	Carbon Stocks	Biodiversity	Ecosites	Agro Diversity	Forest Products	Water Extraction
AMAZONIAN	-8	-15	200	-5	71	0.1
ANDEAN	-7	-8	88	-2	8	0.1
CARIBBEAN	33	-2	213	11	-14	-1.2
SOUTH CONE	-8	-14	128	-13	24	0.2
MESOAMERICA	-15	-7	166	3	39	0.4

Source: own elaboration based on data from FAO (2010), FAO (2013), IUCN (2013) UNESCO (2013)

inverse trend, with a general increase in the provision of regulating and cultural services and a general decrease in the demand of provisioning services.

As Chapter 4 outlines, human well-being indicators have improved for the most part, which raises the question about to what extent the observed loss of ES diversity is a consequence of having improved the living conditions of LAC inhabitants. For instance, Rodrigues et al. (2009) found a boom-and-bust pattern in levels of human development (life expectancy, literacy and standard of living) across the deforestation frontier in the Brazilian Amazon, where human development increased rapidly in the early stages of deforestation and then declined as the frontier advanced. Per capita timber, cattle and crop production also reveal a boom-and-bust pattern across the deforestation frontier.

3.5 What options are available in order to spare land and halt deforestation?

Taking into consideration the different drivers of deforestation across LAC, it is clear that a pool of different measures is needed in order to overcome the existing competition for land and develop regional land use strategies to balance food production, rural development and the maintenance of LAC's ES in the long run.

One possible solution is to unwind land competition in LAC as a further intensification of agriculture. Strategic and sustainable agricultural intensification, in terms of elevating yields of existing croplands of under-yielding nations, might be the solution to meet the global crop demand without causing irreversible ecosystem damage (Tilman et al., 2011). In countries like Honduras, Guatemala and Nicaragua, staple crops such as maize have yields below 2.1t/ha/yr, two and three time smaller than those obtained in Brazil or Argentina at present (FAO, 2012). In order to bridge this yield gap, rural development programmes need to be fostered, together with further investments to modernize agriculture, and ensure greater legal certainty to secure such investments, e.g. a better definition of tenure rights (IICA, 2013).

Despite the existing yield gaps across some countries, LAC's agricultural productivity as a whole has increased substantially during the last few decades (Ludena et al., 2010; Maletta and Maletta, 2011). Soybean yields in major producer centres such as Brazil increased at twice the US rate, from a much lower base since 1990 (FAO, 2012), and the yield of tree plantations for wood and pulp in Chile, Brazil and Uruguay is three to four times the level that can be achieved in Europe (FAO, 2010). Soybean, maize and wood-based fuels are the key actors in the agricultural and livestock sector and industries in LAC, and improvements in their productivity may help to spare land. In fact when assessing the evolution of the agricultural sector, it is clear that in the last decade, agriculture growth is mostly being attributed to increasing efficiency and becoming more and more decoupled from land inputs (Figure 3.7).

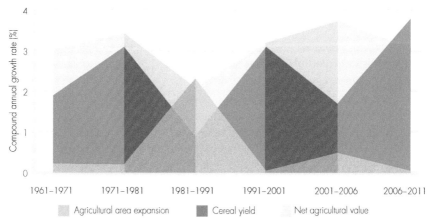

Figure 3.7 Annual growth rates of agricultural land, yields and net production value. *Source: own elaboration, based on FAO (2012)*

Nevertheless, the land-sparing argument, based on modern agriculture, has been criticized for neglecting some important environmental side-effects. It is well known that modern intensive and unsustainable agriculture frequently leads to soil degradation and watershed contamination (Matson et al., 1997; Tilman et al., 2002). Also, natural ecosystems interspersed between highly intensified and productive areas are often forest patches with a low conservation value (Tscharntke et al., 2005; Vandermeer and Perfecto, 2007).

Land sparing through agricultural adjustment has been the predominant land use model followed in Europe and the US. As Tilman et al. (2011) argues, probably the only path to sustain future food demand without causing further ecosystem services losses is through a sustainable intensification of current land use policies, including land use efficiency, together with agricultural practices that avoid depleting soil and biological properties, e.g. agro-forestry practices. Also, a deeper understanding of the environmental implications linked to this land use intensification path is needed (*ibid.*). This will require: determining how land sharing can deliver sufficiently high yields and ecosystem services, assessing trade-offs between increasing yields and environmental benefits across different circumstances and spatial scales, and exploring policy and market mechanisms that enhance sharing initiatives (Garnett et al., 2013).

Nevertheless, Tittonell (2013) recalls on the importance of not falling in to the 'intensification trap', that is the risk of oversimplifying the challenges of feeding a growing population just by intensifying existing agricultural land and balancing environmental trade-offs. He warns against this primarily because intensifying existing agriculture goes hand in hand with larger energy and fertilizer demand, which creates and exacerbates other related societal and environmental problems.

A different argument brought up in support of a less intensive landscape matrix is related to the promotion of organic and wildlife farming agriculture. However, critics argue that organic agriculture may have lower yields and would therefore need more land to produce the same amount of food as conventional farms, resulting in more widespread deforestation and biodiversity loss, and thus undermining the environmental benefits of organic practices. Differences in yields differ greatly depending on the crop type and the region where it is cultivated. According to Seufert et al. (2012), organic to conventional yield ratios of common key LAC products such as soybeans are on average high (0.9). Lower ratios, however, are found for cereals: maize (0.85), barley (0.7) and wheat (0.6).

Overall, and in addition to the pool of measures that can be adopted to overcome land use conflicts between agriculture and nature in LAC, it is important to promote also measures directly aimed at preserving existing nature, e.g. through payment for ecosystem services (see Chapter 14), incentives to reduce deforestation and forest degradation (Box 3.2) and sustainable management of forests and landscape restoration including reforestation. Besides the collection of measures directly targeting at increasing efficient production in the field, off-site efficiency improvements (e.g. along the supply chain) would help to reduce food waste and increase production per unit of land. As IMECHE (2013) highlights we produce about four billion metric tons of food per annum, but it is estimated that 30–50% (or 1.2–2 billion tons) of all food produced never reaches a human stomach due to poor practices in harvesting, storage and transportation, as well as market and consumer wastage. Any such measures should be accompanied by a more transparent food chain with information that will allow consumers to make informed choices.

References

Aide, T.M., Clark, M.L., Grau, H.R., López-Carr, D., Levy, M.A., Redo, D., Bonilla-Moheno, M., Riner, G., Andrade-Nuñez, M.J. & Muñiz, M. (2012). Deforestation and reforestation of Latin America and the Caribbean (2001–2010). *Biotropica*, 0: 1–10.

Armenteras, D., Rodríguez, N., Retana, J. & Morales, M. (2010). Understanding deforestation in montane and lowland forests of the Colombian Andes. *Regional Environmental Change*, 11: 693–705.

Balvanera, P., Uriarte, M., Almeida-Leñero, L., Altesor, L., DeClerck, F., Gardner, T., Hall, J., Lara, A., Laterra, P., Peña-Claros, M., Silva Matos, D.M., Vogl, A.L., Romero-Duque, L.P., Arreola, L.F., Caro-Borrero, A.P., Gallego, F., Jain, M., Little, C., de Oliveira, X.R., Paruelo, J.M., Peinador, J.E., Poorter, L., Ascarrunz, N., Correa, F., Cunha-Santino, M.B., Hernández-Sánchez, A.P. & Vallejos, M. (2012). Ecosystem services research in Latin America: The state of the art. *Ecosystem Services*, 2: 56–70.

Baptista, S.R.(2008). Metropolitanization and forest recovery in Southern Brazil: a multiscale analysis of the Florianópolis city-region, Santa Catarina State, 1970 to 2005. *Ecology and Society*, 13(2): 5.

Bonilla-Moheno, M., Redo, D.J., Aide, T.M., Clark, M.L. & Grau, H.R. (2013). Vegetation change and land tenure in Mexico: A country-wide analysis. *Land Use Policy*, 30: 355–364.

Bray, D.B. & Klepeis, P. (2005). Deforestation, forest transitions, and institutions for sustainability in Southeastern Mexico, 1900–2000. *Environment and History*, 11: 195–223.

Butler, R.A. & Laurance, W.F. (2008). New strategies for conserving tropical forests. *Trends in Ecology & Evolution*, 23: 469–72.

CEPEA (2013). Centro de Estudios Avançados en Economía Aplicada. *Datos PIB do Agronegócio de 1994 a 2011.* ESALQ/USP. [Online] Available from: cepea.esalq.usp.br/pib/. [Accessed June, 2013].

Defries, R.S., Rudel, T., Uriarte, M. & Hansen, M. (2010). Deforestation driven by urban population growth and agricultural trade in the twenty-first century. *Nature Geoscience*, 3: 178–181.

ESA (2010). European Space Agency. Globcover 2009. *Global Land Cover Map.* [Online] Available from: due.esrin.esa.int/globcover/. [Accessed March, 2013].

Farinaci, J.S. (2012). *The new forests of São Paulo State: a multiscale study using the forest transition theory perspective.* PhD dissertation. Instituto de Filosofia e Ciências Humanas, Universidade Estadual de Campinas. Campinas, Brazil.

Farinaci, J.S. & Batistella, M. (2012). Variation on native vegetation cover in São Paulo: an overview of current knowledge. *Revista Árvore* ,36(4): 695–705.

FAO (2009). Food and Agriculture Organisation. *How to Feed the World in 2050.* Rome, FAO.

FAO (2010). Food and Agriculture Organisation. *Global Forest Resource Assessment.* Main Report. Rome, FAO [Online] Available from: www.fao.org/docrep/013/i1757e/i1757e. pdf. [Accessed March, 2013].

FAO (2012). Food and Agriculture Organisation. FAOSTAT. [Online] Available from: faostat3.fao. org/home/index.html. [Accessed May, 2013].

FAO (2013). Food and Agriculture Organisation. AQUASTAT. [Online] Available from: www.fao. org/nr/water/aquastat/main/indexesp.stm. [Accessed March, 2013].

Fischer, G., Hizsnyik, E., Prieler, S., Shah, M. & Velthuizen, H. (2009). *Biofuels and food security. Implications of an accelerated biofuels production.* Summary of the OFID study prepared by IIASA. [Online] Available from: www.ofid.org/LinkClick.aspx?fileticket=O3FeeiHvu7U%3D&t abid=109 [Accessed May, 2013].

Fundação SOS Mata Atlântica & Instituto Nacional de Pesquisas Espaciais (INPE). (2008). *Atlas dos Remanescentes Florestais da Mata Atlântica: Período 2000–2005.* [Online] Available at: mapas.sosma.org.br/. [Accessed June, 2009].

Fundação SOS Mata Atlântica & Instituto Nacional de Pesquisas Espaciais (INPE). (2009). *Atlas dos Remanescentes Florestais da Mata Atlântica: Período 2005–2008.* [Online] Available at: mapas.sosma.org.br/dados/. [Accessed February, 2010].

Fundação SOS Mata Atlântica & Instituto Nacional de Pesquisas Espaciais (INPE). (2010). *Atlas dos Remanescentes Florestais da Mata Atlântica: Período 2008–2010.* [Online] Available at: mapas.sosma.org.br/dados/. [Accessed February, 2011].

Garnett,T., Appleby, M.C., Balmford, A., Bateman, I.J., Benton, T.G., Bloomer, P., Burlingame, B., Dawkins, M., Dolan, L., Fraser, D., Herrero, M., Hoffman, I., Smith,   P., Thornton, P.K., Toulmin, C., Vermeulen, S.J. & Godfray, H.C.J. (2013). Sustainable intensification in agriculture: premises and policies. *Science*, 341: 33-34.

Gloor M., Gatti, L., Brienen, R., Feldpausch, T., Phillips, O., Miller, J., Ometto, J.P., Ribeiro da Rocha, H., Baker, T., Houghton, R., Malhi, Y., Aragão, L., Guyot, J.L.,Zhao, K., Jackson, R., Peylin, P., Sitch, S., Poulter, B., Lomas, M., Zaehle, S., Huntingford, C. & Lloyd, J. (2012). The carbon balance of South America: a review of the status, decadal trends and main determinants. *Biogeosciences*, 9: 5407–5430.

Grau, H.R., & Aide, M. (2008). Globalization and land-use transitions in Latin America. *Ecology and Society*,13 (2): 16.

Gurlit, W., Mencarini, E. & Montealto, R. (2007). *Weighing the risks in South American basic materials.* The McKinsey Quarterly 2007 Special Edition [Online] Available from: commdev. org/files/1730_file_McKinsey_1_.Mining_South_America.pdf. [Accessed October, 2013]

Hecht, S.B., Kandel, S., Gomes, I., Cuellar, N. & Rosa, H. (2006). Globalization, forest resurgence, and environmental politics in El Salvador. *World Development*, 34(2): 308–323.

IBGE (2009). Instituto brasilero de geografia e estatística. Censo Agropecuário 2006. *Brasil, Grandes Regiões e Unidades da Federação.* Rio de Janeiro, IBGE. 777 p.

IICA (2013). Instituto Interamericano de Cooperación para la Agricultura. *Perspectivas de la agricultura y del desarrollo rural en las América: una mirada hacia América Latina y el Caribe.* Santiago, Chile, CEPAL, FAO & IICA.

IMECHE (2013). Institution of Mechanical Engineers. *Global food waste not, want not.* Report No. 31.

INPE (2012). Instituto Nacional de Pesquisas Espaciales. *Projeto PRODES. Monitoramento da floresta amazônica brasileira por satélite.* [Online] Available from: www.obt.inpe.br/prodes/index.php. [Accessed March, 2013].

IUCN (2013). International Union for Conservation of the Nature. *The IUCN Red List of Threatened Species.* Version 2013.1 [Online] www.iucnredlist.org. [Accessed March, 2013].

Klooster, D. (2003). Forest transitions in Mexico: institutions and forests in a globalized countryside. *Professional Geographer*, 55: 227–237.

Kronka, F.J.N, Matsukuma, C.K., Nalon, M.A., Cali, I.H.D., Rossi, M., Mattos, J.F.A., Shin-Ike, M.S. & Pontinha, A.A.S. (1993). *Inventário florestal do Estado de São Paulo.* São Paulo, Brazil, Instituto Florestal-Secretaria do Meio Ambiente. 199 p.

Kronka, F.J.N., Nalon, M.A., Matsukuma, C.K., Kanashiro, M.M., Ywane, M.S.S., Pavão, M., Durigan, G., Lima, L.M.P.R., Guillaumon, J.R., Baitello, J.B., Borgo, S.C., Manetti, L.A., Barradas, A.M.F., Fukuda, J.C., Shida, C.N., Monteiro, C.H.B., Pontinha, A.A.S., Andrade, G.G., Barbosa, O. & Soares, A.P. (2005). *Inventário florestal da vegetação natural do Estado de São Paulo.* São Paulo, Brazil, Instituto Florestal-Secretaria do Meio Ambiente. 200 p.

Lambin, E.F. & Meyfroidt, P. (2010). Land use transitions: Socio-ecological feedback versus socio-economic change. *Land Use Policy*, 27(2): 108–118.

Lambin, E. F., & Meyfroidt, P. (2011). Global land use change, economic globalization, and the looming land scarcity. *Proceedings of the National Academy of Sciences*,108(9): 3465–3472.

Lima-Toivanen, M.B. (2012). The South American Pulp and Paper Industry: The Cases Brazil, Chile, and Uruguay. In: Lamberg, J.A., Ojola, J., Peltoniemi,M. & Särkkä, T. (eds). *The Evolution of Global Paper Industry 1800–2050. A comparative analysis.* Dordrecht, The Netherlands, Springer. pp 243–283.

Ludena, C.E. (2010). *Agricultural Productivity Growth, Efficiency Change and Technical Progress in Latin America and the Caribbean.* IDB Working Paper Series No. IDB-WP-186.

MA (2005). Millennium Ecosystem Assessment. *Ecosystems and Human Well-being: Synthesis.* Washington, DC, Island Press.

Maletta, H. & Maletta, E. (2011). Climate Change, Agriculture and Food Security in Latin America and the Caribbean. Essex, UK, Multi Science Publishing Co Ltd. pp.97–103

Mather, A.S. (1992). The forest transition. *Area*, 24 (4): 367–379.

Matson, P.A., Parton, W.J, Power, A.G. & Swift, M.J. (1997). Agricultural intensification and ecosystem properties. *Science*, 277: 504–509.

Martinelli, L.A. (2012). *Ecosystem Services and Agricultural Production in Latin America and Caribbean*. IDB Environmental Safeguards Unit (VPS/ESG). Technical notes. IDB-TN-382.

Perz, S.G. (2007a). Grand theory and context-specificity in the study of forest dynamics: forest transition theory and other directions. *The Professional Geographer, 59* (1): 105–114.

Perz, S.G.(2007b). Reformulating modernization-based environmental social theories: challenges on the road to an interdisciplinary environmental science. *Society and Natural Resources*, 20: 415–430.

Perz, S.G. & Skole, D.L. (2003). Secondary Forest Expansion in the Brazilian Amazon and the Refinement of Forest Transition Theory. *Society and Natural Resources,*16: 277–294.

Pfister, S., Bayer, P., Koehler, A. & Hellweg, S. (2011). Environmental impacts of water use in global crop production: hotspots and trade-offs with land use. *Environment, Science and Technology*, 45(13): 5761–5768.

Pickersgill, B. (2007). Domestication of plants in the Americas: insights from Mendelian and molecular genetics. *Annals of Botany*, 100: 925–940.

Rademaekers, K., Eichler, L., Berg, J., Obersteiner, M. & Havlik, P. (2010). *Study on the evolution of some deforestation drivers and their potential impacts on the costs of an avoiding deforestation scheme*. European Commission Directorate-General for Environment. Final Report Framework Contract No. DG ENV/G.1/FRA/2006/0073.

Rockström, J., Steffen, W., Noone, K., Persson, Å., Chapin, F. S., Lambin, E., Lenton, T.M., Scheffer, M., Folke, C., Schellnhuber, H. J., Nykvist, B., Wit, C. A. de, Hughes, T., Leeuw, S. van der, Rodhe, H., Sörlin, S., Snyder, P. K., Costanza, R., Svedin, U., Falkenmark, M., Karlberg, L., Corell, R. W., Fabry, V. J., Hansen, J., Walker, B. & Liverman, D. (2009). Planetary boundaries: exploring the safe operating space for humanity. *Ecology and Society*, 14 (2).

Rodrigues, A.S.L., Ewers, R.M., Parry, L., Souza, Jr.C. & Verissimo, A. (2009). Boom-and-bust development patterns across the Amazon deforestation frontier. *Science*, 324: 1435–1437.

Rudel, T., Coomes O.T., Moran, E., Achard, F., Angelsen, F., Xu,J. & Lambin, E. (2005). Forest transitions: towards a global understanding of land use change. *Global Environmental Change*, 15: 23–31.

Rudel, T. (2007). Changing agents of deforestation: From state-initiated to enterprise driven processes, 1970–2000. *Land Use Policy*, 24: 35–41.

Rudel, T.K., Defries, R.S., Asner, G.P & Laurance, W.F. (2009). Changing drivers of deforestation and new opportunities for conservation. *Conservation Biology*, 23: 1396-405.

SAA, CATI & IEA (2009). (Secretaria de Agricultura e Abastecimento/ Coordenadoria de Assistência Técnica Integral/ Instituto de Economia Agrícola). *Levantamento censitário de unidades de produção agrícola do Estado de São Paulo - LUPA*. Available at: www.cati. sp.gov.br/projetolupa. [Accessed May, 2010].

Sánchez-Cuervo, A.M., Aide, T.M., Clark, M.L. & Etter, A. (2012). Land cover change in Colombia: surprising forest recovery trends between 2001 and 2010. PLoS ONE,7: 1–14.

Seufert, V., Ramankutty, N. & Foley, J.A. (2012). Comparing the yields of organic and conventional agriculture. *Nature* 485 (7397): 229–232.

SIFESP (2010). *Sistema de Informações Florestais do Estado de São Paulo*. Available at: www. iflorestal.sp.gov.br/sifesp/. [Accessed July, 2010].

Smith,P., Gregory, P.J., van Vuuren, D., Obersteiner, M., Havlík, P. & Rounsevell, M. (2010). *Competition for land*. Philosophical Transactions of the Royal Society B, 365: 2941–2957.

Tilman, D., Gassman, K.G., Matson, P.A., Naylor, R. & Polasky, S. (2002). Agricultural sustainability and intensive production practices. *Nature*, 418: 671–677.

Tilman, D., Balzer, C., Hill, J. & Befort, B.L. (2011). Global food demand and the sustainable intensification of agriculture. *Proceedings of the National Academy of Sciences of the United States of America*, 108: 20260–20264.

Tittonell, P.A. (2013). Farming Systems Ecology. *Towards ecological intensification of world agriculture*. Inaugural lecture upon taking up the position of Chair in Farming Systems Ecology. Wageningen University, The Netherlands [Online]. Available from: www.wageningenur.nl/upload_mm/8/3/e/8b4f46f7-4656-4f68-bb11-905534c6946c_Inaugural%20lecture%20Pablo%20Tittonell.pdf. [Accessed June, 2013].

Tscharntke, T., Klein, A.M., Kruess, A., Steffan-Dewenter, I. & Thies, C. (2005). Landscape perspectives on agriculturall intensification and biodiversity and ecosystem service management. *Ecology Letters*, 8: 857-874.

UNESCO (2013). United Nations Educational, Scientific and Cultural Organization. *World Heritage Sites and Biosphere Reserve Database* [Online] Available from: www.en.unesco.org/. [Accessed June, 2013].

UNFCCC (2013). United Nations Framework Convention on Climate Change. *Non-Annex I National Communications submitted in compliance with the United Nations Framework Convention on Climate Change*. [Online] Available from: unfccc.int/national_reports/non-annex_i_natcom/items/2979.php. [Accessed May, 2013].

USGS (2008). United States Geological Survey. *Global Land Cover Chracterization Database Version 2*. Reference year 1993 [Online] Available from: edc2.usgs.gov/glcc/glcc.php. [Accessed March, 2013].

Vandermeer, J. & Perfecto, I. (2007). The agricultural matrix and the future paradigm for conservation. *Conservation Biology*, 21: 274–277.

Walker, R. (2012). The scale of forest transition: Amazonia and the Atlantic forests of Brazil. *Applied Geography*, 32: 12–20.

World Bank (2013). *World Development Indicators Database*. [Online] Available from: databank.worldbank.org/data/views/variableSelection/selectvariables.aspx?source=world-development-indicators. [Accessed March, 2013].

SOCIO-ECONOMIC MEGATRENDS FOR WATER AND FOOD SECURITY IN LATIN AMERICA

Coordinator:
Elena Lopez-Gunn, I-Catalist, Complutense University of Madrid, and Water Observatory – Botín Foundation, Spain

Authors:
Rosario Perez-Espejo, Universidad Autónoma de México, México
Elena Lopez-Gunn, I-Catalist, Complutense University of Madrid, and Water Observatory – Botín Foundation, Spain
Manuel Bea, Geosys S.L.,Spain
Guillerno Donoso, Pontificia Universidad Católica de Chile, Santiago, Chile
Pedro Roberto Jacobi, PROCAM /IEE Universidade de São Paulo, Brazil
Julio M. Kuroiwa, Laboratorio Nacional de Hidráulica – Universidad Nacional de Ingeniería, Lima, Peru
Ariosto Matus Perez, Universidad Iberoamericana, México DF, Mexico
Ignacio Pardo, Universidad de la Republica, Uruguay
Andrea Santos, Universidade Federal Fluminense, Rio de Janeiro, Brazil
Bárbara Soriano, CEIGRAM, Technical University of Madrid, Spain
Bárbara A. Willaarts, Water Observatory – Botín Foundation, and CEIGRAM, Technical University of Madrid, Spain
Pedro Zorrilla-Miras, Cooperativa Terrativa, Madrid, Spain.
Ibon Zugasti, Prospektiker, Spain

Highlights

- The chapter provides an overview of the main socio-economic megatrends for Latin American and Caribbean (LAC) countries and how these link to water and food security. Main trends include the demographic transition (population growth, urbanization and migration), development model (income growth, income inequality, poverty and the informal economy), and the impact of globalization (trade liberalization, consumption patterns, food security and health). Other trends are the role of technology and climate change.

- Population will continue to increase, although at a slower pace due to the low fertility rate. LAC is the second most urbanized region in the world. It is a region where the urbanization pattern has been rapid, poorly planned and is causing a growing number of social problems. LAC shows all the signs of international migration processes. Nearly 20 million people live outside the country in which they were born and migrants are especially vulnerable since they are more exposed to risks. Urbanization and migration have changed societies in LAC, their needs and the way the population use their natural resources.

- During the last twenty years LAC's per capita growth rate was 1.6%. High commodity prices are leading to some countries to intensify exports of primary commodities making the region more vulnerable to the global economy. LAC displays poor evidence in terms of reducing poverty given its economic growth. Distribution of wealth is the most important issue for a region which globally is one of the most unequal. The informal economy is growing and informal jobs can reach very high levels.

- Market-oriented reforms adopted during the 1990s have not helped to achieve structural challenges. In many LAC countries the correlation between economic growth and trade openness is weak and trade liberalization has not improved income distribution, neither has it reduced poverty. Trade has changed the dietary patterns of LAC societies thus affecting the use of water. Even though undernourished population has declined, 49 million people are still suffering from hunger.

- LAC is undergoing demographic, epidemiological and nutritional transitions. The latter is characterized by a decrease in malnutrition and an increase in obesity due to dietary changes. The health sector faces two challenges: solving traditional problems of infectious diseases and maternal-child mortality, and combating diseases arising from development: chronic-degenerative, senile and mental illnesses, HIV/AIDS and obesity.

- Information and Communication Technology (ICT) may help to guarantee food and water security in LAC. Agriculture can profit from the use of these technologies, improving water-meters and many areas of the food production chain. The participatory approach of water users connected via ICT may create new pathways for water security. LAC countries must increase investment in Research and Development (R&D) which is on average around 0.6% versus 2.3% for OECD.

- Climate change is another phenomenon to consider for socio-economic trends in LAC, due to the high vulnerability of many regions. Increase of some diseases, food insecurity and a growing perception that access to drinking water might be at stake, are some of the potential impacts from climate change. Floods and droughts are and will continue affecting agriculture in particular countries.

- Latin American trends have to be modified. Measures orientated towards achieving a fair income distribution, public policies oriented towards more vulnerable groups of the population, a model of growth supported by domestic markets, formalization of the informal economy, investment in science and technology and policies for improvement and conservation of natural resources, would be key goals to target and report on future socio-economic megatrends to guarantee water and food security.

4.1   Introduction

Latin America is a continent that has experienced dramatic and largely positive changes over the last twenty to thirty years. Development, political stability and an increased global political role bear witness to these changes. This chapter will review these deep dramatic socio-economic changes, identifying, however, some important pending issues and trends for the future. It therefore provides an overview of the main socio-economic and demographic transformations megatrends of Latin American and Caribbean countries (LAC) and as far as possible how these link to water and food security. It will look at the rapid evolution over the last decades regarding what we consider the main 'megatrends'. First, the demographic transition: population growth, urbanization and migration; second, income growth, inequity, poverty and the informal economy; third, changing lifestyles, trade liberalization, consumption patterns, and health; fourth, scenarios on the role of technology and the emergence of vulnerability due to climate change. Finally, we identify some main challenges in terms of socio-economic megatrends for water and food security.

   Population growth, although slowing down, could place increasing pressure on resource use in general and especially through a change in consumption patterns and an increase in food production, an activity that competes with other economic activities for land and water use. Demographic trends and economic growth patterns have produced large differences in economic and social equity, as well as the sustainability of resource use. In recent years, due to a reduction in external demand, growth in the region has been driven mainly by the expansion of the domestic market, stimulated by subsidy policies in most countries.

Economic growth in LAC was 3.9% in 2013 and is projected to be 4.4% in 2014 (UN projections); in 2012 a significant slowdown ended with a Gross Domestic Product (GDP) increase of 3.1% due to the fall in the export sector of non-food and feed sectors, showing the fragility of the current development model that depends on the demand from uncertain and volatile foreign markets for raw commodities.

The global economic crisis of 2008 affected the terms of trade of the region; with the exception of hydrocarbons (oil), whose prices remained stable, and oilseeds, whose prices increased (CEPAL, 2012a). Most industrialized countries in the region face strong competition from Asian economies, which generates a perverse dependence on the demand for low-value added commodities that affects the development of the manufacturing industry. Additionally, LAC faces environmental problems derived from an extraction of natural resources focused on intensive agriculture (biofuels, food and feed production for export), a model based on the use of high quantities of water and agro-chemicals which has impaired water quality and poses a risk to human health. In addition to soil and water pollution, the loss of biodiversity has also been accelerated due to the pressure from mining, forestry, heavy fishing, urbanization and infrastructure development.

The moderate demographic growth, the relatively steady economic progress mirrored through some indicators of human well-being, can present noticeable differences between countries, regions within the same country or between different levels of income. Some emergent health problems such as obesity affect the population at all different levels of income in most countries of the region. LAC is becoming an exporter of primary materials, principally food (and thus virtual water) that contributes to global food and water security but does not necessarily represent the best development model for the region and for its own food and water security.

4.2 Main drivers

4.2.1 Demographic trends and transitions

This section aims to highlight the trends of the demographic transition in LAC. The tendencies of three main topics are analysed: population growth (fertility and ageing), urbanization and migration. The evolution and tendency of these factors has and will have a crucial influence in the growth rate of the demand for food and therefore, in the scale and intensity of natural resources use. Water, as well as other natural resources, is under the stress from the requirements of an increasingly younger population whose consumption habits are radically changed by urbanization processes and migration.

4.2.1.1 Population growth

In the 20th century, countries of LAC, which currently represent almost 9% of the world population, saw their populations grow at a very high pace. For example, by mid-century most of the countries grew by as much as 3% (Miró, 2006). Later in the 20th century this trend had reversed, and population growth in LAC is slowing significantly.

According to the Economic Commission for Latin America and the Caribbean, ECLAC (CEPAL, 2012b), in 2012 there were 603.1 million inhabitants in LAC and the population is predicted to continue growing despite the sustained low fertility rate. This is due to a relatively high concentration of people of reproductive age, coming from periods when fertility was higher, meaning births exceed the number of deaths.

The regional average of the total fertility rate (TFR) was 5.9 children per woman in the period between 1950 and 1955, but has steadily decreased since the second half of the 1960s to the present. From 1965–1970, the TFR in the region fell by 59% with huge variations among countries, ranging from 20% to 70%. Fertility in LAC in the 1950s and 60s was only surpassed by Africa (6.8 children per woman) and was above the world average of 5.0 children per woman. Currently, the regional value is below the global 2.5 children per woman and resembles the figures seen in Europe forty years ago.

The decline in regional fertility has been sustained, but there are still differences in the current level of fertility between countries. The total fertility rate of Guatemala for the 2005–2010 period is the highest in LAC with an average of 4.15 children per woman and nearly threefold that of Cuba, which is the lowest with an average of 1.49 children per woman

Given the impact of fertility on population projections, the Population Division of the United Nations recommends developing three evolution scenarios for this variable. In the case of mortality and international migration there is only one hypothesis for future changes. In the case of fertility, the most plausible hypothesis is designated as *recommended or media*, while the other two hypotheses for the top and bottom strips of the recommended are also estimated. With the *media or recommended* hypothesis, in 2050 fertility would be 1.85, a figure below the replacement level of 2.2 that is likely be achieved in the period 2015–2020. However, the population would continue growing to 760 million inhabitants in 2050; with the high scenario it would reach 900 million people.

Nowadays the population pyramid of LAC has a rectangular base representing the age group of 40 years (70% of the population of the region). People over 65 (7% of the population) gain relative importance, but the top is still narrow compared with 28% of people under the age of 15. Projections show a decline in varying degrees of the population under 15 years, an increase in the population over 65 years and a thickening of the pyramid between these ages. The region's population will grow older, but despite the decline in the population under 15 years, it will continue presenting a young age structure, allowing for the population to grow as forecasted.

4.2.1.2 Urbanization

LAC is the second most urbanized region in the world, with 79.1% of its population living in cities (UN, 2011); when in 1950 the urban population was less than 42% (CEPAL, 2002). LAC is more urbanized than an average high-income country.

Both megatrends of population growth and urbanization have caused many social changes in recent years. Cities have witnessed and partly helped generate a middle class whose importance has caused substantial changes in transport patterns, habitats and consumption. This urbanization trend is rooted in the early 20th century and is deepening. In Argentina, Chile, Venezuela, Uruguay and Puerto Rico at least nine out of ten people live in urban settings (Table 4.1).

Table 4.1 Evolution of urban population, percentage living in urban areas by region (1925–2000)

REGIONS	1925	1950	1975	2000
World	20.5	29.7	37.9	47.0
Most development regions	40.1	54.9	70.0	76.0
Less development regions	9.3	17.8	26.8	39.9
Africa	8.0	14.7	25.2	37.9
Latin America & Caribbean	25.0	41.4	61.2	75.3
North America	53.8	63.9	73.8	77.2
Asia	9.5	17.4	24.7	36.7
Europe	37.9	52.4	67.3	74.8
Oceania	48.5	61.6	71.8	70.2

Source: year 1925: Hauser and Gardner (1982); years 1950–2000: UN (2011)

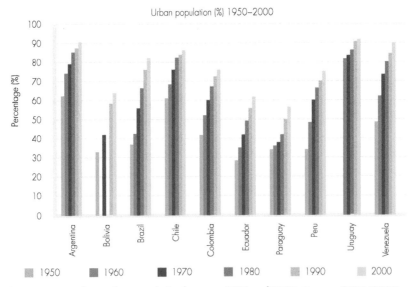

Figure 4.1 Trends in urban population between 1950 and 2000. *Source: CEPAL (2006).*

This pattern of urbanization that has prevailed in LAC has been rapid and poorly planned, not creating an ideal spatial distribution of the population, which is concentrated in large cities. In 2005, there were sixty-seven cities with more than one million inhabitants and four 'megacities' with more than 10 million (Mexico City, São Paulo, Buenos Aires and Rio de Janeiro). These 'megacities' are characterized by inequality, and social problems with a segregated profile in spatial and social terms.

However, the current trend is somehow different. Since 2000, the average annual growth of the urban population is less than 2%, which is a fairly normal population growth (UN-Habitat, 2012). Moreover, the growth of medium-sized cities is an opportunity to overcome the urban problems of the larger cities on the continent.

Often, population growth in urban centres outpaces the ability of utilities to provide adequate services such as water and sanitation. In the absence of piped water systems, communities in these areas meet their water needs through a combination of different sources and means. According to the Global Water Partnership (GWP, 2012) the challenge in LAC is to accelerate the incorporation of the mobile population into informal settlements in order to ensure the formal structure of housing and water and sanitation services. Thus the phenomena of urban transition, the formalization of the economy and water security are all linked. Meanwhile the opposite also holds: rapid urban growth exacerbates the problem (see Box 4.1).

Box 4.1   Slums and access to piped water

Considering that LAC is the second most urbanized region in the world (after North America), the case of slums is extremely pertinent. As can be seen in Figure 4.2 there seems to be a strong correlation between having a high number of people living in slums and overall lack of access to piped water.

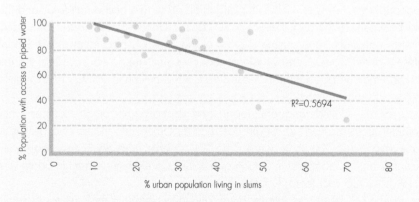

Figure 4.2 Population living in slums and population with access to piped water. *Source: own elaboration based on data from: UN-Habitat, (2012), UN-DESA data (2011) and WHO-UNICEF (2013).*

4.2.1.3 Migration

International migratory processes are motivated by economic, social, cultural and political factors. Recently, studies have also included environmental factors. The total migrant population has been calculated as 3% of the total inhabitants of the planet and 13% of them (about 25 million) were born in LAC. It could be seen as a minor phenomenon but it has a true significance not only in quantitative terms, but also in its impact on social and economic life for both the migrant's country and the host country (CEPAL, 2003).

The Report on Migrations in the World 2010, published by the International Organization for Migration (IOM, 2011), reveals that the number of international migrants increased 11%, from 191 million in 2005, to 214 million in 2010. The Report also indicates that the number of domestic migrants was 740 million in 2009; implying that globally the number of domestic and international migrants is close to 1,000 million, a figure likely to keep increasing (Domínguez-Guadarrama, 2011).

LAC is the scene of intense migration processes that have changed societies in many ways. In this region, all the different types of modern international migration have taken place, from the migration of LAC people (the most visible feature), to immigration, return, irregular migration, forced displacement and the search for shelter, plus the flow of remittances, skilled migration and the presence of dense communities abroad (*ibid.*).

According to the Department of Economic and Social Affairs of the United Nations, in 2010 six out of ten migrants live in developing regions, three-quarters of migrants are concentrated in only twenty-eight countries, and one in every five lives in the United States (UN, 2011). But it is important to point out that only 37% of global migration is from developing countries to developed countries. Most of the displacement takes place between countries in the same category of development.

Furthermore, in LAC nearly 20 million people live outside the country in which they were born and three-quarters of them move to the United States, mainly from Mexico and the Caribbean. From 1970 to 1980, this migration grew two and a half fold, and then duplicated between 1980 and 1990. In 2010, the United States hosted around 43 million foreign nationals, representing 13.5% of the total US population (World Bank, 2011). Results of the 2010 Census indicate that Hispanics made up 16.3% of the total population and that the population increased from 35.3 million in 2000 to 50.5 million in 2010 (Pew Hispanic Center, 2011).

Canada, Spain, United Kingdom, Japan and Australia are other countries where Latin American migrants often go. Spain has recently turned into the second destiny for regional migration; in 2001 there were 840,000 people from South America (mainly from Ecuador) living in Spain and in 2009, one in three foreigners resident in Spain were from LAC (2,479,035 registered) (CEPAL, 2011).

LAC migration has been also intraregional, due to geographic and cultural proximity. In the 1970s, the number of intraregional migrants was near 2 million; in the 1980s and 1990s it grew slowly but by 2000, migrants numbers reached 3 million and by 2005 almost 4 million (3,800,000). At the beginning of the 1990s, most of the immigrants were from outside the region but in 2010, the majority came from the same region, most of them living in Argentina and Venezuela. Costa Rica was the main destination for Central American migrants (CEPAL, 2012b; IOM, 2012).

Throughout its history, Argentina has received immigrants from all its neighbouring countries: Paraguay, Chile, Bolivia, Uruguay and Brazil. This is the case of Venezuela too, where migration was stimulated by the internal conditions such an economic growth and political stability. Immigration in these two countries is higher than emigration. Recent data shows Argentina, Brazil and Chile as the three South American consolidated regional migration receptors. In Brazil, the number of foreigners has experienced strong growth in the past decade: 961,867 in 2010 and 1,510,561 in 2012.

Bolivia, Colombia, Ecuador, Paraguay and Peru maintain a profile mainly for emigration. Within the region, the Bolivians have a strong presence in Argentina and Brazil, Colombians in Ecuador and Venezuela, Paraguayans in Argentina (325,046 in 2001 and 550,713 in 2010), the Peruvian in Argentina (88,260 in 2000 and 157,514 in 2010) and Chile (39,084 in 2002 and 130,859 in 2010) (IOM, 2012).

Inter-urban flows, moving from one city to another, account for the largest volume of population movement within countries of the region. In Mexico, for example, between 1995 and 2000, 70% of the transfers between municipalities were urban–urban type, while rural–urban migration reached 14%. Internal migration is closely related to regional inequalities. In establishing territorial disparities relevant to migration, labour markets play a major role, especially in regard to wages and unemployment in the different zones. There is no evidence, however, that migration reduces the severity of regional inequalities (CEPAL, 2006).

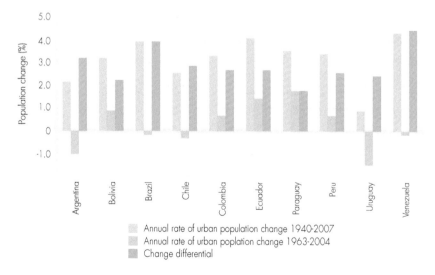

Figure 4.3 Annual rate of urban–rural population change (%). *Source: own elaboration based on data from CEPAL (2006)*

Trade and economic cooperation agreements in LAC (see Chapter 5) such as MERCOSUR (Common Market of the South), CAN (Andean Community of the Nations) and UNASUR (South American Nations Union) have also favoured migration due to its recognition of the importance of the free movement of people (CEPAL, 2012b). Recently, there has been interest in researching the effects of natural disasters; the environment and climate change on migration (see Box 4.4). For example, over 1 million people were estimated to have been displaced due to Haiti's earthquake in 2010 (IOM, 2009).

Even though during the last few years a human rights approach has been progressively introduced into the national and international debate, migrants are very exposed to situations which prevent the exercise of these rights, both during the journey and upon arrival at their destination. These situations include slavery, prostitution, abuse, gender violence, discrimination, expulsion, lack of social support networks and barriers in access to basic health services. In general, this vulnerability is worse in the case of border migrants (CEPAL, 2012b).

4.2.2 Development, income growth, income disparity and poverty

In terms of economic development, LAC was a relatively wealthy region at the start of the 19th century (Millennium Project, 2012). In fact, some countries of LAC were richer than the nascent USA. The Dominican Republic, Mexico and Peru had universities almost one century before Harvard was founded. Haiti was a very wealthy colony in 1800, richer than many parts of the USA. LAC was on a par with most of Europe, and it was richer than Africa, China, India and Japan. At the beginning of the 20th century, Argentina was still one the ten wealthiest countries in the world, and many poor Chinese and Japanese migrated to richer LAC countries like Brazil, Mexico and Peru. However, by the beginning of the 21st century, LAC fell behind, and many countries in East Asia had overtaken it in terms of economic growth. If current trends continue, China will overtake LAC in terms of GDP per capita by the 2020s (Figure 4.4).

Economic growth in LAC in the last thirty years has been modest (in per capita terms) and the varying growth regimes are due to the shocks the region has faced during that period. During the 1970s, shocks were associated with the collapse of the Bretton Woods exchange rate parities and oil-price increases. Throughout the 1980s the region confronted the debt crises and high inflation which was followed by a period of slow and unstable growth and macro-economic instability. Market-oriented reforms were adopted by several LAC countries during the 1990s; however, the slow growth cycle has lasted more than two decades. The different per capita growth rates of seven LAC countries during these periods are presented in Table 4.2.

During the last twenty years (1990–2010) LAC's per capita growth rate has been 1.6% and, of the seven economies studied, only Chile, Costa Rica and Peru exhibit more vigorous growth rates (Figure 4.5). The per capita growth rates observed in LAC throughout the 1980–2010 period also coincide with slower per capita growth in the world economy during the same period.

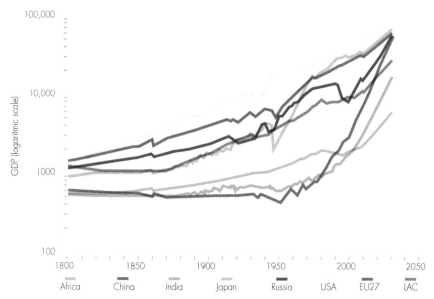

Figure 4.4 Comparative Evolution of GDP per Capita (GDP, logarithmic scale). (Historic and Projections: 1800–2030). Note: The GDP per capita projections are an extrapolation to 2030 using the same growth forecast 2011–2015 by the IMF. *Source: Millennium Project (2012).*

Table 4.2 Real per capita income growth 1960-2010

PERIOD	LESS THAN ZERO	0%–1%	1%–2%	2%–3%	ABOVE 3%
1960–1980			Chile, Peru	Argentina	Brazil, Costa Rica, Colombia, Mexico
1980–2010	Argentina, Peru	Brazil, Costa Rica, Colombia, Mexico			Chile

Source: own elaboration on basis of CEPAL (2005).

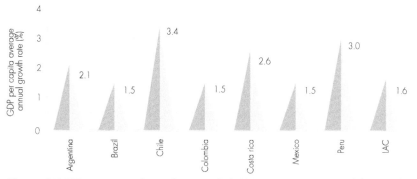

Figure 4.5 GDP mean annual growth rate, period 1980–2010. *Source: own elaboration based on UN data.*

LAC's GDP growth for 2013 was projected to be 3.9% and 4.4% in 2014, but growth volatility is a real possibility. In Brazil, for instance, consensus forecasts for 2012 moved from 3.3% in January to 1.6% in October and drastic corrections are being registered for Argentina (Gurría, 2012). The region still has to tackle many structural challenges in order to turn stability into long-term growth. For example, high commodity prices are leading some countries to favour an economic model based almost exclusively on primary commodities, and this is making the region vulnerable; Chile is an example of this tendency.

Solimano and Soto (2006) found a direct relationship between each country's real GDP per working-age person and 'total factor productivity' (TFP) and the efficiency and rate of the use of capital and labour. Figures 4.6 and 4.7 analyse what appears to be a decoupling in some countries between GDP growth, water consumption and population growth. At LAC scale, this decoupling between population growth and GDP per capita increase and annual freshwater withdrawals seems clear (Figure 4.6). However, a detailed country analysis shows different trajectories (Figure 4.7). While in some countries (Argentina, Peru, Colombia, Costa Rica and Chile) decoupling is a clear, i.e. higher GDP per capita and less water consumption, in other countries (Brazil, Colombia and Mexico) there is no clear trend.

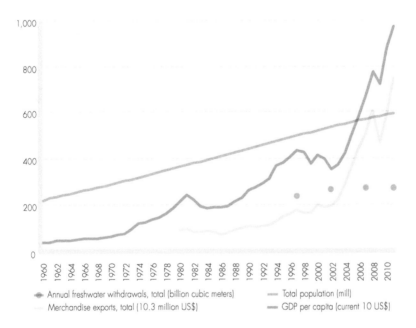

Figure 4.6 Population growth in LAC (1990–2000), Water consumption in LAC (1990–2000) and Evolution of GDP (1990–2000). *Source: own elaboration based on data from World Bank-World Development Indicators database (Population and GDP per capita), FAO-AQUASTAT (Annual freshwater withdrawals) and IMF-World Economic Outlook Database (Total merchandise exports).*

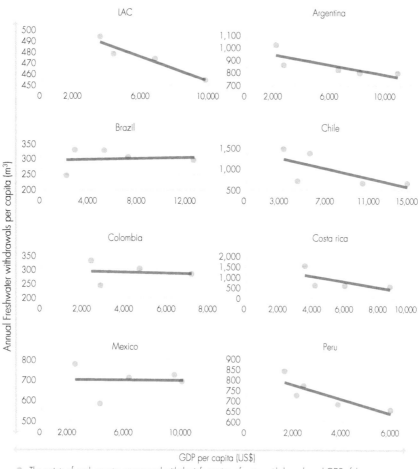

Figure 4.7 Annual freshwater withdrawals per capita vs GDP per capita (1977–2011).
Source: own elaboration on data from FAO-AQUASTAT and World Bank-World Development Indicators database.

4.2.2.1 Beyond GDP: human well-being progress

GDP is an important variable but certainly not the only relevant indicator to measure progress. An analysis based exclusively on GDP would be too simplistic. Thinking beyond GDP we can use a Society–Technology–Economics–Ecology–Politics (STEEP) approach, the Human Development Index (HDI) developed by the United Nations Development Program and other variables. Table 4.3 shows some of the variables included during the Latin America 2030 study on Scenarios (Millennium Project, 2012). It is useful to analyse the, best and worst values for each variable, both in LAC and for the rest of the world.

Table 4.3 Comparative Best and Worst Cases for International Indexes for the year 2010 using the STEEP (Society–Technology–Economics–Ecology–Politics) approach

VARIABLE / INDICATOR/INDEX	WORLD WORST	LATIN AMERICAN WORST	WORLD AVERAGE	LATIN AMERICAN AVERAGE	LATIN AMERICAN BEST	WORLD BEST
Society: HDI (from 0 worst to 1000 best)	0.140 (Zimbabwe)	0.404 (Haiti)	0.624	0.704	0.783 (Chile)	0.938 (Norway)
Technology: E-Readiness index (from 0 worst to 10 best)	2.97 (Azerbaijan)	3.97 (Ecuador)	4.30	5.40	6.49 (Chile)	8.87 (Denmark)
Economics: GDPc (PPP, Thousand US$ 2010)	340 (Congo)	1,121 (Haiti)	10,711	11,188	19,600 (Puerto Rico)	88,232 (Qatar)
Environment: CO_2 emissions per capita (Tons/person)	55.5 (Qatar)	6.0 (Venezuela)	4.6	3.7	0.2 (Haiti)	0.0 (Mali)
Politics: Corruption index (from 0 worst to 10 best)	1.1 (Somalia)	2.0 (Venezuela)	3.3	3.6	7.2 (Chile)	9.3 (Denmark)

Source: Millennium Project (2012).

Notes: (1) The best and worst values correspond to the latest information of the countries with available data in 2010. (2) The Latin American and world averages and based on population-weighted values

The analysis of educational, health or employment indicators also offers relevant information to measure progress in contrast to pure GDP metrics. As Table 4.4 shows, LAC citizens today have greater educational, health and employment opportunities compared to twenty years ago but key issues like wealth distribution and gender equality are pending targets. Regarding health, important progress has been achieved. Life expectancy has increased in all regions, particularly in the Amazonian, Mesoamerican and Caribbean countries. The current rate of life expectancy surpasses 73 years on average. Sanitation facilities and access to safe water source have also increased in most regions but challenges remain in improving access to water and sanitation in rural and peri-urban areas. Schooling rates are also progressing; between 95 to 97% of LAC population complete primary school. Important progress has been reached concerning employment. Nevertheless, the female employment rate (50%) is still far below the men's average (over 70%) and employment vulnerability (unpaid family work or self-employment) has increased for both men and women. Despite this socio-economic progress, income distribution has not improved across all regions. The GINI index has only decreased in the Amazon region and in Mesoamerica. In the other regions it has either increased or remained stable in time. Likewise, the share of the wealth among the richest has increased while it has decreased among the poorest, widening the distance between those that have accrued most of the money and those who have less.

Table 4.4 Trends of human well-being across different regions of Latin America and the Caribbean (LAC) in the last two decades

	DIMENSION	AMAZON		ANDEAN		CARIBBEAN		MESOAMERICA		SOUTH CONE	
	INDICATOR	1988/ 1992	2008/ 2012	1988/ 1992	2008/ 2012	1988/ 1992	2008/ 2012	1988/ 1992	2008/ 2012	1988/ 1992	2008/ 2012
EDUCATION	Rate of males completing primary education (% of relevant age group)	95	91	78	96	95	95	73	95	81	100
EDUCATION	Rate of females completing primary education (% of relevant age group)	88	84	81	97	95	95	71	95	79	100
HEALTH	Life expectancy at birth (years)	63.	71	71	73	70	74	68	75	73	76
HEALTH	Population with access to improved sanitation facilities (%)	68	81	59	66	84	83	66	79	77	89
HEALTH	Rural population with access to improved sanitation facilities (%)	33	64	32	48	81	81	53	71	56	74
HEALTH	Population with access to improved water source (%)	89	94	80	89	91	93	81	92	84	94
JOBS AND EQUITY	Female population employed 15+ (%)	35.8	39.9	35.7	50.6	40.7	48	33.6	43.5	39.4	46.3
JOBS AND EQUITY	Male population employed 15+ (%))	72.1	70.3	75.9	76.5	65.9	67.3	77.9	77.8	75.7	73.3
JOBS AND EQUITY	Vulnerable female employment (% of female employment)	34.1	22.5	37.4	42.1	18.4	31.1	28.6	31.3	27.6	28.4
JOBS AND EQUITY	Vulnerable male employment (% of male employment)	31.4	27.1	31.3	34.5	22.5	40.7	39.9	27.2	22.4	27.8
WEALTH	GDP per capita (constant 2,000 US$)	2,032	2,812	2,258	3,003	6,631	8,498	2,312	3,503	4,020	6,910
WEALTH	GINI index (1–100)	60	55	46	53	45	48	54	51	46	49
WEALTH	Income share held by highest 10%	47	43	35	41	35	38	39	40	36	38
WEALTH	Income share held by lowest 10%	1	1	2	1	2	1	1	1	2	1

Source: own elaboration based on data from World Bank-World Development Indicators database

4.2.2.2 Income growth and poverty reduction

The link between national economic growth and poverty reduction is well known, although it is different for specific countries, each with its own cultural and political history. Due to the negative association between growth and the incidence of poverty,[1] some analysts and international agencies support the recommendation that governments focus on growth in order to alleviate poverty (e.g., Dollar and Kraay, 2001, Ravallion, 2004).

---

1 Statistical analysis has shown that the poverty-reduction elasticity with respect to national income growth has been in the range of 2 to 3.5 percent (Ravallion, 2004)

The debate about growth and poverty is particularly relevant in LAC where countries show poor evidence in terms of poverty reduction over the last fifteen years given their economic growth. Argentina experienced an important increase in poverty during the 1990s, despite having a growing economy during the same period. Poverty significantly decreased in Brazil during the first half of the 1990s, driven by economic growth and improvements in income distribution; however, since 1995, poverty reduction has slowed. There is a positive trend in the reduction of the population below the poverty and indigence line, for the cases of Chile, Brazil, Peru and Colombia, and less constant reduction trends for Mexico, whereas Costa Rica seems fairly stable (Figures 4.8 and 4.9).

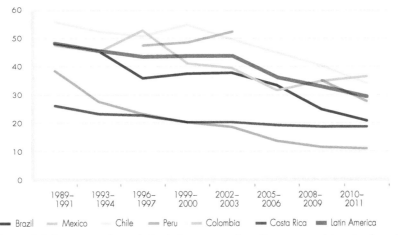

Figure 4.8 **Percentage of population below poverty line.** *Source: own elaboration based on ECLAC (2012).*

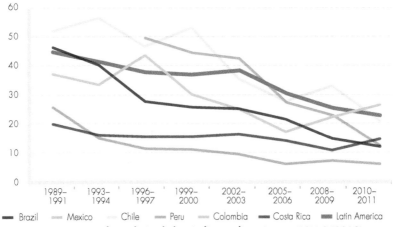

Figure 4.9 **Percentage of population below indigence line.** *Source: ECLAC (2012).*

Chile is a successful story of consistent poverty reduction, from 5 million people below the poverty line in 1990 – approximately 40% of the population – to 2.5 million in 2009 – i.e. about 15%. The rate of extreme poverty also decreased fast, from 13% of the population in 1990 to 3.7% in 2009. The performance of the Andean community economies in terms of poverty reduction has not been so consistent. Poverty has decreased in Peru, whereas in Colombia current poverty levels are equivalent to those of two decades ago. Costa Rica has the lowest poverty rate among Central American countries, and the poverty line is close to 20% while extreme poverty has fallen to almost 7%. Mexico does not have a good track record for poverty reduction; poverty rates recorded during the 2000s are not significantly different from those of the 1990s. Box 4.2 shows how water poverty plays out in LAC, calculated on the basis of research conducted by Lawrence et al. (2003).

For the particular case of Latin America, the most important aspects are not GDP per capita and the reduction or increase in poverty, but the distribution of wealth. GDP per capita has increased in most cases (Figure 4.11) yet things look very different when considering income distribution (Figure 4.12). The income share held by the highest 20% sub-group of population has reduced for the cases of Argentina, Brazil and Colombia, but it is less clear for the cases of Mexico or Chile (Figure 4.12a). In contrast, the income share held by the lowest 20% sub-group of the population has increased in Colombia, Argentina, Brazil and Peru, but it is less marked in countries like Mexico or Chile (Figure 4.12b).

Box 4.2   Water poverty index in LAC

The water poverty index (WPI) is calculated based on a series of parameters related to resources, access, capacity, use and environment. 'Resource availability' is measured taking into account availability but also quality. 'Access' refers to the human access to water, including distance to a safe source, time needed for collection, access for irrigation, etc. 'Capacity' refers to the effectiveness of people's ability to manage water, whereas 'use' refers to the amount of water used for productive uses like agriculture, industry or urban water supply. Lastly, 'environment' accounts for the integrity and flow of ecosystem services provided by freshwater ecosystems. Globally, Finland has the highest WPI score (79) and Haiti the lowest (35). As Figure 4.10 shows, WPI varies across LAC countries, with scores ranging from 55 to 69. Countries with the lowest WPI values are Paraguay, El Salvador, Mexico, Nicaragua and Guatemala. Meanwhile Chile, Ecuador, Uruguay, Costa Rica and Panama have the highest scores. It is striking that countries that do not have the highest rain indexes do not have the lowest WPI. That means that good management is crucial for achieving the best water use given a particularly water resource endowment. For instance, Peru's water resources are slightly more abundant than those of Chile, but Chile has higher levels of the population with access to clean water and sanitation coverage. On the other hand, there are regional differences in each country, especially in the bigger ones: Mexico, Brazil, Argentina or Peru, which have very humid regions and also very dry ones. In these countries water management has to be tailored for each hydrological region to reduce water poverty.

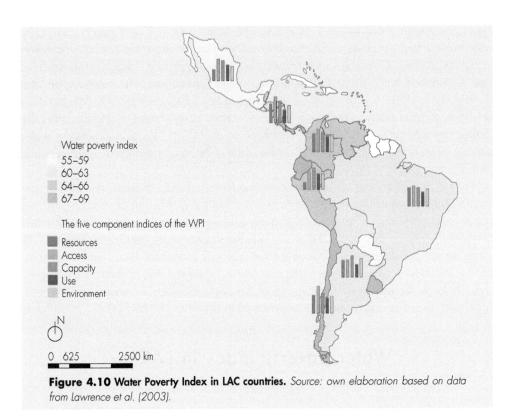

Figure 4.10 Water Poverty Index in LAC countries. *Source: own elaboration based on data from Lawrence et al. (2003).*

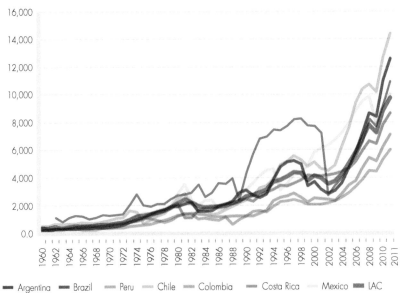

Figure 4.11 Annual GDP per capita growth (expressed in current US$) for the time period 1980–2010. *Source: own elaboration based on data from World Bank-World Development Indicators Database.*

a)

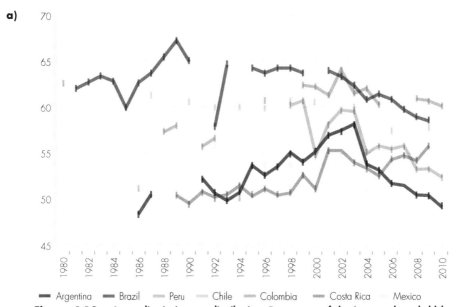

— Argentina — Brazil — Peru — Chile — Colombia — Costa Rica — Mexico

Figure 4.12a Inequality in income distribution. Percentage of the income share held by highest 20% subgroup of population. *Source: own elaboration based on data from World Bank-World Development Indicators Database.*

b)

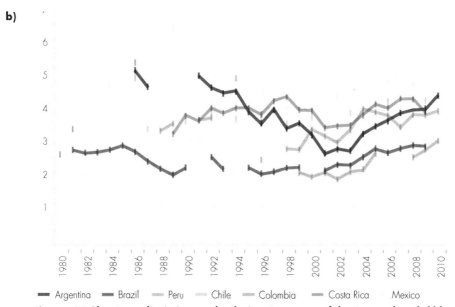

— Argentina — Brazil — Peru — Chile — Colombia — Costa Rica — Mexico

Figure 4.12b Inequality in income distribution. Percentage of the income share held by lowest 20% subgroup of population. *Source: own elaboration based on data World Bank-World Development Indicators Database.*

4.2.2.3 The informal economy in water and food security

The informal economy and its activities are the 'unregistered economic activities that contribute to the officially calculated GDP' (Schneider, 2002). The informal economy contributes to the country's economy but these activities are informal in terms of registration, tax payments, operating licenses, employment conditions and regulations (Becker, 2004). The informal economy represents a transition period that would under normal development paths disappear once countries achieved sufficient levels of economic growth and modern industrial development (Becker, 2004). There is now, however, increased evidence that the informal economy can no longer be considered as a temporary phenomenon. The informal economy has a more fixed character in LAC and this is particularly the case for those countries where incomes and assets are not equitably distributed[2] (Figures 4.12a and 4.12b).

In terms of employment, estimates show that non-agricultural employment as a share of the informal workforce is 57% in LAC. Meanwhile GDP estimates of the contribution of the informal sector (i.e. not the informal economy as a whole, but only informal enterprises) indicate that the contribution of informal enterprises to non-agricultural GDP is significant, representing 29% for LAC (Flodman, 2004). As can be seen in Figure 4.13, the share of informal jobs in total employment can be high, reaching in some cases very high levels like Bolivia, Honduras, Paraguay or Peru, but also pronounced in countries like Colombia, Mexico or Argentina.

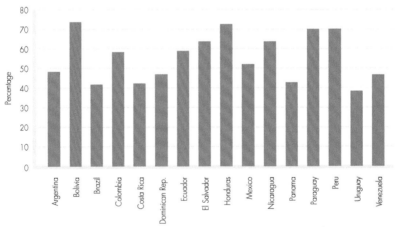

Figure 4.13 Informal employment and the informal economy as part of GDP. *Source: own elaboration based on data from Herrera et al (2004).*

A modern approach to informality does not see informality as an all-or-nothing, but as degrees of informality and formality, along a spectrum where a number of factors can impact its evolution. Informality can become a potential nascent entrepreneurial sector for

---

2 As Becker (2004) states 'if economic growth is not accompanied by improvements in employment levels and income distribution, the informal economy does not shrink'.

growth and innovation or instead a structural problem of under-development and poverty. The majority of informal economy activities provides goods and services, whose production and distribution are legal. LAC has abundant metals, foods and energy resources, and without strong institutions (Millennium Project, 2012), there are risks of criminalization of the economy, and lack of personal safety due to deep social inequalities. Hence the importance of realizing the potential the informal economy has for water and food security in LAC.

4.2.3 Trade liberalization, consumption patterns, food security and health transition

4.2.3.1 Trade liberalization

Over more than four decades, LAC countries have signed agreements and regional integration treaties of different types. The integration aims to remove barriers to free interconnection of the economies in order to increase their production capacity, trade, and investment; that is, to drive economic growth (Guerra-Borges, 2012). There are significant integration structures in the region, including the General Treaty of Central American Integration, the Latin American Integration Association (ALADI), the Southern Common Market (MERCOSUR), the Andean Community and the newly created Union of South American Nations. There are also treaties and other trade agreements of different levels such as the North America Free Trade Agreement signed between Canada, the US and Mexico, or the fifty-eight free trade agreements signed between Chile and other nations. During the 1980s and 1990s, the region undertook deep processes of structural reform: the reduction of state functions, deregulation, and privatization of state enterprises, among others. Trade liberalization (TL), a central component of the reforms, began with a unilateral reduction of tariffs but nowadays includes complex provisions including labour and environmental issues (IICA, 2009).

One question that it is important to discuss is whether trade and trade liberalization benefit the poor in LAC. Trade–poverty linkages are complex and diverse, but according to orthodox mainstream economic theory and empirical findings, they can be divided into a few important pathways: (a) trade-induced growth; (b) effects of trade on prices, income and consumption patterns; (c) effects of trade on wages and employment. Chapter 5 discusses all these pathways in detail. Buitagro (2009) points out that even though trade liberalization has been considered a key element in economic growth and development, indiscriminate trade liberalization strategies have not been beneficial for low- and middle-income countries. Empirical studies conducted in LAC have shown little correlation between trade liberalization, economic growth and poverty reduction (Buitagro, 2009).

Between 1985 and 2000 all economies in the region experienced important TL. Exports increased fivefold in Mexico, tripled in Argentina and doubled in Brazil. A low growth, the unequal distribution of income, and in recent years the high volatility of agricultural prices, has increased society's vulnerability and prevent poverty reduction, especially in the agricultural sector, the most opened in the economy (Table 4.5).

Table 4.5 Merchandise trade in selected countries of LAC (% of GDP)

	1990–2000	2010
ARGENTINA	18	35
BRAZIL	17	19
CHILE	46	61
COLOMBIA	25	31
COSTA RICA	79	64
MEXICO	59	60
PERU	36	45

Source: World Bank-World Development Indicators.

4.2.3.2 Changes in consumption and dietary shifts

Evidence of a change in the dietary patterns of LAC societies have been found since the 1980s. The consumption of fats, animal products, processed foods, fast food and non-alcoholic beverages has increased while cereals, fruits and some vegetables and tuber consumption has diminished. For Regmi (2001), a change in diet occurs gradually and is the result of income growth, urbanization, changing prices, the rise of the processed food sector, changes in the age structure of the population and awareness of food security, among other factors. Morón and Schejtman (1997) and Rastoin (2009) add the 'terciarization' of the agro-food sector and the impact of advertisement.

Regmi et al. (2008) recognize that the global expansion of industries, agribusiness services and supermarket chains, modifies food prices and shapes tastes and diets, tending to standardize how food is produced, distributed and consumed worldwide. For Bermudez and Tucker (2003) food supply is the mechanism by which modernization affects the Latin American diet and the transition or 'convergence' occurs in different social and economic conditions. This has caused a double problem of public health: malnutrition, due to the prevalence of poverty and unequal income distribution, along with obesity and chronic degenerative diseases, result of more 'refined' diets. More use of water and its pollution are collateral problems of dietary changes.

According to the Pan-American Health Organization (PAHO, 2012) at the regional level a detrimental change in nourishing patterns rapidly unfolded. Between 1984 and 1998, the purchasing of refined carbohydrates and sodas increased by 6% and 37% respectively (Rivera et al., 2004).

Box 4.3 The Mexican case

The study by Santos-Baca (2012) corroborates the results of Bermudez and Tucker (2003). She found that the reduction in animal food consumption was created by a reduction in milk intake, a phenomenon that some researchers associate with the increase in soft drink consumption (Rivera et al., 2008: 175). Cereals and legumes particularly beans, which was the basis of the Mexican diet, have reduced their importance. Maize remains the most consumed cereal but the consumption declined 18% from 1992 to 2010, when its price increased 53%. Wheat is the second most important cereal consumed but in much smaller quantities than corn; however, wheat spending is similar to that of corn. The average amounts of consumed fruits and vegetables (sources of essential vitamins and minerals), fell from 0.74kg in 1992 to 0.61kg in 2010; those amounts represent 70% and 66% respectively of the quantity recommended by FAO: 400g per person/day. Beverage consumption, except bottled water, increased 37% between 1992 and 2010, due to the significant increase in the consumption of soft drinks. Consumption of sodas increased 40%, and processed juices and nectars increased 141%. Mexico has the second highest per capita consumption of soft drinks in the world. Unlike in developed countries where changes occurred because of modernization, food transition in Mexico from 1992 to 2010 is characterized by the deterioration in food intake (Figure 4.14).

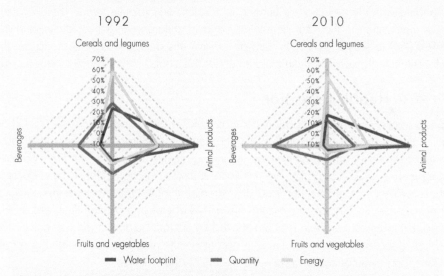

Figure 4.14 Mexican food consumption pattern. Quantity, energy, water footprint of main food products. 1992 and 2010. *Source: own elaboration with data from INEG (1992, 2010), Mekonnen and Hoekstra (2010a, 2010b) and Ercin et al. (2011).*

4.2.3.3 Food security

Hunger currently affects 49 million people living in LAC. This means that 8.3% of the population of the region does not consume the necessary daily calories. Between 1990–1992 and 2010–2012, the undernourished population declined by 16 million people (24.9% over the period), but still an unacceptable number of people are suffering from hunger (FAO, 2012).

The trend in hunger reduction has slowed from 8% between 1990–1992 and 2007–2009, to 2% in 2010–2012. This is the result of the world economic crisis and the slower economic growth in the region but also due to structural problems. The number of hungry people in LAC declined from 57 million in 1990–1992 to 49 million in 2010–2012. The increase in commodity prices and the drought of the last three years have added 3 million people to the category of the poor. Countries like Bolivia and Paraguay, but also Peru, Ecuador and Colombia display hunger problems according to the indicator of chronic malnutrition in children under five years old (FAO, 2012).

On the other hand, adequate calorie consumption does not mean adequate nutritional conditions. Nutrition problems arise from insufficient vitamins and intake of other essential micronutrients. The nutritional problems of the region are not only about hunger in the sense of insufficient energy consumption. Malnutrition caused from inadequate diets which provoke health problems related to nutrients deficiency can also lead to obesity problems (FAO, 2012). According to FAO (2012) Cuba, Argentina, Chile, Mexico, Uruguay and Venezuela have managed to eradicate the scourge of hunger. The case of Brazil is outstanding with the reduction of hunger in absolute and relative terms. The prevalence rate of malnutrition in the total population, and in children under five, diminished from 1999–2000 to 2010–2011. In Peru and Brazil this decrease is very strong but their levels remain high, particularly in children under five years.

4.2.3.4 Health transition

LAC is currently undergoing important demographic, epidemiological (PAHO, 2012) and nutritional transitions (Rivera et al., 2004). The demographic transition is characterized by a reduction in fertility and mortality rates, the increase in life expectancy and by population ageing (PAHO, 2007a and b). The nutritional transition is characterized by a decrease in the prevalence of malnutrition and an increase in obesity, which is a risk factor for chronic diseases such as cardiovascular disease, diabetes and cancer (Pi-Sunyer, 2002). The epidemiological transition is characterized by a triple burden of disease comprising: *infectious diseases*, whose prevalence is declining; *chronic diseases*, registering a rapid increase, and *external causes*, related to accidents and violence.

The drivers of the nutritional transition are complex and multi-causal. Important determinants are the process of urbanization and economic growth, technological changes and innovations that lead to reduced physical activity at work, in leisure and in transportation, and changes in nourishing patterns and dietary intake, with particular emphasis on the increased consumption of processed foods with high-energy content.

The nutritional transition has evolved at different rates in LAC (Barría and Amigo, 2006). Nevertheless, they all display a twofold pattern. On the one hand, there is a diminishing tendency in the prevalence of low weight and height. On the other hand, there is a tendency in the increase in caloric intake, information which has been captured by all surveys of food availability per country (Rivera et al., 2004). Coupled with an increasingly sedentary lifestyle, the result has been a dramatic increase of obesity in many

LAC countries. These are risk factors for morbidity and mortality from diabetes mellitus, hypertension and myocardial infarction, among others.

The health sector in LAC faces a number of challenges: solving problems of infectious diseases and maternal-child mortality, and combating the changes in the disease profiles arising from development and changes in diet; the increase of chronic-degenerative, senile and mental diseases, HIV/AIDS and obesity. Mortality from transmissible diseases and perinatal period decreased while chronic and degenerative diseases linked to external causes (violence, accidents and injuries) have increased. Infant mortality has fallen and the change in age structure has led to an increase of deaths among older adults (Arriagada et al., 2005; CEPAL, 2005).

Obesity is now a widespread growing health problem. Changes in dietary patterns (excessive caloric consumption), sedentary lifestyles, and heavily advertised products with excess fats, salts and sugars, have triggered a rise in obesity (Olaiz-Fernández et al., 2006). Prevalence of hypercholesterolemia found in two cross-sectional samples of adult men and women living in Santiago de Chile increased dramatically in just five years, from 34% in 1987 to 42.5% in men and 46.1% in women in 1992. In Mexico, the mortality rate from diabetes mellitus is 12%. An analysis from 1980 to 1998 of age-adjusted standardized mortality rates for acute myocardial infarction (AMI), diabetes mellitus and hypertension showed a rapid increase of 53% for AMI, 62% for diabetes and 55% for hypertension (Olaiz-Fernández et al., 2006).

In LAC obesity prevalence in adults is high and accounts for over 20% of the population: in Mexico, 33%, Venezuela 31%, Argentina 29% and Chile 29%. The prevalence in children is also concerning. In 2010, more than 2 million under five years old in South America were classified as overweight or obese, more than a million in Central America and approximately 300,000 in the Caribbean. Barría and Amigo (2006) found there is a prevalence of more than 6% of the child population in five countries. Surveys showed that child obesity exceeds 6% in Argentina, Chile, and Peru (Olaiz-Fernández et al., 2006).

The three causes of death (myocardial infarction (AMI), diabetes mellitus and hypertension) have different causes and risk factors. Undeniably, genetics has its influence, but the relationship between these diseases and obesity, poor diet and lack of physical activity is strong and well established in the literature. An important observation about the epidemiological trends in LAC is that obesity and communicable diseases are affecting the populations of all socio-economic levels. Moreover, several studies have found a negative relationship between socio-economic status and prevalence of obesity (PAHO, 2011). Also socio-economic status appears to be positively related to physical activity (Monteiro et al., 2002). These results confront the misconception that obesity is a feature of wealthy populations.

4.3 Other drivers

4.3.1 Scenarios of technological change 2030

A study on potential scenarios for LAC for 2030 was undertaken in 2012. According to the Scenario 1 'Mañana is Today: Latin American Success' of the Latin America 2030 study by the Millennium Project[3] (see Figure 4.15) breakthroughs in science and technology around the world will play a key role. No matter where these advances originated, they will spread quickly throughout the planet. Imagine a scenario in LAC for 2030 where the WTO and Internet 7.0 will help ensure that knowledge moves fast from country to country. Technology will continue improving and synergies among nanotechnology, biotechnology, information technology, and cognitive science (commonly known as NBIC technologies) shall boost technology value and efficiency whilst lowering costs. However, some people could complain about too much technology and unintended consequences like over reliance on technological solutions and furthermore feasible scenarios where socio-political instability and economic constraints become barriers for technology deployment. However, this is one of four possible scenarios for LAC in 2030. These scenarios highlight the role of ICT and technology as game changers, compared to megatrends (NIC, 2008). However, these game changers will probably not materialize without political leadership and vision to address the issues raised earlier in terms of development challenges and opportunities.

4.3.2 Socio-economic impacts of technological change

Information and Communication Technology (ICT) may be the most developed aspect within the NBIC technologies and the best example to analyse their expected impact. ICT has become a key feature in modern life and has proliferated across many sectors, providing new challenges and opportunities. Cell phones that used to be a luxury product can now be bought at an affordable price which has fuelled a rising global ICT market (Figure 4.16).

In 2012, the 2G connection technology (GSM/EDGE) was used by 80% of mobile phones in LAC, although a fast deployment of 3G technologies is foreseen, and the latter should be dominant by 2018 (ERICSSON, 2012). If economic development and consumer demand allow this forecast to be fulfilled, the percentage of individuals using the Internet could skyrocket from 35% in 2011 (16.5% in 2005) to an interval between 70 and 80% in 2018, not far away from the current figures of America's most developed countries, the US and Canada, with approximately 85% (ITU, 2013).[4]

---

3 www.proyectomilenio.org

4 ITU (International Telecommunication Union) is the United Nations specialized agency for information and communication technologies – ICTs.

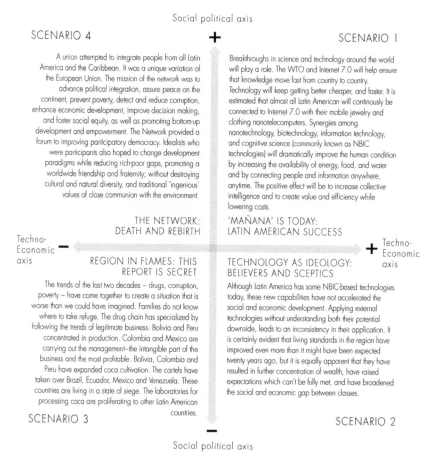

Social political axis

SCENARIO 4 **+** SCENARIO 1

A union attempted to integrate people from all Latin America and the Caribbean. It was a unique variation of the European Union. The mission of the network was to advance political integration, assure peace on the continent, prevent poverty, detect and reduce corruption, enhance economic development, improve decision making, and foster social equity, as well as promoting bottom-up development and empowerment. The Network provided a forum to improving participatory democracy. Idealists who were participants also hoped to change development paradigms while reducing rich-poor gaps, promoting a worldwide friendship and fraternity; without destroying cultural and natural diversity, and traditional 'ingenious' values of close communion with the environment.

Breakthroughs in science and technology around the world will play a role. The WTO and Internet 7.0 will help ensure that knowledge move fast from country to country. Technology will keep getting better cheaper, and faster. It is estimated that almost all Latin American will continously be connected to Internet 7.0 with their mobile jewelry and clothing nanotelecomputers. Synergies among nanotechnology, biotechnology, information technology, and cognitive science (commonly known as NBIC technologies) will dramatically improve the human condition by increasing the availability of energy, food, and water and by connecting people and information anywhere, anytime. The positive effect will be to increase collective intelligence and to create value and efficiency while lowering costs.

THE NETWORK: DEATH AND REBIRTH

'MAÑANA' IS TODAY: LATIN AMERICAN SUCCESS

Techno-Economic axis **—**

+ Techno-Economic axis

REGION IN FLAMES: THIS REPORT IS SECRET

TECHNOLOGY AS IDEOLOGY: BELIEVERS AND SCEPTICS

The trends of the last two decades – drugs, corruption, poverty – have come together to create a situation that is worse than we could have imagined. Families do not know where to take refuge. The drug chain has specialized by following the trends of legitimate business. Bolivia and Peru concentrated in production. Colombia and Mexico are carrying out the management–the intangible part of the business and the most profitable. Bolivia, Colombia and Peru have expanded coca cultivation. The cartels have taken over Brazil, Ecuador, Mexico and Venezuela. These countries are living in a state of siege. The laboratories for processing coca are proliferating to other Latin American countries.

Although Latin America has some NBIC-based technologies today, these new capabilities have not accelerated the social and economic development. Applying external technologies without understanding both their potential downside, leads to an inconsistency in their application. It is certainly evident that living standards in the region have improved even more than it might have been expected twenty years ago, but it is equally apparent that they have resulted in further concentration of wealth, have raised expectations which can't be fully met, and have broadened the social and economic gap between classes.

SCENARIO 3 SCENARIO 2

—

Social political axis

Figure 4.15 Development scenrios for Latin America 2030. *Source: own elaboration based on data from the Millenium Project (2012).*

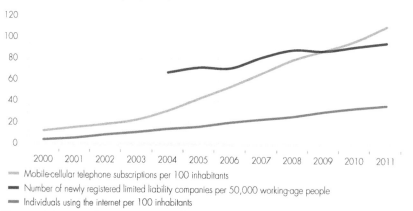

Mobile-cellular telephone subscriptions per 100 inhabitants

Number of newly registered limited liability companies per 50,000 working-age people

Individuals using the internet per 100 inhabitants

Figure 4.16 Trends in **entrepreneurship and access to information and ICT worldwide.** *Source: own elaboration based on data from ITU (2013).*

Thus, in many rural areas of some LAC countries, the breach is greater in basic matters such as access to safe water or water disposal systems than in access to up-to-date ICT. Concerning food security, both public and private sectors are embracing the potential of ICT (World Bank, 2011; Vodafone & Accenture, 2012), concluding that a significant increase in agricultural income can be achieved through their use. The so-called e-agriculture is gaining relevance, taking advantage of ICT to promote new possibilities and alternatives and improve many of the areas related to the food production chain, covering issues as varied as financial services, transportation, commerce and marketing, traceability and quality assurance, storage or training, as well as all activities related to crop management, including the optimization of water management. Farmers can benefit from initial/wider access to credit, logistic and commercial support, building visibility, improving the quality of their products and gaining in capacity and education. Also, farmers can obtain expert agronomic advice, key information on weather forecasts, diseases control or best cultivation practices according to the phenological status of plants.

Concerning water security in LAC, the percentage of people with no access to safe water has successfully decreased from 22.6% in 1980 to 5.8% in 2010. A projection made for 2030 (Millennium Project, 2012) estimates this figure will only reach 3.9%, since most of these people live in rural areas where water plans are difficult to implement. An extended use of ICT could help change this trend. Modern water-meters as well as the fostering of a participatory approach by water users connected via ICT are creating new pathways for water security. As an example, the use of innovative crowd-sourcing approaches via text messages and/or the installation of low-cost performance sensors (Hope, 2012) are allowing the appliance of scale economies to hand-pump construction and maintenance, while increasing transparency of the efficiency and effectiveness of investments.

Technology brings new opportunities and challenges for farmers' capacity building: firstly, promoting online education and training for farmers; secondly, strengthening cooperativism and finally, opening new employment niches, markets and commercialization channels.

However, it would be over simplistic to exalt the role of technology while losing sight of the underpinning structural changes that are needed. For example, the priority of education is fundamental to be able to make the most of these technological opportunities. As can be seen in Figures 4.17, 4.18 and 4.19, the trends in this respect are mixed in relation to primary, secondary and tertiary education, which cautions against the ability to realize the full potential of ICT if no parallel investment is made in education and training.

A further challenge for LAC countries is the need to increase investment in Research and Development (R&D), which is on average around 0.6% of GDP while the same ratio for OECD countries is approximately 2.3%. In order to increase and sustain growth, the region must raise productivity levels to improve competitiveness, which in turn depends on increased innovation and R&D.

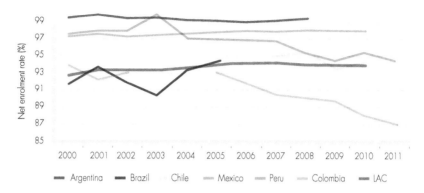

Figure 4.17 Net enrolment rate in first-level education (%). *Source: own elaboration based on data from UNESCO Statistics.*

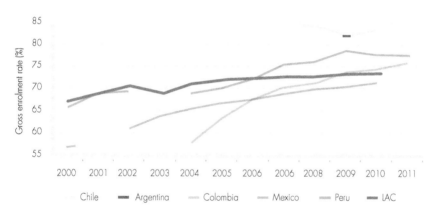

Figure 4.18 Net enrolment rate in second-level education (%). *Source: own elaboration based on data from UNESCO Statistics.*

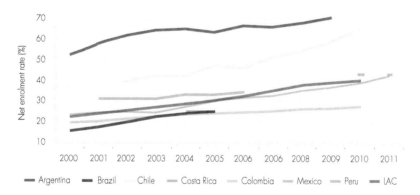

Figure 4.19 Gross enrolment rate in third-level education (%). *Source: own elaboration based on data from UNESCO Statistics.*

4.3.3 Vulnerability to climate change in LAC

Another trend to consider in the future is the potential impact of climate change. Assessing the vulnerability to climate change implies determining the magnitude of adverse effects on the social, economic and ecological systems, its sensitivity to stress factors and its capacity to cope or adapt to the stressor. The Fourth Climate Change Assessment Report of the IPCC (Parry et al., 2007), defines vulnerability as *'The degree to which a system is susceptible to, or unable to cope with adverse effects of climate change, including climate variability and extreme events'.*

The need for improved decision making has motivated an expansion in the number of climate-change impacts, adaptation and vulnerability (CCIAV) assessments and methods in use over the last decades. CCIAVs are undertaken to inform decision making about the degree of risk associated with climate change impacts, so that the most appropriate and cost-effective policy responses can be adopted. The National Communications (NCs) developed by the LAC parties supporting the implementation of the UNFCCC can be classified as a mixed vulnerability and adaptive-based CCIAV. NCs follow mainly a bottom-up approach, where vulnerability to climate is addressed largely as a problem of climate variability within the countries. As Carter and Mäkinen (2011) argue the great majority of assessments that follow this bottom-up approach are found in developing countries, where vulnerability to present-day climatic variability is commonly perceived to be more of a threat than long-term climate change. Table 4.6 summarizes the outcomes of the vulnerability assessments of twenty NCs in LAC countries. The social, economic and environmental risks differ greatly across countries. Nevertheless, social and environmental vulnerability is currently perceived as high in all LAC regions.

In relation to social vulnerability in LAC, the most frequent impacts are related to the increase of diseases, food insecurity and a growing perception that access to drinking water might be at stake. Also, migration linked to worsening climate conditions has been reported, although it is yet unclear whether climate is a driver of such migration flows (see Box 4.4). An increase in the frequency of malaria and dengue fever has also been reported. Food insecurity risks are related to the increased occurrence of the El Niño Southern Oscillation (ENSO) phenomenon and associated with extreme events (e.g. droughts, floods). In 2009/2010 a severe drought affected agricultural areas of Guatemala causing the loss of 100% of the harvest. National production of maize, bean and rice only dropped on average 1.5%; yet over 145,000 people needed emergency food assistance. In 2007/2008, Bolivia was affected by La Niña, which caused floods, droughts, frost and hail-storms, across the whole country. Rice crops losses reached up to 25% and prices of the main basic products rose sharply with a strong negative impact on the price and access to food especially in urban areas. Changing climatic conditions are also affecting water access, like, for example, the case of Chile's glacier Echaurren Norte, one of the most important sources of drinking water for the metropolitan area of Santiago.

Table 4.6 Social vulnerability assessment to climate change in Latin America. Note: Values refer to the percentage of countries reporting vulnerability to the different impacts. Grey cells represent impacts affecting less than 50% of the countries; yellow between 50–75%; and brown cells account for impacts affecting more than 75% of the countries.

	IMPACTS		MESOAMERICA	SOUTH	AMAZONIAN	ANDEAN	FREQUENCY AVERAGE
					VULNERABLE COUNTRIES (%)		
SOCIAL	Higher risk of diseases		75	75	100	100	88
	Unsafe access to drinking water		88	50	33	60	58
	Damages to infrastructures (dwellings, protection areas)		50	25	67	40	46
	Impact on food security		75		67	40	61
	Migrations		50	50	67		56
	Increase in poverty		38				38
	Human losses caused by natural disasters				33		33
ENVIRONMENTAL	Changes in the hydrological cycle and water quality		63	75	33	20	48
	Coastal erosion and coastline retreat		25	60	100	20	49
	Changes in ecosystem productivity and biodiversity loss		75	50	67	60	63
	Higher risk of fires		25	25	67		63
	Salinization		38	25	100		39
	Increase in pests		25		33		29
ECONOMIC	Agriculture	Reduction in water availability	75	25	33	20	38
		Reduction in crop yields	88	25	67	80	65
		Damage to agricultural infrastructures	13			20	17
		Soil erosion and desertification	25			20	23
		Loss of agrodiversity		25		20	23
		Loss of harvests	13			20	17
		Decrease in aquaculture production	38		33	40	37
		Losses in livestock production	38		67		53
	Energy	Lower hydropower generation	38	75	100	20	58
		Damages to energy infrastructures	50	25			8
		Excessive reliance in fossil fuels	13	50			32
		Risk to invest in biofuels		25	33		29
		Increase in energy demand		25			25
	Minery	Reduction in water availability		25			25
	Tourism	Damages to tourism infrastructures	50	25		20	32

Source: own elaboration based on the (NCs) National Communication Strategies of twenty LAC countries (UNFCC, 2013).

Box 4.4 The nexus between climate change and migration

The climate change-migration relationship is as yet unclear. Extreme weather shocks have been associated with migration processes at the micro-level, although no clear macro-trend has been observed. So far no projection exists regarding expected 'environmental migrants' (Wilbanks et al. 2007) and some authors argue that environmental migration might be a 'myth'. Global estimates of environmental migrants –25 million according to Myers and Kent (1995) – are outdated and have been the subject of debate. In LAC no concrete figures exist despite the high frequency of meteorological events that occurred over the last century (see Figure 4.20). In Mesoamerica, storms and floods are the most frequent hazards, whereas in South America, floods prevail. Nevertheless, droughts have caused the largest impacts on the South American population.

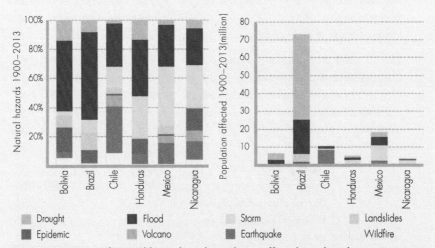

Figure 4.20 Type of natural hazards and population affected in selected countries in Latin America. *Source: own elaboration based on data from the EM-DAT database (2013)*

People often speak of the rural poor as the main victims of migration associated with climate change. In the case of LAC the impact should be observed at a much slower pace given its high degree or urbanization. The highest rural population in LAC is concentrated in the less developed countries, which thus increases the social vulnerability and the risk of migration. The most likely impacts of climate change in LAC may include damage to coastal areas consequently impacting tourism infrastructure and generating migration of local populations to safer areas and the reduction of glaciers, which could affect the Pacific cities and their water reserves. The effects of climate change on migration will be very different in countries like Bolivia or Paraguay, without coasts, than in the Turks and Caicos, the Cayman Islands and the Bahamas, which have a 100% of its population below the 10 metres elevation mark.

4.4 Conclusions

This chapter has identified the main socio-economic trends in LAC and Caribbean (LAC) and their direct and indirect consequences for water and food security. In relation to water security, the LAC population will continue to grow, despite the decrease in the fertility rate, with changes in lifestyles and the growth of the middle class which will increase the demand for water services and food, as well as external demand for producing agricultural commodities for export. As was shown, LAC has experienced a modest level of development in terms of per capita gross domestic product; centred around a fairly intensive use of water resources due to its economic model based on primary goods, recently triggered by the high prices in the international markets.

One of the main issues is whether this economic growth has been decoupled from increased water use. Since the model has been based on a re-primarization of the economy and exports, issues arise regarding the potentially large (green) virtual water of food exports. The other issue relates to poverty and water; water and sanitation in LAC in general has progressed well, in relation to the Millennium Development Goals, particularly in urban areas. LAC is the second most urbanized region in the world, including sixty-seven cities with more than 1 million inhabitants and four 'megacities' with more than 10 million (Mexico City, São Paulo, Buenos Aires and Rio de Janeiro). Sometimes it is hard to provide sufficient water services, with problems emerging from the relatively large informal economy and informal settlements, as well as from dispersed rural areas which could be left behind.

With regard to trade and food sovereignty, related to water security, linkages with price volatility in the case of LAC deserve a more detailed study in order to better understand the inter-linkages due to the economic models adopted in some of the emergent LAC economies. In LAC, the inequitable distribution of income could make it much more difficult to establish potential cross subsidy schemes, e.g. for the urban poor or even in rural areas. Thus equity remains a central pivotal issue for water and food security.

Water and food security confront important challenges imposed by globalization. Trade liberalization, increasing demand from countries such as China for primary goods, compounded with changes in consumption patterns also prompted by urbanization, the increase in per capita income and advertising have changed societal priorities and the way natural resources are used. New dietary patterns based on animal and agriculturally industrialized products require more water and raise issues of food security in terms of the nutritional quality of the 'modern diet'. As a result, health trends reflect the emergence of diseases like obesity and diabetes.

The democratization and adoption of ICTs could present a window of opportunity for water and food security, because of their cost, popularity and access are likely to increase exponentially. Water management in rural areas could be revolutionized through new instruments that can generate more accurate and visible data and information, essential to pinpoint better planning and use of water (and food) resources. The use of ICT can also help to support environmental and biodiversity conservation and further to avoid environmental, social and human harm for future generations.

Rural development could be supported by new technologies in the agriculture and food value chain, access to markets and financial services. There are barriers for ICT uptake in agriculture such as farmers' education and training (illiteracy or lack of technical skills), the lack of awareness and understanding of ICT and also the cost of deployment of some new technologies.

In order to guarantee water and food security faced with the potential impacts of climate change in socio-economic terms, it is important to define long-term targets for CC mitigation, to identify vulnerable regions and groups, prioritize research and adaptation, and to invest in adaptation and mitigation measures. Social and environmental vulnerability is high in all LAC regions where the most frequent impacts are related to increasing diseases such as malaria and dengue fever, food insecurity and a growing perception that access to drinking water may be at risk. Extreme events such as droughts and floods are affecting agricultural areas in many countries and in some cases access to water.

Latin American trends can, however, be modified. Measures orientated towards achieving fair income distribution, public policies orientated towards more vulnerable groups of population, a model of growth sustained on domestic markets, formalization of the informal economy, investment in science and technology and policies for improvement and conservation of natural resources would be key goals to prioritize and thus allow progress to be made on future socio-economic megatrends in order to guarantee water and food security into the future.

References

Arriagada, I., Aranda, V. & Miranda, F. (2005). *Políticas y programas de salud en América Latina. Problemas y propuestas. División de Desarrollo Social.* Santiago, Chile, Comisión Económica para América Latina y el Caribe (CEPAL). Serie Políticas Sociales No. 114.

Barria, R.M.P & Amigo, H. (2006). Transición Nutricional: una revisión del perfil latinoamericano. *ALAN,* 56 (1): 3–11.

Becker, K.F. (2004). *The Informal Economy: a fact finding study prepared for the Swedish International Development Agency.* Stockholm, Sweden, SIDA.

Bermudez, O. & Tucker, K. (2003). Trends in dietary patterns of Latin American populations. *Saude Publica* 19(1): 87–99.

Buitagro, R.E. (2009). Reformas comerciales (apertura) en América Latina. Revisando sus impactos en el crecimiento y el desarrollo. *Revista de la Facultad de Ciencias Económicas: Investigación y reflexión,* 17 (2): 119–132.

Carter, T.R. & Mäkinen, K. (2011). *Approaches to Climate Change Impact, Adaptation And Vulnerability Assessment: towards a classification framework to serve decision-making.* Helsinki, Finland, Finnish Environment Institute (SYKE). MEDIATION Technical Report No. 2.1.

CEPAL (2002). Comisión Económica para América Latina y el Caribe. *Urbanización, redistribución espacial de la población y transformaciones socioeconómicas en América Latina*. Santiago, Chile, CEPAL [Online] Available from: www.CEPAL.org/cgi-bin/getProd.asp?xml=/publicaciones/xml/2/11482/P11482.xml&xsl=/publicaciones/ficha.xsl&base=/publicaciones/top_publicaciones.xsl#. [Accessed June, 2013].

CEPAL (2003). Comisión Económica para América Latina y el Caribe. *La migración internacional en América Latina y el Caribe: tendencias y perfiles de los migrantes*. Serie Población y desarrollo. Santiago, Chile, CEPAL Report No. 35.

CEPAL (2005). Comisión Económica para América Latina y el Caribe. *Panorama social de América Latina 2005*. Documento Informativo. Santiago, Chile, CEPAL.

CEPAL (2006). Comisión Económica para América Latina y el Caribe. *Migración interna y distribución espacial*. CELADE, División de Población de la CEPAL. Brief No. 6.

CEPAL (2011). Comisión Económica para América Latina y el Caribe. *Migración internacional en América Latina y el Caribe. Nuevas tendencias, nuevos enfoques*. Santiago, Chile, CEPAL Report No. LC/R.2170.

CEPAL (2012a). Comisión Económica para América Latina y el Caribe. *Balance Preliminar Economías de América Latina y Caribe 2012*. Santiago, Chile, CEPAL Report No. LC/G.2555-P.

CEPAL (2012b). Comisión Económica para América Latina y el Caribe. *La migración internacional desde una perspectiva regional e interregional*. Santiago, Chile, CEPAL Report No. LC/W.475.

Dollar, D. & Kraay, A. (2001). *Growth Is Good for the Poor*. World Bank Policy Research Working Paper No. 2587.

Domínguez-Guadarrama, R.(2011). El fenómeno migratorio desde una perspectiva global; entendimiento y apuestas teóricas. *Escenarios XXI*, 11(II).

ECLAC (2012). Economic Commission for Latin America and the Caribbean. *Statistical Annex Social Panorama of Latin America 2012*. Santiago, Chile, ECLAC report.

EM-DAT (2013). International Disaster Database. *The OFDA/CRED International Disaster Database*. [Online] www.emdat.be. [Accessed May, 2013].

Ercin, E., Martinez, M. & Hoekstra, A. (2011). Corporate Water Footprint Accounting and Impact Assessment: The case of the Water Footprint of a Sugar-Containing Carbonated Beverage. *Water Resources Management*, 25: 721-741.

ERICSSON (2012). *On the pulse of the networked society*. Ericsson mobility report. [Online] Available from: www.ericsson.com/res/docs/2012/ericsson-mobility-report-november-2012.pdf. [Accessed March, 2013].

FAO (2012). Food and Agriculture Organisation. *Panorama de la Seguridad Alimentaria y Nutricional en América Latina y el Caribe 2012*. Rome, FAO.

Flodman, B.K. (2004). *The Informal Economy. A Fact Finding Study*. SIDA.

Guerra-Borges, A. (2012). *Panorama actual de la integración Latinoamericana y Caribeña*. Mexico, UNAM-Instituto de Investigaciones Económicas.

Gurría, A. (2012). *Perspectives economiques*. Conférence de presse. Paris, 22 May, 2012. OCDE. Report No. 91.

GWP (2012). Global Water Partnership. *Regional Process of the Americas: Access to water and sanitation for all and the Human Right to water and sanitation in the Americas*. Target 1.1. Marseille, France, 6th World Water Forum.

Hauser, P. & Gardner R. (1982). Urban future: Trends and prospects, Honolulu, East-West Population Institute, Reprint. No. 146.

Herrera, J., Roubaud, F. & Suárez Rivera, A. (2004). *El sector informal en Colombia y demás países de la Comunidad Andina.* Instituto de Estadística. Bogotá, Colombia, IRD.

Hope, R. (2012). *Water Security and Mobile Technologies.* University of Oxford, UK. Water Security, Risk & Society brief No. 4.

IICA (2009). Instituto Inter-americano de Cooperación para la Agricultura. *Los Tratados de Libre Comercio negociados por América Latina con La República Popular de China, India, Singapur y Taiwán. Estudio comparativo.* San José, Costa Rica, IICA.

INEGI (1992). Instituto Nacional de Estadística y Geografía. *Encuesta nacional de ingresos y gastos de los hogares* (ENIGH).[Online] Available from: www.inegi.org.mx/est/contenidos/ Proyectos/Encuestas/Hogares/regulares/Enigh/default.aspx. [Accessed May, 2013].

INEGI (2010). Instituto Nacional de Estadística y Geografía. *Encuesta nacional de ingresos y gasto de los hogares* (ENIGH).[Online] Available from: www.inegi.org.mx/est/contenidos/ Proyectos/Encuestas/Hogares/regulares/Enigh/default.aspx. [Accessed May, 2013].

IOM (2009). International Organization of Migration. *Climate, migration, climate change and the environment.* [Online] Available from: publications.iom.int/bookstore/free/migration_and_ environment.pdf [Accessed March, 2013].

IOM (2011). International Organization of Migration. *The State of Environmental Migration 2011.* [Online] Available from: publications.iom.int/bookstore/free/State_Environmental_ Migration_2011.pdf [Accessed March, 2013].

IOM (2012). International Organization of Migration. *Panorama Migratorio de América del Sur, 2012.* [Online] Available from: www.iom.int/.../Panorama_Migratorio_de_America_del_ Sur_2012.pdf [Accessed March, 2013].

ITU (2013). International Telecommunication Union. *Statistics.* [Online] Available from: www.itu. int/en/ITU-D/Statistics/Pages/default.aspx. [Accessed March, 2013].

Lattes, A. (2000). *Población urbana y urbanización en América Latina.* II Jornadas Iberoamericanas de Urbanismo sobre las Nuevas Tendencias de la Urbanización en América Latina. Quito, Ecuador. Conference proceeding. [Online] Available from: www.flacso.org.ec/docs/sfcclates. pdf [Accessed March, 2013].

Lawrence, P., Meigh, J. & Sullivan, C. (2003). *The Water Poverty Index: an International Comparison.* Keele University. Staffordshire, UK, Keele Economics Research Papers.

Mekonnen, M. & Hoekstra, A. (2010a). *The Green, Blue and Grey Water Footprint of Crops and Derived Crop Products.* UNESCO-IHE. Value of Water Research Report Series No. 47.

Mekonnen, M. & Hoekstra, A. (2010b). *The Green, Blue and Grey Water Footprint of Farm Animals and Animal Products.* UNESCO-IHE. Value of Water Research Report Series No. 48.

Millennium Project (2012). Estudio Delphi sobre escenarios para Latinoamérica al año 2030. In: Cordeiro, J.L. (ed). *Latinoamerica 2030.* Caracas, Venezuela, Global Futures Studies and Research.

Miró, C. (2006). La demografía en el siglo XXI en América Latina. *Papeles de Población,* 50 (12): 13–22.

Monteiro, C.A., Wolney, L. & Popkin, B. (2002). Is obesity replacing or adding to under-nutrition? Evidence from different social classes in Brazil. *Public Health Nutrition,* 5.

Morón, C. & Schejtman, A. (1997). *Evolución del consumo de alimentos en América Latina, en Producción y manejo de datos de composición química de alimentos en nutrición.* Rome, FAO.

Myers, N. & Kent, J. (1995). *Environmental Exodus: an Emergent Crisis in the Global Arena.* Washington DC, Climate Institute.

NIC (2008). Network Information Centre. *Global Trends 2025: A transformed world*. November, 2008. [Online] Available from: www.dni.gov/nic/NIC_2025_project.html. [Accessed March, 2013].

Olaiz-Fernández, G., Rivera-Dommarco, J., Shamah-Levy, T., Rojas, R., Villalpando-Hernández, S., Hernández-Avila, M. & Sepúlveda-Amor, J. (2006). *Encuesta Nacional de Salud y Nutrición 2006*. Cuernavaca, Mexico, Instituto Nacional de Salud Pública.

PAHO (2007a). Pan-American Health Organization. *Health in the Americas*. Washington DC. PAHO Scientific Publication No. 622.

PAHO (2007b). Pan-American Health Organization. *Non-communicable diseases in the Americas: building a healthier future*. Washington DC, PAHO.

PAHO (2011). Pan-American Health Organization. *Situación de la salud en las Américas. Indicadores Básicos*. Washington DC, PAHO.

PAHO (2012). Pan-American Health Organization. *Health in the Americas*. Volume II. Washington DC, PAHO.

Parry, M.L., Canziani, O.F., Palutikof, J.P., van der Linden, P.J. & Hanson, C.E. (2007). Climate Change 2007: *Impacts, Adaptation and vulnerability. Contribution of Working Group II to the Fourth Assessment Report of the Intergovernmental Panel on Climate Change*. Cambridge and New York, Cambridge University Press.

Pew Hispanic Center (2011). *Census 2010: 50 Million Latinos Hispanics Account for More Than Half of Nation's Growth in Past Decade*, March 24, 2011. Available at: www.pewhispanic. org/ [Accessed July, 0213].

Pi-Sunyer, F. (2002). The Medical Risks of Obesity. *Obesity Surgery*, 12(1): 6–11.

Rastoin, J.L. (2009). Prospective alimentaire. Serie Systemes Agroalimentariares. 31/2009. *Economies et Sociétés*. p.1725.

Ravallion, M. (2004). Competing Concepts of Inequality in the Globalization Debate. In: Collins, S. and Graham, C. (eds). *Brookings Trade Forum 2004*. Washington DC, US, Brookings Institution. pp.1-38.

Regmi, A. (2001). *Changing Structure of Global Food Consumption and Trade*. Market and Trade Economics Division, Economic Research Service, U.S. Department of Agriculture. Agriculture and Trade Report. WRS-01-1.

Regmi, A., Takeshima, H., & Unnevehr, L. (2008). *Convergence in Global Food Demand and Delivery*. Washington DC, USDA Economic Research Report No. 56.

Rivera, J., Muñoz-Hernández, R., Rosas-Peralta, A. & Popkin, W. (2008). Consumo de bebidas para una vida saludable: recomendaciones para la población mexicana. *Salud Pública de México*, 50 (2).

Rivera, J.A, Barquera, S., González-Cossío, T., Olaiz, G. & Sepúlveda, J. (2004). Nutrition transition in Mexico and in other Latin American countries. *Nutrition Reviews*, 62 (7): 149–157.

Santos-Baca, A. (2012). *Efectos de la apertura comercial en el consumo de alimentos de los hogares urbano-populares, 1992-2010*. MSc Thesis in Social Sciences. Facultad Latinoamericana de Ciencias Sociales. Sede México.

Schneider, F.G. (2002). *The Size and Development of the Shadow Economies of 22 Transition and 21 OECD Countries*. IZA, Institute for the Study of Labour. Discussion Paper No. 514.

Solimano, A. & Soto, R. (2006). Economic growth in Latin America in the late XX century: evidence and interpretation. In: Solimano, A. (ed). *Vanishing Growth in Latin America: The Experience of the Late XX Century*. Massachusetts, US, Edward Elgar Publishers. pp.11–45.

UN (2001). United Nations. *Urban and Rural Population by Age and Sex 1980-2015*. [Online] Available from: www.un.org/en/development/desa/population/publications/dataset/urban/urbanAndRuralPopulationByAgeAndSex.shtml [Accessed June, 2013].

UN (2011). United Nations. *World Urbanization Prospects*. The 2011 Revision, Economic & Social Affairs, New York, UN report.

UN-DESA (2011). United Nations Department of Economics Social Affairs. *World Population Prospects, The 2010 Revision*. Volume I: Comprehensive Tables. UN Report ST/ESA/SER.A/313.

UN-Habitat (2012). United Nations Human Settlements Programme. *Estado de las ciudades de América latina y Caribe 2012. Rumbo a una nueva transición urbana*, Nairobi, Kenia, UN.

UNFCCC (2013). United Nations Framework Convention on Climate Change. *Non-Annex I National Communications submitted in compliance with the United Nations Framework Convention on Climate Change*. [Online] Available from: unfccc.int/national_reports/non-annex_i_natcom/items/2979.php. [Accessed May, 2013].

Vodafone & Accenture (2012). *Connected agriculture. The role of mobile in driving efficiency and sustainability in the food and agriculture value chain*. [Online] Available from: www.vodafone.com/content/dam/vodafone/about/sustainability/2011/pdf/connected_agriculture.pdf. [Accessed March, 2013]. Corporative Report.

WHO-UNICEF (2013). *Joint Monitoring Programme (JMP) for Water Supply and Sanitation data*. [Online] Available from: www.wssinfo.org/data-estimates/table/ [Accessed July, 2013].

Wilbanks, T.J., Romero, P., Lankao, M., Berkhout, F., Cairncross, S., Ceron, J.P., Kapshe, M., Muir-Wood, R. & Zapata Marti, R. (2007). Industry, settlement and society. In: Parry, M.L., Canziani, O.F., Palutikof, J.P., van der Linden, P.J. and Hanson, C.E. (eds). *Climate Change 2007: Impacts, Adaptation and vulnerability*. Contribution of working group II to the fourth assessment report of the intergovernmental panel on climate change. Cambridge, Cambridge University Press. pp. 357–390.

World Bank (2011). *ICT in Agriculture Sourcebook. Connecting Smallholders to Knowledge, Networks, and Institutions* [Online] Available from: www.ictinagriculture.org/content/ict-agriculture-sourcebook. [Accessed March, 2013].

GLOBALIZATION AND TRADE

Authors:

Alberto Garrido, Water Observatory – Botín Foundation, and CEIGRAM, Technical University of Madrid, Spain

Diego Arévalo Uribe, Water Management and Footprint. CTA – Centro de Ciencia y Tecnología de Antioquia, Colombia

Insa Flachsbarth, Water Observatory – Botín Foundation, and CEIGRAM, Technical University of Madrid, Spain

Maite M. Aldaya, Water Observatory – Botín Foundation, and Complutense University of Madrid, Spain

O. Vanessa Cordero Ahiman, CEIGRAM, Technical University of Madrid, Spain

Bárbara Soriano, CEIGRAM, Technical University of Madrid, Spain

Highlights

- The world's economy and agriculture have become ever more intertwined, reinforcing interdependencies, and multiplying network connections. The Latin America and Caribbean (LAC) region has experienced an accelerated growth of imports, exports and inward foreign investment.

- The expansion of the middle class in LAC and Asia and the associated changes in eating habits are adding pressure on agricultural commodities markets, with LAC itself becoming a leading exporter of calories and vegetal proteins required to sustain the expanding livestock sector in the world.

- LAC is still relatively isolated from the rest of the world, in terms of personal air traffic and major port activity, yet well connected through raw materials markets.

- Significant production increases can be obtained in many LAC regions with both rain-fed agricultural practices and farming systems under irrigation.

- LAC's main trading partners are now in Asia, especially China and India, but Central America and the Caribbean still export primarily to North America.

- LAC's increasing exports and imports may have rendered certain social advances in terms of poverty reduction in Argentina, Brazil, Chile, Mexico and Peru. And yet, 174 million people in LAC are considered poor, and 73 million of these are extremely poor (FAO, 2013b). However, causality between trade and poverty cannot be clearly established.

- International trade can make an important contribution to global decoupling (economic growth independent of resource use and impacts) when guided by appropriate environmental and trade policies. These have hitherto been managed separately at country and global levels.

5.1 Introduction

By all accounts, the world has never been more globalized since World War II. Improvements in transportation, logistics, telecommunications and global production systems attest to increasing worldwide economic integration. Furthermore, world food systems have never been as integrated and developed as they are at present (Prakash, 2011) with production specialization, technological advances and the wide dissemination of knowledge. However, doubts exist as to whether agriculture has the potential to feed the world when the population goes beyond 9 billion unless significant improvements are made in production efficiency and food habits change. In 2012, the FAO estimated that 852 million people were undernourished, which is equivalent to 14.9% of the world's population (FAO, 2013b). To this end, the National Intelligence Council of the US has identified the nexus of food, water and energy as one of its four 'megatrends', which are likely to transcend all future scenarios, demonstrating that a growing global population will place more demand on these inextricably linked resources (NIC, 2012).

Globalization is an ambiguous concept without a widely accepted scientific definition. It involves trade relationships and the movement of capital, ideas and even people. It also encompasses the sharing and expansion of risks (such as epidemics or terrorist attacks) and global environmental threats. Agricultural trade has been accelerated by the rapid decline in the costs of cross-border trade of farm produce and other products, driven also by reduction in transportation costs, the information and communication technology (ICT) revolution and major reductions in governmental distortions. As a result, it has altered global agricultural production, consumption and hence trade patterns (Anderson, 2010), not least in the Latin America and Caribbean (LAC) region.

This chapter will review some facts and data that describe what globalization is and what shape it may take in the future. We focus primarily on the LAC region but also provide a global perspective and, in coordination with other chapters of this book (Chapters 4 and 7), we will focus on trade looking in more detail at agricultural commodities. We begin by providing the context of international trade and identify the resulting major trends. In the second section we present the most significant trade data reported at a regional level, again, specifically concentrating on the agricultural context. The third section discusses the main drivers behind the observed trade data and trends in the LAC. The chapter closes with an overview of issues more closely related to social and environmental sustainability.

5.2 Global context and major trends

Recent popular press media has disseminated the trajectory of the world's economic centre of gravity.[1] Two significant points can be identified in the last two millennia. Firstly, the world's economic centre of gravity in 2012 is in almost the same longitude than it was in the year AD 1. Secondly, during almost 2000 years it moved westwards to where, in 1950, it reached the North of Iceland and since then it has moved extremely rapidly to a position in Russia and is projected to continue to a point north of Kazakhstan in 2025. This shift attests to the growth of East Asia and the Pacific area.

Table 5.1 shows the changing percentage of the world's economy and population among the world's regions. LAC's economic importance grew from 5.3% in 1990 to

---

1 Published in *The Economist* (The world's shifting centre of gravity 28 June 2012, 14:34 by The Economist online). Calculated weighting national GDP by each nation's geographic centre of gravity; a line drawn from the centre of the earth through the economic centre of gravity locates it on the earth's surface. (see McKinsey Global Institute, 2013)

8.3% in 2011, whereas its share of the population remained stable at 8.5%. The most significant changes are the increasing share of the world's GDP in South and East Asia, the population decline in North America and the slight decline of Europe and Central Asia.

Table 5.1 Percentage of GDP and population of each region with respect to the world

		1990	2000	2011
SOUTH EAST ASIA	% of GDP/world's GDP	23.2%	26.9%	30.1%
	% of Pop/world's Pop	56.0%	56.3%	55.5%
EUROPE & CENTRAL ASIA	% of GDP/world's GDP	39.0%	29.7%	31.6%
	% of Pop/world's Pop	15.9%	14.1%	12.8%
LATIN AMERICA & CARIBBEAN	% of GDP/world's GDP	5.4%	6.6%	8.3%
	% of Pop/world's Pop	8.3%	8.5%	8.5%
NORTH AMERICA	% of GDP/world's GDP	28.8%	32.9%	23.9%
	% of Pop/world's Pop	5.2%	5.1%	5.0%

Source: World Bank (2012)

A major driving force of the world's economy is the level and instability of commodity markets. Figure 5.1 plots the commodities price indices, showing composite indices of energy, food and metals and minerals, between 1960 and 2011. All three exhibit a long and stable trend between 1978 and 2007, and a rapid escalation after this year. This has been accompanied by increased volatility for all specific products and markets, and the extreme financialization[2] of the commodity markets around the world.

The increasing importance of trade in the world's economy is also a major driver of globalization. Figure 5.2 shows the percentage of trade as a share of the world's GDP. Since 1960, trade value has almost tripled in relative terms in the world and the expansion of trade has been greater in East Asia and the Pacific than in LAC, whose relative trade volume grew less than the world's average.

The growth of trade goes hand in hand with the expansion of transportation. The world's container traffic grew between 2000 and 2010 by a factor of 2.3, and LAC's participation in this traffic augmented from 6.8% in 2000 to 7.4% in 2010. Despite this, the LAC region still lags behind other regions in the world as shown in Table 5.2, which reports the LAC's main port's activity relative to the fifty busiest ports in the world in 2011.

---

2 'Financialisation refers to the increasing amount of liquid funds which have become engaged in agricultural commodity markets over the past years. Often, the role of hedge and index funds is emphasised in particular for price formation on futures markets. Speculation is an even less clearly defined term. Major notions in the literature are speculative bubbles, when asset prices deviate systematically from their fundamental values, speculative hoarding, when stocks are built in the expectation of ever higher prices, and market manipulation, where price movements on less liquid markets are deliberately triggered by some market participants. The economic concept of speculation is yet defined differently; speculators in this meaning are market participants who are willing to take over price risks from hedgers at a premium (and thus fulfil an economically desirable function).' Brummer et al. 2013, p. 3.

Note that Colon and Balboa ports are related to the operation of the Panama Canal and are thus not so involved in operations of loadings and shipments.

The fact that LAC is still weakly connected within world trade circuits is also shown by the statistics of air travel and air passengers. In 2011, only São Paulo-Guarulhos International Airport, ranking 45th, appeared amongst the fifty busiest world airports. In 2010 no LAC airport appears in the list and in 2009 Mexico International Airport is the only LAC present on the list, in 50th position (Airports Council International, 2012).

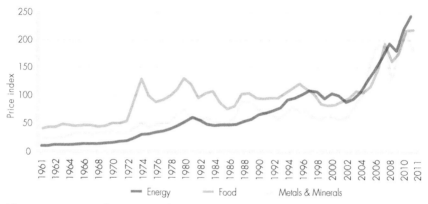

Figure 5.1 Commodities Price Indices (1960–2011) (Average =100). *Source: World Bank (2012)*

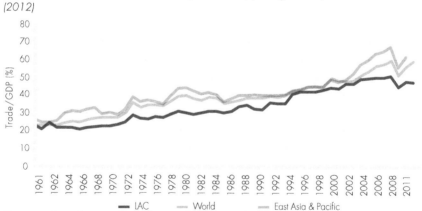

Figure 5.2 Trade as a share of gross domestic product (GDP) (1961–2011). *Source: World Bank (2012)*

Table 5.2 LAC's busiest ports in thousands of TEUs (twenty-foot equivalent units) in 2011

Busiest LACS ports	Rank In world	1000 TEUs	Activity with respect to busiest port (Shanghai) In %
Colon (Panama)	37	3,370	10.6%
Balboa (Panama)	39	3,230	10.2%
Santos (Brazil)	43	2,990	9.4%

Source: Journal of Commerce: The JoC top 50 world container ports (2012)

These data seem to suggest that LAC's growth expansion is primarily based on commodities and industrial products as opposed to the service sector, which could also account for the fact that LAC is relatively less densely populated and the few large populous areas are widely spread across the continent compared to Asia and Europe.

Figure 5.3 shows data for inward foreign direct investment in the region between 1970 and 2011. Foreign direct investment has seen unprecedented growth in the last ten years, but has been very volatile.

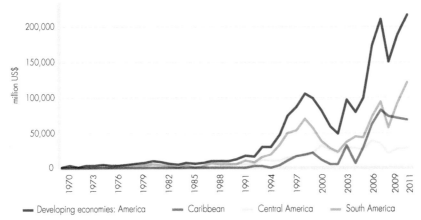

Figure 5.3 Inward foreign direct investment flows, annual, 1970–2011 in million US$. *Source: UNCTAD (2012)*

Figure 5.4 shows that LAC has not been a principal beneficiary of official development assistance in agriculture and infrastructure whereas since 1995 East Asia and Pacific, South Asia and sub-Saharan Africa have received significant aid. However, the region's increasing political stability and economic potential have certainly given an extraordinary push to private investment in infrastructure, making LAC the world's primary recipient in 2002 and 2009 (Figure 5.5).

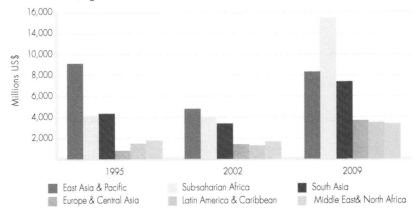

Figure 5.4 Official Development Assistance in agriculture & infrastructure by area, in 1995, **2002 and 2009.** *Source: OECD (2012)*

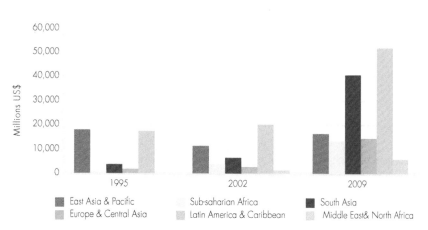

Figure 5.5 Private participation in infrastructure by area in 1995, 2002 and 2009. *Source: PPIAF (2013), World Bank (2012)*

5.3 Trends of trade in the LAC region

LAC has historically established a trade model characterized by intensive goods exports, particularly associated with the region's natural resources (land, minerals, oil and water) and the import of capital-intensive goods and knowledge. While it plays a modest role in world market (10–12%) in recent years some growth has been reported although this is largely due to the increased value of commodities (UNEP, 2010) (Table 5.3).

Table 5.3 Participation of LAC in world agricultural trade in dollar terms

	1980	1985	1990	1995	2000	2005	2010
Agricultural commodities	9.50%	13.58%	11.43%	13.90%	16.29%	16.51%	18.10%
Food commodities	8.40%	9.44%	7.04%	7.98%	9.22%	10.37%	11.55%

Source: WTO (2012)

LAC's trade model has been strengthened by the trend towards an increased integration of the countries with the rest of the world. Indeed, the international integration of the region, especially in South America, is determined by a pattern where natural resources are seen to account for over half of total exports. These are minerals, hydrocarbons (notably natural gas and oil), agricultural, livestock, forestry and fishery products with little or no processing (UNEP, 2010).

Approximately 54% of the region's exports are raw materials. However, there are important sub-regional differences such as Mexico which shows a pattern of exports strongly linked to manufacturing (about 74%). Thus, excluding Mexico, of the remaining

Latin American exports, almost 73% are commodities based on natural resources. In some countries, exports of primary goods exceed 95% of total exports (UNEP, 2010).

A dependence on a few products is also observed. In effect, the ten principal export products of most countries are mining and agricultural goods. At the regional level, the main products exported are crude oil and its derivates (UNEP, 2010).

In the last twenty years (1992–2012), international trade has been conditioned by economic opening based on minimizing the presence of the government through the liberalization of trade forces (Washington Consensus, Williamson, 1990). These policies are considered key tools for development and opening national trade to international competition and elimination of regulations associated with international trade. This has resulted in trade agreements between the countries of the region and has allowed for an open regional market under competitive conditions that promotes trade growth. This situation is clearly shown in the graphs in Figure 5.6.

The composition of exports and imports amongst agricultural and mining products and other merchandises has been quite stable between 1992 and 2011. Agricultural and mining products represented a maximum share of exports of 38% in 1992 and a minimum of 31% in 2001. In terms of imports, 10% were agricultural and mining products in 2011 and 5.8% in 1991 (Inter-America Development Bank, 2012).

Figure 5.7 provides the breakup of agricultural and mining goods exports in 2000, 2005 and 2011 to different world regions. In 2000, North America was the main importer of goods from the three regions of LAC (South America, Central America and the Caribbean). In 2011, Central American and Caribbean exports were still concentrated on North America, but exports from South America were destined primarily to East Asia and the Pacific, followed by the EU and LAC, with North America being the fourth largest importing region.

China and other countries in Southeast Asia are the main importers of Latin American commodities such as copper or soybean. The increasing demand for inputs from emerging economies like India and China has had a noticeable impact on the region's exports. Consumption in Asia, and particularly in China, explains the continued commercial importance of extracting natural resources. In 2007 goods imported from Latin America and the Caribbean were mainly soybean (grain and oil), followed by copper ore (gross and concentrate), copper alloys, fish meal, leather and paper pulp (UNEP, 2010). South American major exports were to East Asia and the Pacific in 2011, whereas Caribbean and Central American major trading partners were in North America. Internal regional trade in LAC's reduced in percentage terms from 2000 to 2011.

The agro-industry has also witnessed strong growth in the region due to increased global demand and international prices for both agro-foods and raw materials to produce biofuels. Agricultural production is being reshaped by an expansion of oilseeds, especially soybean, while there is stagnation in some grains and a reduction in other more traditional products such as coffee and cocoa. There is also an increase in sales of meat, i.e. beef, pork, and poultry, that creates additional demand for grain for animal feed (UNEP, 2010).

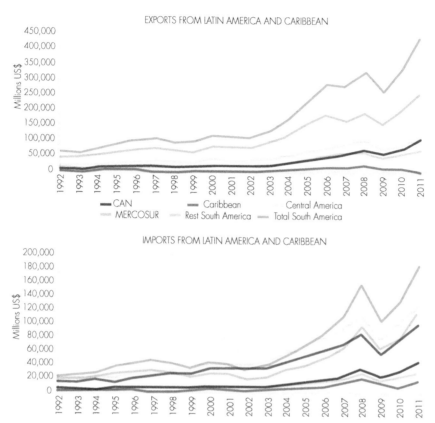

Figure 5.6 Value of imports and exports of agricultural and mining commodities of LAC between 1992 and 2011 expressed in nominal US dollars. *Source: Inter-America Development Bank (2012) Note: CAN (Andean Community of Nations): Bolivia, Colombia, Ecuador and Peru. Caribbean: Antigua and Barbuda, Bahamas, Barbados, Dominica, Dominican Republic, Grenada, Jamaica, Montserrat, Saint Kitts and Nevis, Saint Lucia, Saint Vincent and the Grenadines, Trinidad and Tobago. Central America: Belize, Costa Rica, Guatemala, Honduras, Nicaragua, Panama, El Salvador. MERCOSUR (Southern Common Market): Argentina, Brazil, Paraguay, Uruguay, Venezuela. Rest of South America: Chile, Guyana, Suriname. Total South American (CAN, MERCOSUR, Rest South America): Argentina, Bolivia, Brazil, Colombia, Chile, Ecuador, Guyana, Paraguay, Peru, Suriname, Uruguay, Venezuela.*

At least ten countries – Argentina, Bolivia, Brazil, Colombia, Ecuador, Guatemala, Honduras, Mexico, Paraguay and Peru – produce biofuels, and four countries export biofuels produced from their own crops with Brazil being the largest exporter. There are smaller sales from Bolivia and Guatemala and, recently, from Argentina. However, programmes are underway in almost all countries and so the list of producers is constantly increasing (*ibid.*).

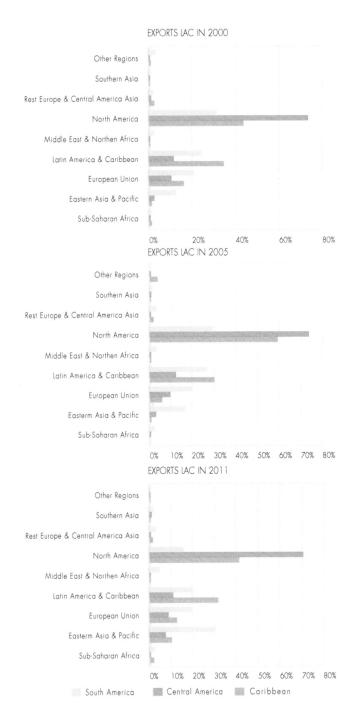

Figure 5.7 Breakup of exports from Latin America and the Caribbean to different world regions in 2000, 2005 and 2011 (%). *Source: Inter-America Development Bank (2012)*

5.4 Main drivers of LAC's increasing globalization and trade

5.4.1 Abundant water and land resources

LAC has the greatest agricultural land and water availability per capita in the world. With 15% of the world's land, it receives 29% of precipitation and has 33% of available renewable resources (Mejía, 2010). In 2007, LAC had about 10.8% of the world's cultivated land, but its growth has lagged behind other regions and the world in general since 1961. Annual growth of LAC's cultivated land between 1961 and 1997 was 1.62% against 4.49% globally, and between 1997 and 2007 annual growth was 0.71% in LAC and 3.04% in the world (Deininger, 2011). This suggests that the expansion of the agricultural frontier has grown at a slower pace in LAC than in other continents. As Chapter 3 explains, the proportion of primary sector activity devoted to agricultural production in LAC remains below the global average despite the net positive deforestation trend.

However, agricultural export growth in LAC has been the result of the expansion of land rather than yields (Deininger, 2011), since most exports come from areas in which yields are at their maximum with little or no yield gaps (Deininger, 2011; Foley et al., 2011). Chapter 10 shows that there are countries in LAC where yield gaps are still significant, and output from cultivated land can still be improved.

5.4.2 Consequences of liberalization and dismantling of tariffs, other trade barriers and the role of Free Trade Agreements

LAC exports represent 6% of world trade (18% of agricultural products), but are affected by 30% of measures of border protection (Giordano, 2012). However, tariffs have been reduced in the region, and freights costs have become more relevant. Tariffs range from 7% in Nicaragua and almost 0% in the Caribbean states, whereas freight costs range from 18% in Ecuador to 7% in Antigua and Barbuda, and between 10% and 15% in most South American countries. Upon the collapse of the third WTO Ministerial Conference of Seattle in 1999 and the standstill of the Doha Round, most LAC countries have adopted free trade agreements (bilateral/plurilateral) in order to pursue export growth and diversification (Dingemans and Ross, 2012).

According to the Information System on Foreign Trade of the Organization of American States (OAS, 2012), there are 111 trade agreements active that involve one or more of the eleven countries with the largest trade volume and population in the region[3] (sixty-three Free-Trade Agreements (FTA) and forty-eight Preferential Trade Agreements (PTA)) with 241 worldwide trade partners (countries).

---

3 Argentina, Bolivia, Brazil, Chile, Colombia, Ecuador, Mexico, Paraguay, Peru, Uruguay, Venezuela.

FTAs are presented as a key policy and economic tool to achieve economic growth and integration. However, the long-term results of these agreements are not wholly positive. Indeed, according to an analysis of the intensive and extensive economics margins,[4] the FTAs have focused on the formalization of existing trade links within natural existing markets, without seeing any real incentive in order to achieve the diversification of production, markets and trade for the region. This situation is clearly shown in Figures 5.8 to 5.10. For instance, Figure 5.10 shows that the large majority of exports have been in the form of increased exports to an already existing market (Intensive Trade Margin). As Dingemans and Ross (2012) conclude, FTAs have not accomplished significant diversification in LAC's exports, despite the significant growth exports shown in Figure 5.8.

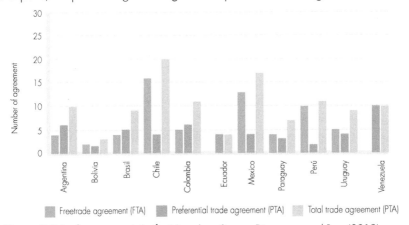

Figure 5.8 Trade agreements in the LA region. *Source: Dingemans and Ross (2012)*

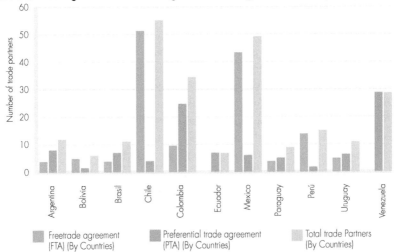

Figure 5.9 Trade partners in the LA region. *Source: Dingemans and Ross (2012)*

4 Intensive margin is the increase of trade with the same products and with the same partners, and extensive margin is the increase of trade of new products with new partners (Brenton et al., 2009).

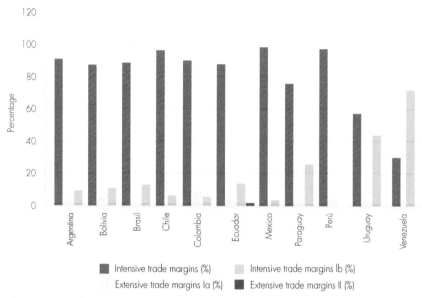

Figure 5.10 Changes in extensive and intensive trade margins in the LA region. *Source: Dingemans and Ross (2012).* Note: Intensive trade margin: expansion of export of a product to an existing market; Intensive Trade Margin Ib, increase of exports to a new market; Intensive Trade Margin Ia, export of a new product to an existing market; Extensive Trade Margin II, increased export of a new product to a new market.

5.4.3 Increasing competitiveness

For many products, LAC exports have been fostered by increasing competitiveness against many other trading partners. In terms of commodity prices and competitiveness, for instance, soybean in Brazil is more efficient than in the US. Annual yield growth in LAC has surpassed others: 2.9% in 1987–2007 and 3.6% 1997–2007, versus 2.1% and 2.2% relative to the world (Bruinsma, 2009). Chapter 10 provides a more detailed overview of trends of agricultural yields for the main products and most of LAC's countries.

5.4.4 Meat demand in emerging countries

More affluent populations have tended to diversify diets towards animal food items which require several multiples of water per calorie of dietary energy. The consumption of calories has also increased significantly in the last four decades in many developing countries. For example, meat demand (including demand for beef, meat, eggs and dairy products) or calorie consumption has grown in the Chinese diet from less than 100kcal/capita/day to more than 600kcal/capita/day between 1961 and early 2003. All of this increase in calorie consumption requires large amounts of grain and fodder (as extensive pastures are relevant for changes in land use) to feed livestock. In addition to calories, meeting growing meat demand requires supplementary production of protein crops for feed. The main growth of protein crop production has been obtained from soybean cultivation, and a significant part of it from soybean produced in Argentina and Brazil.

China alone may account for 43% of additional meat demand worldwide in 2020 compared to 1997, placing higher demand on world water resources and upward pressure on commodity prices in the longer term. But meat demand has also grown in the EU and many other countries. The population of EU27 grew by 20% between 1961 and 2007, whereas consumption of animal protein increased by 80% (Westhoek et al., 2011). It is worth noting that these are the main trading partners of LAC and hence any proposed change in trade patterns should bear this in mind.

Since 2000 there has been a massive change in the composition of the middle class in many world regions. There are various sources to define middle class. With the International Futures model middle-class membership is defined as per capita household expenditures of US$10–50 per day at PPP (power purchasing parity). Goldman Sachs used a comparable GDP per capita of US$6,000–30,000 per year, which yields a similar estimate of 1.2 billion middle-class people in the world in 2010 (NIC, 2012). While in this year, the US, EU and Japan made up 70% of the world's middle class, in 2030 this percentage will shrink to 30%. Except for India, which will make up 25% of the middle class in 2030, in the other growing regions, increased meat consumption is associated with increased affluence. LAC is one of the world's main producers of vegetable protein required to feed the animals, particularly in Argentina and Brazil.

5.4.5 Biofuel use in the EU and US

The expansion of biofuel production and consumption in the EU and US has resulted in significant and sustained pressure on agricultural markets worldwide. World biofuel production increased by a factor of five between 2001 and 2011, reaching 100hm^3/year. About one-third of corn production in the US is used to produce bioethanol. The literature disagrees on the impact of biofuels during the surge of prices between 2007 and 2008. In quantifying the impact of biofuel production on price variability, Mueller et al. (2012) quote sources varying between 3% and 30% (Von Braun, 2008), reaching up to 70% (Mitchell, 2008, who included the indirect consequences on stocks, large use shifts, and speculative activity).

Other evaluations found impacts of 60–70% price increase in corn and 40% in soybeans (Headey and Fan, 2008). About 2% of the world production of grains is used for ethanol production, representing about 14 million hectares in the US, EU, Brazil and China. Rosegrant et al., (2008) found that biofuels were responsible of 30% of the food price increases. Perhaps the most significant effect of the use of biofuels comes from the impact of prices rather than needs of agricultural land, which is expected to grow from 30 to 60 million hectares (Ajanovic, 2011). Pressures to keep on producing more biofuels will grow in the future, to such an extent that oil prices will also grow, in turn further pushing up commodity prices.

However, this prognosis may change if the shale gas expansion in the US were to generate an excess capacity of up to 8 billion barrels, making the OPEC (Organization of the Petroleum Exporting Countries) lose control of oil prices (NIC, 2012). With a breakeven price as low as US$44–68 per barrel, world energy markets may see profound transformations in the upcoming decades, including a reduction of first-generation biofuels.

5.5   Issues of concern

In view of the previously presented data and facts, we have identified five issues of concern for LAC, concerning; (1) implications of trade on water use and access; (2) unregulated access to agricultural land from foreign countries (land-grab); (3) the role of trade on the world's food system; (4) the potential impact of trade on the poor in LAC; and (5) whether trade may hamper the equitable access to land and water resources.

5.5.1   Implications of trade on water use and access

World trade patterns are extremely dynamic and unstable. Specialization, technology adoption and market prices volatility and growth have given rise to fundamental changes in agricultural production and trade worldwide and in LAC. Building on the pioneering analyses of virtual water trade (VWT) (Hoekstra and Chapagain, 2008), a recent literature strand has been analysing trend connections, with a view to observe patterns and draw relevant conclusions for the world's food system today and in the coming decades. Trade connections have been evaluated in physical units (tonnes, monetary units and virtual water), but recently the focus has been placed on the analysis of the networks' formation, stability and configuration. The role of LAC in the world's trade has been presented in Part 3, but Chapter 7 reviews the most recent literature, which offers conflicting views of the role of virtual water trade.

5.5.2   Is land-grab a source of concern for LAC?

Based on the assessment of Rulli et al. (2013), land grabbing is a global phenomenon, which involves at least sixty-two grabbed countries and forty-one grabbers. Africa and Asia account for 47% and 33% of the global grabbed area, respectively. About 90% of the grabbed area is located in twenty-four countries, which includes Argentina, Australia, Brazil and Uruguay, among non-African and non-Asian countries. The grabbed area is often a non-negligible fraction of the country area (up to 19.6% in Uruguay, 17.2% in the Philippines, or 6.9% in Sierra Leone). The countries that are most active in land grabbing are located in the Middle East, Southeast Asia, Europe, and North America.

A key feature of the Latin American case is its intra-regional land grabbing driven by (trans) Latina companies (TLCs). These are companies that involve mainly national capital, as in the case of many Chilean companies, or alliance with companies from different countries in the region, or, finally, alliances of Latino firms with capital from outside the region (Borras et al., 2011). Another particular feature of Latin America is that land grabbing occurs in settings marked by more or less liberal-democratic political conditions, such as those of Brazil, Uruguay and Argentina; not necessarily in fragile states marked by weak governance as is generally believed (ibid.).

However, land-grabbing is a concept that requires finer analyses and conceptualization. The cases of grabbed Latin American countries seem to be closer to foreign direct investments than to actually grabbed land (see Zetland and Möller-Gulland, 2013, for a systematic analysis of land taken as grab or as foreign direct investment). Argentina, Brazil, and other countries are passing legislation in place to control foreign investments to control land grab. Much of the ongoing problem is also related with the land tenure.

5.5.3 Does increased world food demand pose risks for sustainable land and water use in Latin America?

In comparison with other regions, LAC is the region with the greatest land and water resources per capita. As mentioned earlier, the increase of agricultural production has been primarily in terms of land expansion, and less in terms of increased yields. The region's low penetration of irrigation and relatively low utilization rates of fertilizers suggest that agricultural sustainable intensification must be developed in the region, before it can close the gap with North America, Western Europe and South and Southeast Asia (Mueller et al., 2012).

Foley et al. (2011) claim that there are significant opportunities to increase yields across many parts of Africa, Latin America and Eastern Europe, where nutrient and water limitations seem to be strongest. Better deployment of existing crop varieties with improved management should be able to close many yield gaps, while continued improvements in crop genetics will probably increase potential yields into the future. Pfister et al. (2011) show that in LAC the ratio of water and land stress (Relevant for Environmental Deficiency) are clearly less than 1 (being land stress greater than water) in most production regions, except in Chile, Mexico, Peru and northeast Brazil. And yet, the main exporter areas in Argentina, Brazil, Chile and Mexico have yields close to their potential ceiling. This means that agricultural expansion will have to occur in Central America and the Andean region, where intensification still has a lot of potential.

Most serious concerns about continuing agricultural expansion in LAC are related to the reduction of regional and global ecosystem services, chief among these are CO_2 emissions, regulation of the water cycle and biodiversity losses. This is elaborated to a much greater extent in Chapter 3.

5.5.4 Do trade and globalization benefit the poor in LAC?

Trade-poverty linkages are complex and diverse, but according to economic theory and empirical findings, they can be divided into a few important pathways: (a) trade-induced growth; (b) effects of trade on prices, income and consumption patterns; (c) effects of trade on wages and employment.

Trade generally stimulates growth since more open markets lead to access to new technologies and appropriate intermediate and capital goods, which in turn cause increases in production, scale economies and competitiveness. The economy specializes in industries in which it has comparative advantages, meaning that resources are allocated most efficiently (Edwards, 1993; Duncan and Quang, 2002). This is especially important in the agricultural sector, as in LAC a large portion of the poor live in rural areas. If more open agricultural trade generates growth in this sector, it is likely that the rural poor will benefit (Bakhshoodeh and Zibaei, 2007; Cain et al.,2010; Cervantes-Godoy and Dewbre, 2010). Some empirical studies underpin the trade-growth nexus in LAC; for example, Castilho et al., (2012) studied the impact of globalization on household income inequality and poverty using detailed microdata across Brazilian states from 1987 to 2005. Results suggest that trade liberalization contributes to growth in poverty and inequality in urban areas and may be linked to reductions in inequality (possibly poverty) in rural areas. Edwards (1998) analyses comparative data for ninety-three countries, among them ten LAC, and finds that trade openness favours growth and that capital accumulation plays an important role in reducing poverty. Dollar (2005), however, counters that those countries being increasingly integrated into world markets are those where poverty has increased most since the 1980s.

We investigated this relationship in five LAC countries between the mid 1990s of the past century and 2010. Figure 5.11 shows that there is a correlation between the degree of trade openness and GDP growth in the agricultural sector. However, results seem to be very side-specific, depending on each country's development level and on whether it has a net importing or exporting position in agriculture. It seems that in Mexico the correlation between economic growth and trade openness is weak, while Chile even shows a negative correlation. Mexico is a large net importer with comparative advantages in other sectors due to its scarce natural resource endowments. Therefore, open trade in agriculture might not enforce growth in this sector. Chile's is rather shifting away from agriculture, because it is already a developed country in comparison with the other four. Especially in Peru, agricultural trade openness seems to favour GDP growth in agriculture.

Secondly, trade affects agricultural prices and relative prices in an economy, and in turn the real income of poor households, since agriculture represents their main livelihood source and their main consumption expenditure. To what degree price changes transmit to poor household's income depends on market access and their ability to benefit from the trade environment. Hassine et al. (2010) and Taylor et al. (2010) find that lower tariffs reduce nominal incomes for nearly all rural household groups in El Salvador, Guatemala, Honduras and Nicaragua, but they also lower consumption costs substantially leading to a positive net effect on rural households' welfare. Field and Field (2010) and Finot et al. (2011) came to the conclusion that tariff reductions in Chile and Peru between 1994 and 2006 increased total household incomes.

We investigate the relationship between trade openness and income of the 10% poorest population group in five LAC countries. Figure 5.12 shows that the direct effect of liberalizing agricultural markets on the development of the income of the poor is rather small, with the strongest correlation within a 95% confidence interval in Chile and Peru. Both countries have been increasingly exporting high value products and importing lower value staple food. The results show that the poor have probably benefited from these market-driven changes in the sector of agriculture.

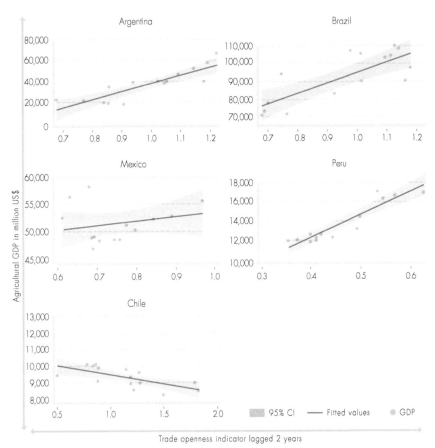

Figure 5.11 Trade and agricultural growth nexus in five LAC countries (1995–2010). *Source: own elaboration, data from WBDI (2012) and FAO (2013a)*

Thirdly, the impact of trade on wages and employment is grounded in the Heckscher-Ohlin model. With labour as an input factor, developing countries will specialize in the production of labour-intensive products which boosts demand for labour and in turn leads to higher wages in these sectors and thus poverty reduction. One of the reasons why agricultural trade liberalization is so important for poverty alleviation is that low-skilled workers in rural areas will benefit through production responses. For example Bussolo et al. (2011) found that the losses and gains in agricultural wages exhibit strong regional patterns: real wages of unskilled farmers rose in Latin America, the Middle East, and East Asia and Pacific, while declining in other developing regions.

Due to missing data, a direct analysis between agricultural trade and wages would not deliver reliable results. Thus we directly view a possible connection between agricultural trade liberalization and poverty rates in five LAC countries (Peru, Mexico, Chile, Argentina and Brazil). Figure 5.13 shows a clear trend between more liberalized trade and declining poverty rates.

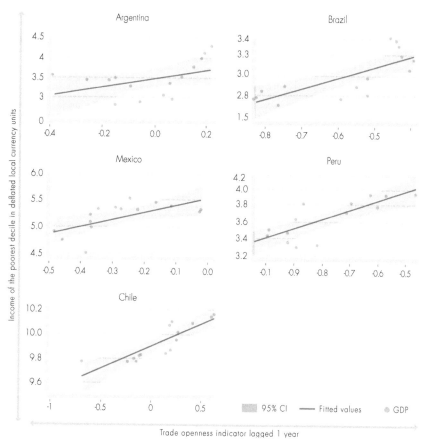

Figure 5.12 Trade and income of the poorest decile in five LAC countries (1996–2010). *Source: own elaboration, data from WBDI (2012) and SEDLAC (2012)*

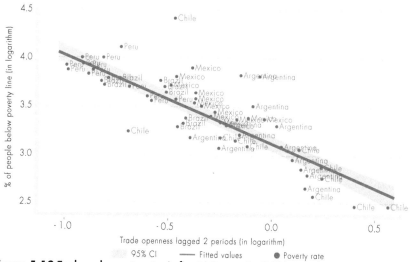

Figure 5.13 Trade and poverty rates in five LAC countries (1996–2010). *Source: own elaboration, data from WBDI (2012) and SEDLAC (2012).*

5.5.5 Do increasing trade and globalization impair or hamper the equitable access to resources (land and water)?

There are conflicting views about the role of trade regulations over water use. Gawel and Bernsen (2011) advise not to look at water footprint differences to support international trade regulations. The need for water governance at the global scale results from growing concerns over, firstly, water security in many parts of the world, and secondly, whether the existing commodity market system can deliver security as well as the necessary stewardship of water resources (Allan, 2011; Sojamo et al., 2012). Even if international trade presently involves products for which a significant part of the production is water-intensive, and virtual water flows are mainly subordinated to world trade rules, the policy linkages between international trade and impacts on freshwater have rarely been analysed. Chico et al. (2014) discuss options to improve global water governance through trade.

It is well known that water is seldom the dominant factor determining trade in water-intensive commodities. Many factors influence virtual water trade, such as direct or indirect subsidies, availability of land, labour, technology, level of socio-economic development, national food policies and international trade agreements (Aldaya et al., 2010; Rogers and Ramirez-Vallejo, 2011). Currently, virtual water flows are mainly subordinated to world trade rules (Hoekstra et al., 2011). The European Single Market and WTO frameworks are potentially suited to address the link between international trade and sustainable water use. Hoekstra et al. (2011) identifies several mechanisms to better ensure that trade and sustainable water use go hand in hand, such as product transparency or an international water pricing protocol. Trade will never contribute to optimal production and trade outcomes, from a water perspective, as long as water remains so underpriced (*ibid.*). There is a need to arrive at a global agreement on water pricing structures that cover the full cost of water use, including investment costs, operational and maintenance costs, a water scarcity rent and the cost of negative externalities of water use. Without an international treaty on proper water pricing it is unlikely that a globally efficient pattern of water use will ever be achieved. However, finding a harmonized water pricing mechanism may be so elusive that second-best solutions may be more feasible.

More recently, the WTO has started looking at the trade interventions that can influence water-related policies on either the production side (irrigation subsidies) or the consumption side (water footprint labelling) (Jackson et al., 2014). More work is needed to clarify key concepts and to enhance transparency in order to have a more comprehensive understanding of the ways in which these rules alter water resource outcomes.

Even if it is not yet widely recognized, the private sector has also a vital role to play in ensuring food-water security. Food supply chains operate beneath a complex pact between the state and the market. The agents in these food supply chains – mainly farmers – determine whether food-water is managed sustainably and securely (Allan, 2011; Sojamo et al., 2012).

Water and food security is today much more related to economic capacity and trade, than to physical water scarcity. Knowledge about the virtual water flows entering and leaving a country can cast a completely new light on the actual water scarcity of a country. This shift in perception forces a reconsideration of what are the main problems of food security, away from pure physical scarcity and technological fixes. The main issues that have to be addressed globally in relation to food security are: the hidden monopolies that currently exist in the WTO, the potential threat of political embargoes and the need for domestic social changes to be fulfilled.

References

Airports Council International (2012). *Passengers data*. Several years. [Online] Available from: www.aci.aero/Data-Centre/Annual-Traffic-Data/Passengers/2010-final. [Accessed January, 2013].

Ajanovic, A. (2011). Biofuels versus food production: Does biofuels production increase food prices? *Energy*, 36 (4): 2070–2076.

Aldaya, M.M., Allan, J.A. & Hoekstra, A.Y. (2010). Strategic importance of green water in international crop trade. *Ecological Economics*, 69 (4): 887–894.

Allan, T. (2011). *Virtual Water: Tackling the Threat to Our Planet's Most Precious Resource*. London, IB Tauris Publishers.

Anderson, K. (2010). Globalization's effects on world agricultural trade, 1960–2050. *Philosophical Transactions of the Royal Society of London – Series B: Biological Sciences*, 365 (1554): 3007–3021.

Bakhshoodeh, M. & Zibaei, M. (2007). No agriculture trade openness and poverty reduction: cross-country analysis. *Iran Agricultural Research*, 25(2).

Borras, S.J., Franco, J., Kay, C., Gómez, S. & Wilkinson, J. (2011). Land grabbing in Latin America and the Caribbean. *The Journal of Peasant Studies*, 39: 1–54.

Brenton, P., Newfarmer, R. & Walkenhorst, P. (2009). Avenues for export diversification: issues for low-income countries. Germany, University Library of Munich. MPRA Paper No. 22758.

Bruinsma, J. (2009). The resource outlook to 2050: by how much do land, water and crop yields need to increase by 2050?. In: *Conference on Expert Meeting on How to Feed the World in 2050*. Rome, FAO.

Brummer, B., Korn, O., Jaghdani, T.J., Saucedo, A. & Schlüßler, K. (2013). Food price volatility drivers in retrospect. Policy Brief 1. EU ULYSSES Project. [Online] Available from: www.fp7-ulysses.eu/publications/ULYSSES%20Policy%20Brief%201_Food%20price%20volatility%20drivers%20in%20retrospect.pdf. [Accessed August, 2013].

Bussolo, M., De Hoyos, R. & Medvedev, D. (2011). Free trade in agriculture and global poverty. *The World Economy*, 34: 2019–2043.

Cain, J., Hasan, R. & Mitra, D. (2010). Trade liberalization and poverty reduction: new evidence from Indian states. In: Bhagwati, J., & Panagariya, A. (eds). *'India's Reforms: How They Produced Inclusive Growth'*. New York, Oxford University Press. pp. 91–169.

Castilho, M., Menéndez, M. & Sztulman, A. (2012). Trade liberalization, inequality, and poverty in Brazilian states. *World Development*, 40 (4): 821–835.

Cervantes-Godoy, D., & Dewbre, J. (2010). *Economic importance of agriculture for poverty reduction*. OECD Publishing. Report No. 23.

Chico, D., Aldaya, M.M., Flachsbarth, I. & Garrido, A. (2014). Virtual water trade, food security and sustainability: lessons from Latin America and Spain. In: Martínez-Santos, P., Aldaya, M.M. & Llamas, R. (eds). *Integrated water resources management in the 21st Century: Revisting the paradigm*. London, CRC Press. Forthcoming.

Deininger, K. (2011). Challenges posed by the new wave of farmland investmente. *The Journal of Peasant Studies*, 38 (2): 217–247.

Dingemans, A. & Ross, C. (2012). Free trade agreements in Latin America since 1990: an evaluation of export diversification. CEPAL *Review*, 108W: 27–48.

Dollar, D. (2005). Globalization, poverty, and inequality since 1980. T*he World Bank Research Observer*, 20 (2): 145–175.

Duncan, R. & Quang, D. (2002). *Trade Liberalisation, Economic Growth and Poverty Reduction Strategies*. Canberra, Australia, AusAID. Technical report.

Edwards, S. (1993). Openness, trade liberalization, and growth in developing countries. *Journal of Economic Literature*, 31: 1358–1393.

Edwards, S. (1998). Openness, productivity and growth: what do we all know? *The Economic Journal*, 108: 383–398.

FAO (2013a). Food and Agriculture Organization of the United Nations. FAOSTAT. Available from: faostat.fao.org/. [Accessed January, 2013].

FAO (2013b). Food and Agriculture Organization of the United Nations. FAO Statistical Yearbook 2013. Rome, Italy, FAO.

Field, A.J. & Field, E.M. (2010). Globalization, crop choice, and property rights in rural Peru, 1994–2004. In: *The Poor Under Globalization in Asia, Latin America, and Africa*. Oxford, Oxford Univ. Press. pp. 284–323.

Finot, A., LaFleur, M., & Lima, J. D. (2011). *Analysis of the effects of trade opening on household welfare: an application to Chile, 1999–2006*. CEPAL Working Paper No. 405.

Foley, J.A., Ramankutty, N., Brauman, K.A., Cassidy, E.M., Gerber, J.S, Johnston, M., Mueller, N.D., O'Connell, C., Ray, D.K., West, P.C., Balzer, C., Bennett, E.M., Carpenter, S.R., Hill, J., Monfreda, C., Polasky, S., Rockström, J., Sheehan, J., Siebert, S., Tilman, D. & Zaks, D.P.M. (2011). Solutions for a cultivated planet. *Nature*, 478 (7369): 337–42.

Gawel, E. & Bernsen, K. (2011). Global trade, water scarcity and the limited role of virtual water. *Gaia*, 3: 162–167.

Giordano. P. (2012). Food Security and Trade in LAC. Necessarily unstructured and unrelated comments to David Laborde. *Workshop Drivers of Food Security in Latin America and the Caribbean* CIAT/IDB, Cali, Nov. 8–9, 2012.

Hassine, N. B., Robichaud, V. & Decaluwe, B. (2010). *Agricultural Trade Liberalization, Productivity Gain, and Poverty Alleviation: A General Equilibrium Analysis*. Cairo, Egypt, Economic Research Forum Working Paper No. 519.

Headey, D. & Fan, S. (2008). Anatomy of a crisis: the causes and consequences of surging food prices. *Agricultural Economics*, 39: 375–391.

Hoekstra, A.Y. & Chapagain, A.K. (2008). *Globalization of Water: Sharing the planet's freshwater resources*. Oxford, Blackwell Publishing.

Hoekstra, A.Y., Aldaya, M.M. & Avril, B. (2011). *Proceedings of the ESF Strategic Workshop on Accounting for Water Scarcity and Pollution in the Rules of International Trade*, Amsterdam, 25–26 November 2010. UNESCO-IHE. Value of Water Research Report Series No. 54.

IDB (2013). Inter-America Development Bank. Statistics and databases. [Online] Available from: www.iadb.org/en/research-and-data/statistics-and-databases,3161.html#.UkK8JsY0xZo [Accessed February, 2013].

Jackson, L.A., Pene, C., Martinez-Hommel, M.B., Hofmann, C. & Tamiotti, L. (2014). Water policy, agricultural trade and WTO rules. In: Martinez-Santos, P., Aldaya, M.M. and Llamas, M.R. (eds). *Integrated water resources management in the 21st century: Revisting the paradigm.* London, UK, CRC Press. Forthcoming.

Journal of Commerce (2012). *The Joc Top 50 World Container Ports.* [Online] Available from: www.joc.com/sites/default/files/u48783/pdf/Top50-container-2012.pdf.[Accessed February 2013].

Mejía, A. (2010). *Water For The Americas: Challenges & Opportunities.* Rosenberg International Forum on Water Policy. Buenos Aires, Argentina.

Mitchell, D. (2008). *A note on rising food prices.* World Bank Policy Research Working Paper Series No. 4682. [Online] Available from: www-wds.worldbank.org/external/default/WDSContentServer/WDSP/IB/2008/07/28/000020439_20080728103002/Rendered/PDF/WP4682.pdf [Accessed February 2013].

Mueller, N.D., Gerber, J.S., Johnston, M., Ray, D.K., Ramankutty, N. & Foley, J. A. (2012). Closing yield gaps through nutrient and water management. *Nature,* 490 (7419): 254–257.

NIC (2012). *National Intelligence Council. Global Trends 2030: Alternative Worlds.* Washington DC. [Online] Available from: globaltrends2030.files.wordpress.com/2012/12/global-trends-2030-november2012.pdf. [Accessed March, 2013].

OAS (2012). Organization of American States. *Foreign Trade Information System of the Organization of American States.* [Online] Available from: www.sice.oas.org. [Accessed February, 2013].

Pfister, S., Bayer, P., Koehler, A. & Hellweg, S. (2011). Environmental impacts of water use in global crop production: hotspots and trade-offs with land use. *Environmental Science Technology,* 45 (13): 5761–5768.

PPIAF (2013). Public-Private Infrastructure Advisory Facility. Data sources. [Online] Available from: www.ppiaf.org. [Accessed March, 2013].

Prakash, A. (2011) *Safeguarding food security in volatile global markets.* Rome, FAO.

Rogers, P. & Ramirez-Vallejo, J. (2011). Failure of the virtual water argument, in: Hoekstra, A.Y., Aldaya, M.M. and Avril, B. (eds). (2011) *Proceedings of the ESF Strategic Workshop on Accounting for Water Scarcity and Pollution in the Rules of International Trade,* Amsterdam, 25–26 November 2010: UNESCO-IHE. Value of Water Research Report Series No. 54.

Rosegrant, M.W., Zhu, T., Msangi, S., Palazzo, A., Miroslkav, B., Magalhaes, M., Valmonte-Santos, R., Ewing, M., Nelson, G.C., Koo, J., Robertson, R., Sulser, T., Ringler, C. & Lee, D. (2008). Global scenarios for biofuels: impacts and implications. *Review of Agricultural Economics,* 30 (3): 495–505.

Rulli, M.C., Saviori, A. & D'Odorico, P. (2013). Global land and water grabbing. *Proceedings of the National Academy of Sciences* 110 (3): 892–897.

SEDLAC. (2012). Socio-Economic Database for Latin America and the Caribbean. SEDLAC and The World Bank. [Online] Available from: data.worldbank.org/data-catalog/sedlac [Accessed March, 2013].

Sojamo, S., Keulertz, M., Warner, J. & Allan, J.A. (2012). Virtual water hegemony: the role of agribusiness in global water governance. *Water International,* 37(2):169–182.

Taylor, J. E., Naude, A.Y. & Jesurun-Clements, N. (2010). Does agricultural trade liberalization reduce rural welfare in less developed countries? The case of CAFTA. *Applied Economic Perspectives and Policy*, 32: 95–116.

The Economist (2012). *The world's shifting centre of gravity*. June 28, 2012. [Online] Available from: www.economist.com/blogs/graphicdetail/2012/06/daily-chart-19. [Accessed march, 2013].

UNCTAD (2012). United Nations Conference on Trade and Development. UNCTADSTAT. [Online] Available from: unctadstat.unctad.org/ReportFolders/reportFolders.aspx?sRF_ActivePath=P,5,27&sRF_Expanded=,P,5,27 [Accessed May 2013].

UNEP (2010). United Nations Environment Programme. *Environment Outlook for Latin America and the Caribbean*: GEO LAC 3. Panama City, Panama, UNEP.

Von Braun, J. (2008). *Responding to the World Food Crisis Getting on the Right Track: IFPRI 2007–2008*: International Food Policy Research Institute (IFPRI). Annual Report Essay 2007–2008. Essay No. 1.

Westhoek, H.J., Rood, G.A., Berg, M, Van Den, Jansen, J., Nijdam, D., Reudink, M. & Stehfest, E. (2011). The protein puzzle: the consumption and production of meat, dairy and fish in the European Union. *European Journal of Food Research Review*, 1 (3):123–144.

Williamson, J. (1990). *Latin American Adjustment: How Much Has Happened?*. Washington DC, Instituto de Economía Internacional.

WBDI (2012). World Bank Development Indicators (WBDI). [Online] Available from: data.worldbank.org/data-catalog/world-development-indicators. [Accessed February, 2013].

WorldBank (2012). World Bank Statistics. Available from: data.worldbank.org/ [Accessed February, 2013].

WTO (2012). World Trade Organization. Statistics database. Available from: stat.wto.org/Home/WSDBHome.aspx?Language=. [Accessed February, 2013].

Zetland, D., & Möller-Gulland, J. (2013). The political economy of land and water grabs. In: Allan, T.,Keulertz, M., Sojamo, S., and Warner, J. (eds). *Handbook of Land and Water Grabs in Africa. Foreign direct investment and food and water security*. Abingdon, Routledge. pp.257–272.

TRACKING PROGRESS AND LINKS BETWEEN WATER AND FOOD SECURITY IN LATIN AMERICA AND THE CARIBBEAN

Authors:
Bárbara Willaarts, Water Observatory – Botín Foundation, and CEIGRAM, Technical University of Madrid, Spain
Alberto Garrido, Water Observatory – Botín Foundation, and CEIGRAM, Technical University of Madrid, Spain
Bárbara Soriano, CEIGRAM, Technical University of Madrid, Spain
Marcela Molano, CEIGRAM, Technical University of Madrid, Spain
Olga Fedorova, CEIGRAM, Technical University of Madrid, Spain

Highlights

- The concepts of water security (WS) and Food Nutritional Security (FNS) have evolved from narrow and well-defined goals of guaranteeing citizens' access to sufficient water and food resources into much broader concepts, embracing health, sustainability, efficiency and social equity aspects. Such wide visions go beyond the physical availability or productive value of water and food, and testify to its importance as key elements to human well-being.

- During the last decade, significant progress has been achieved across many Latin American and Caribbean (LAC) countries on essential WS fronts, such as improving access to drinking water and sanitation, reducing social vulnerability to water hazards and water use efficiency. These achievements have contributed to improving health, physical protection and material needs, but important challenges remain. Water pollution is now one of the most important water security threats to LAC and requires greater attention at all levels.

- Efforts to improve basic WS goals are still needed in most countries, particularly in the low income countries of the Caribbean, Mesoamerica and Andean regions. Wealthier countries such as Mexico, Chile, Argentina, Brazil or Uruguay have higher WS standards, although physical water scarcity is becoming a growing problem, particularly in som of these countries.

- As with WS, most countries in LAC have improved basic food security indicators, predominantly in terms of food availability and access. However, the food crisis of 2007–2009 slowed down progress or even worsened some indicators for a few countries like Haiti, Paraguay and Guyana. Others like Bolivia, Ecuador, Peru, Honduras, Guatemala, Nicaragua and El Salvador have made significant progress, but still have a considerable gap to bridge.

- The most important challenge in LAC regarding food and nutritional security (FNS) is to overcome malnutrition rather than a physical lack of food. Currently, there are still 49 million people undernourished (8% of LAC population), but obesity now affects 20% of the LAC population (> 110 million people) and overweight up to 35% (> 200 million people).

- Between 2000 and 2010 WS and FNS indicators have progressed more rapidly and consistently in the wealthiest half of LAC countries. Progress among the poorest countries has been more erratic, inconsistent and inadequate. Per capita income is a good predictor of the levels of WS and FNS standards but there is considerable variation of performance amongst countries with similar incomes. This suggests that setting the right priorities and implementing the right policies can make a difference.

6.1   Introduction

The concept of security has long been understood as a country's safety faced with external aggression (e.g. wars or conflicts) and the defence of national interests in foreign policies (UNDP, 1994). Yet, human security has a much wider interpretation as it is focused on improving human well-being within countries, beyond defending strategic interests between nations. As the 1994 Human Development Report states 'Human security is concerned with how people live in societies, how freely they exercise their many choices, what access do they have to material well-being, and whether they live in a climate of political stability and peace' (*ibid.*). Because of the many dimensions included in the notion of human well-being, different security branches have emerged since the early 1990s, including food and nutritional security (FNS), water security (WS) and/or environmental security (ES).

WS and FNS are particularly concerned with those issues surrounding water and food, e.g. access, availability, quality and stability, which are critical to human well-being. Both *securities* imply that people have sufficient and stable access to food, enjoy a healthy diet, have access to drinking water and improved sanitation facilities and are physically protected from water hazards, among many other aspects. Not being deprived of these conditions is also a necessary condition for living a dignified life and being morally resilient. The future prospects of a foetus, a new-born or a child are to a great extent conditioned on the mother's and the household's material well-being. A child with adequate access to drinking water, sanitation and food security will have a better chance of surviving and progressing to a mature age. Further, being physically protected against natural disasters and diseases are fundamental conditions for human security and societal resilience.

The extent to which a country is water and food secure depends on the physical environment but predominantly it is the level of poverty and the constrained socio-economic context that really dictates their degree (Grey and Sadoff, 2007). As Allan (2013) states '(…) poverty determines water poverty: water poverty does not determine poverty' (p. 2) When both these circumstances are aligned, harsh natural conditions and widespread poverty, options to improve water and food gaps are rather complex. In Latin America and the Caribbean (LAC), water and land resources are for the most part abundant, and what lies behind existing water and food insecurities is the prevailing poverty (OECD, 2013a). While LAC is on good track to meet many of the Millennium Development Goals (MDGs) ahead of 2015, poverty and inequality are still widespread in the region, and basic indicators of human material well-being remain below minimum standards. Currently, LAC still has 49 million undernourished people, 33 million lacking access to an improved clean water source and 20 million still practice open defecation (FAO, 2012a; WHO-UNICEF, 2013). In addition to this, the region also faces serious nutritional problems, with 20% of the population being obese (equivalent to over 110 million people) and 13.5% of pre-school children with stunted growth (FAO, 2012b; Finucane et al., 2011; Onis et al., 2011).

Improving WS and FNS within countries requires a wide range of different policies, as well as a clear definition of priorities based on their socio-political and economic statuses. However, in spite of these differences, there are also numerous interrelated aspects of water and food within countries that call for a joint analysis, since both securities are inextricably linked. Currently, 95% of the water consumed in LAC is used for producing food (Mekonnen and Hoekstra, 2011); therefore improving FNS inevitably requires having secure access to sufficient and stable water resources. Also, other important components of FNS in LAC like food safety, acceptable cooking conditions and personal hygiene require a minimum set of water quality standards to be in place. The importance of water for food production is what led Allan (2013) to distinguish between 'food-water', i.e. 90% to 95% of total water consumption which is invested in agricultural production, and 'non-food water', i.e. the remaining 5% to 10% of water resources needed to sustain all the other economic activities beyond agriculture.

The aim of this chapter is to explore the progress achieved in WS and FNS in LAC countries during the last decade, outline the main challenges ahead and assess the relevance of the food-water security link in this region. Accordingly, this chapter is organized as follows: Section 6.2 provides a conceptual discussion of the concepts of WS and FNS, reviewing how these two concepts have been defined and refined over time by different authors and institutions; Section 6.3 quantitatively synthesizes the trends and progress of both securities over the last decade; Section 6.4 assesses the links between both securities outlining the different synergies found in the LAC context; and lastly, Section 6.5 includes some final remarks.

6.2 Evolving concepts of water and food security

6.2.1 Water security: concept and metrics

The concept of WS was introduced in the early 1990s and it has evolved significantly ever since (Cook and Bakker, 2012; López-Gunn et al., 2012). Originally WS was approached from a physical perspective, linked to the idea of national security, and the threat that physical water scarcity and conflicts-over-water could represent for neighbouring countries (Starr, 1992). Under this framework, WS was closely linked to the goal of ensuring sufficient water resources and guaranteeing access in order to maintain political stability within and outside national borders.

Over time, the concept has further evolved to include other economic, social and environmental aspects of water important to human well-being beyond its physical availability. These include protection against water hazards, safeguarding human health, maintenance of healthy aquatic ecosystems as well as cultural and spiritual values linked to water (see Table 6.1). One of the most recent definitions proposed by UN-Water (2013) defines WS as *'the capacity of a population to safeguard sustainable access to adequate quantities of acceptable quality water for sustaining livelihoods, human well-*

being, and socio-economic development, for ensuring protection against water-borne pollution and water-related disasters, and for preserving ecosystems in a climate of peace and political stability' (p.2).

The increasing use of this concept raises fundamental questions for water policy, including whether or not it overlaps or aligns with IWRM (see Chapter 15). As Cook and Bakker (2012) argue, both approaches are complementary since water security is focused on the end goal (being water secure) whereas IWRM is process-orientated (the path and steps required to become water secure).

Table 6.1 Human well-being dimensions considered under different approaches to water security

DIMENSIONS	Starr (1992)	GLOBAL WATER PARTNERSHIP (2000)	Grey and Sadoff (2007)	UNESCO IHE (2010)	UN WATER (2013)
Access to drinking water and sanitation					
Protection from water hazards					
Ecosystem protection (water quality and quantity)					
Adequate livelihoods (e.g. health, material goods, education)					
Preserve non-material aspects of water (e.g. cultural and ethical values)					
Maintain peace and political stability (e.g. transboundary water cooperation, public participation, etc.)					

Source: own elaboration

Improving WS of LAC citizens will require a pool of measures, including hard-path solutions, i.e. technological responses based on infrastructure development, as well as soft-path solutions, i.e. an institutional response including legal framework development and enforcement, greater transparency or economic instruments to improve water management. The type of measures as well as the implementation sequence will largely depend on the socio-economic context and the degree of development within countries, above any favourable hydrological condition. Foremost because improving WS and reducing people's vulnerability to water risks largely relies on the capacity of nations to make investments and develop infrastructures and policy tools (Grey and Sadoff, 2007; Allan, 2013).

Nevertheless, having a favourable hydrological situation is an advantageous factor to become more water secure. As described in Chapter 2, LAC is extremely well endowed in terms of water resources; however, it also has a high hydro-climatic variability (e.g. floods and droughts linked to the El Niño and La Niña phenomena). Such inherent variability often affects the most vulnerable and poorest, but also LAC's richest countries, such as Chile or Mexico. In fact, droughts in Chile represent a major water risk since they are highly frequent in the centre-north part of the country, where the majority of the population lives and most agriculture takes place (UNESCO, 2010).

An inherent characteristic of countries' WS is that it is a scale-dependent goal (Cook and Bakker, 2012). In fact, national WS assessments can mask significant variations compared to those performed at the more regional or local scale (Vörösmarty et al., 2010). Moreover, WS goals are likely to change over time, depending on the priorities countries have at a given time or stage of development. For instance, in Europe conventional approaches to water management have for a long time prioritized the need for building infrastructures and attending to the increasing demands of competing users. However, the goal of the current European water policy i.e. the Water Framework Directive (WFD 2000/60/EC) represents a radical shift with respect to this previous approach since it considers environmental sustainability of aquatic ecosystems as a priority to ensure WS in Europe.

The benefits gained by LAC countries when improving their WS and reducing their water risk to tolerable levels entail inevitable *trade-offs*, e.g. guaranteeing water access to big urban areas requires the constructions of dams, and even large inter-transfer schemes, which often have large social and environmental implications. However, some of these *trade-offs* are avoidable, such as reducing water pollution, and these will depend to a larger extent on the priorities defined by governments. The path followed by developed regions such as Europe to achieve WS has brought about serious environmental degradation, and yet there is no full understanding of the costs and the actions needed to reverse this problem despite ongoing efforts. Hence, developing countries striving for WS would need to make large investments in water management and infrastructure at all levels, but they can benefit from the experience gained in regions like Europe of the need to pay greater attention to institutional development, environment sustainability and social inclusion to avoid unintended and avoidable costs.

In order to keep track of regional progress in WS, a number of operational frameworks have been developed over the last few years (see Figure 6.1). The overall purpose of these frameworks is to determine whether countries or regions are on the right path to increase resilience to water risks and what are the main challenges. As Figure 6.1 shows, the majority of existing operational frameworks propose a different set of indicators to measure the hydrological status within countries (resource physical availability and environmental status), as well as the use and access of water from a socio-economic perspective (access, sanitation and economic water efficiency). The existence of water institutions to ensure WS stability is barely considered under these frameworks, partly because of the lack of robust metrics to measure institutional progress, and also because of the difficulty of quantifying what is good governance. Neither the risks related to water hazards, nor those associated with natural disasters, are explicitly considered in most of the cases despite the importance they have in regions like LAC.

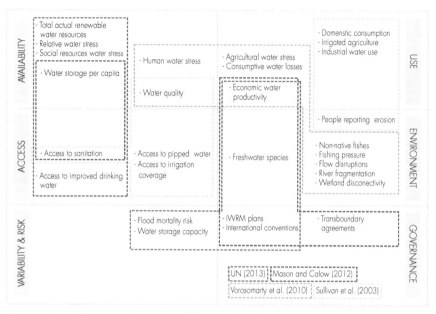

Figure 6.1 **Indicators and operational frameworks for measuring water security.** *Source: Own elaboration based on UN (2013), Mason and Calow (2012), Sullivan et al. (2003) and Vörösmarty et al. (2010)*

6.2.2 Food and nutritional security: concepts and dimensions

Similarly to the WS concept, the notions of FNS have evolved significantly in the last sixty years. Table 6.2 synthesizes the major milestones of the concept since the 1940s.

The notion of food security has generated tremendous attention in the last years, and it is now a well-established concept. According to FAO (1998) food security (FS) exists when (a) all people at (b) all times have (c) both physical and (d) economic access to sufficient food to (e) meet their dietary needs for (f) a productive and healthy life. Often, FS is framed in four dimensions: availability, access, stability and utilization.

According to FAO (1998):

> Food insecurity exists when people are undernourished due to the physical unavailability of food, their lack of social or economic access, and/or inadequate food utilization. Food insecure people are those individuals whose food intake falls below their minimum calorie (energy) requirements, as well as those who exhibit physical symptoms caused by energy and nutrient deficiencies resulting from an inadequate or unbalanced diet, or from the inability of the body to use food effectively because of infection or disease. An alternative view would define the concept of food insecurity as referring only to the consequence of inadequate consumption of nutritional food, considering the physiological utilization of food by the body as being within the domain of nutrition and health. Vulnerability refers to the full range of factors that place

Table 6.2 Evolving definition and scope of the food security concept

	FOOD AND NUTRITIONAL SECURITY
1940–1980	Food security and nutrition security (WW II), 43 countries met in Hot Springs, Virginia, 1943
	'Freedom from want' meaning a secure, adequate and suitable supply of food for every man, woman and child, where 'secure' referred to the accessibility, 'adequate' referred to the quantitative sufficiently of the food supply and 'suitable' referred to nutrient content.
1980–1990	'Concept of entitlement' Sen (1982). Food problems associated to agricultural production and food supply, but also with the governing economies and societies.
1940–1980	1996 World Food Summit 'All people at all times have physical and economic access to sufficient, safe and nutritional food to meet …'
	'A person is considered nutritionally secure when he/she has a nutritional diet and the food consumed is biologically utilized… resisting or recovering from disease, pregnancy, lactation and physical work' Frankenberger et al. (1997)
	Joint use of FS and NS concepts IFPRI, UNICEF and FAO (mid-1990s)
2000–PRESENT	Road Map for Scaling-Up Nutrition 'NS is achieved when secure access to an appropriately nutritious diet is coupled with a sanitary environment, adequate health services and care, to ensure a healthy and active life for all household members' 2010
	Weingärter (2010), Food and Nutritional Security is a condition under which adequate food (quantity, quality, safety, socio-cultural acceptability) is available and accessible for and satisfactorily utilized by all individuals at all times to live a healthy and happy life.
	FAOs 'FNS is a condition when all people at all times consume food of sufficient quantity and quality in terms of variety, diversity, nutrient content and safety to meet their dietary needs and food preferences for an active and healthy life, coupled with a sanitary environment adequate health and care' (CFS, 2009)

Source and quotes from: Pangaribowo et al. (2013)

people at risk of becoming food insecure. The degree of vulnerability for an individual, household or group of persons is determined by their exposure to the risk factors and their ability to cope with or withstand stressful situations.

Hoddinott (1999) claims that there are 200 definitions and 450 indicators of food security. As we will review in section 6.3.2 below, dozens of indicators are identified as having a direct and indirect influence on food security assessments. Less straightforward and evident are the drivers of food insecurity. Consider one of the factors that have been mentioned as having a crucial impact on the number of people suffering from hunger or being vulnerable to food insecurity: agricultural prices levels and volatility. Swinnen and Squicciarini (2012) found contradictory statements from two leading institutions, FAO and OXFAM, in relation to the role of agricultural prices in explaining rural poverty and food insecurity. The difficulty of ascertaining the impact of food prices on food security is due to the fact that people in poor rural areas are often producers and consumers, a factor whose complexity escalates as some households could be net buyers under some price situations and net sellers under others.

Recently, the notion of FS has also been expanded to include nutritional security, the two now being commonly addressed as 'Food and Nutritional Security' (FNS). The G8 supported the New Alliance for Food Security and Nutrition, which included the endorsement of the 'Scaling Up Nutrition movement' and 'welcome the commitment of African partners to improve the nutritional well-being of their populations, especially during the critical 1,000 days window from pregnancy to a child's second birthday'. This attests to the fact that, while calorie intake may be sufficient to cover body-energy demands, many other dietary elements are also required, especially for pregnant women and children, to ensure a healthy life and growth.

And yet, well-known experts still puzzle at the low adoption rates of a number of crucial habits for health improvement and income generation among the world's poorest, e.g. application of fertilizers, use of anti-malaria nets, application of chlorine to drinking water, vaccinations and routine medical checks to name but a few (Banerjee and Duflo, 2011). Another unresolved query is the increasing prevalence of obesity among the poorest households in some developed and developing countries alike. Ultimately, having a healthy diet requires not only sufficient access to food under all FNS dimensions, but also the willingness to adhere to it and minimum knowledge of its components and sources.

What the above comments may suggest is the following. First, whilst an increase in agricultural production is fundamental in order to increase FNS, it may not guarantee it. This is one of the blurring elements of the linkages between water and food security, in the sense that more water (or land) available for agriculture does not necessarily improve FNS indices, although increasing agricultural production among the poorest rural households improves their nutritional outlook. Second, the new approach of FNS places more emphasis on nutritional aspects than FS, but in order to monitor them there is a need for data which is much harder to obtain and of which we do not have historical records. Furthermore, the consequences of reduced FNS could have delayed effects which may only become evident as children become adolescents and young adults. Third, as this book shows, virtually all the variables directly related to FNS in the LAC region have been changing rapidly in the last decade, in the course of which commodities prices have become very volatile and followed an upward trend (see Chapters 4 and 5). Thus, FNS performance indicators co-vary with other major drivers; with which it has only an indirect relationship, meaning causality is almost impossible to establish (see Table 6.5).

As Barrett (2010) mentions, the FNS concept is elusive because a single indicator cannot summarize its complexity. It is thus necessary to analyse a set of indicators in order to capture all its relevant dimensions. Some of the existing composite food security indicators that focus on macro levels are: the FAO Indicator of Undernourishment (FAOIU); the Global Hunger Index (GHI); the Global Food Security Index (GFSI); the Poverty and Hunger Index (PHI); the Hunger Reduction Commitment Index (HRCI)). Some indicators that focus on micro level are the anthropometric indicators (measure nutritional outcomes) and the medical and biomarkers indicators (measure micronutrient deficiencies) (Pangaribowo et al., 2013). Many of the different FNS frameworks or compound indicators developed complement each other because they refer to different critical dimensions of food security (see Figure 6.2). Dimensions such as access, use and utilization are well represented by most composite indicators, only stability is clearly under-represented. Pangaribowo et al. (2013) recommend including two outcome indicators to capture the short-term FNS stability: per capita food supply variability and food price variability.

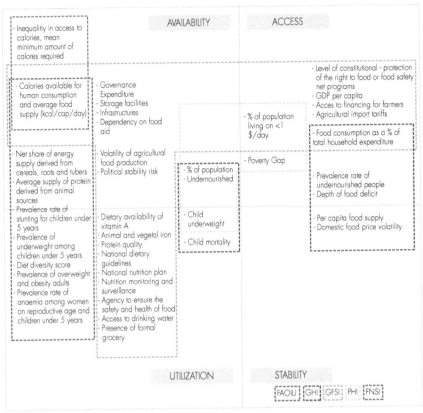

Figure 6.2 Existing food and nutrition indicators. *Source: own elaboration based on Panga-ribowo et al. (2013).* Note: the FAO Indicator of Undernourishment (FAOIU); the Global Hunger Index (GHI); the Global Food Security Index (GFSI); the Poverty and Hunger Index (PHI); Food and Nutrition Security Indicators proposed by EU project 'Food Secure' (FNSI).[1]

6.3 Water and food security status and trends in LAC

6.3.1 Water security performance

Table 6.3 summarizes the WS status of LAC countries in 2010 and the progress achieved since 2000. The framework used in this assessment to measure WS is a mixture of the ones presented in Figure 6.1. An imposed pre-requisite was to choose only those indica-tors for which it was possible to track changes over time, as well as selecting a pool of indicators capable of reflecting the different dimensions involving WS.

In terms of blue water availability (runoff), LAC countries have a privileged status, only the Caribbean islands of Dominican Republic and Haiti show a total actual renewable water resources (TARWR) below 3000 m³/cap/yr (Table 6.1). Despite this overall water

---

1 www.foodsecure.eu/

richness, physical blue water scarcity exists due to the spatial mismatch between where water is naturally available and where it is demanded. For instance, more than 75% of Mexicans live in basins where water consumption is at least twice the volume of water renewed naturally every year (blue water scarcity index ≥ 2) (see Figure 6.3 and Table 6.3). The northern part of Chile also faces serious blue water stress, with current demand being three times more than the natural available flow. In the northern part of Argentina and northeast Brazil, blue water scarcity problems are currently affecting 14% and 13% of their national populations respectively, and this trend has grown since the year 2000. Along the Peruvian coast, blue water scarcity is approaching a critical threshold, which poses an important risk for Peru's development since the majority of the population and agricultural activity is concentrated along the coastal basin.

Green water (soil moisture) plays a fundamental role in LAC's agriculture (see Chapter 7) and it is a key asset for achieving regional and global food-water security. Green water availability (measured in terms of arable land per capita) in LAC is high (0.26 ha per capita per year in 2010), and only the Caribbean islands, Costa Rica and Colombia have lower ratios. These punctual green water shortages are mostly compensated through regional agricultural trade and do not represent a major water risk for the above mentioned countries. The most important risk from a food-water security perspective in LAC is related

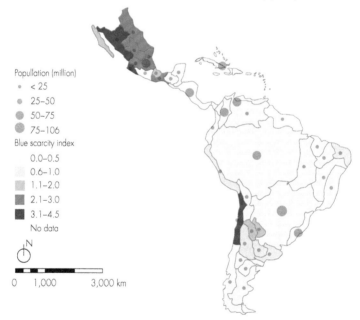

Figure 6.3 **Blue water scarcity and population distribution estimates for 2010 in Latin America.** *Source: own elaboration with data from Hoekstra and Mekonnen (2011) and CIESIN-FAO-CIAT (2005).* Note: The blue water scarcity index as defined by Hoekstra and Mekonnen (2011) is the ratio between the annual blue water consumption and the naturally available runoff minus the environmental flow requirements.

to the intra- and inter-annual variability of green water, i.e. the high frequency of droughts and floods linked to El Niño and La Niña, and the impacts these phenomenon have on rain-fed agriculture and food security. Alongside this, the high reliance on green water for food production has associated large environmental trade-offs, since the expansion of arable land calls for the extension of the agricultural frontier over natural ecosystems (see Chapter 3).

With regard to access to water, significant improvements have been achieved across most LAC countries. Approximately 90% of the households in LAC have access to an improved water source and 76% to sanitation facilities (see Table 6.3). Only Bolivia and Haiti remain below these rates, particularly regarding sanitation facilities. This positive trend nevertheless masks an important gap between urban and rural access, particularly in the Andean region, Brazil and some Mesoamerican countries such as Nicaragua (see Figure 6.4). According to the latest figures of the Joint Monitoring Program on water (WHO-UNICEF, 2013) in 2011, 32.7 million people in LAC still have no access to an improved drinking water source and 21 million still practice open defecation, the majority of these in rural areas.

Assessing the productive use of water determines nations' dependency on water resources for its economic development. Table 6.3 summarizes the trends in green water productivity. Overall, the majority of countries show a positive increase in the efficiency of green water use (measured in terms of improvements in rain-fed agricultural yields), particularly the most important agricultural producers like Chile, Argentina, Brazil and Paraguay. In Mesoamerica, green water productivity has increased, but to a lesser extent. Only the countries Dominican Republic and Cuba, together with Belize, have experienced a reduction in their agricultural yields. These results evidence a progressive decoupling of agricultural growth from agricultural area expansion, which is a positive sign to increase food-water security. Regarding blue water efficiency use in agriculture, no data exists to track progress over time, which prevents a detailed analysis. However, as discussed in Chapter 10 and detailed in Figure 6.5, irrigation efficiency in LAC remains low compared to the global average (39% of LAC average compared to a global efficiency of 56%). Mesoamerican countries and the Caribbean islands show the lowest rates of irrigation efficiency.

Concerning the environmental status of aquatic ecosystems, the indicator on freshwater diversity status shows a clear trend of environmental degradation across the entire region (see Table 6.3). Countries whose rivers are most degraded include Brazil, Colombia, Peru and Mexico. Overall, and despite the lack of robustness of this indicator, it seems clear that averting environmental degradation and reduced water quality is probably the next most important challenge LAC needs to face in order to avoid unintended environmental but also social and economic side effects. Figure 6.6 shows the trends in public investments in LAC countries on water resources management. Since 2000 a large fraction of the public investments in LAC (either as Official Development Aid (ODA) or as Other Official Flows (OOF)) have been directed to mixed projects of water supply and sanitation. Wastewater treatment investments still represent less than 1% of total public investments.

Table 6.3 Water security progress between 2000 and 2010 in LAC

Region	Country	AVAILABILITY — Total Actual Renewable water resources (TARWR) m³/cap./yr [1] 2000	2010	AVAILABILITY — % Pop. living in water scarce basin 2000	2010	AVAILABILITY — Green water availability (ha of arable land/cap./yr) [1] 2000	2010	ACCESS — % Pop with access to an improved water source 2000	2011	ACCESS — % Pop with access to sanitation 2000	2011	UTILIZATION — (Green water) Arable productivity (ton/ha/yr) [1] 2000	2010	STATUS — Freshwater biodiversity (number of threaten species) [2] 2000	2008	RISK — Flood risk Index (% population affected by water floods, storms and landslides) [3] 2000	2010	GOVERNANCE — State recognition on the human right to water and sanitation [3] 2000	2010	GOVERNANCE — Water laws (national laws) [4] 2000	2010
AMAZONIAN	Brazil	45,920	41,886	12	12	0.33	0.31	93	98	75	79	2.6	3.6	70	737	<1	<1	no	yes	1	3
AMAZONIAN	Guyana	326,558	318,783	<1	<1	0.61	0.56	89	94	79	84	2.6	2.8	3	23	2	14	yes	yes	no data	no data
AMAZONIAN	Suriname	254,167	230,624	<1	<1	0.12	0.11	89	92	81	83	2.5	2.8	2	16	<1	3	no data	no data	no data	no data
ANDEAN	Bolivia	71,900	61,707	<1	<1	0.36	0.38	79	88	37	27	1.5	2.0	23	57	<1	<1	no	yes	1	1
ANDEAN	Ecuador	33,242	28,938	<1	<1	0.11	0.11	84	94	63	92	1.6	2.4	no data	118	<1	<1	yes	yes	1	2
ANDEAN	Peru	71,974	65,068	<1	<1	0.14	0.13	80	85	63	71	2.6	3.1	45	229	<1	<1	yes	yes	1	1
ANDEAN	Colombia	51,901	45,432	<1	<1	0.07	0.04	91	92	73	78	2.7	3.3	148	101	<1	<1	no	yes	1	2
ANDEAN	Venezuela	48,787	41,886	<1	<1	0.11	0.10	92	no data	89	no data	3.0	3.1	43	no data	<1	<1	no	no data	0	2
MESOAMERICA	Belize	70,532	58,333	<1	<1	0.26	0.20	85	98	83	90	2.6	2.5	9	25	14	6	no	yes	no data	no data
MESOAMERICA	Costa Rica	27,529	45,432	<1	<1	0.05	0.04	95	97	91	95	2.5	2.6	31	69	3	<1	no	yes	1	1
MESOAMERICA	El Salvador	4,213	4,052	<1	<1	0.11	0.11	83	88	61	87	2.1	2.7	4	19	<1	<1	no	yes	0	1
MESOAMERICA	Guatemala	9,432	7,542	<1	<1	0.12	0.10	87	90	71	78	1.8	2.1	34	73	<1	<1	no	yes	0	2
MESOAMERICA	Honduras	14,809	12,370	<1	<1	0.17	0.13	81	87	65	77	1.3	1.5	27	56	4	<1	no	yes	1	2
MESOAMERICA	Mexico	4,455	3,983	77	78	0.25	0.22	89	96	75	85	2.8	3.5	205	335	<1	<1	yes	yes	2	3
MESOAMERICA	Nicaragua	37,663	33,492	<1	<1	0.38	0.33	80	85	48	52	1.5	1.7	6	26	3	<1	yes	yes	1	2
MESOAMERICA	Panama	48,224	41,445	<1	<1	0.19	0.16	90	94	65	69	3.3	4.2	no data	no data	<1	<1	no	yes	1	1
SOUTH CONE	Argentina	21,616	19,968	13	14	0.76	0.77	96	96	92	96	4.4	6.0	40	86	<1	<1	yes	yes	6	6
SOUTH CONE	Chile	58,414	84,483	62	63	0.11	0.07	95	96	92	96	4.4	6.0	7	57	<1	<1	yes	yes	1	2
SOUTH CONE	Paraguay	60,337	51,157	<1	<1	0.57	0.59	74	86	58	71	2.0	2.8	9	27	1	<1	no	yes	no data	no data
SOUTH CONE	Uruguay	41,805	41,124	<1	<1	0.42	0.55	98	100	97	100	2.9	3.6	10	35	<1	<1	yes	yes	no data	no data
CARIBBEAN	Cuba	3,411	3,387	no data	no data	0.32	0.32	91	94	87	91	2.1	1.8	16	no data	<1	4	no	no	1	1
CARIBBEAN	Dom. Republic	2,370	2,088	no data	no data	0.07	0.09	86	83	78	83	2.8	2.3	no data	no data	2	<1	no	no	no data	no data
CARIBBEAN	Haiti	1,580	1,386	no data	no data	0.10	0.11	62	69	21	17	0.8	3.3	no data	65	2	3	no	yes	1	1
	LAC weighted average	34,917	32,465	22	23	0.28	0.26	90	90	75	76	2.6	3.3	34	65	3	3	no	yes	1	2

Legend: ■ Improvement above the regional average growth · ■ Improvement below the regional average growth · ■ Deterioration or no improvement · □ No risky change

Source: own elaboration using data from EM-DAT (2013), FAO (2013b; 2013c; 2013d), Hoekstra and Mekonnen (2011), IUCN (2013), World Bank (2013) and WHO-UNICEF (2013)

1 Data for 2000 represent an average for the values of 1999-2001, whereas data for 2010 represent also an average for values of 2008–2010.
2 The inventory of freshwater threatened species was for the first time conducted in 2004 and updated in 2008.
3 State recognition of the human right to water and sanitation acknowledged in national constitutions, laws or policies.
4 Includes national or regional water legislation, laws on natural resources with a specific section on water, domestic supply legislation and specific groundwater law in selected LAC countries.

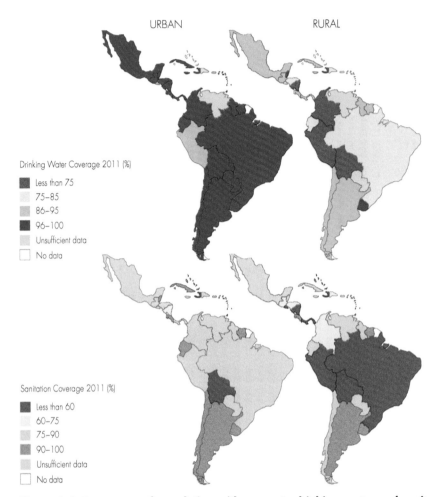

Figure 6.4. Percentage of population with access to drinking water and sanitation coverage in urnban (left) and rural (right) areas in LAC. *Source: own elaboration based on data from WHO-UNICEF (2013).*

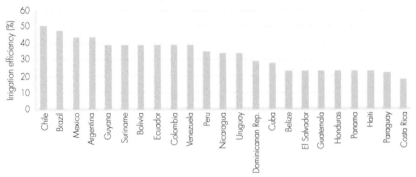

Figure 6.5 Irrigation efficiency (measured in terms of water requirement ratios) for Latin American countries, average for the period 1990–2012. *Source: FAO (2013a)*

The high hydro-climatic variability across many LAC countries represents an important water risk. Floods and droughts have large impacts on WS and FNS as they have large social and economic implications. As Table 6.3 shows, the social impacts of floods (measured in terms of the percentage of the population affected) are relatively low (<3%) for the entire LAC, but in countries like Belize, Guyana or Cuba they have larger impacts. Figure 6.7 summarizes the economic impacts attributed to natural hazards in LAC since 1980. Even though variability is a constant over time, economic impacts related to water hazards are still high, for instance in 2010 they peaked to almost 2 % of LAC's GDP. These trends shows that the region's vulnerability to water hazards is still high, and may not subside, in relative terms, as more growth is seen in terms of infrastructures, the economy, population density and the concentration of said population, thus increasing exposure to these risks (Berz, 1999; Mills, 2009).

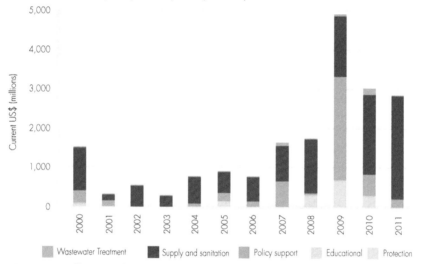

Figure 6.6 Allocation of public investments in water supply and sanitation in LAC, 2000–2010. *Source: based on data from OECD (2013b).*

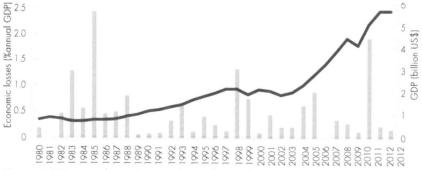

Figure 6.7 Economic losses (expressed in % of annual GDP, bars) attributed to water-related hazards (storms, floods and droughts) and GDP evolution (in USD, line) in Latin America and the Caribbean, 1980–2012. *Source: EM-DAT (2013) and World Bank (2013)*

Good governance and the development of a basic legal framework is a pre-requisite for ensuring countries' WS in the long run (Cook and Bakker, 2012). Several countries such as Costa Rica have made significant progress towards WS despite lacking a national water act. Nevertheless, the existence of a basic legal framework should facilitate the road to improve WS within countries. The recognition of water as a human right, either in their constitutions or under different legislations, and the number of existing water laws (national or regional water acts, groundwater, urban water supply) were used here as a proxy-indicator to ascertain the extent to which legal baseline conditions are in place in LAC countries to reach WS goals and minimize water risks (see Chapter 11). As Table 6.3 shows, water governance overall seems to have progressed substantially more than some WS goals. There is a close correlation between progress achieved in water access and sanitation and the development of legal frameworks. However, these legal frameworks have not been effective at reducing other important water risks associated with increased water pollution and vulnerability to hydro-meteorological events, probably because policy goals were mostly oriented towards securing access to citizens.

The above results can be summarized into two major trends. First, government priorities to improve WS (mostly those concerned with securing access and sanitation) have been effective and remarkable progress has been accomplished. Still, greater efforts are required among the low- and middle-income countries of Mesoamerica and Andean region (see Figure 6.8). The second trend is that upcoming water challenges will most likely require addressing the growing water pollution problem, particularly in megacities, because of the high threat such a trend could represent for LAC's development.

Figure 6.8 Water security performance in LAC countries. *Source: own elaboration based on the data from Table 6.3.*

6.3.2 Food security indicators in LAC

In this section we review a selection of food and nutritional indicators across LAC countries in order to track progress in FNS during the period 2000–2010. As with WS indicators, only those available for the majority of LAC countries and for which it was possible to track temporal changes were considered in this analysis. The selected indicators are shown in Table 6.4.

Table 6.4 Food and nutritional security indicators selected to assess Food and Nutritional Security (FNS) performance in Latin America and the Caribbean (LAC)

DIMENSION	INDICATOR	UNITS
AVAILABILITY	Per capita total amount of net calories available in a given country	kcal/person/day
	Average supply of protein derived from animal resources	g/cap/day
ACCESSIBILITY	Prevalence rate of undernourished people	% of population
	Depth of food deficit (how many calories would be needed to lift the undernourished from their status)	kcal/person/day
UTILIZATION	Prevalence rate of stunting for children under five years old (height-for-age < two standard deviations of the WHO Child Growth Standards Median)	%
	Body Mass Index [BMI < 18.5 Low BMI (chronic energy deficiency)/ BMI > 25 Overweight]	kg/m^2
STABILITY	Per capita food supply variability (Variability of the net food production value between 2004 and 2006 in constant $ divided by the population from UN 2010 estimates.)	%
	Cereal import dependency (Cereal imports/(cereal production+cereal import-cereal export)	%

Source: FAO (2012c)

Table 6.5 reports the progress of the indicators between 2000 and 2010. The indicators that show the best performance in LAC are those related to availability and access. 'Energy supply' improved in most countries and in those where it worsened, only slight reductions were experienced. Among these, Paraguay has the lowest levels and worsened over the specified time period. Ecuador, Guatemala and Haiti stood at fewer than 2,500 kcal/cap in 2010. Also, availability of 'energy from animal protein' improved in most countries. It ranges from 63 grams of protein per capita per day in Argentina to 9 in Haiti. It is below 30 in Belize, Bolivia, Cuba, Guatemala, Haiti, Honduras, Nicaragua, Paraguay, Peru, El Salvador and Suriname. In addition to Haiti's low score, the availability of animal protein is also particularly low in Nicaragua and Guatemala (19), although 35% higher than in 2000. It decreased in Paraguay (reaching 29), Uruguay and Argentina, but in these last two it is still above the regional average.

Overall it is interesting to note that food availability has improved the most among the Andean and Mesoamerican countries and the Caribbean. All Andean countries have improved their availability and access indicators (cells in green). Some countries have recorded increases higher than the average regional growth in these indicators. Some examples of this remarkable positive performance are: Peru and Venezuela in the Andean Region, Dominican Republic in the Caribbean and Nicaragua, Panama and

Honduras in the Mesoamerican region. Although this last region has exhibited significant growth, the case of Guatemala ought to be highlighted. In this country the prevalence of undernourishment ratio is still above 20%. Nicaragua has reduced this indicator from 37.5% to 22.7%, but still this percentage is notably high. Paraguay has seen all of its availability and access indicators go down between 2000 and 2010.

Food access indicators such as 'prevalence rate of undernourished people' and 'depth of food deficit' behaved well in the region. However, a few countries (Argentina, Costa Rica, Guatemala, Paraguay, El Salvador, and Uruguay) worsened in one or the other. Guatemala and Paraguay experienced significant worsening indicators. But Peru, Honduras, Nicaragua, Panama and Venezuela improved significantly. The depth of food deficit was still above 150 kcal in Guatemala, Haiti and Nicaragua in 2010.

Trends for food utilization vary across the LAC region. The prevalence of stunting for children under five has improved in most of the cases, except in Guyana, Dominican Republic and Haiti. Although the largest improvements were concentrated among Andean countries, these countries still have a high percentage of children likely to have stunted growth (more than 20% of children under five years old). For the year 2010, Bolivia, Ecuador and Peru also displayed this ratio above 20%. Considering the relative number of stunted children under five, in 2010 the prevalence rate was 8.2% in Argentina, 7.1% in Brazil, 12.7% in Colombia, 15.5% in Mexico, and 28.2% in Peru, to mention only the most populous countries.

Regarding food, stability indicators vary across the region. In terms of stability, the indicator 'variability of food supply' exhibits a mixed performance in the region. Some countries reduced it significantly, mainly in the Mesoamerican region (Belize, Costa Rica, Nicaragua and Panama) and in the Andes (Bolivia, Ecuador and Peru). Others saw it worsened, including Chile, Paraguay, and Brazil. Most of the countries show a greater cereal imports dependency ratio in 2010 than in 2000, predominantly among Mesoamerican and Caribbean countries where it ranges from 12% in Paraguay to more than 100% in Haiti.

The role of international trade as a means of achieving improved food security has been at the centre of numerous discussions, both in the academic world and at the top international political arena because of the 2007–2009 food price crises. The G20[2] wrote:

(1). Under the Food Security pillar of the Seoul Multi-year Action Plan on Development, the G20 request that FAO, IFAD, IMF, OECD, UNCTAD, WFP, the World Bank and the WTO work with key stakeholders to develop options for G20 consideration on how to better mitigate and manage the risks associated with the price volatility of food and other agriculture commodities, without distorting market behaviour, ultimately to protect the most vulnerable. … [This report] has been prepared

---

2 G20 Agricultural Ministers agreed in June 2011 on an 'Action Plan on food price volatility and agriculture', www.g20-g8.com/g8-g20/g20/english/news/news/declaration-of-the-ministers-of-agriculture.1401.html.

Table 6.5 Food security progress between 2000 and 2010 in LAC countries

| | | AVAILABILITY | | | | | | ACCESS | | UTILIZATION | | | | STABILITY | | | |
	COUNTRY	ENERGY SUPPLY (kcal/cap/day)		ENERGY FROM ANIMAL PROTEIN (gr/cap/day)		PREV. UNDER-NOURISHMENT (%population)		DEPTH OF FOOD DEFICIT (kcal/cap/day)		PREV. STUNTING CHILDREN UNDER FIVE YEARS (%)		BODY MASS INDEX (kg/m²)		VARIABILITY FOOD SUPPLY (kcal/cap/day)		CEREAL IMPORTS DEPENDENCY RATIO (%)	
		2000	2009	2000	2009	2000	2010	2000	2010	2000[1]	2010[1]	2000	2008	2000	2010	2000	2010
AMAZONIAN	Brazil	2,882	3,173	40	45	13	8	83	60	14	7	25	26	18	18	19	14
	Guyana	2,814	2,718	38	31	8	6	52	37	14	18	24	26	46	29	36	34
	Suriname	2,457	2,548	24	23	18	13	121	89	15	11	25	27	30	28	34	30
ANDEAN	Bolivia	2,121	2,172	23	26	30	26	200	175	33	27	24	27	64	13	29	24
	Colombia	2,662	2,717	29	34	13	12	86	82	18	13	24	26	24	21	54	59
	Ecuador	2,221	2,267	26	32	20	19	126	124	33	29	25	27	46	19	37	37
	Peru	2,379	2,563	21	25	23	14	151	92	31	28	25	26	36	14	46	49
	Venezuela	2,484	3,014	34	44	15	5	102	20	17	16	27	27	47	41	54	48
MESOAMERICA	Belize	2,560	2,680	26	26	9	7	52	45		22	26	30	36	30	30	30
	Costa Rica	2,825	2,886	36	39	5	5	27	35	9	6	26	26	60	33	87	95
	Guatemala	2,096	2,244	14	19	27	30	167	192	50	48	24	27	48	16	45	49
	Honduras	2,435	2,694	22	27	17	11	100	60	43	30	25	26	8	16	46	52
	Mexico	3,158	3,146	37	42	<5	<5	21	6	22	16	27	29	23	16	35	34
	Nicaragua	2,148	2,517	14	19	38	23	265	151	31	23	25	27	43	16	31	39
	Panama	2,195	2,606	36	40	25	12	175	79	22		26	27	86	41	64	70
	El Salvador	2,561	2,574	18	25	11	12	64	74	29	21	25	28	30	20	47	54
SOUTH CONE	Argentina	3,268	2,918	67	63	<5	<5	7	25	17	8	27	28	54	35	1	0
	Chile	2,808	2,908	38	47	<5	<5	31	26	3	2	26	28	20	39	45	52
	Paraguay	2,596	2,518	42	29	13	19	85	132	18	18	25	25	12	25	15	12
	Uruguay	2,844	2,808	56	39	<5	<5	27	33		14	25	27	43	35	27	16
CARIBBEAN	Antigua & Barbuda	2,155	2,373	49	60	39	21	293	156					48	60	99	99
	Bahamas	2,785	2,750	56	57	6	7	42	50			26	29	41	64	99	99
	Barbados	2,832	3,021	48	54	<5	<5	32	24			26	28	68	16	109	112
	Cuba	3,046	3,258	23	26	<5	<5	21	6	7		25	26	88	111	72	76
	Dominica	3,081	3,147	50	54	<5	<5	24	18					35	36	97	98
	Dom. Republic	2,322	2,491	23	29	23	16	156	107	8	10	25	26	37	31	76	75
	Grenada	2,220	2,456	37	48	31	21	228	156			25	27	45	26	178	129
	Haiti	1,931	1,979	8	9	52	45	429	375	28	30	23	23	55	35	58	59
	Jamaica	2,729	2,807	36	40	7	9	49	58	7	6	23	27	14	7	101	102
	Saint Kitts&Nevis	2,513	2,546	43	43	20	17	143	121					34	54	101	100
	Saint Lucia	2,720	2,710	55	54	11	14	74	97			24	27	30	39	100	101
	S.Vincent	2,528	2,914	35	48	14	6	99	39			25	27	32	14	178	188
	Trinidad & Tobago	2,696	2,751	25	31	13	10	94	73	5		25	29	16	26	111	116

■ Improvement above the regional average growth   ■ Improvement below the regional average growth   ■ Deterioration

Source: FAO (2012c)

by the listed organisations, with the addition of IFPRI and the UN HLTF, in response to the G20 request. (2). The approach taken in this report reflects the view of the collaborating international organisations that price volatility and its effects on food security is a complex issue with many dimensions, agricultural and non-agricultural, short and long-term, with highly differentiated impacts on consumers and producers in developed and developing countries.

Timmer (2013) indicated that:

Macro food security refers to a society-wide sense that food is reliably available in urban markets and that adequate purchasing power is a sufficient condition for accessing this food. 'Micro' food security requires that all households (urban and rural) have access to sufficient food, but that is only possible when poverty has been eliminated. 'Macro' food security is often confused (especially politically) with food self-sufficiency, but imported food often plays a key role in providing macro food security. (p.12)

Openness and increasing reliance on trade to import food staples is both a necessity and source of serious concern. Primarily, while 16% of the world's population today relies on food imports, Fader et al. (2013) conclude that 50% of the population will be dependent on imports in 2050 because of land and water constraints, even if food productivity in these countries reached its maximum potential. The OECD (2013b) reports that the net agricultural trade of all the developing countries, excluding Brazil, worsened significantly after the food crisis of 2007–2009.

It has been concluded by numerous authors that the food crisis in 2007–2009 worsened food security indicators in many countries (de Schutter, 2012; and OECD, 2013b). In Table 6.6 it is clear that the rate of improvement of food security indicators was much slower between 2007–2009 and 2010–2011 than it had been between 1990–1992 and 2007–2009. In some countries, including Colombia, Costa Rica, El Salvador, Guatemala, and Paraguay the proportion of people that suffered from hunger increased during the last comparison periods.

Table 6.6 Percentage of people suffering from hunger

	1991–92	2007–09	2010–12	CHANGE BETWEEN 1990–92 AND 2007–09	CHANGE BETWEEN 2007–09 AND 2010–12
LAC	14.6	8.7	8.3	-5.9	-0.4
Caribbean	28.5	18.6	17.8	-9.9	-0.8
Cuba	11.5	<5	<5		
Dominican Rep.	30.4	15.9	15.4	-14.5	-0.5
Haiti	63.5	46.8	44.5	-16.7	-2.3
Latin America	13.6	8.1	7.7	-5.5	-0.4
Argentina	<5	<5	<5		
Bolivia	34.6	27.5	24.1	-7.1	-3.4
Brasil	14.6	7.8	6.9	-6.8	-0.9
Chile	8.1	<5	<5		
Colombia	19.1	12.5	12.6	-6.6	0.1
Costa Rica	<5	5.0	6.5		1.5
Ecuador	24.5	19.6	18.3	-4.9	-1.3
El Salvador	15.6	11.3	12.3	-4.3	1.0
Guatemala	16.2	30.2	30.4	14.0	0.2
Honduras	21.4	11.6	9.6	-9.8	-2.0
Mexico	<5	<5	<5		
Nicaragua	55.1	23.9	20.1	-31.2	-3.8
Panama	22.8	13.1	10.2	-9.7	-2.9
Paraguay	19.7	16.8	25.5	-2.9	8.7
Peru	32.6	15.9	11.2	-16.7	-4.7
Uruguay	7.3	<5	<5		
Venezuela	13.5	<5	<5		

■ Improvement above the regional average growth   ■ Deterioration

Source: FAO (2012b)

The case of Paraguay has special relevance for our study. In 2011, it exported 48% of the soybean production (FAO, 2012b), reaching US$2.23 billion in exports revenues, 44% more than in the period 2009–2010. And yet, food security indicators worsened significantly in the period of measurement.

In Table 6.7 we report the ratio of imports over national utilization of wheat and maize in several LAC countries. Note that among the worst performing countries in terms of food security indicators, all except Paraguay had dependency rates of 99% or 100%.

De Schutter (2012) highlights some of the improvements being achieved in LAC on implementing the right to food, including: (1) the increased recognition of the right to food in the constitutions of many countries – rich and poor alike – with the development of an expansive legal framework on FNS (e.g. Ley Sistema de Seguridad Alimentaria y Nutricional in Guatemala (2005), Ley de Soberanía y Seguridad Alimentaria in Ecuador (2006), Ley Orgánica de Seguridad Alimentaria y Nutricional in Brazil (2006), Ley Orgánica de Seguridad y Soberanía Agroalimentaria in Venezuela (2008), Ley de Soberanía y Seguridad Alimentaria y Nutricional in Nicaragua (2009), or Ley de Seguridad Alimentaria y Nutricional in Honduras (2011)); and (2) the development of FNS strategies and plans of action (e.g. the Plan Nacional de Seguridad Alimentaria 2009–2015 of Paraguay, the Política Nacional de Seguridad Alimentaria y Nutricional of Nicaragua, the Política de Seguridad Alimentaria y Nutricional 2006–2015 of

Table 6.7 External dependencies of wheat and maize in LAC, (average 2007–2008 and 2011/2012)

	WHEAT		MAIZE	
	Ratio Imports/ Utilization	Consumption (kg/cap/yr)	Ratio Imp/ Utilization	Consumption (kg/cap/yr)
Mesoamerica & Caribbean				
Costa Rica	100	50	·	·
Dominican Rep	99	29	·	·
El Salvador	100	31	38	116
Guatemala	99	34	28	85
Haiti	100	25	·	·
Honduras	98	32	40	79
Mexico	54	50	28	144
Nicaragua	100	21	19	57
Panama	100	43	83	24
South America				
Bolivia	70	55	·	·
Brazil	61	52	·	·
Chile	35	114	52	17
Colombia	100	27	38	41
Ecuador	99	35	37	17
Peru	91	57	54	19
Venezuela	96	56	39	49
Uruguay	·	·	26	32

Source: FAO (2012b)

Honduras, the Política Nacional de Seguridad Alimentaria y Nutricional 2008 in Colombia, the Estrategia Nacional de Reduccion de la Desnutrición Crónica 2006–2016 of Guatemala, the Política Nacional de Seguridad Alimentaria y Nutricional (2003 and 2011) of El Salvador or the Plan Nacional de Seguridad Alimentaria y Nutricional 2009–2015 of Panama). Furthermore, a series of national social programmes also aim explicitly at combating hunger and food and nutrition insecurity, such as the 'Fome Zero' in Brazil, the 'Vivir mejor' in Mexico, 'Bogotá sin Hambre' in Colombia, 'Desnutrición Cero' in Bolivia, or 'Hambre más urgente' in Argentina.

Underlying the general improvement of the LAC region in most FNS indicators, the other side of the coin of food insecurity and probably the greatest challenge this region needs to face in relation to malnutrition is obesity. As shown in Table 6.5, most countries' body mass index indicates worrying levels of overweight (i.e. are above 25 kg/m^2). LAC is the second region in the world, after the US, with the highest percentage of its population obese or overweight (Finucane et al., 2011). Obesity today affects 20% of the Latin American population (> 110 million people) and overweight up to 35% (> 200 million people) (FAO, 2012b). In countries such as Belize, Mexico, Venezuela, Argentina and Chile obesity affects almost 30% of the countries' population, whereas in Brazil and most Andean countries it affects closer to 20% of the population (ibid.). Yet, the highest rates of overweight and obesity are found in those countries which are at a stage of nutritional post-transition (FAO, 2010, see also Box 6.1). The underlying reasons behind this type of food insecurity are diverse and include economic, as well as cultural factors. As claimed by Cuevas et al. (2009), 'the increase of overweight and obesity [has] been

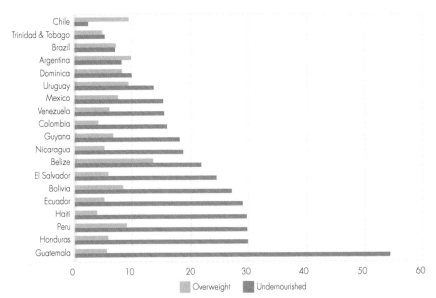

Figure 6.9 Percentage of undernourished and overweight children under five years old (2000–2009). *Source: FAO (2010) using data from Global Health Observatory-WHO 2010*

attributed to lifestyle changes occurring in recent decades related to rapid socioeconomic development, including a more Westernized diet, physical inactivity, urbanization, rural-urban migration and some maternal-fetal factors' (see Box 6.1). Obesity is a serious sign of malnourishment, stands in contrast to the hunger pandemic and has consequences for future generations. Countries that have eradicated hunger are those in a stage of nutritional post-transition and have the highest rates of child obesity. Among them, Argentina, Uruguay and Chile show obesity rates above 9 % (see Figure 6.9).

Box 6.1 The nutritional transition of emerging economies: the case of Brazil

Population growth, economic globalization, improving living standards and urbanization are causing important changes in the global food system in addition to modifying the dietary habits in many parts of the world (CAWMA, 2007; Godfray et al., 2010). As countries develop and populations become wealthier, the nutritional transition occurs. This transition implies a shift away from traditional staple foods such as roots and tuber vegetables and a rise in the consumption of meat and milk products, refined and processed foods as well as sugars, oils and fats (Ambler-Edwards et al., 2009).

In Brazil important changes have occurred to the food consumption patterns since 1987 (see Figure 6.10). In absolute terms, food consumption per capita has decreased over time from 360kg per capita in 1987 to 315kg per capita in 2009. However, most importantly the composition of the diet has experienced significant changes. In 1987 Brazilians had a balanced diet with an intake of predominantly vegetables, fruits, cereals and legumes (around 90 per capita per year of each). Rice, native tubers such as cará, potatoes, beans and tropical fruits like bananas and citruses were fundamental components of the diet prior to 1990. Animal protein consumption in the late 1980s was relatively high (> 50kg per person per year), equivalent to the average intake of richer regions like Europe (≈ 60kg per person per year in 1990) (Westhoek et al., 2011). However, since 1987 noteworthy changes have taken place in the composition of the food pyramid. Overall, the intake of vegetables, fruits and dairy products has decreased significantly (between 36 and 38%), whereas the consumption of processed food, stimulants and sugary products has experienced a dramatic increase (80%). Brazilians eat twice as much sugar as they did in 1987, 30% more processed food and almost 50% more non-alcoholic drinks and mineral water. The largest reduction in fruit and vegetable consumption is due to the lower intake of citruses and local tubers. Among the dairy products, the largest reduction is due to the lower intake of milk (from 68 litres per capita in 1987 to 40 litres in 2009). Overall, a nutritional transition in Brazil occurred in the late 1990s and early 2000s, overlapping with the economic takeoff of the country. Nevertheless, and compared to the prevailing trend in other developed regions, diet changes in Brazil have not translated into a greater consumption of animal protein, simply of food items linked to urban lifestyles.

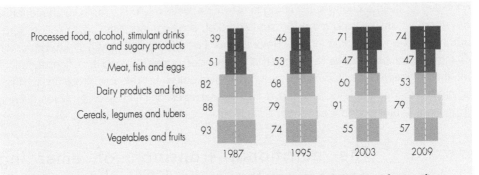

Figure 6.10 Food consumption pyramids (in consumed kg per capita per year) for Brazilians during the last two decades. *Source: own elaboration based on household survey – Pesquisa de Ornamentos Familiares: POF – for the years 1987/1988, 1995/1996, 2002/2003 and 2008/2009 (IBGE, 2013).*

6.4 Linking water and food security in Latin America

The purpose of this final section is to assess whether improvements and progress in water and food security indicators correlate across countries and to what extent they are interrelated. As shown in previous sections, economic development to a large extent explains part of the trends and current status. Therefore, in order to carry out the joint analysis of water and food security indicators we grouped the countries according to per capita income (as measured in 2010). The four figures (6.11 to 6.14) all have three panels, each with the set of countries belonging to the corresponding quartile of per capita income. Lastly, for each country and panel we present two points, corresponding to the pairs of selected WS and FS indicators measured in 2000 and 2010. Note that the scale differs across the three panels of each graph. This way data in this section shows five dimensions: time, country, per capita income, one WS indicator and one FS indicator.

The following pairs of indicators are plotted in Figures 6.11, 6.12, 6.13 and 6.14: prevalence of undernourishment (%) against access to improved sanitation (%); prevalence of stunting in children under five (%) against access to improved sanitation in rural areas (%); and finally prevalence of stunting in children under five (%) against access to drinking water (%).

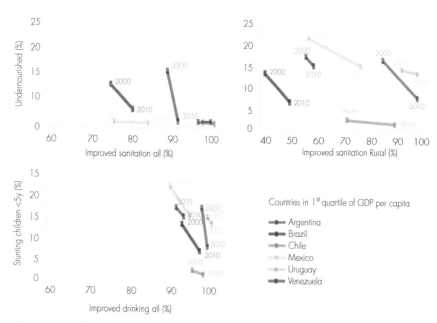

Figure 6.11 Three pairs of water and food security indicators measured in 2000 and 2010 (countries of the first quartile of per capita income in 2010). *Source: FAO (2010) using data from Global Health Observatory.*

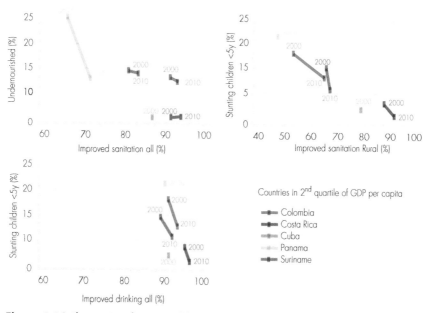

Figure 6.12 Three pairs of water and food security indicators measured in 2000 and 2010 (countries of the second quartile of per capita income in 2010). *Source: FAO (2010) using data from Global Health Observatory.*

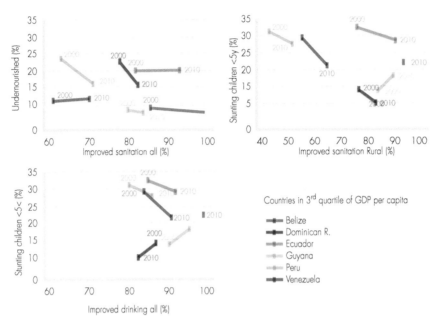

Figure 6.13 Three pairs of water and food security indicators measured in 2000 and 2010 (countries of the third quartile of per capita income in 2010). *Source: FAO (2010) using data from Global Health Observatory.*

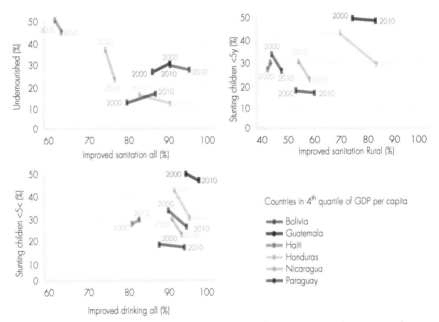

Figure 6.14 Three pairs of water and food security indicators measured in 2000 and 2010 (countries of the fourth quartile of per capita income in 2010). *Source: FAO (2010) using data from Global Health Observatory.*

By examining the Figures 6.11 to 6.14, we can draw the following conclusions. First, per capita income largely explains the pattern of improvements of the five indicators represented in these figures. From countries with the highest (Figure 6.11) to the lowest per capita income (Figure 6.14), the direction and slope of the segments overall become less homogeneous and more chaotic. In the groups of countries of the two lowest quartiles, some segments are upwardly sloped, and some hardly show any improvement between 2000 and 2010. Therefore, income growth and per capita income is fundamental for improving both WS and FS indicators.

Second, the reduction of the prevalence of stunting in children under five is closely correlated to the improvement of access to sanitation in rural areas. Except for Guyana, the remaining twenty-two countries exhibit downward sloping segments whose slopes tend to be similar within groups of countries. This would indicate that improved sanitation and the reduction of stunting in children evolve in parallel, although causation cannot be established.

Third, based on the different improvements and base levels of the percentage of undernourished people and stunted children across groups of countries, it seems clear that the reduction of undernourishment precedes the reduction of stunting in children. This would suggest that countries find it easier to reduce undernourishment rates than reducing the proportion of stunted children. We would thus conclude that ensuring nutritional security is more complex than simply reducing undernourishment, such as these concepts are defined by FAO. NS requires more specific programmes, population targets and a strong focus on pregnant women and children, especially amongst the most vulnerable.

Fourth, improving sanitation is for the most part preceded by improvements in access to drinking water, especially in rural areas. The consequences of not improving sanitation infrastructure and delaying its deployment to further stages of economic development are found in impaired water quality and ecosystems, reduced biodiversity and a greater prevalence of water-borne diseases.

Last, there is still a huge gap in terms of improving sanitation in the region, especially in rural areas. The investments required to bridge this gap are reviewed in Chapter 13, and the institutional challenge is the focus of Chapter 1.

6.5 Final remarks

The overview of a wide range of variables for most LAC countries within a span of a decade tells three overall stories. First, that the consequences in coping with the problems of insufficient sanitation have eventually materialized in increasing costs to reverse its impacts and in moving towards more sustainable economic development. It is true that the investment needs are, for some countries, overwhelming. For others with growing economies and rapid poverty alleviation, ensuring proper sanitation in rural areas and water treatment in both large cities and rural areas should be an affordable priority.

Second, it seems that common patterns of nutritional transition in the prosperous LAC countries show growing rates of overweight and obesity. This has worrying negative effects, in both impaired human health and pathologies, but also in the larger footprints of the diets that are behind this emerging pandemic. In the case of LAC, the 49 million people suffering from undernourishment coexist with 110 million obese people, and with 200 million overweight. Only by educating people at the basic level can this trend be curbed and a worse disaster averted. It is important that the nutritional transition does not follow this path, but solutions are far from clear.

Last, while the performance in LAC countries of most WS and FNS indicators can be explained by the relative level of per capita income, there are significant differences amongst countries even within the same income quartile. National policies are thus crucial to rapidly improve the situation and reach the poorest and more vulnerable members of society.

References

Allan, J.A. (2013). *Food-water security: beyond hydrology and the water sector.* Working paper. [Online] Available from: www.ditchley.co.uk/assets/media/Allan%20-%20water%20 security%20chapter%20v3.pdf [Accessed May, 2013].

Ambler-Edwards, S., Bailey, K., Kiff, A., Lang, T., Lee, R., Marsden, T., Simons, D., & Tibbs H. (2009). *Food Futures: Rethinking UK Strategy.* The Royal Institute of International Affairs. London, Chatham House Report.

Banerjee, A.V., & Duflo, E. (2011). *Poor economics: a radical rethinking of the way to fight global poverty.* New York, Public Affairs.

Barrett, C.B. (2010). Measuring food insecurity. *Science, 327*: 825–828.

Berz, G.A. (1999). Catastrophes and climate change: concerns and possible countermeasures of the insurance industry. *Mitigation and Adaptation Strategies for Global Change, 4*: 283–293.

CAWMA (2007). Comprehensive Assessment of Water Management in Agriculture. *Water for Food, Water for Life: A Comprehensive Assessment of Water Management in Agriculture.* London, Earthscan, and Colombo, International Water Management Institute.

CIESIN-FAO-CIAT (2005). Center for International Earth Science Information Network-Columbia, University United Nations Food and Agriculture Programme, Centro Internacional de Agricultura Tropical. *Gridded Population of the World, Version 3 (GPWv3): Population Count Grid, Future Estimates.* Palisades, NY, NASA Socioeconomic Data and (SEDAC). [Online] Available from: sedac.ciesin.columbia.edu/data/set/gpw-v3-population-count-future-estimates [Accessed May 2013].

CFS (2009). Committee on World Food Security. *Reform of the Committee on World of Food Security: Final Version.* [Online] Available from: ftp://ftp.fao.org/docrep/fao/meeting/017/ k3023e3.pdf [Accessed September, 2012].

Cook, C., & Bakker, K. (2012). Water security: Debating an emerging paradigm. *Global Environmental Change, 22*(1): 94–102.

Cuevas, A., Alvarez, V., & Olivos, C. (2009). The emerging obesity problem in Latin America. *Expert Review Cardiovascular Therapy, 7*(3): 281–288.

De Schutter, O. (2012). *United Nations Special Rapporteur on the Right to Food*. September, 2012. Briefing Note. [Online] Available from: www.srfood.org/images/stories/pdf/otherdocuments/note06-septembre2012-en-v2.pdf [Accessed May, 2013].

EM-DAT (2013). Emergency Events Database on Disasters. World Health Organization and Centre for Research on the Epidemiology of Disasters Database. [Online] Available from: www.emdat.be/ [Accessed May, 2013].

Fader, M., Gerten, D., Krause, M., Lucht, W., & Cramer, W. (2013). Spatial decoupling of agricultural production and consumption: quantifying dependences of countries on food imports due to domestic land and water constraints. *Environmental Research Letters*, 8(1): 014046.

FAO (1998). Food and Agriculture Organization. *Guidelines for national food insecurity and vulnerability information and mapping systems (FIVIMS): background and principles*. Rome, FAO. [Online] Available from: www.fao.org/docrep/meeting/W8500e.htm [Accessed May, 2013].

FAO (2010). Food and Agriculture Organization. *Panorama de la Seguridad Alimentaria y Nutricional en América Latina y el Caribe 2010*. Rome, FAO.

FAO (2012a). Food and Agriculture Organization. *Coping with water scarcity. An action framework for agriculture and food security*. Rome, FAO.

FAO (2012b). Food and Agriculture Organization. *Panorama de la Seguridad Alimentaria y Nutricional en América Latina y el Caribe 2012*. Rome, FAO. [Online] Available from: www.rlc.fao.org/es/publicaciones/panorama-2012/.[Accessed May, 2013].

FAO (2012c). Food and Agriculture Organization. *Food security indicators*. Rome, FAO. [Online] Available from: www.fao.org/fileadmin/.../foodsecurity/Food_Security_Indicators.xlsx. [Accessed May, 2013].

FAO (2013a). Food and Agriculture Organization. *Irrigation water requirement and water withdrawal by country*. Rome, FAO. [Online] Available from: www.fao.org/nr/water/aquastat/water_use_agr/index7.stm [Accessed May, 2013].

FAO (2013b). Food and Agriculture Organization. FAOSTAT. [Online] www.faostat.fao.org/ [Accessed May, 2013].

FAO (2013c). Food and Agriculture Organization. AQUASTAT. [Online] Available from: www.fao.org/nr/water/aquastat/main/index.stm [Accessed May, 2013].

FAO (2013d). Food and Agriculture Organization. WATERLEX Statistics. Rome, FAO. [Online] Available from: faolex.fao.org/waterlex/ [Accessed May, 2013].

Finucane, M.M., Stevens, G.A., Cowan, M.J., Danaei, G., Lin, J.K., Paciorek, C.J. & Ezzati, M. (2011). National, regional, and global trends in body-mass index since 1980: systematic analysis of health examination surveys and epidemiological studies with 960 country-years and 9.1 million participants. *The Lancet*, 377 (9765): 557–567.

Frankenberger, T.R., Oshaug, A. & Smith, L.C. (1997). *A definition of nutrition security*. Atlanta, CARE.

GWP (2000). Global Water Partnership. *Towards Water Security: A Framework for Action*. [Online] Available from: www.gwp.org/ [Accessed July, 2013].

Godfray, H.C.J., Beddington, J.R., Crute, I.R., Haddad, L., Lawrence, D., Muir, J.F. & Toulmin, C. (2010). Food security: the challenge of feeding 9 billion people. *Science*, 327 (5967): 812–818.

Grey, D. & Sadoff, C.W. (2007). Sink or swim? Water security for growth and development. *Water Policy*, 9(6): 545.

Hoddinott, J. (1999). *Choosing outcome indicators of household food security. Technical Guide No. 7*. Washington DC, IFPRI. [Online] Available from: afghanlivelihoods.com/virtual-library/livelihoods/householdfoodsecurity-ifpr.pdf [Accessed July, 2013].

Hoekstra, A.Y. & Mekonnen, M.M. (2011). Global water scarcity: the monthly blue water footprint compared to blue water availability for the world's major river basins. Delft, The Netherlands, UNESCO-IHE. Value of Water Research Report Series No. 53.

IBGE (2013). Instituto Brasileiro de Geografia e Estatística. *Pesquisa de Ornamentos Familiares (POF) 1987/1988, 1995/1996, 2002/2003 & 2008/2009*. [Online] Available from: www.sidra.ibge.gov.br. [Accessed July, 2013].

IUCN (2013). International Union for Conservation of the Nature. *The Red List Database of Threatened Species*. Version 2013.1. [Online] Available from: www.iucnredlist.org. [Accessed July, 2013].

Lopez-Gunn, E., Willaarts, B.A., Dumont, A., Niemeyer, I. & Martinez-Santos, P. (2012). The concept of water and food security in Spain. In: De Stefano and Llamas. *Water, agriculture and the environment in spain: can we square the circle?* CRC – Press and Botín Foundation.

Mason, N. & Calow, R. (2012). *Water Security: from abstract concept to meaningful metrics. A initial overview of options*. London, Overseas Development Institute.

Mekonnen, M.M. & Hoekstra, A.Y. (2011). *National water footprint accounts: The green, blue and grey water footprint of production and consumption:* UNESCO-IHE. Value of Water Research Report Series No. 50. [Online] Available from: www.waterfootprint.org/Reports/Report50-NationalWaterFootprints-Vol1.pdf [Accessed May, 2013].

Mills, E. (2009). A Global Review of Insurance Industry Responses to Climate Change. *The Geneva Papers*, 34: 323–359.

OECD (2013a). Organization for Economic Development and Cooperation. *Global food security. challenges for the food and agricultural system*. Paris, OECD publishing.

OECD (2013b). Organization for Economic Development and Cooperation. *OECD StatExtracts*. [Online] Available from: stats.oecd.org/Index.aspx?DatasetCode=CRSNEW [Accessed July, 2013].

Onis, M.D., Blössner, M. & Borghi, E. (2011). Prevalence and trends of stunting among pre-school children, 1990–2020. *Public Health Nutrition*, 14 (7): 1–7.

Pangaribowo, E.H., Gerber, N. & Torero, M. (2013). *Food and Nutrition Security Indicators: A Review* (No. 147911). FOODSECURE working paper No. 5. [Online] Available from: www.foodsecure.eu. [Accessed March, 2013].

Sen, A. (1982). *Poverty and famines: an essay on entitlement and deprivation*. London, Oxford University Press.

Starr, J. (1992). Water security: The missing link in our Mideast strategy. *Canadian Journal of Development Studies/Revue canadienne d'études du développement*, 13(4): 35–48.

Sullivan, C.A., Meigh, J.R., & Giacomello, A.M. (2003). The Water Poverty Index: Development and application at the community scale. *Natural Resources Forum*, 27 (3): 189–199.

Suweis, S., Rinaldo, A., Maritan, A. & D'Odorico, P. (2013). Water-controlled wealth of nations. *Proceedings of the National Academy of Sciences*, 110 (11): 4230–4233.

Swinnen, J. F. & Squicciarini, P. (2012). Mixed messages on prices and food security. *Science*, 335(6067): 405–406.

Timmer, P. (2013). Preface. In: Gaultier, F. (ed). *Managing food price instability in developing countries: A critical analysis of strategies and instruments*. Paris, CIRAD.

United Nations (2012). *Food and nutrition security for all through sustainable agriculture and food systems*. Note from the United Nations System High Level Task Force on Global Food Security. [Online] Available from: www.un-foodsecurity.org. [Accessed July, 2013].

UNDP (1994). United Nations Development Programme. *New dimensions of human security. human development Report 1994*. New York, Oxford, Oxford University Press.

UNESCO (2010). *Guía metodológica para la aplicación del Análisis Regional de Frecuencia de Sequías basado en L-momentos y resultados de aplicación en América Latina*. Montevideo, Uruguay, CAZALAC. Documentos Técnicos del PHI-LAC No. 27.

UN-Water (2013). United Nations Water Programme. *Water security & the global water agenda: a un-water analytical brief*. [Online] Available from: www.unwater.org/downloads/watersecurity_analyticalbrief.pdf [Accessed July, 2013].

Vörösmarty, C.J., McIntyre, P.B., Gessner, M.O., Dudgeon, D., Prusevich, A., Green, P. & Davies, P.M. (2010). Global threats to human water security and river biodiversity. *Nature,467* (7315): 555–561.

Westhoek, H., Rood, T., Van den Berg, M., Janse, J., Nijdam, D., Reudink, M. & Stehfest, E. (2011). *The protein puzzle*. The Hague, PBL Netherlands Environmental Assessment Agency.

WHO-UNICEF (2013). *Joint Monitoring Programme (JMP) for water supply and sanitation data*. [Online] Available from: www.wssinfo.org/data-estimates/table/ [Accessed July, 2013].

World Bank (2013). *World Bank Development Indicators Database*. [Online] Available from: data.worldbank.org/data-catalog/world-development-indicators [Accessed July, 2013].

Part 3

Water for food and non-food

WATER AND AGRICULTURE

Coordinator:
Maite M. Aldaya, Water Observatory – Botín Foundation, and Complutense University of Madrid, Spain

Authors:
Erika Zarate, Good Stuff International, Switzerland
Maite M. Aldaya, Water Observatory – Botín Foundation, and Complutense University of Madrid, Spain
Daniel Chico, Water Observatory – Botín Foundation, and CEIGRAM, Technical University of Madrid, Spain
Markus Pahlow, Department of Water Engineering & Management, University of Twente, The Netherlands
Insa Flachsbarth, Water Observatory – Botín Foundation, and CEIGRAM, Technical University of Madrid, Spain
Gabriela Franco, Departamento de Economía Agraria Pontificia Universidad Católica de Chile,Santiago, Chile
Guoping Zhang, Water Footprint Network, The Netherlands
Alberto Garrido, Water Observatory – Botín Foundation, and CEIGRAM, Technical University of Madrid, Spain
Julio M. Kuroiwa, Laboratorio Nacional de Hidráulica – Universidad Nacional de Ingeniería, Lima, Peru
Julio Cesar Pascale Palhares, Embrapa Cattle Southeast, São Carlos, Brazil
Diego Arévalo Uribe, Water Management and Footprint. Centro de Ciencia y Tecnología de Antioquia, Colombia

Contributors:
Mesfin Mekonnen, University of Twente, The Netherlands
Barbara Soriano, CEIGRAM, Technical University of Madrid, Spain
Laurens Thuy, Utrecht University, The Netherlands
Luis F. Castro, School of Civil Engineering,Universidad Nacional de Ingenieria, Lima, Peru

Highlights

- This chapter shows the strong links between water, agriculture and the economy in Latin America and Caribbean (LAC). Both green and blue water are vital for LAC's economies and for its food security. Awareness of LAC's virtual water trade volumes and water footprints alone will not solve the local or global water problems. However, the awareness gained increases the likelihood that optimized water allocation decisions, which consider the hydrological and economical aspects of water resources, are made.

- Agriculture is a significant economic sector for many LAC countries with some being major world players in the agricultural commodities world markets, such is the case for Brazil and Argentina who contribute to 13% of the global green water export. At the micro level, agriculture still plays a significant role for the food security of the population.

- The consumptive water use of agricultural production was on average 1,057Gm^3/yr for the period 1996–2005; of which, 95% corresponds to the green water footprint, whereas 5% refers to the blue component. This indicates that LAC relies heavily on green water for agricultural production, i.e. rain-fed agriculture.

- Maize is a fundamental crop in Argentina, Brazil, Chile, Mexico and Peru, representing 15% of the total agricultural blue and green water footprint (773,408hm^3/yr) and contributing to 35% of the agricultural nitrogen pollution, estimated as grey water footprint, in Argentina, Brazil, Chile, Colombia, Mexico and Peru. Only in Mexico, maize contributes 60% of the agricultural grey water footprint.

- Grazing represents 24% of the total green water footprint of agriculture in these countries. The blue water consumption by the animal water supply is very significant in Argentina, Brazil, Chile, Mexico and Peru, which amounts to 13% (38,825hm^3/yr) of the total consumption.

- Concerning agricultural products, the LAC region was a net exporter of green virtual water (14Gm^3/yr) and a net importer of blue virtual water (16Gm^3/yr) during the period 1996–2005.

- Export-oriented industrial agriculture has become the main driver of South American deforestation.

- Sustainable water management should not be seen as a barrier for the development of the region, but rather as the way to develop and grow as a region.

- Understanding the magnitude of overlap and interactions between poverty, conservation and macro-economic processes is crucial in order to identify possible win–win solutions for the LAC region. Access to agricultural water has secondary effects on poverty through output, employment and prices.

7.1 Introduction

The Latin American and Caribbean region (LAC) as a whole is increasingly becoming a major source of agricultural commodities for the world market and thus influencing food security. As such, improving resource management in the region promises to have important benefits for both the inhabitants of LAC and the world.

Agriculture is essential to food security. However, food production requires substantial amounts of water, both stored in the soil as soil moisture from rain (green water) and as water for irrigation (blue water). FAO (2012b) estimated an annual blue water use in LAC of 262,800hm³/yr. Globally, agriculture is the sector with the largest water withdrawal by far, with about 70%. This percentage compares to 73%, (192,700hm³/yr) in LAC, whereas 19% and 9% correspond to the domestic and industrial sectors respectively (ibid.). The Guyana sub-region (Guyana and Suriname) and Southern Cone (Argentina, Chile, Paraguay and Uruguay) have the highest level of agricultural water use, with values of 96% and 91% respectively (ibid.). Agriculture is also central to economic growth in LAC. For the period 2000–2007, it contributed an average of 9.6% to its GDP and exports of agricultural commodities accounted for 44% of total export value in 2007 (Bovarnick et al., 2010). Notably the agricultural sector provides employment for about 9% of LAC's population (UNEP, 2013).

Globally, a substantial part of the most fertile land is already being used for agriculture. According to FAO (2012a), much of the remaining arable land is located in LAC and sub-Saharan Africa, however, it is in remote locations, far from population centres and agricultural infrastructure, and cannot be converted into productive land without investments in infrastructure development. In LAC, agricultural production increased by more than 50 % from 2000 to 2012, with Brazil expanding production by more than 70 %. Most food is produced by rain-fed agriculture in LAC, with 87% of the cropland being rain-fed (Rockström et al., 2007). The irrigation potential for the region is estimated at 77.8 million hectares (FAO, 2013), whereas in 2009 the LAC region had 13.5 million hectares of irrigated agriculture. The gap between the irrigation potential and actually irrigated agriculture is due to increasing costs of construction, limited government support for large-scale irrigation investments and concerns about the negative social and environmental impacts of irrigation (UNCTAD, 2011). Most of the regional irrigation potential (66%) is located in four countries: Argentina, Brazil, Mexico and Peru (ibid.). Figures on irrigation potential usually only take into account climatic conditions and land irrigation sustainability, while studies including surface- and groundwater availability are considered scarce (FAO, 2013).

Water quality deserves as much attention as water quantity. Local and regional physical water scarcity problems are exacerbated by severe water quality problems in LAC; leading to the frequent usage of wastewater for irrigation. Many countries in LAC have been facing increasing challenges in water quality management. The world's major water quality issues as identified by United Nations (UN, 2003) are organic pollution, pathogens, salinity, nitrate, heavy metals, acidification, eutrophication and sediment load either in surface water bodies or in groundwater.

LAC is relatively well endowed with water resources. However, the spatial and temporal variability of water, coupled with rapid urbanization and inadequate water governance is putting considerable pressure on the available water resources (see Chapter 2 and 6 for an analysis of water scarcity in LAC). Ironically, in the water abundant LAC, almost 20% of its nearly 600 million inhabitants do not have access to drinking water, 20% do not have any kind of access to a sewage system, and less than 30% of the wastewater receives treatment (Proceso Regional de las Américas, 2012). In addition almost 18 million of children under five suffer from chronic malnutrition (FAO, 2012b). This elevated distributive inequity is a notable element in the reality of LAC.

This chapter analyses the challenges and opportunities of water management in the region from the perspective of the agricultural sector. First, water is accounted in terms of quantity and quality. Virtual water trade in the LAC region is also analysed and, finally, a productivity analysis is presented taking into account social and economic aspects.

7.2 Methodology and data

In this chapter we use the water footprint (WF) (Hoekstra et al., 2011) to calculate water consumption. The 'water footprint' is a measure of humans' appropriation of freshwater resources. Freshwater appropriation is measured in terms of water volume consumed (evaporated or incorporated into a product) or polluted per unit of time. A water footprint has three components: green, blue and grey. The blue water footprint refers to consumption of blue water resources (surface and ground water). The green water footprint is the volume of green water (rainwater stored in the soil as soil moisture) consumed, which is particularly relevant in crop production. The grey water footprint is an indicator of the degree of freshwater pollution and is defined as the volume of freshwater that is required to assimilate the load of pollutants based on existing ambient water quality standards.

In the context of the countries considered, the water footprint accounting is applied from two perspectives: the water footprint of agricultural production and the water footprint of agricultural consumption. The water footprint of agricultural production for a given country refers to the blue, green and grey water footprints of all the agricultural processes, that is, crop and livestock production, taking place within the political borders of the country. The water footprint of agricultural production is equivalent to the agricultural 'water footprint within the area of the nation' (Hoekstra et al., 2011), and is defined as the total freshwater volume consumed or polluted within the territory of the nation as a result of activities within the different sectors of the economy, in this case agriculture.

The water footprint of agricultural consumption refers to the quantification of the water consumed and polluted to produce the agricultural products consumed by the population of a country. It consists of two components: the internal and external water footprint of national consumption. The internal water footprint is defined as the use of domestic water resources to produce goods and services consumed by the population of the country. It is the sum of the water footprint within the nation minus the volume of virtual-water exported to other nations through the export of products produced with domestic water resources. The external water footprint is defined as the volume of water resources used in other nations to produce goods and services consumed by the population in the nation under consideration. It is equal to the virtual water import into the nation minus the volume of virtual water export to other nations as a result of re-export of imported products. The virtual water export from a nation consists of exported water of domestic origin and re-exported water of foreign origin. The virtual-water import into a nation will partly be consumed, thus constituting the external water footprint of national consumption, and may in part be re-exported (Mekonnen and Hoekstra, 2011).

The grey water footprint data used refer to the nitrogen pollution alone and are based on Mekonnen and Hoekstra (2011), who estimated the grey water footprint based on nitrogen leaching-runoff from fertilizer use. The fraction of nitrogen that leaches or runs off multiplied by the nitrogen application rate represents the load of nitrogen reaching the surface and subsurface water bodies. Some 10% of the applied nitrogen fertilizer is assumed to be lost through leaching-runoff. In order to estimate the grey water footprint, an ambient water quality standard of 10mg/l measured as Nitrate-nitrogen (NO_3-N) was used, following the guidelines of the US Environmental Protection Agency (US-EPA).

The countries analysed in this chapter as LAC correspond to the thirty-three countries of the Economic Commission for Latin America and the Caribbean (ECLAC) plus Puerto Rico. Data from other non-sovereign Caribbean islands are included in tables whenever available.

7.3   Water accounting

7.3.1   Water quantity

7.3.1.1   Water withdrawal in agriculture

In the majority of the countries of the region, irrigation is seen as an important means to increase productivity, and enable and intensify crop diversification, an objective of most agricultural policies of governments in the region (FAO, 2013). Irrigated areas increased steadily during the 20th century and particularly from the 1950s onwards (*ibid.*). These increases are, however, modest in comparison to Asia and sub-Saharan Africa. Mexico has by far the largest irrigated area with over 6.5 million hectares; and Brazil is next with 3.2 million hectares, followed by Chile, Argentina, and Bolivia (UNCTAD, 2011). About

0.5 million hectares in Brazil are located in the semi-arid northeast region – an area with the lowest social and economic indicators (Oliviera et al., 2009).

Figures on irrigation water use (non-consumptive) are expressed in cubic metres per hectare per year, and show certain homogeneity for the whole of South America and the Greater Antilles, varying between 9,000m³/ha/yr and 12,000m³/ha/yr. Figures for Mexico are slightly higher, 13,500m³/ha/yr, and for Central America even higher. In the case of Mexico, the higher value is probably due to its climatic characteristics (higher potential evapotranspiration), while Central America is dominated by its permanent crops (banana, sugar cane, etc.) and its high cultivation intensity in temporary crops such as rice (FAO, 2013).

Concerning the irrigation techniques, surface irrigation is by far the most widespread irrigation technique in LAC. Table 7.1 presents information on irrigation techniques by sub-region for the countries in which information was available. It is worth noting the importance of localized irrigation in the Lesser Antilles (32.1%), where water scarcity and farm characteristics have induced an extensive utilization of localized irrigation, and in Brazil (6.1%). Sprinkler irrigation covers significant areas in Cuba (51%), Brazil (35%), Panama (24%), Jamaica (17%) and Venezuela (16%).

According to FAO (2013), the major source of irrigation water in the region is surface water, with the exception of Nicaragua and Cuba where groundwater is the source for respectively 77% and 50% of the area under irrigation.

Mexico, Brazil, Argentina, Chile, Venezuela and Peru have the highest irrigation water withdrawal (FAO, 2013) and account for 81% of the total irrigation water withdrawal in the region. It is worth noting that from these six countries, Mexico, Chile and Peru have the highest levels of water scarcity in the region.

7.3.1.2 Blue and green water consumption of agricultural production

Quantifying actual crop water consumption is crucial to understanding real water needs for agriculture. The consumptive water use of agricultural production (crops and livestock) for the LAC region, i.e. the green and blue water footprints of agricultural production, was on average 1,057Gm³/yr for the period 1996–2005, corresponding to 13.9% of the global water footprint of agricultural production (Mekonnen and Hoekstra, 2011). Of these 1,057Gm³, 95% corresponds to the green component of the water footprint, whereas only 5% corresponds to the blue component. Brazil alone accounts for 42.4% of the total (green and blue) water footprint in the region, followed by Argentina (17.1%), Mexico (11.7%), Colombia (4.9%) and Paraguay (3.1%) (Figure 7.1). These five countries account for 79.2% of the total water footprint of the region. This data points towards two fundamental issues: (i) LAC relies heavily on green water (95%) for agricultural production, i.e. rain-fed agriculture; (ii) Brazil and Argentina alone account for 60% of agricultural water consumption in LAC. This provides an indication of the global significance of these two countries in terms of agricultural water consumption and virtual water trade.

The total blue water footprint of agricultural production in the region was 50.9Gm³/yr. In this case, the country with the biggest contribution is Mexico (29.2%), followed by

Table 7.1 Irrigation techniques in the LAC region

| SUB-REGION | IRRIGATION TECHNIQUES | | | | | | |
| | SURFACE | | SPRINKLER | | LOCALIZED | | TOTAL |
	ha	%	ha	%	ha	%	ha
MEXICO	5,802,182	92.7	310,800	5.0	143,050	2.3	6,256,032
CENTRAL AMERICA	418,638	93.0	17,171	3.8	14,272	3.2	450,081
GREATER ANTILLES	746,894	63.6	407,075	34.6	21,256	1.8	1,175,225
LESSER ANTILLES	2,890	53.8	761	14.2	1,725	32.1	5,376
GUYANA SUB-REGION	201,314	100	0.0	0.0	0.0	0.0	201,314
ANDEAN SUB-REGION	3,379,637	95.6	122,364	3.5	34,536	1.0	3,536,537
BRAZIL	1,688,485	58.8	1,005,606	35.0	176,113	6.1	2,870,204
SOUTH SUB-REGION	3,445,068	95.6	95,730	2.7	62,153	1.7	3,602,951
LAC REGION	15,672,050	86.7	1,960,365	10.8	453,105	2.5	18,097,720[1]

Source: FAO (2013).

1 This is an approximate figure of land under irrigation, which represents the physical area with irrigation infrastructure. It is not the area that is actually irrigated in a given year. As a global figure provided by FAO, 80% of the area under irrigation is actually irrigated. Given the problems in operation, maintenance and rehabilitation of the irrigation districts, it is estimated that the real figure must be lower (see section 7.1 for estimated numbers of area under irrigation in LAC).

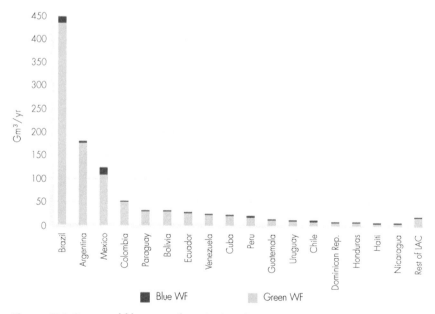

Figure 7.1 Green and blue water footprint (in cubic Gigametres per year) of agricultural production for the LAC region (average 1996–2005). *Source: own elaboration based on data from Mekonnen and Hoekstra (2011).*

Brazil (23.7%), Argentina (10.0%), Peru (8.4%) and Chile (4.9%). These five countries are responsible for 76.2% of the total blue water footprint in the LAC region and for 75% of the total (green and blue) water footprint of the region.

Not surprisingly, countries with fewer available water resources in the areas of important economic activity, like Mexico, Peru and Chile, rely more on blue water resources compared to the other countries. Brazil and Argentina occupy together 55% of the LAC area and therefore contribute with a significant blue water footprint. These five countries with the greatest blue water footprint of agricultural production, namely Mexico, Brazil, Argentina, Peru and Chile, together cover 75% of the LAC area.

Figure 7.2 shows the distribution of agricultural green and blue water footprints for Mexico, Brazil, Argentina, Peru and Chile, according to their main agricultural uses.

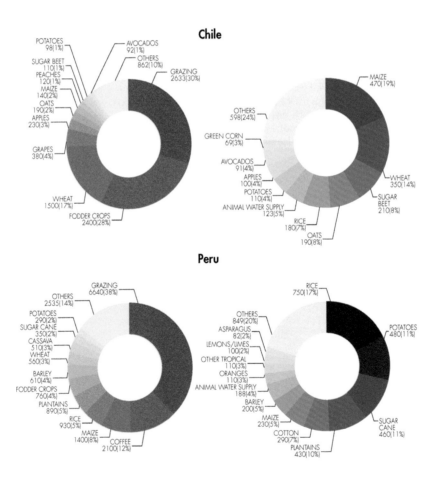

(Figure 7.2 continues in the next page)

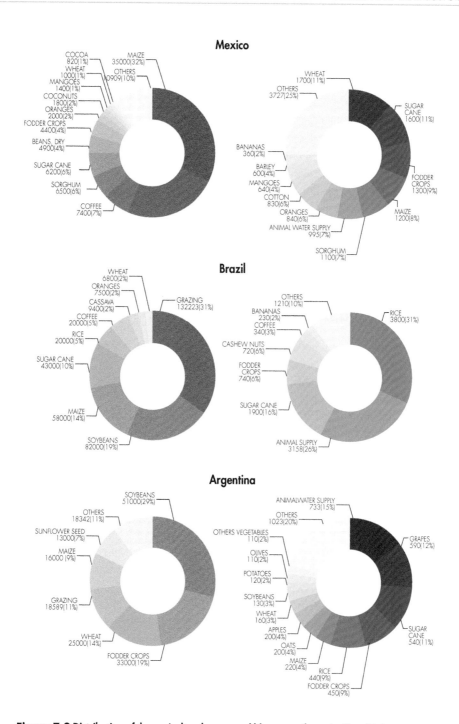

Figure 7.2 Distribution of the agricultural green and blue water footprint (in cubic hectometres per year) of Mexico, Brazil, Argentina, Peru and Chile (average for the years 1996–2005). *Source: own elaboration based on Mekonnen and Hoekstra (2011) and the Water Footprint Assessment Tool (WFN, 2013b).*

Maize is a fundamental crop in all five countries as shown in Figure 7.2. It represents 15% of the total agricultural (blue and green) water footprint (WF) of these five countries equivalent to 773,408hm³/yr. Soybean is especially important in Brazil and Argentina, and accounts for 17% of the total agricultural blue and green WF of these five countries. Grazing contributes significantly with 24% of the total green WF of agriculture in these countries. The blue water consumption for the animal water supply in the five countries, which amounts to 13%, or 8,825hm³/yr, is also noteworthy. In the context of water policy, being aware of water allocation for livestock is essential when considering food security for LAC (Box 7.1). Sugar cane is also an important crop for all the above-mentioned countries except Chile (for climatic reasons), which shows a stronger production of cash crops such as grapes, apples and avocados. Rice makes up a significant part of the blue WF for all the countries except Mexico (14% of the total blue WF of the five countries). Potatoes constitute a very important crop in Peru (Box 7.2).

7.3.1.3 Water footprint agricultural products' consumption: externalization of the water footprint

The average global water consumption of agricultural products was 1,156m³/capita/yr (88% green, 12% blue) for the period 1996–2005 (Mekonnen and Hoekstra, 2011). The equivalent value for the LAC region was 1,473m³/capita/yr (94% green, 6% blue). Figure 7.3 shows that water footprints range between 3,420m³/capita/yr (98% green, 2% blue) for Bolivia and 833m³/capita/yr (95% green, 5% blue) for Nicaragua. Chile, Peru, Mexico and Dominican Republic have the highest percentage of blue water in their water footprints of consumption, with values of 16, 15, 10 and 10% respectively. Countries with the lowest blue water proportion are Bolivia (2%), and Brazil, Uruguay, Paraguay and Dominica (3%).

The virtual water import dependency of a nation is defined as the ratio of the external to the total water footprint of national consumption, whereas the national water self-sufficiency is defined as the ratio of the internal to the total water footprint of national consumption. The Lesser Antilles and Mexico have the highest virtual water dependency in the LAC region. Saint Lucia, Trinidad and Tobago and Bahamas show virtual water dependencies above 90%, whereas Mexico's corresponding value is approximately 45%. This means that these countries import most of the virtual water required to cover the agricultural needs of its population, meaning they have a notable dependency on external water resources. Chile and Peru, both countries characterized by significant levels of water scarcity (see Chapter 2), show virtual water import dependencies of 37 and 34% respectively. Conversely, Paraguay, Argentina, Bolivia and Brazil have very low virtual water import dependency values (2, 3, 9 and 9 % respectively) indicating high self-sufficiency. This means that these countries use their own available resources to supply most of the agricultural products consumed by their inhabitants.

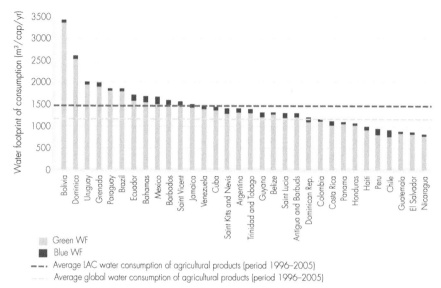

Figure 7.3 Water footprint (in cubic metres per inhabitant per year) of the consumption of agricultural products (green and blue) in the LAC region (average 1996–2005). *Source: own elaboration based on data from Mekonnen and Hoekstra (2011).*

Box 7.1   Water footprint of poultry and swine production per Brazilian state

Brazil is one of the major producers of animal products in the world and also a large exporter. The country is rich in water sources, which are mostly located in the Amazon Basin. Swine and poultry production are concentrated in different regions, mainly in the south, one of the most urbanized and industrialized parts of the country. Therefore, studies that aim to calculate the water footprint are extremely important to the society to inform upon water security, elaborate discussions on the topic and ensure the future of the production.

We calculated the water footprint of pigs slaughtered in 2008 in south-central states of Brazil. Calculations considered indirect water consumed in grain production (corn and soybean), and direct water, drinking and washing water consumed on the farm. Rio Grande do Sul was the state with the largest water footprint (2,702,000hm³, 99.9% green and 0.09% blue), followed by Santa Catarina (2,401,000hm³, 99.88% green and 0.12% blue), and Parana (1,089,000hm³, 99.85% green and 0.15% blue). These are the states where slaughter is practised most. Although, Rio Grande do Sul is the second in terms of animals slaughtered, its water footprint was the largest due to dry climatic conditions, which require more water to produce the same amount of corn and soybean. States with high corn and soybean productivity had a lower ratio of water volume consumed per kg of meat, namely Distrito Federal (2.49m³/kg), Parana (2.53m³/kg), and Goias (2.77m³/kg).

The water footprint of broiler chicken slaughtered in the decade 2000–2010 in each of Brazil's south-central states was also calculated. Similarly the calculation considered indirect water, consumed in grain production, and direct water, consumed on the farm. South states had the largest water footprints and the largest number of animals slaughtered during the period. The average footprint for Parana in the decade in question (2000–2010) was 4,334hm³ (99.7% green and 0.3% blue) and Rio Grande do Sul 4,216hm³ (99.8% green and 0.2% blue). Slaughters increased and/or remained constant in all states. Annual variation was determined by productivity of corn and soybeans.

Results show that water management in animal production should not only address the farm; but also include related agricultural supply chains, where most of the water consumed is green. Blue and grey water footprints, most notable in the direct water use of the farm, are also important as they are consumed in watersheds with an increased potential for water use conflicts (Palhares, 2012).

Box 7.2 Importance of potatoes in the Peruvian diet

Potato (*Solanum Toberusum*) is a South American tuber that grows in a wide variety of environments, ranging from cold to temperate climates, and in altitudes ranging from sea level to 4,700m. It is the fourth most important crop in the world behind rice, wheat and maize and the third most important in human consumption, feeding more than one billion people worldwide (CIP, 2010).

FAO (2008) indicates that potatoes are very productive from the nutritional viewpoint. For each m³ of water applied to potato crops, 5,600 calories are produced. By comparison, 1m³ of water applied to corn produces 3,800 calories and only 2,000 calories if it is applied to rice. In addition, 1m³ of water applied to potatoes produces 150g of proteins and 540mg of calcium. Therefore, potatoes' protein content per cubic metre is more than double that of maize and wheat and offers twice the calcium provided by wheat and four times that of rice.

The average European consumption is 87.8kg potatoes/year/person. By comparison, per capita consumption of potatoes per year is 60kg in North America, 13.9kg in Africa, 23.9kg in Oceania and 20.7kg in Latin America, although its consumption is steadily growing in the latter region (FAO, 2008).

In Latin America, the highest yields are obtained in Argentina (28.7t/ha) and the lowest yields are obtained in Bolivia (5.6t/ha). In the Andean countries potato cultivation is mostly in hands of small farmers. Higher yields are related to improved technology, sufficient water supply and better management.

The Andean population uses productive domesticated species to overcome the limitations of poor productivity of wild plants, although these do not grow at altitudes greater than 4,500m. *Solanum jozepozukii* and *Solanum curtilobum* are frost-resistant potatoes that grow at high elevations where agriculture is practised (Moran, 1982).

An ongoing study (LA-Peru, 2012) indicates that, on average, production of 1kg of potatoes requires only 469 litres of water. Mekonnen and Hoekstra (2011) provide a lower global average WF figure of 290litre/kg: 66% related to green, 11% to blue and 22 % to grey WF. Potato cultivation is concentrated in the mountainous area of the Andean region and the Pacific Basin. Crops are rain-fed during the wet season (January–March) and during the rest of the year in which precipitation is negligible, flood or furrow irrigation is used. In some cases, water is not applied in the last months of the vegetative period, and the yield is very low (Egúsquiza, 2000). Initial watering appears to be sufficient to achieve an acceptable growth and even with a low yield potatoes help to cover part of the basic nutritional needs of poor communities in the Andean Highlands.

Further population growth and shortage of water resources in some areas in the near future may force a substantial change in crop cultivation patterns. For instance, rice is grown in a number of valleys where water is scarce. It might be more advantageous from the water conservational, nutritional and even economic point of view to grow potatoes instead. In addition, potato productivity ought to be increased, particularly in the Andean countries.

7.3.2 Water quality

The most well-known effects of agriculture on water quality are due to chemical contamination by fertilizers and pesticides that accumulate in water sources. Additionally the reuse of sewage effluent for irrigation, known to transmit a number of pathogens even after secondary water treatments, can seriously affect the quality of the water used in agriculture. Significant water pollution due to irrigation has been reported in Barbados, Mexico, Nicaragua, Panama, Peru, Dominican Republic and Venezuela (Biswas et al., 2006). In addition, the problem of salinity caused by irrigation is a serious constraint in Argentina, Cuba, Mexico, and Peru and, to a lesser extent, in the arid regions of northeastern Brazil, north and central Chile and some small areas of Central America (*ibid.*).

This section focuses mainly on the agricultural grey water footprint caused by nitrogen pollution in LAC due to the use of fertilizers. The total of which amounted to 44,412hm^3/yr for the period 1996 to 2005. This value corresponds to 46% of the total grey water footprint in the region; 96,649hm^3/yr including the industrial and domestic sectors (17% and 37%, respectively). The countries contributing the most to the total agricultural grey WF of the region are Brazil, Mexico, Argentina, Chile, Colombia and Peru. The total agricultural grey WF of these six countries was 39,017hm^3/yr, corresponding to 88% of the agricultural grey WF in the LAC region. Brazil and Mexico alone already constitute 61% of the agricultural grey water footprint in the region (and 51% of the LAC area).

7.3.2.1 Most important corps contributing to the grey water footprint in the LAC region

Figure 7.4 shows the crops contributing the most to the grey WF for Brazil, Mexico, Argentina, Chile, Colombia and Peru.

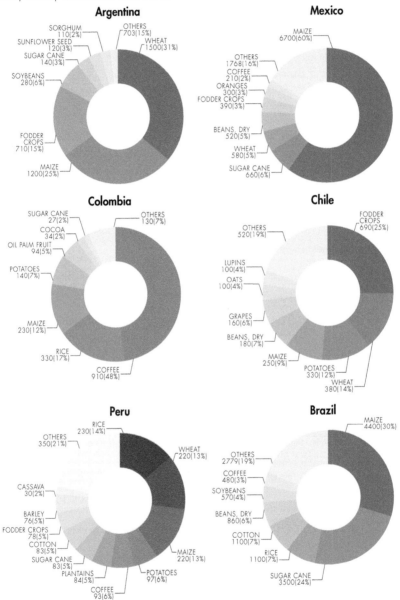

Figure 7.4 Composition of the agricultural grey water footprint (in cubic hectometres per year) by crops in Brazil, Mexico, Argentina, Chile, Colombia and Peru. *Source: own elaboration based on Mekonnen and Hoekstra (2011) and the Water Footprint Assessment Tool (WFN, 2013b).*

These figures show that maize is a heavily fertilized crop and contributes significantly to the grey WF in all six countries: 35% of the agricultural grey WF of these six countries corresponds to this crop. In Mexico alone it contributes to 60% of the agricultural grey WF. Sugar cane contributes 12% of the total agricultural grey water footprint of these six countries, whereas coffee, rice and fodder crops contribute 5%. Notably coffee contributes 48% of the agricultural grey WF of Colombia.

These above-mentioned grey water footprint results are only with respect to nitrogen, for which the grey water footprint for all the countries and products is publicly available (Mekonnen and Hoekstra, 2011). This allows for straightforward comparisons, however, a large number of agrochemicals are used in the LAC region. For example, Costa Rica tops the list of Latin American countries using multiple agrochemicals, which counterbalances many of their environmental policies seeking to improve environmental quality in the country (LA-Costa Rica, 2012). Costa Rica annually imports about 13,000t of some 300 active ingredients, many of which are restricted and/or prohibited in other countries and are even included in international disposal agreements (*ibid.*). A portion of the active ingredients is repackaged and re-exported. Although there are no precise data on the exported quantities, it is estimated that around 20–25% of total imports are re-exported (Ramirez et al., 2009). The import data therefore does not accurately reflect the quantities used in the fields, but they serve to check usage trends (LA-Costa Rica, 2012).

7.3.2.2 Grey water footprint of consumption of agricultural products in LAC

The average world WF of consumption of agricultural products was $1,268m^3$/capita/yr during the period 1996–2005, with $1,156m^3$/capita/yr corresponding to the blue and green WF and $112m^3$/capita/yr to the grey WF, equivalent to 91 and 9% of the total respectively (Mekonnen and Hoekstra, 2011). For the LAC region, the average was $1,560m^3$/capita/yr, with $1,473m^3$/capita/yr corresponding to the blue and green WF and $87m^3$/capita/yr to the grey WF, equivalent to 94 and 6 % respectively. Grey WF values range from $272.4m^3$/capita/yr for Belize and $19.5m^3$/capita/yr for Bolivia.

The externalization of the grey WF is equivalent to the externalization of pollution due to importing of agricultural products. Argentina has the lowest external grey water footprint as a proportion of their total grey WF (6%), together with Paraguay and Belize (9%). On the other hand, countries like Bahamas, Saint Lucia, Grenada, Trinidad and Tobago, Saint Vincent and the Grenadines, Antigua and Barbuda and Dominican Republic have a 100% external grey water footprint. This indicates that while for Argentina, Paraguay and Belize the pollution caused by consumption of agricultural products (in this case due to nitrogen) is mostly internal, i.e. caused within the borders of the countries, pollution caused due to consumption of agricultural products in the Antilles is borne by other countries.

7.3.3 Virtual water flows related to trade of agricultural products

The net virtual water import of a country or region during a given period of time is defined as the gross import of virtual water minus the gross export. A positive net import of virtual water implies net inflow of virtual water to the country or region. A negative net import of virtual water implies net outflow of virtual water, which means that the country is a net exporter of virtual water (Hoekstra et al., 2011). LAC was a net exporter of virtual water in terms of agricultural products during the period 1996–2005 (Mekonnen and Hoekstra, 2011). The net virtual water import for LAC was 125.4Gm³/yr. This means that for agricultural products, LAC was a net exporter of green virtual water (141.5Gm³/yr) and a net importer of blue virtual water (16.1Gm³/yr).

Figure 7.5 shows the countries with the largest virtual water flows of agricultural products in the region. Mexico is the largest virtual water importer, followed by Trinidad and Tobago, Venezuela, Peru and Chile. The countries with the largest virtual water exports related to agricultural products are Argentina, Brazil, Paraguay, Uruguay and Honduras.

Argentina and Brazil primarily produce for world markets under rain-fed conditions, which indicates an increased use of green water instead of blue water. This is reflected in the scale differences used for blue and green virtual water exports in Figure 7.6. According to Mekonnen and Hoekstra (2011), these two countries contribute with 13% of the total green water exported in the world (whereas LAC contributes with 19%), which constitutes an indication of the global importance of green water provided to the world food market by Argentina and Brazil, notably as green water is generally associated with lower opportunity costs than blue water (Albersen et al., 2003). Following the notion of opportunity costs, it has been argued that the use of green water in crop production

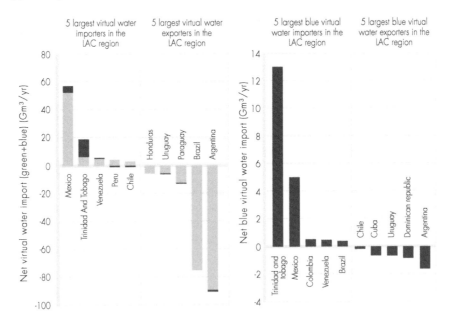

Figure 7.5 Largest total (green and blue) net virtual water importers and blue net virtual water importers (in cubic Gigametres per year) of agricultural products in the LAC region (average 1996–2005). *Source: own elaboration based on data from Mekonnen and Hoekstra (2011).*

is considered more sustainable than blue water use, except when replacing high-value ecosystems (Yang et al., 2006; Aldaya et al., 2010; Niemeyer and Garrido, 2011). On the other hand, expanding rain-fed agriculture is often associated with massive land use changes. Especially in Brazil where increasing virtual water exports contained in soybeans has led to a threefold land footprint increase.

Green virtual water exports

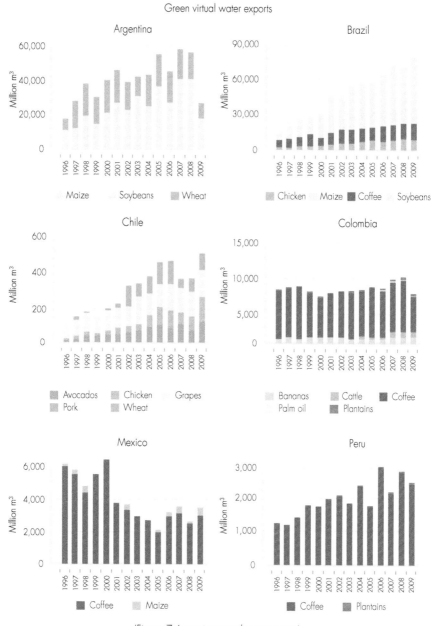

(Figure 7.6 continues in the next page)

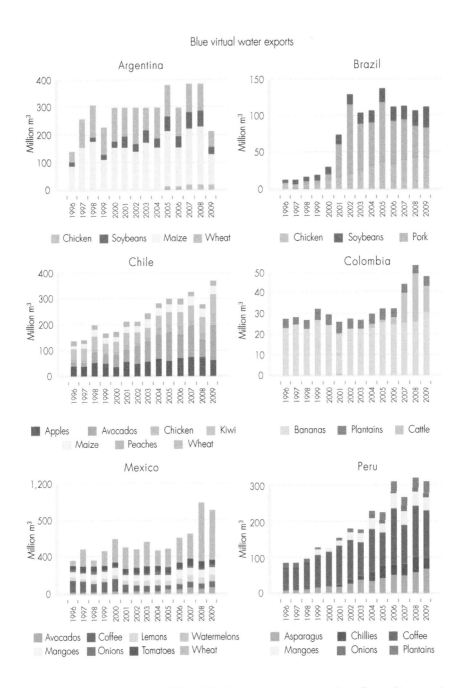

Figure 7.6 Green (above) and blue (below) virtual water exports (in million cubic metres) per country and main products (1996–2009). *Source: own elaboration based on data from Mekonnen and Hoekstra (2011) and FAO (2012d).*

Green virtual water imports

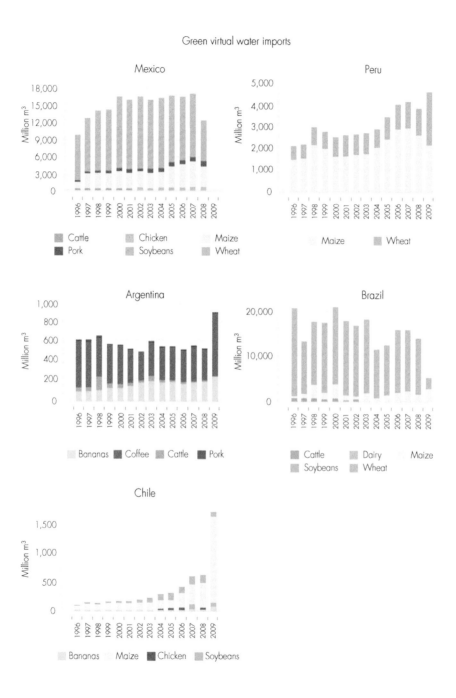

(Figure 7.7 continues in the next page)

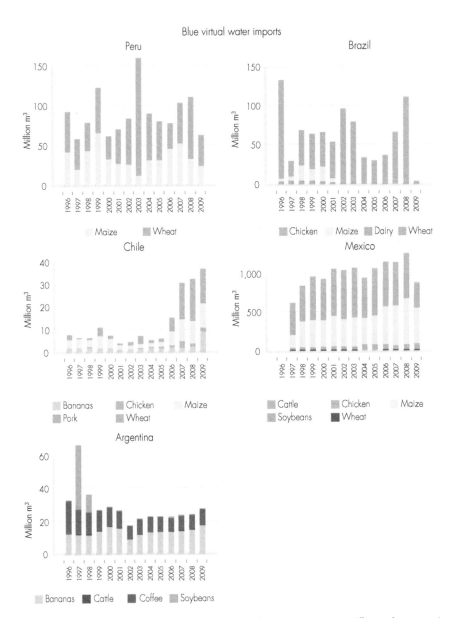

Figure 7.7 Green (above) and blue (below) virtual water imports (in million cubic metres) per country and main products (1996–2009). *Source: own elaboration based on data from Mekonnen and Hoekstra (2011) and FAO (2012d).*

Mexico is a large agricultural net importer. This country must cope with green water constraints and thus highly depends on irrigated agriculture. The substitution of domestic staple food production by imports has led to a shift in agricultural production towards higher value fruits and vegetables as well as livestock production (Figure 7.7). Fruits and vegetables are mostly produced under irrigated conditions leading to higher blue water use. Furthermore, agricultural production has increased substantially due to global market forces. This has resulted in accelerating blue water depletion rates. For example, the Rio Grande river basin has already reached or surpassed sustainable extraction rates during some months of the year (Chapter 6). A similar trend can be observed in Chile and Peru. In Argentina and Brazil blue water exports play a rather minor role.

Trade patterns are extremely dynamic and unstable. Specialization, technology adoption and market prices volatility and economic growth have given rise to fundamental changes in agricultural production and trade worldwide and in LAC (Figure 7.8). From Figure 7.8, one can see that the Caribbean economies are increasingly dependent on virtual water imports while the South Cone and Amazonian region are increasing their virtual water exports the majority of which are green virtual water exports.

Deforestation continues to be the dominant land-use trend in LAC, and subsistence agriculture, an important part of many local economies, is one of the major contributors (Grau and Aide, 2008). But, socio-economic changes related to globalization are promoting a rapid change towards agricultural systems oriented to local, regional, and global markets. The Amazon basin is the region that has lost the largest area to deforestation, with the greatest impacts on biodiversity and biomass loss, but other biomes have also been and continue to be severely affected by conversion to agriculture and pastures (see Chapter 3). Export-oriented industrial agriculture has become the main driver of South American deforestation. In Brazil, Bolivia, Paraguay, and Argentina, extensive areas of seasonally dry forest with flat terrain and enough rainfall for rain-fed agriculture are now being deforested for soybean production, which is mainly exported to China and the European Union.

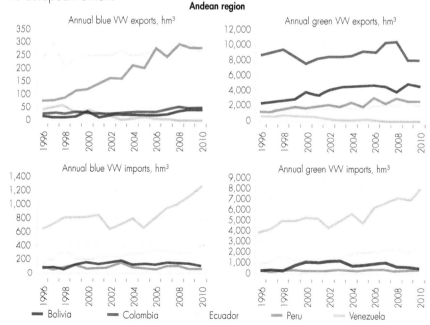

Andean region

(Figure 7.8 continues in the next page)

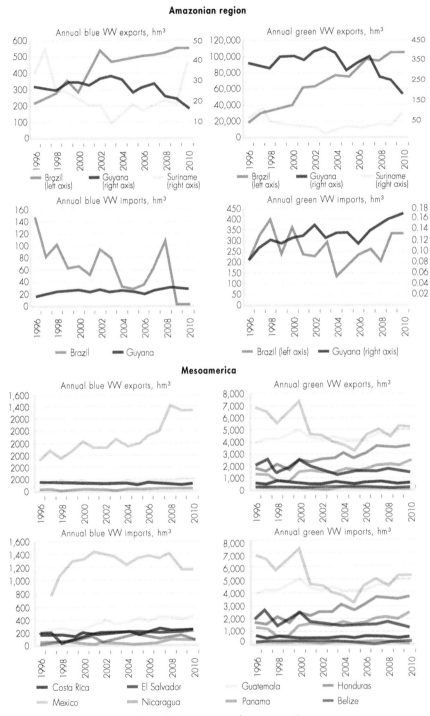

(Figure 7.8 continues in the next page)

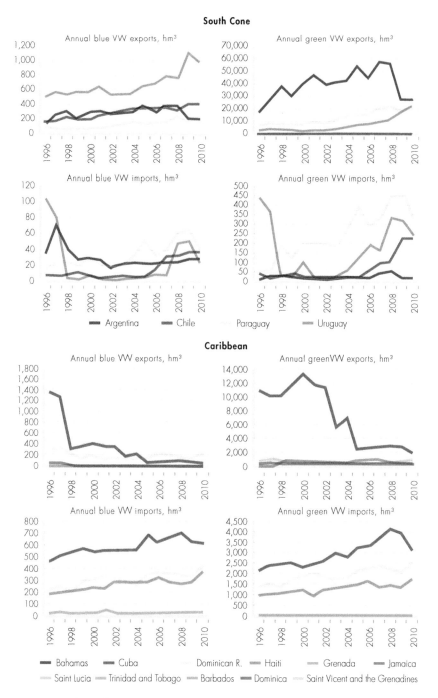

Figure 7.8 Blue and green virtual water exports and imports (in million cubic metres) between 1996 and 2010 in LAC. Note the difference in scales for the vertical axes in the plots. *Source: own elaboration based on data from the Water Footprint Network WaterStat Database (WFN, 2013a).*

7.4 Trends in agriculture: physical, economic and social aspects

7.4.1 Land accounting

The evolution of arable lands in LAC since 1995 (Table 7.2) shows that arable land use has particularly increased for the countries in the Amazonian region, in the South Cone and in Mesoamerica. It has remained constant in the Andean region, and decreased in the Caribbean region. In 2011, average arable land values ranged between 3.2% for the Andean region and 14.9% for the Caribbean. However, the arable land per capita shows a decrease for all the LAC regions between 1995 and 2011, except for the South Cone region, which increased from 0.47ha/person in 1995 to 0.53ha/person in 2011. The lowest regional average of arable land per capita is registered for the Caribbean region (0.08ha/person), and the highest for the South Cone (0.49ha/person).

7.4.2 Productivity analysis

7.4.2.1 Yield

According to the CAWMA (2007), part of the increase in food production can be achieved by improving crop yields and increasing crop water productivity through appropriate investments in both irrigated and rain-fed agriculture. There is good scope for improved productivity in LAC rain-fed areas but less so in irrigated areas. Rain-fed agriculture holds great under-exploited potential for increasing water productivity through better water management practices – gaining more yield and greater value from water. This is an effective means of intensifying agricultural production and reducing environmental degradation (*ibid.*).

LAC is globally important in a number of crops and often achieves yields significantly above the developing world average (Hall, 2001). As shown in Table 7.3, the major cereal yields (e.g. maize, wheat, rice) have increased in line with their production, during the period 1995–2005. The average regional yield per unit of land for wheat in LAC is similar to the average yield output of 2.5–2.7t/ha in North America, while wheat yield in Western Europe is approximately twice as large (5t/ha) and in sub-Saharan Africa it remains below 2t/ha. Yield increases have also happened in tuberous crops (principally potato).

However, yield gaps are still significant in the region, though not so pronounced for the main exporters, such as Argentina or Brazil. Closing the yield gap on a large scale requires investments in rural infrastructure and institutions as well as technology transfer. In LAC, public sector agencies together with the private sector have made some headway in closing the yield gap.

Table 7.2 Evolution of the arable land (in % of countries' land area) in Latin American and Caribbean countries, for the years 1995, 2002 and 2011

	1995	2002	2011
AMAZONIAN REGION			
BRAZIL	6.86	7.27	8.50
GUYANA	2.44	2.29	2.13
SURINAME	0.37	0.29	0.38
ANDEAN REGION			
BOLIVIA	2.31	2.86	3.54
COLOMBIA	2.16	1.99	1.89
ECUADOR	5.69	5.48	4.65
PERU	2.81	2.85	2.85
VENEZUELA, RB	2.93	2.83	2.95
CARIBBEAN			
ANTIGUA AND BARBUDA	9.09	9.09	9.09
BAHAMAS, THE	0.60	0.70	0.90
BARBADOS	37.21	32.56	27.91
CUBA	34.30	35.70	33.35
DOMINICA	4.00	6.67	8.00
DOMINICAN REPUBLIC	18.63	17.96	16.56
GRENADA	5.88	5.88	8.82
HAITI	29.03	32.66	36.28
JAMAICA	14.59	12.47	11.08
PUERTO RICO	3.72	7.67	6.76
ST KITTS AND NEVIS	26.92	26.92	19.23
ST LUCIA	8.20	3.28	4.92
ST VINCENT AND THE GRENADINES	12.82	12.82	12.82
TRINIDAD AND TOBAGO	7.80	5.85	4.87
MESOAMERICA			
BELIZE	2.72	3.07	3.29
COSTA RICA	4.31	3.92	4.90
EL SALVADOR	28.09	33.30	32.09
GUATEMALA	12.64	13.30	14.00
HONDURAS	14.30	9.55	9.12
MEXICO	12.91	12.91	13.11
NICARAGUA	13.71	16.62	15.79
PANAMA	6.73	7.37	7.26
SOUTH CONE			
ARGENTINA	9.90	10.18	13.90
CHILE	2.85	2.22	1.77
PARAGUAY	6.54	8.08	9.82
URUGUAY	7.37	7.43	10.32

Source: World Bank (2013).

Table 7.3 Yield compound annual growth rate by crop and country, period 1995–2001[1]

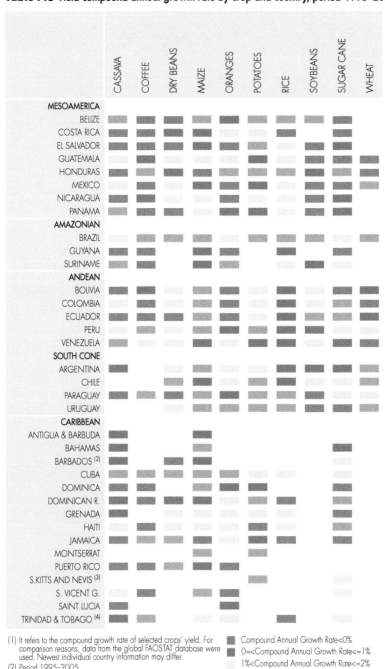

(1) It refers to the compound growth rate of selected crops' yield. For comparison reasons, data from the global FAOSTAT database were used. Newest individual country information may differ.
(2) Period 1995–2005
(3) Dry Beans: Period 1998–2010
(4) Sugar: Period 1995–2007

■ Compound Annual Growth Rate<0%
■ 0=<Compound Annual Growth Rate<=1%
□ 1%<Compound Annual Growth Rate<=2%
■ Compound Annual Growth Rate>2%
□ No data

Source: FAO(2012d)

7.4.2.2 Economic

Agricultural economic productivity (US$/ha)

Agriculture is a significant economic sector for many of the LAC countries. It is so at the macro level, with some of the countries being major world players in the agricultural commodities markets, or at the micro level, with agriculture playing a significant role in terms of food security.

In the last decade, the largest producers in the Southern hemisphere have responded to demand by increasing their cultivated areas, especially that of cereals, oil crops and sugarcane, and most significantly the share of those products that are irrigated. However, the countries production differs greatly. Some countries have highly specialized production (Argentina, Brazil), while others rely on a wider array of products (Mexico, Colombia, Peru, Chile). Consequently the economic effects of world markets on each country's agricultural sector will differ substantially.

On average, yields in the region have improved in the period 2000–2010 by 9% whereas economic productivity of land grew a 19% (constant US$/ha, own calculations based on FAO, 2012d). As reported by FAO (2012a), the increase in production, productivity and income vary between the countries. Figure 7.9 shows the compound growth rate in agricultural land productivity in physical productivity, that is, yield (t/ha), and in economic productivity (US$/ha) between the average of the years 1991–1993 and 2008–2010 for the countries in Central and South America, for some specific products. Economic productivity growth rates are consistently higher than physical productivity growth rates. Particularly potatoes, coffee, wheat and maize have shown in average higher growth rates. Nevertheless, the behaviour of each product shows great variations among countries, as in the case of sugarcane or cassava.

Economic blue water productivity: surface and groundwater

For selected countries Figure 7.10 shows the area harvested and the economic water productivity per crop alongside the share of blue WF related to the total (green and blue) WF. These data are averages for the period 1996–2005. The cultivated surface data was obtained from FAO (2012d). Economic water productivity was calculated using the average producer's price per crop (US$, constant prices) from FAO (2012d) divided by the green and blue water footprint. Data on green and blue water footprints was obtained from the respective countries report or, in the absence of a specific national figure, from Mekonnen and Hoekstra (2011).

Some countries show low economic water productivity, such as Argentina, Brazil, Nicaragua, Bolivia, Uruguay and Mexico. In very general terms, these countries dedicate significant areas for the cultivation of cereals, coffee, cocoa and sugarcane, which have lower economic productivity. Peru, Ecuador and Chile, and to a lesser extent Colombia and Costa Rica, do have a notable amount of area dedicated to crops with medium-high economic productivity, like grapes, onions, pineapples and potatoes. On average, Chile, Venezuela and Costa Rica show higher average productivities (0.57, 0.54 and 1.21US$/m^3 respectively), whereas Bolivia, Argentina and Brazil show lower ones (0.13, 0.12 and 0.11US$/ m^3).

Figure 7.9 Compound growth rate (%) of yield (t/ha) and economic productivity (US$/ha) between av. 1991–1993 and av. 2008–2010 for selected countries and crops. *Source: own elaboration based on FAO (2012d).*

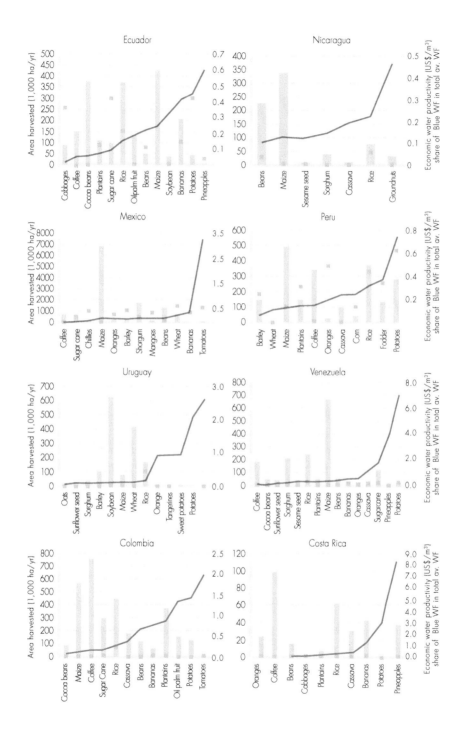

(Figure 7.10 continues in the next page)

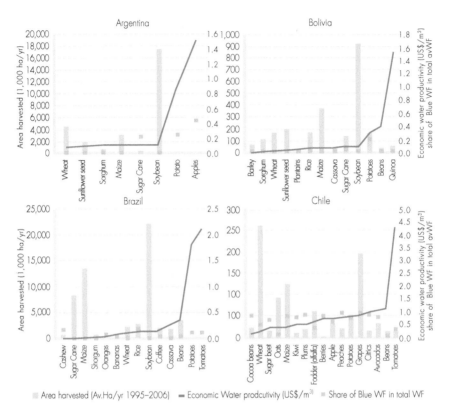

Figure 7.10 Average cultivated area (1,000ha/yr), economic water productivity (US$/m³) and share of blue WF in crop WF for selected countries and crops. The data shown corresponds to an average of the years 2007-2010. Note the difference in scale for each country. *Source: Own elaboration based on FAO (2012d) and Mekonnen and Hoekstra (2011).*

7.4.2.3 Social

Insecure access to reliable, safe, and affordable water keeps hundreds of millions of people from escaping poverty. Most of them rely directly on agriculture for their food and income. According to the CAWMA (2007), poverty could be reduced by improving access to agricultural water and its use. Livelihood gains of smallholder farmer could be obtained by securing water access (through water rights and investments in water storage and delivery infrastructure), improving value obtained by water use through pro-poor technologies, and investing in roads and markets.

Increased productivity by improving irrigation has a multiplier effect on the economy (Table 7.4). Improved agricultural water management boosts total farm output. Increased output may arise from improved yields, reduced crop loss, improved cropping intensity, and increased cultivated area. Reliable access to water enhances the use of complementary inputs such as high-yielding varieties and agrochemicals, which also increases output levels (Hasnip et al., 2001; Bhattarai and Narayanamoorthy, 2003; Hussain and Hanjra,

2003; Smith, 2004; Huang et al., 2006). FAO (2003) data show that the major sources of growth in crop production for all developing countries during 1961–1999 were yield increase (71%), area expansion (23%), and cropping intensity (6%). Empirical evidence for a sample of forty countries shows that for a 1% improvement in crop productivity poverty – in terms of those living on less than US$1 a day – fell by about 1% and the human development index rose by 0.1% (Irzet al., 2001). There seems to be a solid link between yield growth, poverty reduction, and human development. Access to agricultural water has secondary effects on poverty through output, employment and prices. Two factors contribute to output fluctuations: rainfall variability and the relative prices of outputs. Food grain output is sensitive to variations in rainfall (Lipton et al., 2003; Smith, 2004) and as such reliable access to agricultural water not only raises crop output levels, but also usually reduces variance in output across seasons and years.

Finally, stabilization of farm output cannot be achieved merely through a reliable system of agricultural water management. Reducing risk and uncertainty for farmers requires the general improvement of the farming environment (Smith, 2004).

Table 7.4 Impact of irrigation by type of system

	IMPACT	LARGE-SCALE PUBLIC, DRY ZONE	LARGE-SCALE PUBLIC, PADDY-BASED	SMALL- OR MEDIUM- SIZE COMMUNITY-MANAGED	PRIVATE, COMMERCIAL	SMALLHOLDER, INDIVIDUAL
ECONOMIC	Production	Low positive	Low positive	Low positive	High positive	High positive
	Food security	High positive	High positive	High positive	Low positive	High positive
	Rural employment	High positive	High positive	High positive	Low positive	High positive
SOCIAL	Settlement strategies	Mixed	Mixed	High positive	None	None
	Social capital	None	Low positive	High positive	None	None
	Health	Mixed	Mixed	Mixed	Low negative	Mixed
ENVIRONMENTAL	Biological diversity	Mixed	Mixed	Mixed	Mixed	None
	Social and water conservation	Mixed	Mixed	Mixed	Mixed	None
	Water quality	High negative	Mixed	Mixed	High negative	Low negative
CULTURAL	Religious ceremonies	Low negative	None	Low positive	None	None
	Landscape, aesthetics	Mixed	High positive	High positive	Low negative	None
	Cultural heritage	Mixed	Mixed	High positive	None	None

Source: CAWMA (2007)

7.5 Conclusions and recommendations

The LAC region's economy is on average growing rapidly. With its green water and land availability, LAC could potentially represent a good opportunity to produce and supply more food for itself and for other parts of the world. This option also denotes the chance to boost economies in some of these emerging countries. This is the general case for the whole continent; however, particular areas, such as the Antilles, show severe water scarcity levels at the country level, with high levels of dependency on external water resources for food supply.

In spite of the positive agricultural development perspectives and the satisfactory water availability in most areas of the LAC region, if not carefully planned, using local water resources to satisfy this food demand may exert more pressure on water and land resources and increase the already severe water quality problem in the region. The combination of rapid urbanization over the past fifty years and more importantly weak governance are crucial factors affecting water scarcity in a water-rich region.

As economies emerge and there is more investment for natural resources exploitation and use, competition among sectors increases, such as in the case of biofuels and mining versus agriculture for food in the LAC region. The domestic, industrial and hydropower sectors also compete with agriculture. The complex trade-offs across sectors and across water users can best be managed through integrated water management at the river basin level, developed in agreement with the national policies and planning – but establishing appropriate institutions for inter- and intra-sectorial water allocation remains an important challenge under the fragmented management structure in most of LAC. Appropriate water accounting systems, including the green, blue and grey water footprint and the related socio-economic and environmental impacts can inform decision-makers, planners and developers at different levels (river basin, departmental, national) on the sustainability of different water management options. These water accounting systems can also inform about crop water consumption and its economical and social benefits to optimize the allocation of water resources when planning irrigation development (Box 7.3). Sustainable water management should not be seen as a barrier for the development of the region, but rather as the way to develop and grow as a region.

Overall, this chapter shows the strong links between water, agriculture and economy in LAC. Both green and blue water are a vital fuel for LAC's economies and for its food security. Awareness of LAC's virtual water trade volumes and water footprints will not alone solve the local or global water problems. However, the awareness gained increases the odds that optimized water allocation decisions, which consider the hydrological and economical aspects of water resources, are made (Allan, 2011).

Box 7.3 Water footprint assessment of Porce River Basin, Colombia

The Water Footprint Assessment (WFA) of Porce River Basin (2012) included the five main productive sectors in the basin (crop and livestock, industry, domestic, hydropower and mining) and the four phases of the WFA were analysed.

The total WF of crop production was 250hm^3/yr, (93% green – 5% blue – 2% grey). Coffee is the crop that contributes the most to the WF (green and blue, 31%), followed by sugar cane with 19%, potatoes 15% and plantain 8%. In terms of the grey WF, coffee is the crop with the highest impact in the watershed followed by potatoes (based on nitrogen). The water footprint of livestock is 700hm^3/yr, (66% green – 32% blue – 2% grey). Cattle contribute with more than 80% to the total WF of livestock, followed by horses, poultry and pigs respectively. Cattle equally occupy the first place (76% blue and 65% grey), followed by poultry (11% blue and 21% grey), pigs (10% blue and 9% grey) and horses (3% blue and 5% grey).

Table 7.5 The green, blue and grey water footprint in the Porce River Basin

SECTOR	GREEN WF m^3/yr	BLUE WF m^3/yr	GREY WF m^3/yr	CRITICAL POLLUTANT
CROP PRODUCTION	231.0	13.5	4.8	N
LIVESTOCK	463.0	12.4	215.8	N
HOUSEHOLD	-	27.8	11,788.2	BOD
INDUSTRIAL	-	8.0	4,078.5	BOD
HYDROPOWER	-	24.4	-	-
MINING	-	3.7	3,059.1	TSS

Source: CTA (2013)

The environmental, economic and social components of the WF sustainability assessment were included. The biggest environmental problem identified is the lack of pollution assimilation capacity, especially in the upper basin (city of Medellin). This region presents critical pollution indexes, according to the maximum allowed concentration criteria used. For the economic analysis, apparent water productivities were analysed for each of the productive sectors. For the social analysis indicators on public health, coverage in water supply and sanitation were taken into account.

The complex WF sustainability assessment (environmental, economic and social) identifies the basin's hotspots, enabling the formulation of responses in terms of public policy and public–private partnerships.

References

Allan, T. (2011). *Virtual Water: Tackling the Threat to Our Planet's Most Precious Resource*. London, IB Tauris Publishers.

Albersen, P.J., Houba, H.E.D. & Keyzer, M.A. (2003). Pricing a raindrop in a process-based model: General methodology and a case study of the upper-zambezi, *Physics and Chemistry of the Earth*, 28: 183–192.

Aldaya, M.M., Allan, J.A. & Hoekstra, A.Y. (2010). Strategic importance of green water in international crop trade, *Ecological Economics*, 69 (4): 887–894.

Bhattarai, M., & Narayanamoorthy, A. (2003). Impact of Irrigation on Rural Poverty in India: An Aggregate Panel-Data Analysis. *Water Policy*, 5 (5–6): 443–458.

Biswas, A.K., Tortajada, C., Braga, B., & Rodriguez, D.J. (eds) (2006). *Water Quality Management in the Americas. Water Resources Development and Management*. The Netherlands, Springer.

Bovarnick, A., Alpizar, F. & Schnell, C. (eds) (2010). United Nations Development Programme. *The Importance of Biodiversity and Ecosystems in Economic Growth and Equity in Latin America and the Caribbean: An economic valuation of ecosystems*. UNDP.

CAWMA (2007). Comprehensive Assessment of Water Management in Agriculture. *Water for Food, Water for Life: A Comprehensive Assessment of Water Management in Agriculture*. London, Earthscan.

CIP (2010). Centro Internacional de la Papa. *Facts and Figures about Potato*. Lima, Peru, CIP. [Online] Available from: www.cipotato.org. [Accessed February, 2013].

CTA (2013). Centro de Ciencia y Tecnología de Antioquía. Evaluación de la huella hídrica de la cuenca del río Porce. Resumen de resultados. Mayo 2013. pp.102.

Egúsquiza, B.R. (2000). *La Papa, Producción, Transformación y Comercialización*. Convenio MSP ADEX-USAID. Lima, Perú, Universidad Nacional Agraria La Molina.

FAO (2003). Food and Agriculture Organization. *World Agriculture: Towards 2015/2030. An FAO Perspective*. FAO and Earthscan.

FAO (2008). Food and Agriculture Organization. *Potato International Year*. [Online] Available from: www.potato2008.org/es/mundo/america_latina.html. [Accessed February, 2013].

FAO (2012a). Food and Agriculture Organization. *The State of Food and Agriculture*. Rome. [Online] Available from: www.fao.org/docrep/017/i3028e/i3028e.pdf [Accessed February, 2013].

FAO (2012b). Food and Agriculture Organization. *Panorama de la Seguridad Alimentaria y Nutricional en América Latina y el Caribe 2012*. Rome, FAO. [Online] Available from: www.rlc.fao.org/es/publicaciones/panorama-2012/. [Accessed May, 2013].

FAO (2012c). Food and Agriculture Organization. *The Outlook for Agriculture and Rural Development in the Americas: A Perspective on Latin America and the Caribbean*. Santiago, Chile, CEPAL, FAO & IICA. [Online] Available from: www.rlc.fao.org/en/publications/outlook-2013 [Accessed February, 2013].

FAO (2012d). Food and Agriculture Organization. FAOSTAT. *Production and Trade Statistics*. [Online] Available from: faostat.fao.org/site/291/default.aspx. [Accessed December, 2012].

FAO (2013). Food and Agriculture Organization. AQUASTAT. [Online] Available from: www.fao.org/nr/water/aquastat/main/index.stm [Accessed February, 2013].

Grau, H.R., & Aide, M. (2008). Globalization and land-use transitions in Latin America. *Ecology and Society*, 13 (2): 16.

Hall, M. (ed.) (2001). Food and Agriculture Organization. *Farming Systems and Poverty. Improving farmers' livelihoods in a changing world.* FAO-World Bank.

Hasnip, N., Mandal, S., Morrison, J., Pradhan, P. & Smith, L.E.D. (2001). *Contribution of Irrigation to Sustaining Rural Livelihoods.* Wallingford, UK, HR Wallingford.

Hoekstra, A.Y., Chapagain, A.K., Aldaya, M.M. & Mekonnen, M.M. (2011). *The Water Footprint Assessment Manual: Setting the global standard.* London, Earthscan.

Huang, Q., Rozelle, S., Lohmar, B., Huang, J. & Wang, J. (2006). Irrigation, agricultural performance and poverty reduction in China. *Food Policy,* 31 (1): 30–52.

Hussain, I. & Hanjra, M.A. (2003). Does irrigation water matter for rural poverty alleviation? Evidence from South and South-East Asia. *Water Policy,* 5 (5–6): 429–42.

Irzet, X., Lin, L., Thirtle, C. & Wiggins, S. (2001). Agricultural productivity growth and poverty alleviation. *Development Policy Review,* 19 (4): 449–66.

LA–Costa Rica (2012). *Report of Costa Rica.* Contribution to the Water and Food Security in Latin America Project. Water Observatory Project Madrid, Spain: Fundacion Botín.

LA–Peru (2012). *Report of Peru.* Contribution to the Water and Food Security in Latin America Project. Water Observatory Project Madrid, Spain: Fundacion Botín.

Lipton, M., Litchfield, J. & Faurès, J.M. (2003). The effects of irrigation on poverty: a framework for analysis. *Water Policy,* 5 (5–6): 413–27.

Mekonnen, M.M. & Hoekstra, A.Y. (2011). *National Water Footprint Accounts: The Green, Blue and Grey Water Footprint of Production and Consumption. Volume 2: Appendices.* Delft, The Netherlands, UNESCO-IHE. Value of water, Research Report Series No. 50.

Moran, E.F. (1982). *Human Adaptability. An Introduction to Ecological Anthropology.* Colorado, Westview Press.

Niemeyer, I. & Garrido, A. (2011). International farm trade in Latin America: does it favour sustainable water use globally? In: Hoekstra, A.Y., Aldaya, M.M. & Avril, B. (eds). *Proceedings of the ESF Strategic Workshop on Accounting for Water Scarcity and Pollution in the Rules of International Trade.* Delft, The Netherlands, UNESCO-IHE. Value of Water Research Report Series No. 54. pp. 63–84.

Oliviera, A.S., Trezza, R., Holzapfel, E., Lorite, I. & Paz, V.P.S. (2009). Irrigation water management in Latin America. *Chilean Journal of Agricultural Research,* 69: 7–16.

Palhares, J.C.P. (2012). Pegada Hídrica das aves abatidas no Brasil na década 2000–2010. In: Macari, M. & Soares, N.M. (Org.). *Água na avicultura.* Campinas, Brazil, Facta. pp. 40–53.

Proceso Regional de las Américas (2012). Agenda del Agua de las Américas. *Agenda del agua de las Américas: Metas, soluciones y rutas para mejorar la gestión de los recursos hídricos.,* Marseille, France, 6th World Water Forum.

Ramírez, F., Chaverri, F., de la Cruz, E., Wesseling, C., Castillo, L. & Bravo, V. (2009). *Importación de plaguicidas en Costa Rica.* Periodo 1977–2006. Instituto Regional de Estudios en Sustancias Tóxicas. Facultad de Ciencias de la Tierra y el Mar. Heredia, Costa Rica, Universidad Nacional,. Serie Informes Técnicos IRET No. 6.

Rockström, J., Hatibu, N., Oweis, T. Y., Wani, S., Barron, J., Bruggeman, A., Farahani, J., Karlberg, L. & Qiang, Z. (2007). Managing water in rainfed agriculture. In: Molden, D. (ed.). *Water for Food, Water for Life: A Comprehensive Assessment of Water Management in Agriculture.* London, Earthscan. pp. 315–352

Smith, L.E.D. (2004). Assessment of the contribution of irrigation to poverty reduction and sustainable livelihoods. *Water Resources Development,* 20 (2): 243–257.

UN (2003). United Nations. *Water for People, Water for Life*. The first UN edition of the World Water Development Report (WWDR1).

UNCTAD (2011). United Nations Conference on Trade and Development. *Water for Food. Innovative water management technologies for food security and poverty alleviation*. [Online] Available from: http://unctad.org/en/Docs/dtlstict2011d2_en.pdf. [Accessed February, 2013]. Current Studies on Science, Technology and Innovation report no 4.

UNEP (2013). United Nations Environment Programme. *The State of Natural Resource Management in LAC - Opportunities for Sustainable Practices and Prioritization of Resources*. UNEP - Division of Technology, Industry and Economics (UNEP DTIE) and UNEP Regional Office for Latin America and the Caribbean (PNUMA ROLAC).

WFN (2013a). Water Footprint Network. WaterStat Database. [Online] Available from: www.waterfootprint.org/?page=files/WaterStat. [Accessed March, 2013].

WFN (2013b). Water Footprint Network. Water Footprint Assessment Tool. [Online] Available from: www.waterfootprint.org/?page=files/waterfootprintassessmenttool [Accessed February, 2013].

World Bank (2013). *World Bank Development Indicators*. [Online] Available from: data.worldbank.org/indicator. [Accessed May, 2013].

Yang, H., Wang, L., Abbaspour, K.C. & Zehnder, A.J.B. (2006). Virtual water trade: An assessment of water use efficiency in the international food trade, *Hydrology and Earth System Sciences*, 10: 443–454.

WATER SECURITY AND CITIES

Authors:
Enrique Cabrera, ITA, Universitat Politècnica de València (UPV), Spain
Emilio Custodio, Dept. Geo-Engineering, Universitat Politècnica de Catalunya (UPC), Barcelona, Spain

Contributors:
Ramón Aguirre, Sistema de Aguas de la Ciudad de México, México
Emilia Bocanegra, Universidad Nacional de Mar del Plata, Argentina
Gerson Cardoso da Silva Jr, Universidade Federal do Rio de Janeiro, Brazil
Manuel Cermerón, Aqualogy, Spain
Javier Dávara, Aqualogy-SEDAPAL, Peru
Maria Josefa Fioriti, Subsecretaría de Recursos Hídricos, Argentina
Ricardo Hirata, Universidade de São Paulo, Brazil
Joaquim Martí, Aguas Andinas, Chile

Highlights

- Latin America has achieved good progress in urban water supply and sanitation, although gaps have to be bridged and efficiency has to be improved, especially in what refers to sanitation.

- Urban water supply quality presents deficiencies in some urban areas, some due to poor natural water quality and to inadequate functioning of supply networks.

- A lot could be socially gained by investments in better functioning of urban water supply and sanitation networks.

- Water users have to cover the full cost of urban water supply and sanitation, including maintenance and renovation of infrastructure, most probably the main bottleneck in the way to achive sustainable urban water systems.

- The most deprived population has the right to receive drinking water and sanitation at affordable price for them, but the additional cost has to be covered by the other users and the society in general, without compromising the needed investments.

- The dilemma between public and private water services is not the key issue.

8.1   Introduction

A large and growing fraction of humanity currently lives in urban areas, many of which are, so-called, megacities. Table 8.1 shows some of the most relevant cities in Latin America (LA). In this chapter we consider only continental Latin American countries, Mexico being the most northerly country down to Argentina and Chile in the south. The focus is on water security for large urban areas, with particular emphasis on water services. Various case studies are provided by local experts with short comments on water issues in specific LA large cities. These represent some of the best managed cities in the area, so to some extent the sample is biased. However, what is presented helps to understand the current situation even though they do not necessarily correspond to the general picture of the continent as a whole.

The urban water cycle is a relatively new concept (Cabrera and Custodio, 2013). Even if all early civilizations had large waterworks in order to secure good access to water for their citizens, the generalized establishment of urban supply did not start worldwide until the middle of the 19th century. Thus, available data and studies are based on recent history. In particular, the introduction of water chlorination by English physician John Snow to control water-borne diseases was a real turning point. Although it was not until the early

Table 8.1 Data on some of the largest cities in LA

TOWNS	POPULATION 2011 (millions)	WORLD RANKING IN URBAN POPULATION	SUPLIED SURFACE AREA (km²)	% POPULATION OF LA	PER CAPITA DAILY SUPPLY (L/day/cap)	PER CAPITA DAILY SUPPLY (hm³/cap)
SÃO PAULO	20.6	9	3173	3.43	180	3,702,240
MEXICO D.F.	20.0	10	2046	3.34	360	7,211,520
BUENOS AIRES	13.8	21	2642	2.30	370	5,097,120
RIO DE JANEIRO	12.6	26	4026	1.94	190	2,207,040
LIMA	9.4	30	648	1.57	254	2,387,600
BOGOTÁ	9.0	36	414	1.50	136	1,225,224
SANTIAGO DE CHILE	6.2	53	984	1.03	196	1,209,516
MONTERREY	4.2	88	894	0.70	300	1,266,000
CARACAS	3.3	124	272	0.55	400	1,311,600

Source: own elaboration.

20th century that this technique was widely applied (McGuire, 2013), it is still credited for being the primary cause of last century's increase in life expectancy. In terms of public health, it could be considered the greatest advancement of the millennium.

By the early 20th century most of the world's developed cities had introduced urban water supply networks. However, a few decades later these networks were stressed due to the much higher demand produced by rapid urban growth.

Apart from some pioneering examples of sewage water treatment (the first plant was established in 1890 in Worcester, Massachusetts, USA), they were not at all common in the USA until the end of the Second World War; when in 1948, under the Federal Water Control Act,[1] funds were made available for the construction of treatment plans to ensure water quality. In developed countries, almost 100% of urban and industrial wastewater is currently treated, albeit large improvements can still be made. In Europe the main impetus came from new regulations relating to water quality and pollutants; chief amongst these is the Water Framework Directive (OJEU, 2000). There are still some challenges given that tertiary wastewater treatment to eliminate organic load is still not sufficient to eradicate some worrisome contaminants, which appear in relatively low concentrations.

Latin America has partly followed the path of developed countries, albeit delayed in time and with a long way to go. This is reflected in the America's Water Agenda (Regional Process of the Americas, 2012), a report backed by the Inter-American Development Bank and produced by relevant institutions of the Americas involved in water affairs.

---

1 www.fws.gov/laws/lawsdigest/FWATRPO.HTML (accessed in June, 2013)

The challenges for America are grouped into fourteen points, of which the first four refer specifically to water supply and sanitation:

- Implementation of the human rights; access to water and sanitation, in response to the July 2010 United Nations resolution A/RES/64/292
- Water, sanitation and service quality deficiencies
- Universal coverage for current urban and marginal areas
- Water and sanitation in rural areas.

A large number of the remaining points also refer to urban water, albeit not so directly, like the 5th point (climatic change), 6th point (governance), 7th point (need for integrated water resources management), 10th point (increasing water contamination, largely due to urban and industrial pollution), and 14th point (creation of a political environment willing to make the necessary investments into urban water planning and infrastructure).

All of these aspects are relevant and important since humans need water to live and yet the way to provide this water is becoming progressively more complex. Towns and the associated water systems expand, while their supporting infrastructures are neither upgraded nor modernized at the same pace, and thus they age and their service quality decreases. This is an issue that must be specifically addressed as often citizens believe that, water being a universal right, the government ought to provide it at no cost to them. This explains their reluctance to pay the true cost of a service that is becoming increasingly costly and complex. As a consequence, politicians responsible for urban water management, who depend largely on citizens votes in many countries, and LA is not an exception, are unlikely to charge the full water service cost. The consequence is that water services often become economically collapsed. The antidote is full, transparent information.

Available documents show that top-down points of view alongside socio-economic and political analysis tend to be prioritized whilst technical aspects are largely disregarded, when they could in fact be the key to better understanding of the problems. To some extent, what follows is an attempt to better understand this discrepancy.

8.2   Water sources for urban supply

Most large cities in LA depend on surface water from river basins (Santiago, Chile, Box 8.1). However groundwater is also important for many of them, occasionally the only source, as is the case of Mar del Plata, Argentina (Box 8.2), or sometimes groundwater is a key complement that cannot easily nor quickly be substituted (São Paulo, Brazil, Box 8.3). At times it is used as a necessary backup as in Lima, Peru, or simply for the supply of fast expanding and poor peri-urban areas (Rio de Janeiro, Brazil, Box 8.4; Lima, Peru, Box 8.5, and Buenos Aires, Argentina, Box 8.6). Mexico City, in terms of urban water management, is one of the most complex ones in the world. It is served by a mixture of imported, local and further afield groundwater resources (Box 8.7 and SACM, 2012). Even though many of the large cities on the continent are near or next to the sea, seawater desalination is seldom employed as an urban water source in LA, with a few exceptions in medium-size towns (e.g. Antofagasta and Iquique, Chile).

Box 8.1   Water supply to Santiago, Chile

[By Joaquim Martí, Aguas Andinas, Santiago, Chile, and Manuel Cermerón, Aqualogy, Barcelona]

Santiago, the capital city of Chile, is located in the Intermediate Chilean Depression, the depression between the Andes and the Chilean coastal range. It receives an annual average precipitation of 320mm mainly concentrated in the May–August period; the summers are long and dry. The population is about 6.5 million inhabitants, of which 95% have their water supply and sanitation needs provided by the Aguas Andinas group, the remaining 5% are served by a municipal company.

Most of the supplied water (86%) is surface water from the Maipo and Mapocho rivers, complemented by some 14% groundwater. To guarantee a continuous drinking water supply, large reservoirs exist, such as El Yeso (220hm^3), Laguna Negra (600hm^3) and Laguna Lo Escañado (50hm^3). Two main plants, Las Vizcachas (15.8m^3/s) and La Florida (4.0m^3/s), produce drinking water alongside twelve smaller plants. The water is then distributed via a 12,094km piping network. Unaccounted water amounts to 29% and includes technical, commercial and measurement components.

The sanitation network covers 100% of the area with 10,501km of sewers. Sewage water is treated in thirteen plants, the two main ones being, La Farfana (8.8m^3/s) and Mapocho–Trabal (6.6m^3/s). All treated waste waters are disposed of downstream from Santiago, in the Mapocho River.

Drinking water price is 0.595US$/m^3, plus 0.465US$/m^3 for sewage service and 0.307US$/m^3 for wastewater treatment. However, low-income citizens may ask for half the price to be covered. All the revenue generated is used to cover the operation, maintenance, replacement and any future works needed to improve the service and ensure water supply security.

Box 8.2   Water security in the supply of Mar del Plata, Argentina

[By Dr Emilia Bocanegra, Universidad Nacional de Mar del Plata, Mar del Plata, Argentina]

The coastal town of Mar del Plata, Argentina, is approximately 450km south of Buenos Aires. It spans an area of 80km^2 and has 620,000 residents, a figure which is doubled during the summer period. The local economy depends on tourism, harbour activities (mostly fisheries), textiles and leather, and fruit trees and orchard agriculture. All water resources are supplied by groundwater coming predominantly from rural areas in the north and northeast of the city but also abstracted inside the urban and peri-urban area. Approximately 129hm^3/yr of good quality groundwater is abstracted from 274 wells. Intensive groundwater abstraction in the urban area has induced seawater intrusion into the aquifer, forcing forty wells to be abandoned. Other wells have also been taken out of service due to high level of nitrates. The limit of 45mg/L of nitrate concentration is exceeded in 117 wells. Water is periodically chlorinated and meets drinking water standards in 96.6% of cases. Some 93% of the population is connected to the distribution network. The network loses 40% and suffers 0.38breaks/km each year. Most residences have tanks to store water since in summer months water pressure may be insufficient to reach some neighbourhoods.

Regarding the sanitation network 92% of population is connected to it, and it suffers 0.17breaks/km each year. Four main servers feed a final pre-treatment plant where solids are separated, dried and aerobically stabilized to produce a soil conditioner that is principally used for ornamental plant cultivation. The effluent is disposed of in the sea.

Drinking water is served at a cost of 0.17US$/m³, plus 0.12US$/m³ for sanitation. The joint average yearly charge for connection is 98US$/yr. Commercial and industrial establishments are often metered amounting to approximately 20% of the water provided. The domestic charge is calculated according to the surface area of the dwelling and the area of the city where it is located.

Urban water use rate is high (296L/day/cap). Despite public campaigns to reduce water use, results have not been very effective. In order to encourage water savings there is a surcharge for indoor pools.

The contribution from customers is sufficient for the efficient maintenance and operation of the water and sanitation services. Network expansion and major infrastructure projects are financed by subsidies received from the government. Expansion is planned in the future including the incorporation of a further seventy-four wells into the network in order to supply the western main and the construction of a submarine outfall in order to improve local coastal seawater quality. The municipal company Obras Sanitarias de Mar del Plata (OSMP, Mar del Plata Waterworks) is responsible for supplying water and sanitation services and regulates the different uses under the Water Code of the Province of Buenos Aires thus helping water governance. Currently there is no known conflict between OSMP and rural and industrial users.

Box 8.3 Water security in the metropolitan area of São Paulo, Brazil: the key role of groundwater

[By Ricardo Hirata, CEPAS-Universidade de São Paulo, São Paulo, Brazil]
The megacity of São Paulo, including the surrounding thirty-five municipalities, reunites 20.6 million inhabitants. They are located in the Alto Tietê Watershed (ATW), which has a surface area of 5,720km². The public water supply system is operated by SABESP, a state-owned company that provides 2,144hm³/yr (68m³/s) of water coming from eight surface water sources, half of them imported from another water basin. More than 95% of the total population is provided for in this way. Additionally, a total of 347hm³/yr (11m³/s) of water comes from approximately 9,000 privately owned tube-wells. Even though this supply is only 17% of total demand, if it fails due to water contamination or excess of abstraction, it will seriously compromise water security in the ATW. SABESP

currently has no more water resources available to substitute for this groundwater use. Water transfers from further water basins, although planned, will not be available in the near future. Average groundwater abstraction in the area is about 32% of total recharge (400mm/yr), including rainfall infiltration (45%). Losses from the supply and sewage network are 55%. However, abstraction is irregularly distributed in the area and concentrates especially in the central area of Penha-Pinheiros sub-basin where abstraction exceeds 80% of the recharge.

Water supply security in the ATW is also a concern with regard to well conditions as 60–70% of them lack operation permits and, as a consequence, there is no control on them. Well drilling and construction does not follow acceptable standards, so the wells and the aquifer itself are under risk of contamination. More than 50% of declared contamination cases in São Paulo State are located in the ATW. This is the result of intense industrialization and unplanned, dense and rapid urban expansion.

Box 8.4   Water security in the metropolitan area of Rio de Janeiro, Brazil: groundwater, the unknown resource

[By Gerson Cardoso da Silva Jr, IGEO, Universidade Federal do Rio de Janeiro, Brazil] The Metropolitan Area of Rio de Janeiro (MARJ), Brazil, with a land area of 5,292km^2 plus 400km^2 of the Guanabara Bay, comprises seventeen municipalities with approximately 12.6 million inhabitants. The significant demographic and economic growth in recent decades is the cause of a notable increase in the consumption of water resources. The Hydrographical Region of Guanabara Bay (HRGB) includes the MARJ and other territories, but predominantly, river headwaters. About 50m^3/s are transferred from the Paraíba do Sul River to the Guandu Water Plant, the main source of water for public supply. The price for the final user is approximately 0.65US$/m^3 which includes sewage treatment. This is complemented by water from other minor sources, groundwater being approximately 2% of the total. In general, high-quality water is supplied by the water plants. The gross revenue accounts for financial costs and investments, as well as for operational and maintenance costs. Water losses due to leaks in pipelines and 'social losses' (e.g. unpaid water for shantytowns) represent 30% of the total distributed water.

In many situations a significant use of groundwater resources as a supplementary source of water in the region is made, even though the aquifer characteristics are poorly known. Most wells lack any kind of register or permit. The Rio de Janeiro State has implemented in recent years a programme for well legalization, raising the number of permitted wells from a few hundred to thousands in the last five years alone, although official federal statistics point to approximately 0.5 million shallow wells in the area, which are mainly used by low-income populations as a complementary source and sometimes as the sole water source. Aquifer overexploitation is not reported. Groundwater quality is sometimes poor due to pollution or salinity.

About 44% of sewage water (13.2m³/s) is treated, and most of the remaining discharge flows to the sea through a pipeline or to the Guanabara Bay. In general the situation is improving.

Box 8.5 Urban water in Metropolitan Lima, Peru

[By Javier Dávara, Aqualogy-CEDAPAL, Lima, Peru, and Manuel Cermerón, Aqualogy, Barcelona]

The metropolitan area of Lima, which comprises Lima (Peru's capital) and Callao (main Peru's harbour) has approximately 9.4 million inhabitants. It is in an arid zone which depends fully on river water from the high Andes Range. Urban water supply is tapped from two of the three local rivers and the local aquifer. In this arid area aquifer recharge is through river and urban water infiltration.

The state-owned Servicio de Agua Potable y Alcantarillado de Lima (SEDAPAL, Lima's Drinking Water and Sanitation Service) fully supplies forty-three out of the forty-eight city districts and partially two others, with 83% of surface water and 17% of groundwater, which is treated in three plants. Some 89% of the inhabitants are supplied through 13,700km of pipes, and 85% of sewage water is collected through 12,000km of sewers, 20.6% of which is treated in seventeen plants.

Water prices are 0.67US$/m³ plus 0.29US$/m³ for sanitation. There is fixed base rate plus a charge proportional to the volume of the water supplied, with different rates for domestic, commercial, industrial and state demands.

Box 8.6 Urban water in the city of Buenos Aires, Argentina

[By María Josefa Fioriti, Under-Secretariat of Water Resources, Buenos Aires, Argentina]

The water supply and sanitation of the Autonomous City of Buenos Aires, the capital of Argentina, and its seventeen neighbouring administrative areas ('partidos') is the responsibility of AySA (Agua y Saneamientos Argentinos S.A.). The very flat area covers over 18,11km², it has 10.2 million inhabitants, of which 90.6% receive drinking water and have sanitation. 4.53hm³/day of surface water from the Rio de la Plata and 0.25hm³/day of groundwater from 238 wells is supplied, about 600m³/day/cap. The water supply network exceeds 18,000km. Supplied water quality complies with the fifty-eight values of the Regulatory Framework, which is based on the Argentinean Food Code and Word Heath Organization recommendations.

Some 62.6% of population is served by the 10,600km sewage water network and five treatment plants, discharging to an outfall that takes and diffuses the 2.25hm³/day of effluents to a point near the end of the estuary of La Plata River, 2.5km offshore.

Water is charged to domestic, commercial and industrial establishments, and non-occupied land according to the metered volume for those who have a water meter and according to building or house characteristics for the other ones. Social tariffs are applied to low-income citizens and also they participate in local water management affairs as regulated by national authorities.

The Director Plan Framework for water aims to have 100% coverage by 2018. This will require a new drinking water treatment plant, modernization of the existing networks, five new sewage water treatment plants and the enlargement of four of the existing ones.

Quality of service, the protection of human rights and the delivery of concession conditions to the water company depends on the Water and Sanitation Regulatory Entity (ERAS). Water planning for water service and sanitation works depends on the Planning Agency (APla).

Box 8.7 Urban water in Mexico City

[By Ramón Aguirre Díaz, Director General of Waters Systems, Mexico City, Mexico]. The metropolitan area of Mexico City exceeds 20 million inhabitants and is located in the Valley of Mexico, a closed basin 2,240m above sea level. Current water outflows from the area are artificial. Water availability is 160m^3/yr/cap, while for the whole country is 4,090m^3/yr/cap. The main water source is the aquifer, which has been over-drafted by a factor of two for more than a decade. This is a non-sustainable situation. Some studies indicate that the aquifer could be completely exhausted in two or three decades. Complementary water supply from the Cutzamala water surface dams is not enough in the event of a serious drought. About 10% of the population receives water once or twice per week, another 15% will suffer from low water pressure during five to six hours per day and 5% does not receive good quality drinking water. Current average urban water use is 530L/day/dwelling, which is a very high figure under current circumstances. This is the result of water prices being well below the real cost of water and the almost non-existent water meters in the dwellings.

To deal with this huge and concerning problem and to take care of the future generation, an Integrated Water Resources Management Program (PGIRH) has been prepared and launched, for the coming twenty years. Among the many actions, 225 new wells, the protection of 111 springs, the substitution of 5,700km of pipes to reduce leakages, 1,326km of new pipes, seventy-four improved or new drinking water treatment plants and the use of reclaimed treated sewage water for non-drinking uses are included. Aquifer recharge with reclaimed sewage water treated up to drinking water quality is foreseen to increase storage and to control subsidence problems, especially where thick clay layers exist. The programme aims at raising water prices to cover real cost and to install in-house water metering to reduce water use, but considering social tariffs for the more deprived. Although the programme is very ambitious and involves very high investment of 13 billion US$ investment in twenty years, it is feasible and it is what the City of Mexico requires to achieve a sustainable, quality water service.

Water quantity is not generally a limitation in LA (see per capita consumption in Table 8.1). However, per capita consumption is often high, particularly in Argentina and Mexico. The limitations come mainly from the needed infrastructure associated to water storage and transport. Sometimes water transportation may involve important energy consumption when natural barriers have to overcome, as in Peru for the transport of water resources from the water-rich and relatively low-lying upper Amazonas basin to supply the arid western coastal area, where most of the large urban areas are located. Mexico City and other urban areas at high altitude require costly water pumping from low-lying areas. Furthermore conflicts with local residents arise as they compete for the resources or do not accept wastewaters for agriculture in exchange of handling over the water resources they already have. Here groundwater plays an important role but it is also a source of conflicts with other users. Deeper wells are drilled in order to tap up-to-now little exploited aquifers and offered as a new water resource when it is in fact part of the same system and their exploitation will only worsen current problems of over-draft and also serious land subsidence in some areas.

The water quality of urban supply sources is often an important issue, and will be more in the future. Buenos Aires (Argentina) takes a large part of its water from the urban and industrially polluted Rio de la Plata, which can suffer large quality fluctuations. The urban areas in Colombia and Venezuela cannot use the numerous local rivers, lakes and aquifers due to intense pollution from important urban, industrial and mining areas upstream. Groundwater is commonly of enough good quality, but there are important exceptions. In coastal urban areas seawater intrusion has forced the closure of part of the supply wells (Mar del Plata, Argentina; Recife, Brazil). High natural groundwater salinity due to aridity is found in northern Chile and Peru, and in parts of northeastern Brazil. An excess of nitrates is also a common groundwater quality problem, mostly caused by the activity of the urban area itself, for example when collective sanitation is insufficient (Conurbano Bonaerense, Buenos Aires, Argentina, and Lima, Peru) and in some cases of agricultural origin, although this does not occur as intensely as it does in North America and Europe. This last aspect is important around urban areas in Mesoamerica (Costa Rica, Nicaragua, and Guatemala). Unwanted and noxious solutes, such as relatively high concentrations of fluoride and arsenic, can be found in groundwater. Although this problem rarely affects people in large cities connected to public water supply, it is a serious issue in many rural areas of Argentina, Paraguay, Chile, Peru and Mexico. Some large towns do often have to blend the different water sources in order to dilute waters that do not meet the standards or take out of service some others – an often-used situation when sanitary authorities intervene or use them only as backup in emergencies. Blended water may not always be available to all citizens. Treatment for natural poor water quality is not common in LA, where looking for new water sources is preferred, even if this is more expensive.

8.3 Current situation of urban water services

The initial euphoria at establishing water services throughout the 20th century, their advantages being so obvious, made sure that the funds needed to initiate the projects were provided. However, this euphoria has not been maintained, and in fact many water services are currently in dire need of modernization. This and the great population growth, especially in urban areas, are responsible for the existing problems, which are especially pronounced in LA. The most important of them are presented below.

8.3.1 Insufficient water service cover

Reducing the percentage of people lacking water services is one of the main Millennium Development Goals, which also include halving by 2015 the number of people lacking drinking water. The World Summit for Sustainable Development in Johannesburg in 2002 added sanitation to these goals (GTAS, 2003). These goals have been globally accomplished and exceeded for water supply, but not for sanitation that is likely to fall 85% short of the objective (Regional Process of the Americas, 2012). Prospects are good for LA (ONU, 2011; UNICEF/OMS, 2012), although in rural areas the situation is not as positive since 20% of these populations do not have adequate water supply and 50% of them lack appropriate sanitation (Pearce-Oroz, 2011) (see Chapter 6 for a detailed presentation of these indicators).

8.3.2 Lack of urban and industrial wastewater treatment

This seems to be the main mid-term problem to be addressed by developing countries and by LA in particular. By prioritizing the Millennium Development Goals to increase water supply and sanitation networks, wastewater treatment has been put to one side. The Americas' Water Agenda report (Regional Process of the Americas, 2012) highlights the effect of increasing water contamination but does not deal with the causes; the problem is simply considered a service deficiency. The IV World Water Forum of Mexico (CNA/WWC, 2006) provided the impressive figure that more than 86% of wastewater is disposed of into the environment without any treatment, and irrigating with untreated wastewater is a common practice, highly risky for citizens' health.

Good wastewater treatment before disposal is a key component of sustainable water management, even if it requires costly investments, expensive maintenance and modernization of the existing network. After a global analysis avoiding contaminated water in the urban environment, treatment is much cheaper than importing water resources from elsewhere. However, in the real world short-term goals overcome the mid- and long-term points of view. This problem will never be solved if water prices do not allow a reasonable cost recovery since Governments do not have enough economic resources to subsidize water treatment, especially for a population that is polluting more and more and increasingly using more household polluting chemicals.

8.3.3 Economic unsustainability

Economic sustainability is a necessary although not sufficient condition to solve the above mentioned problems. Efficient management needs not only technological capacity to identify the best cost-benefit actions, but also economic resources to carry them out. Without them the system advances towards economic collapse, as too often happens in LA urban water services.

In practice, the efficient and accurate administration of economic resources is difficult, as dealt with extensively in the OECD (2012) report, although this will be not addressed here. Up to seven key coordination gaps are identified for better water governability: administration, information, policy, capacity, financing, objectives and accounting. Economic resources are needed and they should be wisely managed.

Water service economic unsustainability refers not only to the lack of funds needed to improve the facilities and systems of the fast growing cities – they may come from international organizations in the case of poor countries – but also to the more complex problem of obtaining the funds needed for the correct functioning of the existing, expensive infrastructures. These funds ought to be secured from the water services users themselves. This opens two permanently debated points: 1) tariffs that allow for cost recovery, which is a complex matter in countries with deep social inequalities, and 2) corruption, which is especially concerning in LA. In poorly consolidated democracies corruption is a rather frequent temptation in front of the increasing economic resources being used and attracted, as shown in an OECD website[2] on this topic.

8.3.4 Ageing of existing water infrastructures

The ageing of water infrastructures is a very serious problem that passes almost unnoticed due to the more conspicuous problems commented above. This problem even affects the USA, the richest of all the American countries, as underlined by the EPA (Environmental Protection Agency). In 2007 the EPA estimated that 334,800 million US\$ has to be invested during twenty years in order to modernize drinking water supplies (EPA, 2009). In other recent studies it is estimated that about 1,600,000km of drinking water pipes have to be modernized in the USA, which means about 3000 billion US\$ over the coming twenty-five years (AWWA, 2012). Figure 8.1 depicts the magnitude of this great problem by showing the evolution of the median age of the infrastructures and the median age of the population in the USA.

In LA this problem is even more serious. This is a direct consequence of the economic unsustainability mentioned previously, worsened by the fact that most new investments are for new infrastructures to progressively extend water services and approach the MDGs, while existing infrastructures are poorly maintained. Some of them are many decades old. All of them have an expiry date. Maintenance, rehabilitation or replacement failures will have serious consequences in the long term. In some way they are the main cause of the deficient quality of supplied services (Regional Process of the Americas, 2012).

---

2 www.oecd.org/corruption/latinamerica

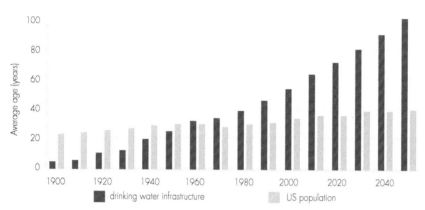

Figure 8.1 Median age of population and of water pipes in the USA. *Source: Buchberger (2011)*

8.3.5 Poor quality of water services

This is the natural consequence of economic unsustainability, caused by poor maintenance and the lack of infrastructures renewal. The Regional Process of the Americas (2012) report explicitly highlights the most relevant deficiencies to be insufficient water disinfection, poor surveillance of water abstractions, discontinuous service, insufficient pressure, high leakage percentage and the low wastewater treatment degree. These are big challenges.

The problems identified by the Regional Process of the Americas (2012) report can be grouped into water quality (treatment and monitoring) and distribution (pressure, leakage, continuity of service). Something similar has been identified by Aqua Rating, a project sponsored by the IDB with the help of the IWA (International Water Association). The project aims at developing a system to qualify water and sanitation suppliers (Krause et al., 2012). For water service quality the project considers drinking water quality, water distribution for use and consumption, wastewater collection, and care of the users (this last point being more commercial than structural).

8.4 The enormous cost of poor water service quality

In any country urban water services should and can be economically, environmentally and socially sustainable. In spite of the big financial resources required, scale economies in cities allow to provide these water services at a reasonable cost, which is more difficult in rural areas (Pearce-Oroz, 2011). This is not discussed here. Sustainability demands governance and long-term vision, which is the bottleneck of implementing these systems. The other aspects are easier to solve. The fact that providing a low-quality service is more costly to citizens than attaining the adequate standards needs to be clearly shown and argued. This can be easily understood from a social point of view by showing the savings that would be obtained if the individual no longer had to support the failings of the poor-quality service. However, the supplier will always look to their own short-term economy if not pushed to do otherwise by well-informed and organized citizens.

Each of the six deficiencies described below is accompanied by the associated costs, which are seldom considered and often transferred as a burden for future generations. However when all these costs are included, good management of the water service is much cheaper. The problem comes down to the need to involve society much more in the water services through appropriate institutions.

8.4.1 Insufficient disinfection and drinkability

According to the World Health Organization (OMS, 2002), inadequate water and sanitation are the main causes of illness, such as malaria, cholera, dysentery, schistosomiasis, infectious hepatitis and diarrhoea, which are related to 3,400 million deaths in the world and LA is not an exception. Inadequate water and sanitation are also a main cause of poverty and of the growing gap between the rich and the poor. Figure 8.2 shows the close link between child mortality and access to improved water and sanitation (Robinson et al., 2006).

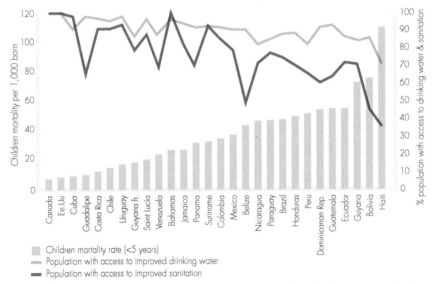

Figure 8.2 Access to water and sanitation (%) and child mortality (deaths per 1,000 born) in different American Countries. *Source: Robinson et al. (2006)*

8.4.2 Lack of quality control

There is a long path through pipes from the drinking water plant to the citizen's tap. The pipes may be sometimes over one hundred years old. Even if at the inflow point to the system water quality is guaranteed, along the distribution path it deteriorates, especially when the residence time in the distribution network can be up to several days. A control of water at the outflow of the drinking water plant is needed but is not sufficient as the water quality will deteriorate, the older the network is.

Much attention has been paid to this aspect in the last two decades in developed countries. One example is the EPANET programme for analysis of water networks, sponsored by the US Environmental Protection Agency (Rossman, 2000). It is a free access programme aimed at facilitating the study of water quality evolution through pipes.

Urban water quality is of great concern in all developed countries (Kastl et al., 2003; MHLS, 2010) due to trihalomethane generation, which is a carcinogenic compound resulting from the progressive interaction of organic matter in water – mostly that from surface resources – and chlorine (Aieta and Berg, 1986). This is the reason why in some countries, like The Netherlands and many areas in Germany, the use of chlorine as a disinfectant has been abandoned and substituted by other disinfection processes, which do not impair the taste, but they are often more expensive. As a result, these countries have drastically reduced the consumption of bottled water (den Blanken, 2009).

8.4.3 High percentage of losses

Unreliable or inadequate pipes for the soils in which they are laid, careless assembly, lack of maintenance and scarce – sometimes, inexistent – network rehabilitation or replacement are the main factors that cause the water distribution system deterioration. Water supply reliability depends on pipe proofing, which decreases progressively over time.

Network water losses in LA are often very high. According to ADERASA (Association of Regulating Entities for Drinking Water and Sanitation of the Americas) the average value is 42% (Figure 8.3), and this value is possibly quite conservative as it refers to a set of the best managed towns in the area (ADERASA, 2012). According to the Cooperativa Andina de Fomento (Andean Cooperative for Enterprenery, CAF, 2011) losses are 40%, which is also quite optimistic as it is estimated from data contributed by the own water distribution utilities. In a later publication (CAF, 2012) the fraction of unaccounted water (real leakage, measurement errors and thefts) had risen to 50%. Consequently, it seems that in LA about half of outflow water from drinking water plants does not get to the users, which is economically, environmentally and socially unsustainable, even if the water resource is not entirely lost for later use.

The consequences of high network losses are well known. The most relevant are:

- Existing infrastructures (pumps, drinking water plants, reservoirs, networks) become insufficient to secure the water supply and they have to be enlarged. This means doubling current capacity in LA.
- High probability of pathogenic intrusion into the water network. If water pressure drops below atmospheric pressure in a network – something that happens often when water service is not continuous – leaked water may get back into the system (Kirmenyer and Mantel, 2001). This poses a more serious problem than quality deterioration during distribution.
- Water supply may be intermittent in dry periods, when less water is available. This has a series of other drawbacks that will be commented upon below. Should the water losses be small, there is no need to interrupt water supply. There are other more reasonable methods to save water without recourse to temporal interruption of the supply (Cabrera, 2007).
- Water users' poor trust in the water utility, thus reducing willingness to pay.

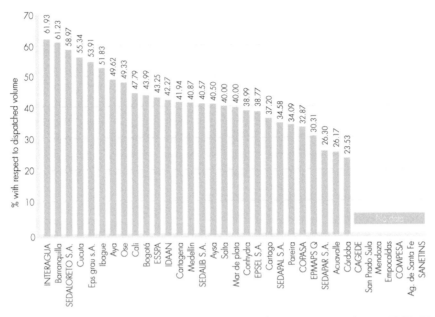

Figure 8.3 Water network losses in representative urban areas in LA. *Source: ADERASA (2012)*

8.4.4 Low water supply pressure and domestic storage tanks

Leakages aside, a network becomes insufficient if it cannot supply the demanded flow at water demand peaks at the established pressure – at least approximately 20m of water head (two atmospheres) – when water demand peaks. Problems appear when the network does not improve its capability at the same rate as population growth. Often pipe diameters are those required by the city before a significant increase of connected people. In order to solve supply problems the manager of an insufficient network forces or recommends the construction of domestic water storage tanks and butts, to decouple water demand from users from actual supplied water. They store water when demand is low (e.g., from 0:00 to 6:00 am) and release it in peak hours. The method has clear disadvantages:

- Water quality can be significantly affected due to storage time, mostly because storage time is unknown, but also to the high possibility of a pathogenic intrusion as pressure – the guarantee against intrusion – is lost. Besides, tanks and cisterns are barely protected and cleaned; this contributes to a worsened water quality guarantee, as shown in the experimental studies carried out in Tiquipaya, Bolivia (Schafer and Mihelcic, 2012).
- Energy is wasted since water is depressurized to later be pressurized again for the users.
- From a more technical point of view, water demand timing is quite distorted and the system loses a great deal of its characteristics, thus making it more difficult to conduct mathematical simulation and leakage surveys.

Temporal water service interruption shows that the network is clearly inadequate to supply a growing population. Looking globally at the problem, this implies a very important economic loss (Cabrera et al., 2013): the money the system manager saves by not investing is much less that the set of expenses transferred to all connected customers.

8.4.5 Intermittent service

Often inefficiency – high leakage – and insufficiency – lack of pressure – go together and then the water supply manager has no other recourse than intermittent service by interrupting water supply periodically. This is frequent in LA, where it has been observed in almost 30% of the systems (CAF, 2012). The most serious drawbacks of water quality loss and the causes of the system becoming no more secure are well documented (Yepes et al., 2001; Totsuka et al., 2004). One of the most serious problems is faster network deterioration as a result of alternatively pressurizing and depressurizing the network, which leads to increasingly more frequent and lasting breakdowns and water losing points. According to Charalambous (2011), the number of pipe breaks each year triples with respect to a continuous water supply.

8.4.6 Lack of wastewater treatment

In LA most urban and industrial water discharges lack treatment, and this has important consequences due to the serious sanitary, environmental and the not always fully recognized economic consequences. If the town pollutes the near-by waters, it has to look for more distant water supply sources and consequently more expensive ones. Looking for water sources further afield is not a new phenomenon: in 1890 in order to supply Paris, it was proposed that water be taken from Lake Leman, 450km away (Barraqué, 2004). What at that time was a fashion, declined by end of the 19th century and beginning of the 20th century after the discoveries of Pasteur and Koch that lead to disinfection thus allowing the use of local water. However, in some of America's large cities, importing water seems a real necessity, as is the case of Mexico City (Perlo and González, 2009), and probably of Lima.

8.4.7 Conclusion on poor water quality supply

From any point of view, making use of economies of scale is the most logical and feasible solution to provide a good quality urban water service to the citizen at the lower possible cost when all involved costs (e.g. the bottled water to substitute for tap water, see Box 8.8) and a mid- and long-term economic balance are considered (Cabrera et al., 2013). However, this needs a good administrative structure and political support, in addition to a large capacity for leadership amongst decision makers. Currently, most or all urban water cycles in LA are unsustainable.

Box 8.8 Bottled water and soft drinks for domestic consumption

In many LA countries the production of bottled water and soft drinks is not only quite developed but has a special economic and social relevance, with the trade mostly in the hands of a few large firms. Mexico is a leading country. One of the main drivers seems to be the once unreliability and poor quality of in-house water supply, without guarantee of being pathogen free, or too saline, or containing an excess of toxic natural solutes for sustained consumption, such as fluorine or arsenic. This is the case in many rural and small urban areas of Mexico, Argentina and Paraguay, most of them using groundwater without adequate hydrogeological studies. Large efforts are devoted to better understand the origin of these natural pollutants and on practical means to reduce economically their content. Interruptions in domestic supply in the past, as explained in the main text, can be considered the main cause of the widespread use of small domestic storage tanks on the roofs of the houses, as can be seen in many of the low-rise residential areas of several towns in LA. Some water-related diseases are not uncommon, often with mild results to locals, not always to travellers; there is a real risk of epidemic spreading. Another important driver of bottled water and beverages consumption is the increasing living standards. Drinking bottled products is often considered a sign of affluence, which becomes a social issue and a display of personal status. This is widely propagated by effective advertising by large companies and good distribution logistics. Even though this represents a large part of income to individuals and to society, it is difficult to be reversed.

Production of bottled water and drinks uses about two to three times the water in the final product and about 25% of its volume as fuel consumption. These amounts include the water used to produce the bottles and the fuel to distribute them which has to be added as well as that needed to produce the additives. Furthermore used plastic bottles is becoming a serious environmental problem.

Bottled water price in the market ranges from 500 to 5,000 times higher than the average tap water price. It is one of the main hidden personal, social and environmental costs of a deficient water supply.

8.5 Causes of the current water services situation

After describing the main problems of urban water services and the associated serious consequences, the present situation is taken into account in order to define measures aimed at improving it. This is a complex task. Economic and social–political causes have been described in detail by some organizations (CAF, 2011; Regional Process of the Americas, 2012; OECD, 2012) as well as by some authors that have a great wealth of knowledge about LA (Jouravlev, 2004; Hantke-Domes and Jouravlev, 2011), and will not be repeated here. As an overview, key phrases from three of these works that explain current situation are given opposite.

In the report 'Water Governability in Latin America and the Caribbean: A multilevel point of view', the OECD (2012) concludes that [translation from Spanish]:

Key challenges are institutional and territorial fragmentation and badly managed multi-level governance, as well as limited capacity at the local level, unclear allocation of roles and responsibilities and questionable resource allocation. Insufficient means for measuring performance have also contributed to weak accountability and transparency. These obstacles are often rooted in misaligned objectives and poor management of interactions between stakeholders. (p.15)

The report on Americas' Water Agenda (Regional Process of the Americas, 2012) provides a similar diagnosis:

It is important to emphasize that in contrast to the institutional strength and stability in Canada and the USA, the social perception studies on the role of public institutions in Latin America and the Caribbean (LAC) demonstrate low credibility. Different factors not always attributable to the institutions have had an influence on this: the magnitude of the challenges faced, institutional weakness, the scarcity of economic resources, preconceived judgments and ideological notions with respect to the role of the State, the regulation and participation of the private sector, the weakness of civil society organizations, the perception of seizure of the institutions by interested sectors and the problems associated with globalization.(p.6)

Some years before, the same problems were pointed out in Jouravlev (2004) on the status of drinking water and sanitation services on the 21st century [translation from Spanish]:

Despite the differences to be expected in a region that is home to many different countries, reforms have many common features, such as; institutional separation between the roles of sectorial policymaking, economic regulation and the management of the systems; the deepening and consolidation of decentralization in the provision of services; the general interest in promoting private participation, the development of new regulatory frameworks, and the demand, born out of the 1980s crisis, that services should aim to be self-financed, and when in place, subsidy plans should be set up to help low-income groups.

These coinciding diagnoses can be summarized as weak governance and a lack of leadership. In short a public urban water service for a big city is a project common to all its inhabitants and can only be carried out successfully with sound governance and leadership. Given that water and the nature of human beings are the same regardless of country boundaries, the differences between national water policies depend on the strength of their governing institutions and political structures as well as on other cultural and economic factors. From a strictly technical point of view the problem is not unduly difficult in the big cities, however, the story is different in rural areas where a priori scale savings are not possible to make water supply economically feasible.

In order to easily overcome these handicaps, some existing problems have to be solved, the difficulty of which will vary according to the country, although some already have plans in place. Even when plans exist, it helps in taking into account these handicaps, which are reviewed over.

8.5.1 Lack of professional capacity of political decision makers

Decision makers, besides applying common sense and being honest, must give priority to sound solutions for citizens and especially for future generations. They ought not to be guided by short-term goals and self-interest, but will require training and good foundations on which to build. With this in mind they will boldly identify decisions for the future, abandoning the short-term visions that have become commonplace, and will strive for mid- and long-term solutions.

8.5.2 Lack of training for managers and engineers

In the urban water cycle three different decision levels exist: political, managerial and mechanical. These last two refer respectively to those controlling finances and human resources and those carrying out technical decisions. However, current training opportunities for staff are not sufficient to deal with the complexity of urban water supply. Since economic resources are often lacking, these managers and engineers with little training have to identify optimal cost-benefit solutions. This is highlighted in the OECD (2012) report, in which it is said that in two-thirds of the LAC countries studied, the capacity gap is a big handicap to the effective implementation of water policies.

8.5.3 Lack of environmental knowledge amongst citizens

State financing of large water infrastructures and the principle that water is a public good have fostered the belief amongst citizens that they have a universal right to good-quality tap water at little or no expense. Water as such has only an environmental and opportunity cost, but the infrastructures needed to carry the water to the citizen has high economic costs which must be paid by water users. Without this, recurrences continue to happen: a progressive deterioration of the infrastructure that benefits nobody. Another aspect to be considered is that the tariffs ought to be progressive to help the economically weakest gain access to water they need with any loss in water quality.

The lack of public knowledge about real water costs makes the need to increase water tariffs a poorly understood issue. This is regrettably part of a political debate when it should be a 'state' affair given that the survival of the city is at stake. The average citizen is intelligent and, if given the right information, is likely to be able to see the difference between opportunistic and self-interest policies, and those which are honestly beneficial to the population. This is of crucial importance in order to separate water from the political arena as much as possible.

8.5.4 Allocating responsibility

Fragmented institutions, with diverse points of view of the same issue, make for difficult decision making. This problem is highlighted in all reports where water problems are analysed in any detail. As water is such a key issue, all politicians try to include it in their agendas, or avoid it as much as possible in the case of serious conflict. Drinking water's increasing complexity has favoured an increase in the number of government departments involved in managing the issue: water, environment, health, industry, mining, etc. This makes for cumbersome and bureaucratic governance. More nimble and operative structures are needed in order to solve the plethora of water problems.

Involving users in water affairs may be an interesting approach. An approach is that adopted in Buenos Aires, where AySA has had considerable success doing specific work with lower-income sectors, through the methods of 'participatory models of governance' (MPG) and the 'Water+Work Plan' (Plan agua más trabajo), and to a lesser extent to 'Sanitation+Work Plan', which is reflected by joint action with neighbourhood communities, municipalities, government agencies, and social organizations. The company performs most of the financing of projects and provides technical supervision and neighbours help with labour, receive technical training, participate in workshops on proper use, and receive a discount on their bills (Lopardo and Lentini, 2010).

8.5.5 Lack of water service standards

In order to have a good-quality water service, objectives are needed and have to be made explicit. Management indicators (Alegre et al., 2006) allow them to be set with relative accuracy. For example, when defining the minimum water pressure that has to be guaranteed at demand peak, one must take into account the acceptable percentage of losses. This is needed to evaluate the present situation and set future targets. Independently of objectives set by financing organizations (Krause et al., 2012), which are establishing water service quality evaluation systems, the corresponding country authority (or regulator if it exists) should be the one to set the detailed objectives.

8.5.6 Unclear pay rules when the service is externalized

The public–private urban water management debate is intrinsic to the service and continues to cause debates. In 1875, the city of Birmingham (UK) bought the service from the private enterprises that had started them, arguing that 'the quantity and quality of water to be supplied to the public are matters of greater importance than making profits and should be controlled and managed by representatives of the people and not by private speculators' (Thackray, 1990). Shortly afterwards the story was repeated. In 1898 the city of Amsterdam bought the private company because of problems with financing the necessary extensions to the network arguing that the private company was more interested in making profits than in supply' (Swemmer, 1990). After more than a full century the debate continues, with those who defend public involvement bringing forward similar arguments.

Circumstances in LA have made the above commented issue a conflictive one (Castro, 2007; Ducci, 2007; ISF, 2008). At the same time that public management is preferred, there is also a need to attract and obtain private investment. This explains why prestigious institutions revisited the problem and proposed guidelines for developing reasonable agreements (Solanes and Jouravlev, 2007; OECD, 2009). This will not be elaborated upon further here, but will undoubtedly continue to be debated in all countries for many years to come (Jones et al., 2004; Boland, 2007).

8.5.7 Political prices and criteria

The lack of clear criteria used to establish the final price to be paid by citizens for the water service is probably the largest Achilles' heel of the urban water cycle. Since not all costs are recovered, a large part of the infrastructure has to be financed with public funds of diverse origins or with subsidies. Since this means large money sums passing through a few hands it is not rare that corruption occurs (Solanes and Jouravlev, 2005). Thus, clear criteria for self-financing are needed, in such a way that the sustainability of these water services is guaranteed by honouring the cost recovery principle as promoted by the Water Framework Directive in the European Union (OJEU, 2000), and also with an equitable tariff system to protect the weakest, without compromising efficiency. This equity is crucial given the large inequalities in LA.

8.5.8 Poor transparency in urban water management

Since water is a public good that is essential for life, it has to be flawlessly managed. There is an ethical obligation for transparent management, and this is even more serious for water supply. Countries have to establish the means needed to accomplish this goal. It would be recommendable that all money coming from urban water should be invested in water, and not be used for other needs, as often happens due to biased political interference. An efficient regulation service is one of the best ways in which to accomplish this goal, much as efficiency is a key strategy in order to reach a more sustainable future (BNA, 2012)

8.6 Conclusions

Providing a quality drinking water service, adequate sanitation and correct treatment of wastewater is a very complex problem, especially in LA where the population, especially in urban areas, has expanded rapidly. It is even more difficult considering that the current situation is far from desirable. The root of the problem is neither economic (scale economies allow for the provision of good quality and sustainable water services at a reasonable cost) nor technical (current engineering can deal with the most complex problems). The handicaps to be overcome are a severe lack of governance and of institutional leadership which are currently unable to deal with the rapid evolution. Since the path to be followed is well known, it is hoped that improvement of the current situation and avoidance of the serious consequences of not doing anything could be seen in the near future. LA citizens deserve the best quality service at the lowest possible cost though the differing characteristics of each country must be taken into account before any plan is implemented.

Acknowledgements

Special thanks to Andrés Benton for his help obtaining information on Mexico City. Complementary data are found in other chapters of this book and in the supporting reports contributed by the different country teams.

References

ADERASA (2012). Asociación de Entes Reguladores de Agua y Saneamiento de América. *Informe anual 2012.* Asunción, Paraguay: Grupo de Trabajo Regional de Benchmarking.

Aieta E.M. & Berg J.D. (1986). A review of chlorine dioxide in drinking water treatment. *Journal American Water Works Association, 78* (6): 62–72.

Alegre, H., Baptista, J.M., Cabrera Jr., E., Cubillo, F., Duarte, P., Hirner, W., Merkel, W. & Parena, R. (2006). *Performance Indicators for Water Supply Services.* Second edition. London, IWA Publishing.

AWWA (2012). *American Water Works Association Buried no Longer: Confronting America's water infrastructure challenge.* Denver, Colorado.

Barraqué, B. (2004). Water policies in Europe after the Framework Directive. In: Cabrera, E. (ed.). *Challenges of the New Water Policies for the XXI Century.* pp. 39–50. Balkema.

BNA (2012). Business News Americas. *Efficient Water Use in Latin America: Good business for all.* Infraestructure Intelligence Series.

Boland, J.J. (2007). The business of water. *Journal of Water Resources Planning and Management.* ASCE, May–June: 189–191.

Buchberger, S. (2011). *Research & Education at the Nexus of Energy and Water.* Environmental Engineering Program, University of Cincinnati.

Cabrera, E. (2007). Los cortes de agua y el racionamiento racional en épocas de escasez. In: Cabrera, E & Babiano, L. (eds). *'la sequía en España. Directrices para Minimizar su Impacto'.* pp.212–231, Ministerio de Medio Ambiente de España.

Cabrera, E., Cobacho R., Espert V. & Soriano J. (2013). *Assessing the full economic impact of domestic water tanks.* International Conference on Asset Management for Enhancing Energy Efficiency in Water and Wastewater Systems. Marbella, Spain, IWA.

Cabrera, E. & Custodio, E., (2013). Urban water, an essential part of integrated water resources. In: Martínez-Santos, P., Aldaya, M.M. & Llamas, M.R. (eds). *Integrated Water Resources Management in the 21st Century: Revisting the paradigm.* Boca Ratón, FL, CRC-Press. Forthcoming.

CAF (2011). Cooperación Andina de Fomento. *La infraestructura en el desarrollo integral de América Latina. Diagnóstico estratégico y propuestas para una agenda prioritaria.* XXI Cumbre Iberoamericana de Jefes de Estado.

CAF (2012). Cooperación Andina de Fomento. *Agua potable y saneamiento en América Latina y el Caribe: metas realistas y soluciones sostenibles.* Propuestas para el 6° Foro Mundial del Agua.

Castro, J.E. (2007). *La privatización de los servicios de agua y saneamiento en América Latina.* Buenos Aires, Nueva Sociedad.

Charalambous, B. (2011). The hidden costs of resorting to intermittent supplies. *Water* 21: 29–30.

CNA/WWC (2006). Comisión Nacional del Agua and World Water Council. *Water Problems in Latin America.* Mexico, 4th World Water Forum.

den Blanken, M. (2009). *Asset Management. A necessary tool for a modern water company.* Miami, LESAM. Leading-Edge Conference on Strategic Asset Management.

Ducci, J. (2007). *Salida de operadores internacionales de agua en América Latina.* Banco Interamericano de Desarrollo.

EPA (2009). Environmental Protection Agency. *Drinking Water Infrastructure Needs Survey and Assessment.* Washington DC, Office of Ground Water and Drinking Water. Fourth Report to Congress. EPA 816-R-09-001.

GTAS (2003). Grupo de Trabajo sobre Agua y Saneamiento. *Cómo lograr los objetivos de desarrollo del milenio con respecto al agua y saneamiento: ¿Qué será necesario?* Proyecto Milenio. New York, UN.

Hantke-Domas, M. & Jouravlev, A. (2011). *Lineamientos de política pública para el sector de agua potable y saneamiento.* Santiago de Chile, Chile, CEPAL.

ISF (2008). Asociación Catalana de Ingeniería sin Fronteras. *Con el agua al cuello. Ascenso y caída de las multinacionales en Sudamérica: El proceso de mercantilización de la gestión del agua. El caso de la inversión española en el Cono Sur.* Barcelona.

Jones, S.A., Houghtalen, R.J., Jones, D.T. & Niblick, B. (2004). Privatization of municipal services: Sustainability issues of production and provision. *Journal of Infrastructure Systems ASCE:* 139–148.

Jouravlev, A. (2004). *Los servicios de agua potable y saneamiento en el umbral del siglo XXI.* Santiago de Chile, Chile, CEPAL. Recursos Naturales e Infraestructura número 74.

Kastl, G., Fisher, I., Sathasivan, A., Chen, P. & van Leeuwen, J. (2003). *Modelling water quality from source water to tap by integrated process models.* Townsville CD-ROM: 1799–1804: MODSIM, Integrative Modelling of Biophysical, Social and Economic Systems for Resource Management Solutions.

Kirmeyer, G.J. & Martel, K. (2001). *Pathogen Intrusion into the Distribution System.* Denver, Colorado, American Water Works Association.

Krause, M., Cabrera, E. Jr. & Cubillo, F. (2012). AquaRating: An innovative system for assessing utility practice and performance. *Water* 21: 65–66.

Lopardo, R.A. & Lentini, E. (2010). *Supply and sanitation: serving the urban unserved in Latin America, with a special focus on Argentina.* Presented in the Seventh Biennial Rosenberg International Forum on Water Policy. Buenos Aires. [Online] Available from: http://rosenberg.ucanr.org/documents/argentina/Lopardo%20Final%20Paper.pdf [Accessed September, 2013].

McGuire, M.J. (2013). *The Chlorine Revolution: water disinfection and the fight to save lives.* Denver. American Water Works Association.

MHLS (2010). Ministry of Healthy Living and Sport. *Comprehensive Drinking Water Source-to-tap Assessment Guideline.* Victoria, British Columbia, Canada.

OECD (2009). Organisation for Economic Cooperation and Development. *Private Sector Participation in Water Infrastructure.* Checklist for Public Action. Paris.

OECD (2012). Organisation for Economic Cooperation and Development *Gobernabilidad del Agua en América Latina y el Caribe: Un enfoque multinivel.* Paris.

OJEU (2000). Official Journal of the European Union. Directive 2000/60/EC of the European Parliament and of the Council of 23 October 2000 establishing a framework for Community action in the field of water policy.

OMS (2002). *Organización Mundial de la Salud. Agua para la salud: un derecho humano:* Centro de Prensa Organización Mundial de la Salud. [Online] Available from: www.who.int/mediacentre/news/releases/pr91/es/index.html. [Accessed May, 2013].

ONU (2011). Organización de las Naciones Unidas. *Millennium development goals: 2011 progress chart.* [Online] Available from: www.un.org/milleniumgoals/pdf(2011)¬MDReport201. [Accessed May, 2013].

Pearce-Oroz, G. (2011). *Los desafíos del agua y saneamiento rural en América Latina para la próxima década.* Washington, World Bank.

Perlo, M. & Gonzalez, A. (2009). *¿Guerra por el agua en el Valle de México.* 2° Edición. Mexico DF, Mexico, Universidad Nacional Autónoma de México.

Regional Process of the America (2012). *Americas' water agenda: targets, solutions and the paths to improving water resources management.* February 2012. Available at: www.gwp.org/Global/GWP-CAm_Files/Americas'%20Water%20Agenda.pdf

Robinson, K., Infante, R. & Trelles J. (2006). *Agua, saneamiento, salud y desarrollo. Una visión desde América Latina y el Caribe.* Documento Preparatorio del IV Foro Mundial del Agua. Mexico.

Rossman, L. A. (2000). EPANET 2: *User's Manual.* Cincinnati, Environmental Protection Agency.

SACM (2012). Sistema de Aguas de la Ciudad de Mexico. *El gran reto del agua en la ciudad de México. Pasado, presente y prospectivas de solución para una de las ciudades más complejas del mundo.* [Online] Available from: issuu.com/politicaspublicas/docs/aguadf [Accessed July, 2013].

Schafer, C.A. & Mihelcic, J.R. (2012). Effect of storage tank material and maintenance on household water quality. *Journal American Water Works Association,* 104 (9): E521–E529.

Solanes, M. & Jouravlev, A. (2005). *Integrando economía, legislación y administración en la gestión del agua y sus servicios en América Latina y el Caribe.* Santiago de Chile, Chile, CEPAL.

Solanes, M. & Jouravlev, A. (2007). *Revisiting Privatization, Foreign Investment, International Arbitration and Water.* Santiago de Chile, CEPAL.

Swemmer, F.F. (1990). Water supply and water resources management. In: Schilling, K.E. & Porter, E. (eds). *Urban Water Infrastructures.* Kluwer Academic Publishers. pp. 173–188.

Thackray, J.E. (1990). Privatization of water services in the United Kingdom In: Schilling, K.E. & Porter, E. (eds). *Urban Water Infraestructures.* Dordrecht, The Netherlands, Kluwer Academic Publishers. pp. 33-41.

Totsuka, N., Trifunovic, N. & Vairavamoorthy, K. (2004). *Intermittent Urban Water Supply under Water Starving Situations.* Proc.30th WEDC International Conference. pp. 505–512. Vientiane, Lao.

UNICEF/OMS (2012). *Progresos en materia de saneamiento y agua potable. Informe de actualización 2012.* Programa Conjunto UNICEF/WHO de Monitoreo del Abastecimiento de Agua y Saneamiento.

Yepes, G., Ringskog, K. & Sarkar, S. (2001). The high cost of intermittent water supplies. *Journal of Indian Water Works Association,* 33(2).

WATER, ENERGY, BIOENERGY, INDUSTRY AND MINING

Authors:
Emilio Custodio, Dept. Geo-Engineering, Universitat Politècnica de Catalunya (UPC), Barcelona, Spain
Alberto Garrido, Water Observatory – Botín Foundation, and CEIGRAM, Technical University of Madrid, Spain

Highlights

- Hydropower is the main energy source in the Latin American (LA) region as a whole (52%), although not in all countries and its relative weight has decreased. It still has growth potential, but new projects will face growing physical, economic and social barriers, including environmental restrictions, the rights of native and local inhabitants, and biodiversity conservation. In some locations hydropower may reduce water availability and security for other uses.

- Fuel energy production may compete for a large part of available water resources in some of the highly populated and semi-arid and arid areas, even when they use closed-cycle water cooling or are placed at coastal areas. Nuclear energy and other forms of energy production are less developed and their effect on water resources is local. Some interesting geothermal energy production exists.

- Mining and related industrial sectors stand amongst the fastest-growing industries in the region. They may be water resource intensive and consequently affect water availability and security when located close to urban areas or in arid areas. So their water needs and consumption, together with their wastewater and mining residues, are becoming stressful factors and a significant source of pressures in numerous basins in the region. In some cases non-renewable groundwater reserves are consumed. In some areas artisanal and small mining activities cause some serious pollution problems to downstream water resources, as is the case of gold mining.

- Industry, energy production and mining together consume 8 to 15% of water resources in the seven considered LA countries.[1] Water consumption relative to available water resources is respectively 4 to 9% for industry, 2 to 5% for energy production (hydroelectric energy water consumption not included) and up to 6% for mining. Water economic productivity in Chile may range from 3 to 10US$/m^3.

- Crops grown for biofuels are increasingly becoming major export products, with a large production in Brazil, and Argentine being now the second largest exporter. Many countries in LA are promoting the cultivation of crops for biodiesel and bio-alcohol. Water resources and land used for these crops compete with land for food production. Especially concerning are projects in dry areas where intensive irrigation is needed.

---

1 This chapter focuses mainly on Argentina, Brazil, Chile, Colombia, Costa Rica, Mexico and Peru.

9.1 Introduction

This chapter focuses on two topics that are usually considered separately. First, the amount of water required by energy-producing plants. Second, two economic sectors – mining and industry – are also water-demanding and are experiencing tremendous growth in the Latin American (LA) region. A detailed treatment of both would request at least a chapter for each of these topics, but being the focus of the volume water and food security, a review of data and their analysis from this narrower perspective has been chosen.

For the most part this chapter refers to continental LA, and specifically to the seven Ibero-American countries (Argentina, Brazil, Chile, Colombia, Costa Rica, Mexico and Peru) that have provided data from their country reports.[2] However, the discussion goes beyond political boundaries in order to consider different geographical and climatic areas. Except for the large islands of Cuba, Hispaniola (the Dominican Republic and Haiti) and Puerto Rico, the Caribbean area consists of small islands with specific characteristics that introduce quite different circumstances for water resource security, which in some cases involve seawater desalination as an important complementary source. For this reason, they will not be included in what follows. Unless otherwise indicated, in this chapter only mobile water resources (blue water) are considered. Water use refers to water supplied to the activity and water consumption is the part of water used that is not available afterwards due to evaporation, impaired quality or disposal into the sea or a water body that has no further possible use downstream.

Water is needed for many human activities beyond drinking purposes, urban services and the production of food and fibres. It is also used for industrial processes – including food processing and fuel production – and for mining. Additionally, water resources are used for energy production and consumed whilst energy is also needed to make water available for use and for the treatment and safe disposal into the environment thereafter. There is competition for using and securing water resources between all these demands. The important related topic of water security in urban areas is considered in Chapter 8. Except for agriculture, bio-fuel production – a particular form of water intensive agriculture – and hydroelectricity – which uses large water flows but consumes a small part – the other activities often demand a moderate fraction of LA countries' water resources.

Although water consumption for energy, industry and mining may only be a small percentage of the countries consumption, it could be locally significant, especially in small basins and in the arid and hyper-arid areas of LA. This consumption may also be economically and socially important, and therefore water quantity and quality should be guaranteed. Industrial and mining activities may generate wastewater and by-products that could have a large negative impact on water resources quality and on the environment. Circumstances are quite different from one country to another and even inside a given country, as shown along this book, so generalizations may be meaningless. Thus, country

---

2 These country reports include data and analyses carried out by the project partners. A summary of the consortium and specific represence is include in the volume's foreword and introductory section.

comments give only a coarse and blurred picture – sometimes too biased – that has to be afterwards considered in more detail taking into account actual territorial circumstances and local situations.

Data and analysis provided here serve to give a general overview of the situation in the first decade of the 21st century. Most LA countries are under fast development and, in spite of fluctuations and some political instability in some of them, conditions have and will continue to improve. The associated industrial and mining development – partly fostered by current high prices for minerals in many of the LA countries – contributes to increased water resources use and consumption and energy demand but also to a more efficient use in economic terms. Estimates of future evolution trends can be found in WEC (2010) and Jiménez-Cisneros and Galizia-Tundisi (2012).

This chapter does not try to present a detailed bibliographical review of energy – including bio-energy, industry, mining activities and water use but instead to contribute to an evaluation of how the associated water needs may affect water and food security in LA. This is done by examining existing data in the reports of the partner countries amongst other sources of information and personal experience. Partners' reports are cited as LA-country (2012) except when specific data are attributed to particular authors (see footnote 2). A general overview is first presented, followed by specific comments on water use and consumption in energy production, industry and mining, which leads to some considerations on water security from the point of view of the activity undertaken and subsequently of general human water needs.

9.2 General overview

Table 9.1 shows some basic data on water use for energy, industry and mining. Hydroelectricity has a large water use (demand) but only a small fraction is consumed. While traditional mining may demand rather large water quantities that are disposed of, modern mining is predominantly a water consumer since internal recycling is important. However, the fraction of water that is disposed of may be unusable and, what is more, may impair water quality downstream, which is equivalent to additional water consumption.

The economic productivity of water is an important driver for water rights acquisition in areas where the resource is scarce, but there are other factors to be considered such as legal restrictions, existing rights – local customs and rights may already be in place amongst natives – and social pressure.

The average economic productivity of water use has been estimated for Chile (Table 9.2) and is markedly higher in industry and the mining sector, e.g. for copper production the value may exceed $50US\$/m^3$. This in turn explains the pressure they impose in order to get water resources when they are scarce, as is the case in the northern part of the country. This may also account for the fact that the companies involved have a greater capacity to purchase water rights in order to secure their supply (see Chapter 13). The very low productivity of water for energy production derives from the fact that in Chile a large fraction of energy production is hydroelectric power that uses large water flows but only consumes a small fraction. If actual water consumption is considered, which is highly variable from one plant to another, the economic productivity of water becomes much higher. Besides, many hydroelectric plants use water resources in remote areas, where no other significant productive uses exist except forestry, fishing and landscaping. Associated externalities are a cost that is often not accounted for. Jiménez-Cisneros and Galizia-Tundisi (2012) gives an average water use productivity in Mexico of 78US$/m^3, varying from 1.3US$/m^3 in agricultural states to 100US$/m^3 in the Federal District, where urban and industrial uses dominate.

Table 9.1 General data for the first decade of the 21st century (values rounded up)

	SURFACE AREA 1000km²	POPULATION Million	USABLE WATER km³/yr	USABLE WATER mm/yr	USABLE WATER % GW	WATER USED km³/yr	WATER USED mm/yr	WATER CONSUMED km³/yr	WATER CONSUMED mm/yr	WATER CONSUMED m³/cap./year	% OF WATER CONSUMED FOR ENERGY [3]	% OF WATER CONSUMED FOR INDUSTRY	% OF WATER CONSUMED FOR MINING	% OF WATER CONSUMED FOR TOTAL
MEXICO	1,973	117	550	279	29	—	—	80	40	683	5	4	0.07	9
COSTA RICA	51	4.7	110	2,157	30	25	490	0.5	10	106	—	—	—	
COLOMBIA	1,142	46	2,640	2,312	20	1,200	1,051	11	10	239	2	5	—	7
BRAZIL	8,515	197	10,110	1,188	19	—	—	58	7	294	—	—	—	
PERU	1,285	29	2,046 [1]	1,592	13	32	25	20	16	690	2	4	2	8
CHILE	756	17	1,060	1,402	14	140	185	15	20	882	[2]	9	6	15
ARGENTINA	2,780	40	1,750	629	14	650	233	190	68	4,750	[2]	6		6

1 1.8% of this amount in the Pacific area, where most human and industrial activities are done
2 Included in industry
3 Consumption by evaporation in surface reservoirs for hydroelectricity in generally not considered
Source: own elaboration based on LA–Argentina (2012), LA–Brazil (2012), LA–Chile (2012), LA–Colombia (2012), LA–Costa Rica (2012), LA–Mexico (2012), LA–Peru (2012).

Table 9.2 Economic productivity of used water in Chile

USE	AGRICULTURE AND FORESTRY [1]	HUMAN SUPPLY	INDUSTRY	MINING	ENERGY
US$/m³	0.3	2.4	7.4	4.5	0.02

1 Soil (green) water not included
Source: LA–Chile (2012).

9.3 Energy and water

9.3.1 Energy for water

Energy is needed to abstract, pump, transport and treat surface and groundwater. The specific energy consumption (kWh/m^3) may become significant due to:
- pronounced altitude differences, such as sloping land in urban areas, physical barriers to be overcome in mountainous areas, or when groundwater levels are deep in highly transmissive and intensively exploited aquifer systems, especially under arid conditions
- long distance transportation to supply large urban and industrial areas from remote sites

- high salinity or presence of unwanted components, such as nitrates, fluoride or arsenic; energy-consuming treatments such as membrane processes are used to reduce their concentration
- sewage water treatment before discharge, to a degree that depends on the assimilation capacity of the surroundings and the environmental requirements; energy consumption increases with treatment intensity, and especially when the water is to be reused via water reclamation
- improved irrigation methods that need pressurized water.

In order to produce the energy needed, water resources are also used and consumed. In order to have a complete picture, the energy consumed during the construction of the water works and that for their maintenance and repair should also be included and distributed along the life time of the works. This includes energy to produce the cement and iron consumed and for excavation, drilling and earth movement. Generally this energy is a small fraction of that spent on the cumulative water production over time, but not always, and should appear in the energy footprint, albeit seldom known or considered.

9.3.2 Water to produce energy

Large water flows are needed to produce any source of energy except for wind and sea energy and direct conversion of solar radiation. Results from a wide-ranging survey are shown in Table 9.3. Some comments on the common energy sources follow.

Table 9.3 Energy and water in Latin America and the Caribbean (LAC) in 2005

ENERGY SOURCE	PRIMARY PRODUCTION (10^{15} J)	ELECTRICITY GENERATION (10^3 GWh)
TOTAL WORLD	460	10,100
TOTAL LAC	25.4	909
COAL AND LIGNITE	1.3	29
OIL [1]	14.6	88
NATURAL GAS	5.0	105
NUCLEAR	0.2	20
HYDROPOWER AND GEOTHERMAL	—	629
BIOMASS [2]	4.3	36
SOLAR AND WIND	—	2

1 1.2×10^{15} J of production are from non-conventional fossil sources using 4.0 km^3 of water
2 2.7×10^{15} J of production is non-traditional
Source: modified from WEC (2010).

9.3.2.1 Hydroelectricity

Hydroelectricity is a renewable energy that still has a large potential for further development in LA. It is often considered environmentally friendly but there are important side effects to be evaluated. Hydro-energy plants modify the natural water regime and this has quantity, quality, environmental and health consequences – externalities – and may imply a loss of other opportunities for water use, as well as human displacements and the creation of territorial barriers. On the beneficial side, dams may facilitate the interconnection of otherwise isolated areas and evaporated water helps to stabilize the local climate.

Although hydroelectricity is often presented as an example of a non-consumptive water use, there is an associated consumptive use of water that should be taken into account. This refers mostly to the fraction of the used water lost to evaporation, especially when extensive water storage areas are needed in flat areas located in warm and arid climates. Evaporation rates up to 1m/year are common and may be exceed 2.5 m/year in some areas. In flat areas this may be significant for downstream river basin resources. These water losses can be added to the water consumption footprint of energy production.

Water resources consumption due to evaporation associated with hydroelectric production range from 0.04 to 210m^3/GWh, with median values ranging from 2.6 to 5.4m^3/GWh (Torcellini et al., 2003; Freedman and Wolfe, 2007; WEC, 2010), depending mostly on the surface area exposed to evaporation, relative to the stored volume, climate, timing of storage, and the altitude difference between the reservoir level and the turbine discharge point. Water consumption increases from mountain environments to lowlands. From data in Mekonnen and Hoekstra (2011) considering thirteen hydroelectric plants in South America, covering very large to medium-size ones and from deep storage reservoirs to shallow ones, the following water specific consumptions in m^3/MWh produced can be gathered: 22 to 36 in Argentina (flat areas), 2 to 111 (median 12) in Brazil (from flat areas to narrow valleys), 0.1 in Chile (a narrow valley in a cold area), and 0.1 to 1.0 in Colombia (deep valleys). Comparing with total usable water resources, evaporation from these dams is a negligible quantity in Chile and Colombia, 0.13% in Brazil and 0.3% in Argentina, at the country level. Larger percentages refer to the river basin where the dams are located. Other results derived from other sources are: 0.9m^3/MWh (70m^3/yr/GW installed) in Colombia (Jiménez-Cisneros and Galizia-Tundisi, 2012), which means a river flow loss of approximately 0.5 to 1%; 14 and 24m^3/MWh for the large hydroelectric plants of Itaipú (95,000GWh, 14GW) and Tacuruí (8.4GW) in Brazil; 0.6m^3/GWh (4% of usable water and 25% of water used) in Costa Rica (LA–Costa Rica, 2012). For reversible hydroelectric plants used for energy regulation these values can be higher.

9.3.2.2 Thermoelectricity

Electricity production in thermal plants may demand and consume large water flows. Thermal efficiency depends on thermodynamics – predominantly on the maximum temperature – and may vary from approximately 0.30–0.35 for nuclear and old coal plants, 0.40–0.45 for critical state coal and oil plants and 0.45–0.50 for combined-cycle gas turbines. The heat that is not converted into electricity is transferred to the environment by means of water. Air cooling greatly reduces water consumption but is expensive and thus it is mostly restricted to areas with scarce water flows, such as in some geothermal plants and isolated coal mines.

Thermoelectric plants may use an open cooling water cycle when a large water flow is available, generally a river, or sea water for plants on the coast, as is the case in Mexico.

Waste heat is transferred to the water with a temperature increase of a few degrees, which has to be compatible with ecological restrictions at the disposal site. Discharged water slowly cools to environmental temperature by evaporation, which implies some flow decrease and an increase in salinity. Water use may vary from 30 to 60m³/s/GWe (We=W of electrical power), depending on the admissible temperature increase in the outflow water, or a little less for thermally efficient nuclear plants. Flows can be halved in sea water cooled plants when a higher temperature increase in discharged water is allowed.

When water resources are scarce, the other commonly used cooling method is the closed water cycle, in which heat is transferred to an external water flow closed circuit that is cooled by water evaporation in natural or forced convection, high cooling towers. Water consumption is the sum of evaporated water, leakages and the renewal of water in the circuit to prevent salinity build-up via evaporation. Approximately 0.5 to 0.7L/s/MWe are consumed, depending on plant thermal efficiency, which equates to 1.8 to 2.5m³/MWhe, and 8m³/MWhe for older plants. Some average water consumption values are given in Table 9.4. In most cases the presence of a thermoelectric plant may produce a significant decrease of resources in small river basins or aquifers. Water disposed of may affect local water salinity and also carry with it corrosion products and in-plant treatment chemicals that may be of some concern if not duly treated before discharge. Water recycling in the cooling system is generally three to ten times the water use for a low salinity water supply. These figures are within the same range of results from other studies carried out in the US (Averyt et al., 2011) and Spain (Hardy et al., 2012).

Table 9.4 Average water consumption rates for thermoelectric plants with closed cooling

TYPE OF PLANT	COAL	OIL	COMBINED CYCLE TURBINE	NATURAL GAS AND MIX	NUCLEAR	HYDROPOWER
m³/MWhe	1.8 [1]	1.2	1.2	0.7	2.7 [2]	1.0 [3]

1 Part of water used for handling of ashes
2 Lower operating temperatures
3 The value for hydropower is actually highly variable while the other figures have a small range

Source: modified from WEC (2010).

9.3.2.3 Geothermal electricity generation

Geothermal electric energy production is significant in some countries such as Mexico, Costa Rica, El Salvador, Nicaragua and Guatemala, and is currently receiving a push in Chile, even though some former projects were not completed due to legal and environmental restrictions. Geothermal production uses closed water cooling units that require water. Air cooling is used when geothermal water is too saline and no other sources of fresh water are available. Waste heat per unit of power is greater than in conventional thermoelectric plants due to the lower thermodynamic efficiency. Most of the water produced is re-injected to avoid pollution problems and to recharge the geothermal aquifer. Some production data are shown in Table 9.5.

Table 9.5 Geothermal energy in some representative LA countries

COUNTRY	MEXICO	EL SALVADOR	NICARAGUA	COSTA RICA	GUATEMALA
EXISTING POWER (MWe)	850	105	70	125	24
% OF COUNTRY ELECTRICITY	3.2	20	17.2	10.2	3.7

Source: own ellaboration.

Water consumption is highly variable. As a reference, in the Salton Sea geothermal fields (3.2GWe) in the USA, water use is $2.2hm^3$/yr or approximately 1.2L/MWe. Studies carried out in Australia indicate that this consumption may be about 2.5L/s/MWe (Clark et al., 2010). Thermo-solar energy production also needs cooling water, which ranges from 0.26 to $0.9m^3$/MWh.

9.3.2.4 Energy from biomass

Biomass is generally used for heating, including industries, especially those that produce it, such as sugar, paper and cellulose factories, but in small quantities it is also used for electricity production. Part of this bioenergy is derived from forest products, mostly consuming soil (green) water, and vegetal wastes from agricultural production. Water is needed for in-plant energy production, mostly to generate process heat and electricity, with little thermodynamic efficiency compared to large thermal plants. This is due to the often low working temperatures and the small, non-optimal units. Energy is also needed for collection, transportation and temporal storage of wastes and products. Similar processes are applied in the less common plants used to transform forest bio-matter into gas and liquid fuels. World biofuel (biodiesel and ethanol) production reached $100hm^3$ in 2010 (HLPE, 2013). This source claims that seventeen LA countries have adopted biofuel policies with specific targets and mandates for transport fuels. Biomass transformation is reported to produce 2% of energy in Costa Rica, 4% in Chile and is significant in Mexico.

9.3.3 Water and land needs to produce biofuels

Renewable energy production is a priority for the 21st century and an important part of it is solar energy captured through biosynthesis. This requires vegetal biomass that consumes large quantities of water, besides land and nutrients. In humid climates crop production is mostly rain-fed but in semi-arid and arid areas irrigation is needed, and consequently this is a source of conflict for the often scarce available water resources.

The planned production of bio-matter to be transformed into fuel – biodiesel and bioethanol amongst others – is currently important in some LA countries (Balat and Balat, 2009; Saulino, 2011). It has been well established in Brazil since the 1970s, and has recently received a push in Argentina (Nass et al., 2007; CADER, 2010, 2011), Colombia, Peru and there is some small production in Paraguay. This refers mostly to the intensive cultivation of maize, sugar cane, other grains, sunflower, and soybeans. Besides water for irrigation, when it is needed, and that used to produce the nutrients – often imported from other areas – water is also needed for the production process in a factory, plus the energy embedded in the facilities and the machinery. The social and economic benefits and the energy balance are not always clear, even if private gains are obtained and there is a good prospect for exporting, mostly to the United States, Europe and Japan. Many different interests and points of view are involved. For many there is a threat to food security and income on a national scale and worldwide, and further the energy used in the production is considered a waste of fossil fuel.

Bioethanol (bioalcohol) was considered as an alternative motor fuel in Brazil as early as the 1930s. Its industrial production started in the 1970s with the programme PROAL-COOL. Up to 10% of ethanol can be mixed with gasoline without modifying the motor or it can be used directly or with up to 10% gasoline in modified engines. In LA it is predominantly produced in Brazil, but also in Argentina since 2009 (Babcock and Carriquiry, 2012) for domestic consumption, and is starting in Peru, Colombia and Costa Rica, mostly from sugar cane. Current production in hm^3/yr for Argentina, Brazil, Colombia and Peru are respectively 0.28, 29, 0.3 and 0.14 (USDA, 2011). The characteristic sugar cane's specific mass production is 75,000kg/ha/yr, and yields 6–8 $m^3/ha/yr$ of alcohol. Approximate data for sugar cane is provided in Table 9.6.

Bio-alcohol can also be produced from corn, other grains and lignocellulose, but at a higher cost. Some data on prices are given in Table 9.7.

Table 9.6 Sugar-cane production and crop area

COUNTRY	No.mills	CANE PRODUCTION (10^6 t/yr)	SURFACE (10^3 ha)	% CROP AREA	BIO-ALCOHOL (hm^3/yr)
BRAZIL	350	460	9,000	4	20 [1]
ARGENTINA	30		50	1.5	0.2
PERU	1	9.3	69	1.3	0.1 [2]

1 85% for internal use, of which 90% is as biofuel
2 ongoing project; 90% for export to the EU
Source: own elaboration based on technical unpublished data.

Table 9.7 Approximate costs of producing bio-alcohol and comparative cost of oil

	SOURCE BIOALCOHOL					OIL
	SUGAR-CANE	CORN	BEAT	WHEAT	LIGNO-CELLULOSE	
US$/L	25-50	60-80	60-80	70-95	80-110	80

Source: own elaboration based on technical unpublished data.

Costs and possibilities are country specific and depend on factors ranging from rainfall and water availability to soil value and the calculation methods used. Crops need nutrients and may produce important externalities, so the real gain and sustainability is open to debate. Brazil claims the energetic value of the bio-alcohol they produce is approximately

8–10 times that of the fuel used in the production. Crop yield has improved by a factor of 1.6 and fossil fuel consumption in the production has decreased by a factor of 0.75.

Biodiesel is produced from oleaginous plants (mostly soybeans) in Argentina (Hilbert el al. 2012), and from palm oil and castor oil in Brazil. It is added to diesel fuel at 5%–7%, with the prospect of attaining up to 10%. Current production in hm^3/yr in Argentina, Brazil, Colombia and Peru is 2.90, 2.65, 0.54 and 0.03 respectively (USDA, 2011). Although Brazil started production earlier (in 2005), Argentina having only started in 2009 (Hilbert et al., 2012) is currently the world's second largest producer after the USA and the main world exporter.

Water consumption in the factories producing biofuel varies between 4 and 6L/L (volume of water/volume of biofuel), which could be cut down to 2.5 by improvement in production. These data can be compared to 2.5 to 5.5L/L to produce petrol, and the 1.9L/L minimum thermodynamic requirements to produce inorganic alcohol. However, the main water consumption is due to irrigation, which ranges from 0 in fully rain-fed areas to 800 to 2000L/L for irrigated crops in arid areas. National water values often only consider water resources (blue water) consumed and thus results vary greatly according to the country or region.

A key element of biofuels production is related to the land and soil (green) water needs. HLPE (2013) compiled the following ranges in $ha/m^3ge/yr$ (ge = gasoline equivalent): for ethanol it is required 0.300 from sugar cane, 0.465 from corn and 0.470 for cellulosic material; for biodiesel, 1.540 from jatropha and 0.310 from palm oil. This means that to produce 1 hm^3/yr of sugar cane approximately 300,000ha of cropland is needed.

9.4 Water for industry

Industry covers a large and variable group of activities, many of which depend on the specific economy of the country or region. Industrial areas are highly variable in LA, from heavily industrialized zones, such as São Paulo (Brazil), Mexico City and Monterrey (Mexico), where the metal sector and petrochemical industries are present, to other areas in which industry is relatively less important and a large proportion of the factories are for food processing. Water needs and the environmental impact are thus quite different. Often thermo-power plants and treatment plants for minerals that are not in the mining area, such as smelters, are considered as industrial plants.

A large proportion of industrial factories are small to medium size, in or around towns, and thus water demand and use is generally included in urban water and the disposal of used water goes to the municipal sewage system. Whilst large self-supplied factories and industrial areas can be found, most of them are oil refineries, chemical plants (which include fertilizer production and natural and artificial textiles), sugar factories ('ingenios') and biofuel production plants processing rain-fed or irrigated agricultural production. In some cases tanneries (leather factories) may be important, as is the case in some areas of

Mexico, Peru and Argentina. Food-processing industries often use water from the supply network while the production of bottled water and refreshments is partly supplied by the municipal network and partly self-supplied, as commented in Box 8.8 of Chapter 8.

Even if factories generally demand a small fraction of the total resources, they may pose important burdens on their surroundings since they are competing for the scarce local resources. This may become locally unpopular and provoke reactions from citizens and the mass-media. Furthermore, in the absence of strict environmental regulations or when the enforcement of such regulations fails, whether it be due to powerful lobbying groups or public administration weaknesses, factories are likely to pollute both surface and groundwater.

Water use data vary from country to country and over time due to continuous improvements in water use efficiency, to reduce production costs and due to environmental pressure to save scarce water resources. Some industrial processes are especially water intensive, such as the production of paper, cellulose, petrochemicals and artificial fibres. In Chile, a 40 m^3/t water demand for paper production is mentioned, where it was formerly of 110m^3/t (LA–Chile), and this value can still be greatly reduced further, as shown by the experience in Spain. In Mexico, about 50% of water for industrial use is for cooling and 35% for industrial processes, and an important fraction of it is wasted. Also in Mexico, the main oil-related industry uses approximately 230hm^3/yr, about half surface water and half groundwater. The water/product ratio is 1.0 for refining, 0.6 for basic gas and oil products and 4.7 for petrochemical products. The water needs for fuel production and processing are shown in Table 9.8. The industrial water use in Mexico for the principal water demanding industries is given in Table 9.9. Self-supplied industries use 3,100hm^3/yr (45% groundwater) and thermoelectric units use 4,100hm^3/yr (12% groundwater), and both of them use 9% of the country's water resources.

Table 9.8 Water needs for fuel production, including processing

FUEL MINERAL	COAL	URANIUM	CRUDE OIL	NATURAL GAS
m³/MJ content	335	184	3,809	218

Source: WEC (2010)

Table 9.9 Industrial water use in Mexico for the main water-intensive sectors.

INDUSTRY	SUGAR	CHEMICALS	OIL	PAPER AND CELLULOSE
% of industrial water use	40	22	7	8
% of industrial water consumption	35	21	6	8

Source: Jiménez-Cisneros and Galizia-Tundisi (2012)

River pollution due to the combined effect of wastewater from urban centres and industry produces important problems in some areas. Worldwide known problems are those of Lerma River and Chapala Lake (Mexico DF), Tieté River (São Paulo, Brazil), the highly polluted Tigre, Matanza–Riachuelo and Reconquista river stream systems around Buenos Aires (Argentina) where a special organization has been formed to try to control it (Autoridad de Cuenca Matanza–Riachuelo, ACUMAR), and downstream Bogotá (Colombia). Except for coastal Buenos Aires the other urban areas are continental and their effect on water resources is therefore greater.

In Mexico, in order to treat 2,500hm^3/yr of wastewater 1,650GWh/yr are used; of this total 900hm^3/yr are from factories, consuming 600GWhe/yr (Jiménez-Cisneros and Galizia-Tundisi, 2012). Total energy consumption for the water cycle is approximately 13,500GWhe/yr, or 7.1% of Mexican energy consumption.

9.5 Water for mining

Mining is an important sector in many LA countries. It is a key source of income and employment and is a sector which is on the rise given the increasing world demand for some metals (see Chapter 5). LA countries are very important world producers of silver, copper, molybdenum, zinc, aluminium, strontium, gold, iron and nickel. In Chile, copper contributes 90% of the economic value of the country's mining, US$ 9 billion to the GNP and produces US$ 45 billion in exports. LA countries supply 51% of the world's silver, 45% of its copper and overall 25% of the world's metal market. The production of lithium, a series of secondary metals and coal are also important, as well as gems. Classical mining areas are those of San Luis Potosí (Mexico), Zacatecas (Mexico), Ouro Preto (Minas Gerais, Brazil), and several Andean areas of Chile, Peru, Bolivia, Argentina and Colombia. La Guajira (Colombia) is an important world coal producer. There are large companies but also numerous small, even artisanal ones, especially for precious metals and gems. They attract 32% of the world's economic investments in mining. Mining activities can be seen as both producers and consumers of water, this second aspect being a serious problem in some areas. Most new mines exploit large and deep pits. Some new mining activities exploit existing natural brines in 'salares' (salt pans) to extract dissolved substances such as lithium and potassium, and also nitrates in some cases.

Mining by means of underground galleries or deep open pits may intersect aquifers or induce the infiltration of river or lake water. This is avoided as much as possible, sometimes with artificial impermeable barriers, but often water drainage cannot be controlled or is the result of operation failures. Pumping out this water is often a costly, energy intensive activity and water has to be disposed of. This produces desiccation problems in some areas and inundation in others, alongside quality problems since pumped water may be acidic or have excessive loads of some undesirable components. This water and that produced inside the mine area, including tailings (mine dumps) drainage, has to be disposed of. A fraction of mine water production is often used for mine operation and dust control.

Open pits become evaporation surfaces that may consume 1 to 2m/yr of water depth, depending on the area. Rainfall may be scarce in many of the arid mining areas of the Americas, often less than 100mm/yr or even as low as a few mm/yr, and thus this may compromise during a long time and even forever the future water resources, as has been observed in the arid and hyper-arid areas of Peru, northern Chile, western Bolivia and northwestern Argentina.

Water is needed for the operation of the mines, mostly to supply mineral leaching areas and mineral processing, but also for dust control. This is a moderate quantity but may become a serious demand in arid and hyper-arid areas. Mineral concentrates are often transported from inside the mining plants and to further away facilities in order to process the final product or to ship it. This transportation can be done by means of pipelines as slurries, thus using large water flows that are often not returned to the mine. This increases mining water needs – a serious challenge in arid areas – and may be a water quantity and quality disposal problem at the processing plant.

Numerous improvements for in-mine water use efficiency through recycling have been introduced to reduce water use. However, mining continues to be a serious challenge in many arid areas where it is necessary to provide enough water to the mining sector whilst preserving human supply, protecting the local environment and avoiding the spread of air and water-borne contamination. Long water transfers have been or are being planned to make mining possible, although excess water disposed of by the mine may become an added problem to the local environment. Current use of water in mining is given in Table 9.10.

Table 9.10 Current water consumption in mining (values rounded up)

COUNTRY	MINING	WATER CONSUMPTION		
	% of GNP	hm³/yr	% [3]	Comments
CHILE	12	260	8.8	growing; mostly for copper [1]
PERU	6	210	2	growing
MEXICO	1.6	55	0.07	[2]
ARGENTINA	3.2			

1 Economic productivity of used water: 4.4US$/m³; 1950hm³/yr used; other sources show up to 300hm³/yr
2 27hm³/yr consumed, 26hm³/yr disposed of; 74hm³/yr recycled; 2% of employment
3 Percentage of available water resources
Source: Jiménez-Cisneros and Galizia-Tundisi (2012), LA–Chile (2012); LA–Mexico (2012); LA–Peru (2012).

The most important supply problems appear in the arid and hyper-arid western coastal areas of South America, especially in the Tarapacá (Region I) and Antofagasta (Region II) areas of Chile. Important water rights purchases have been made at prices between 75,000 and 225,000US$/L/s (see Chapter 13). In 2006 this prompted one of the large companies operating in the area to invest approximately 160 million US$ to obtain 500L/s of fresh water at a coastal seawater desalination plant for leaching sulphide mineral concentrates. With a total investment of 870 million US$, water is pumped up to an altitude of 3200m through a 170km pipeline.

Mining carried out by large companies is generally much less water consuming and produces less water quality degradation per unit of production and per unit value of production than small-scale and artisanal (informal) mining. This last point is common practice in many areas, especially in the Andean region and to exploit secondary mineral accumulations ('placeres') in alluvial deposits and other sediments, or through small underground mining. They are widespread in the wet areas of eastern Peru and Colombia. This mining contributes more employment per unit of production and per unit value (but under poor working and health conditions) than large mines, but it may be highly detrimental to the local environment and water resources. To extract gold, amalgamation with mercury (quicksilver) and cyanide treatment is carried out, and consequently serious mercury and cyanide pollution is produced in rivers, lakes and groundwater. These small-scale and artisanal activities are often poorly controlled, and become important environmental (Hajeck and Martínez Anguita, 2012) and social problems to which governments often turn a blind eye, especially if the native population and poor people are involved in the mining.

Although environmental restoration is possible and mining permits are currently under consideration, the current situation shows that the impact of past activities, failings and unaccounted situations often appear during and after mine operation. Post-mining correction activities carry the risk of not being executed since in many cases the responsibility is passed from the mining companies to governments as money transfers.

One of the largest water resources problem, affecting especially groundwater resources and their relation with the water cycle, is the lack of knowledge and trained personnel, during the mine's operation and especially after its closure. Trained persons are scarce in many countries and are employed preferably to support direct mining activities, which is the priority and are much better paid posts. Thus, it is not rare that governmental organizations in charge of environmental control and regulation are not able to keep a stable workforce due to the higher salaries offered by mining companies. This is a common situation in LA.

Water consumption in copper mining is currently 0.3 to $1.2m^3/t$ of treated mineral, with an average value of $0.75m^3/t$. This is a clear improvement compared to 2 m^3/t some years ago; there are hopes that this will be reduced to $0.05m^3/t$. Current consumption is approximately 75 to 100 L/kg of refined copper and the apparent water economic productivity is approximately 80US$/$m^3$.

Gold production in Colombia is 56t/yr. In the Porce River basin, in the highlands (LA–Colombia, 2012), with 4,000 hm^3/yr of water resources, gold mining uses $0.5hm^3$/yr to produce 3t/yr by using 80–100t/yr of mercury, but actual water consumption is from 0.5 to $1.5hm^3$/kg of gold produced when the flow needed to dilute the pollutants is considered. Water productivity is around 460US$/$m^3$.

Oil extraction is an important mining activity in LA, mainly in the large basins on the eastern side of the Andes Range, from Peru to Colombia–Venezuela, including the central Amazonia in Brazil, as well as in a series of formations in Mexico and southern

South-America, in Argentina and Chile. Water use for abstracting the oil is generally small and highly variable, depending on the circumstances. Oil is abstracted jointly with large flows of often saline and highly contaminated water, which is mostly re-injected to enhance production or is just disposed of safely. Failings or accidents may contaminate groundwater resources and later surface water resources too, for a long time. Secondary and tertiary oil recovery is done by water injection, generally using small flows. Also small flows are needed for advanced gas recovery by 'fracking', which is currently being considered in LA. Chemicals used are an environmental concern and a poorly understood source of pollution. CO_2 injection into deep formations to reduce its emissions into the atmosphere is being considered in Brazil. This needs water for treatment and cooling, and especially to produce energy for the capture process at the plant. The water resources impact of these small amounts will likely be important in the future in water scarce areas.

As is the case of diverse regions of the world and especially in arid and semi–arid areas, as discussed in Chapter 2, in some of the dry areas of LA groundwater reserves in some of the large aquifers are being depleted due to intensive exploitation, at a rate much higher than renovation (Custodio, 2010, 2011). This groundwater withdrawal due to mining activities is happening in the hyper-arid areas of the Andean Region, comprising coastal Peru, northern Chile, southwestern Bolivia and northwestern Argentina, where groundwater renovation is scarce or nil. Groundwater abstraction is for the most part to supply the mining of metal ores and also for brine extraction in salt pans ('salares') used to exploit some solutes such as lithium, potassium and nitrates. The sustainability of small springs and groundwater discharges that are important for some human settlements and of ecological and touristic value, such as high altitude wetlands ('bofedales'), is of special concern. Rainfall in the intermediate depressions is a few mm/yr on average and the scarce recharge is produced occasionally by some sporadic floods in gullies whose headwaters are in the highlands ('altiplano'). Even though rainfall in the altiplano is scarce, a combination of almost bare soil of low humidity retention (mostly young acidic ignimbrites) and rainfall retention in the seasonal snow cover favour some recharge that manages to sustain some springs which yield water with a very long turnover time (Acosta et al., 2013).

9.6 Discussion and conclusions on water security for energy production, industry and mining activities and for human uses in LA

What have been presented in the preceding sections are general considerations on water use and consumption in the different sectors of energy production, industry and mining, with specific references to LA countries and regions, and especially to the seven countries that have contributed reports. For many aspects data have not been found and an in depth bibliographical search has not been performed. Thus, part of the comments and warnings are qualitative and their relative importance remains speculative. Additionally it should be noted that part of the data was obtained from reports that have not been checked or are not always well defined.

Not all of the sectors – energy production, industry and mining – are similarly present in all LA countries. In Argentina, Brazil, Paraguay, Chile, Colombia and Costa Rica hydroelectricity is an important energy source, while in Mexico coal, oil and thermoelectricity contribute a larger fraction of the country's needs. Only Argentina has operating nuclear plants, although their contribution to the country's total energy needs is small.

Hydroelectricity may consume water by evaporation in the storage reservoirs, which is often a small fraction of river flow, but in some cases it may be large enough to affect downstream water security by reducing flow, increasing salinity and modifying seasonality. Specific water consumption for energy production varies over a wide range, from less than $1m^3/GWh$ to more than $100m^3/GWh$, depending on local conditions. At the national scale this amounts to 0.1% to 4% of total water resources, although in some cases, particularly in warm, flat areas it can be up to 25%.

In thermoelectric plants, cooling – in open and closed cycles – is generally done with river water, but in Mexico marine water cools important power plants located in coastal areas. Geothermal plants along the western mountainous areas of LA are in arid regions and use closed cycle cooling fed with groundwater or air cooling. These cooling needs water consumption may be a significant fraction of local surface and groundwater resources, which in arid areas can compete against other water demands for a large share. Thus, water security may become an important consideration for the plant operation, for the downstream local population and for the environment.

The production of biofuels may introduce an important water demand where irrigation is needed, which may in turn have a great impact on local and downstream local water security. This would be especially true in semi-arid and arid areas and furthermore in the areas from where the water resources are to be taken. It seems that some projects on the Pacific side of South America may create important local water imbalances or require expensive water conveyance systems and energy-consuming water imports from further afield areas for the sake of income from biofuel exportation.

Water security considerations for industry are as varied as the involved activities. In many cases they are connected to urban water supply and share their water security circumstances, as explained in Chapter 8. This includes part of the production of bottled water and refreshments that are common in LA, Mexico being a world leader in per capita production and consumption. Large industrial establishments, which include thermoelectricity production, have their own water supply. Other important water-independent industries are those related to refineries, large chemical plants, smelters for iron, aluminium and other metals, textiles, leather and large sugar plants, amongst others. Comments made above for energy water security also apply here. Additionally water security for populations and the environment has to consider the pollution generated by these plants – something which is highly dependent on the types of activity and technology – and also on the existence and enforcement of legislation and civil society action. Circumstances vary largely in LA. Large industrial concentrations are found in several places in Mexico, Brazil and

Argentina, and large sugar plants ('ingenios') in Colombia. The impact of water security on the population also depends on the location of these industrial plants. Many of them are close to the coast – large lakes do not exist – and have less downstream water security impact, but others are far inland and are often at high altitude (São Paulo, Bogotá, Mexico City) and thus have a higher impact on downstream water security.

Mining is an important activity and a great source of income in many LA countries such as Mexico, Colombia, Venezuela, Brazil, Peru, Chile, Argentina and Bolivia. Some mines are in areas with plenty of water – where the problem is how to get rid of it – but others are in semi-arid areas with water supply problems (e.g. central Mexico, northern Colombia, northeastern Brazil) and in arid and hyper-arid areas (northern Chile, northwestern Argentina, eastern Bolivia, southern Peru) where water resources are very scarce and groundwater with very slow renovation (up to several thousand years) is used and partly mined. Water security for mining is an important concern, so in some cases seawater desalination at the coast has been introduced. For example in northern Chile costly desalinated seawater is pumped to the highlands where the copper mines are located. In the case of mining, water security can be solved when mining can support the involved cost of procuring and producing water given the current high prices of metals.

From the point of view of human water needs, mining may become an important threat in arid and semi-arid areas, but may also generate large benefits. Mining may seriously interfere with water security of locals by reducing river and spring flow, even exhausting them, or in other cases damage wetlands. This is a complex situation as changes in the groundwater resources are slow and delayed, which may pass unnoticed for years. Detailed hydrogeological studies are therefore needed to measure this impact over time. Thus, it is important to know the pace of recovery after a mine closes; it may be that this rate is too slow to be significant. It is relatively common that open pit mines are not refilled as they may be conceivably re-opened in the future or is not considered in their mining permit; thus this can leave a large and deep lake capable of evaporating large water flows if groundwater seepage is enough or if surface water gets in when barriers fail. This may reduce local and downstream water resources and even exhaust springs and small streams. There is little information on this issue, especially due to poor monitoring since many large mining activities are relatively young and the evaluation is complex owing to weather and climate variability.

From the water quality point of view, mining may affect the water security of inhabitants and of the environment, both local and downstream. This is due to the disposal of water with high salinity, acid and/or containing diverse unwanted and noxious solutes derived from minerals – diverse heavy metals – or from concentration and processing, such as flotation compounds, and quicksilver (mercury) and cyanide in the case of the many gold mines in LA, especially the small and artisanal ones. This is a common situation in Colombia – where the supply and even agricultural use of water from many rivers is jeopardized – and the Amazonian side of Peru. The situation is less acute in the case of well-operated modern mining, where wastewater disposal is relatively small and controlled.

Industry, energy production and mining together consume 8 to 15% of water resources in the seven considered LA countries. Water consumption is respectively 4 to 9% for industry, 2 to 5% for energy production (hydroelectric energy consumption not included) and up to 6% for mining. Water economic productivity for these uses may range from 3 to 10US$/m³. It is a high value when considering direct costs and benefits but if externalities are considered the economic picture may change, depending on the social discount rate that is applied.

Acknowledgements

Several experts have explicitly contributed data: Bárbara Soriano (CEIGRAM/UPM, Madrid), Maria-Josefa Fioriti (Subsecretaría de Recursos Hídricos, Buenos Aires), Jorge Benites Agüero (Autoridad Nacional del Agua, Lima), Luis Alberto Pacheco-Gutierrez (UNAM, Coyoacán, Mexico). Lucia de Stefano (FB/UCM) and Bárbara Willaarts (FB/UPM, Madrid), Enrique Cabrera (UPV, Spain), and Blanca Jiménez-Cisneros (UNESCO-PHI, Paris) have contributed useful comments.

References

Acosta, O., Guimerà, J., Custodio, E., Ansón, I. & Delgado, J.L. (2013). *Contribución al conocimiento de la hidrogeología de las cuencas intraandinas del N de Chile*, Congreso Hidrogeológico Argentino, La Plata, Argentina.

Averyt, K., Fisher, J., Huber-Lee, A., Lewis, A., Macknick, J., Madden, N., Rogers, J. & Tellinghuisen, S.(2011). *Freshwater use by U.S. power plants: Electricity's thirst for a precious resource. A report of the Energy and Water in a Warming World initiative*, Cambridge, Union of Concerned Scientists.

Babcock, B.A. & Carriquiry, M. (2012). *Prospect for corn ethanol in Argentina Center for Agricultural and Rural Development.* pp. 1–30. Ames, Iowa, Iowa State University.

Balat, M. & Balat, H. (2009). Recent trends in global production and utilization of bio-ethanol fuel. *Applied Energy*, 86: 2273–2282.

CADER (2010). Cámara Argentina de Energías Renovables. *State of the Argentine biofuels industry*, Buenos Aires, Argentina. pp. 1–24. Report 4° Trim.

CADER (2011). Cámara Argentina de Energías Renovables. *Estado de la industria argentina de biodiesel.* Buenos Aires, Argentina. Report 4° Trim.

Clark, C.E., Harto, C.B., Sullivan, J.L. & Wang. M.Q. (2010). *Water use in the development and operation of geotermal power plants*, Environmental Science Division, Argonne National Laboratory. [Online] Available from: www.osti-gov/bridge. [Accessed June, 2013].

Custodio, E. (2010). Intensive groundwater development: A water cycle transformation, a social revolution, a management challenge. In: Martínez–Cortina, L., Garrido, A. and López–Gunn, E. (eds). *Rethinking Water and Food Security.* pp. 259–298. Botín Foundation–CRC Press.

Custodio, E. (2011). Hidrogeología en regiones semiáridas y áridas, VII Congreso Argentino de Hidrogeología–V Seminario Hispano–Latinoamericano, Temas Actuales de la Hidrología Subterránea/Hidrogeología Regional y Exploración Hidrogeológica. *Salta*: 1–17.

Freedman, P.L. & Wolfe, J.R. (2007). *Thermal Electric Power Plant Water Uses; Improvements Promote Sustainability and Increase Profits*, Presented at the Canadian-US Water Policy Workshop, Washington, DC, 2 October 2007; pp. 1–11.

Hajek, F. & Martínez de Anguita, P. (2012). ¿Gratis? Los servicios de la naturaleza y cómo sostenerlos en el Perú. *Servicios Ecosistémicos Perú:* 1–436.

Hardy, L., Garrido, A., Juana, L. (2012). Evaluation of Spain's water–energy nexus. *International Journal of Water Resources Development*, 28 (1): 151–170.

Hilbert, J., Sbarra, R. & López Amorós, M. (2012). *Producción de biodiesel a partir de aceite de soja: contexto y evolución.* Buenos Aires, Ediciones INTA.

HLPE (2013). *Biofuels and food security: A report by the High Level Panel of Experts on Food Security and Nutrition of the Committee on World Food Security*, Rome, FAO.

Jiménez-Cisneros, B. & Galizia-Tundisi, J (Coord.) (2012). *Diagnóstico del Agua en las Américas*, Mexico.,Red Interamericana de Academias de Ciencias (IANAS) and Foro Consultivo Científico y Tecnológico (FCCyT). [Online] Available from: www.foroconsultivo.org.mx/home/index.php/libros-publicados/diagnosticos-y-analisis-de-cti/991-diagnostico-del-agua-en-las-americas. [Accessed February, 2013].

LA–Argentina (2012). *Report of Argentina.* Contribution to the Water and Food Security in Latin America project. Water Observatory Project Madrid, Spain, Fundacion Botín.

LA–Brazil (2012). *Report of Brazil.* Contribution to the Water and Food Security in Latin America project. Water Observatory Project Madrid, Spain, Fundacion Botín.

LA–Chile (2012). *Report of Chile.* Contribution to the Water and Food Security in Latin America Project. Water Observatory Project Madrid, Spain, Fundacion Botín.

LA–Colombia (2012). *Report of Colombia.* Contribution to the Water and Food Security in Latin America project. Water Observatory Project. Madrid, Spain, Fundacion Botín.

LA–Costa Rica (2012). *Report of Costa Rica.* Contribution to the Water and Food Security in Latin America Project. Water Observatory Project Madrid, Spain, Fundacion Botín.

LA–Mexico (2012). *Report of Mexico.* Contribution to the Water and Food Security in Latin America Project. Water Observatory Project Madrid, Spain, Fundacion Botín.

LA–Peru (2012). *Report of Peru.* Contribution to the Water and Food Security in Latin America Project. Water Observatory Project Madrid, Spain, Fundacion Botín.

Mekonnen, M.M. & Hoekstra, A.Y. (2011). *The water footprint of electricity from hydropower.* Institute for Water Education, UNESCO-IHE. Research Report Series number 51.

Nass, L.L., Pereira, P.A.A. & Ellis, D. (2007). Biofuels in Brazil: an overview. *Crop Science*, 47(6): 2228–2237.

Saulino, F. (2011). *Implicaciones del desarrollo de los biocombustibles para la gestión en el aprovechamiento de agua.* pp. 1–67. Santiago de Chile, CEPAL.

Torcellini, P., Long, N. & Judkoff, R. (2003). *Consumptive water use for US power production*, Golden. Technical Report number NREL/TP-550-33905.

WEC (2010). World Energy Council. *Water for energy.* pp. 1–51. London.

Part 4

Economic, legal and institutional factors for achieving water and food security

10

WATER EFFICIENCY:
STATUS AND TRENDS

Coordinator:
Maite M. Aldaya, Water Observatory – Botín Foundation, and Complutense University of Madrid, Spain

Authors:
Guillermo Donoso, Pontificia Universidad Católica de Chile, Chile
Maite M. Aldaya, Water Observatory – Botín Foundation, and Complutense University of Madrid, Spain
Wilson Cabral de Sousa Junior, Aeronautics Technology Institute, São José dos Campos, Brazil
Xueliang Cai, International Water Management Institute (IWMI), Pretoria, South Africa
Daniel Chico, Water Observatory – Botín Foundation, and CEIGRAM, Technical University of Madrid, Spain
Angel de Miguel, IMDEA Agua – Madrid Institute for Advanced Studies, Spain
Aurélien Dumont, Complutense University of Madrid ,and Water Observatory – Botín Foundation, Spain
Luis Gurovich, Pontificia Universidad Católica de Chile, Chile
Jonathan Lautze, International Water Management Institute (IWMI), South Africa
Elena Lopez-Gunn, I-Catalist, Complutense University of Madrid, and Water Observatory – Botín Foundation, Spain
Markus Pahlow, Department of Water Engineering & Management, University of Twente, The Netherlands
Julio Cesar Pascale Palhares, Embrapa Cattle Southeast, Brazil
Erika Zarate, Good Stuff International, Switzerland

Contributors:
Barbara Soriano, CEIGRAM, Technical University of Madrid, Spain
Laurens Thuy, Utrecht University, The Netherlands

Highlights

- Latin America may well be water rich, but economic and urban growth from the last two decades has polluted freshwater resources of many countries.

- Several factors such as population growth, rapid urbanization, water contamination and pollution, and increased water demands due to increased economic growth are putting considerable pressure on available water resources. Decoupling economic growth from water use is at the core of innovation strategies for sustainable consumption and production and ultimately for resource efficiency.

- In LAC, as in other regions of the world, agriculture is the main user of freshwater. Within this sector about 90% of the water consumption is based on green water – rainwater stored in the soil as soil moisture.

- The greatest opportunity for improvement in water productivity and efficiency is in rain-fed agriculture through enhanced and known management practices.

- In general, irrigation efficiency of the existing systems in LAC countries is medium to low; the average irrigation efficiency for the region is reported at 39%, varying between 30 and 40%, whereas the world average is 56%.

- Urban water use in LAC also shows low technical water efficiency relative to developed countries; on average, water conveyance efficiency is reported to be 59%.

- Water efficiency in the electricity sector also shows significant room for improvement.

- Thus LAC countries must improve water use efficiencies in order to increase water and food security as well as protect aquatic ecosystems. LAC countries must consider water policy changes that provide adequate incentives to use water resources efficiently and ultimately achieve a more sustainable use of water in all sectors.

10.1 Introduction

10.1.1 Rationale for water efficiency in Latin America and the Caribbean

Latin America and the Caribbean (LAC) is graced with an abundance of fresh water, holding 31% of the world's freshwater resources (UNEP, 2010). However, several factors such as population growth, rapid urbanization, water contamination and pollution, and increased water demands due to increased economic growth are putting considerable pressure on available water resources.

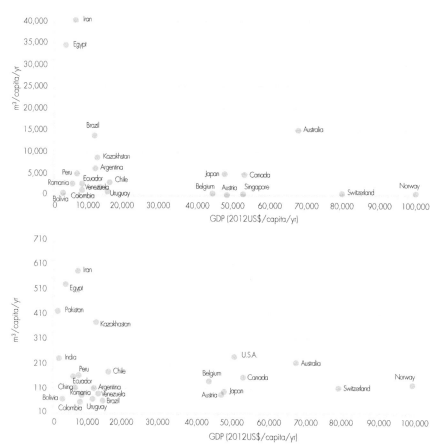

Figure 10.1 The relation between the blue water footprint of production (upper) and consumption (lower) and the level of economic development. *Source: own elaboration based on data from Mekonnen and Hoekstra (2011) and World Bank (2013).*

Available empirical evidence suggests a dubious relationship between the rates of water consumption and GDP growth in many countries (Figure 10.1).

Some developed countries (e.g. USA) and developing countries (e.g. India and China) have high water consumption rates per unit of GDP, i.e. a high water intensity ratio. Other developed countries (e.g. Singapore, Switzerland, Norway) and many developing countries have a low water consumption rate per unit of GDP (e.g. Uruguay) (Figure 10.1).

These examples suggest that relative decoupling of economic growth from water use is already happening in some countries. However, these assessments do not take into account the increases in burden shifting through virtual water flows (Gilmont, 2013). For example, OECD countries may have achieved the 'decoupling' by shifting water intensive production activities towards non-OECD countries (Figure 10.1). Decoupling should be assessed for Latin America and the Caribbean in light of the evidence that it is a net virtual water exporting region.

Empirical case studies of selected countries confirm that decoupling of economic growth from water uses and water pollution is not an automatic by-product of growth in national incomes but requires dedicated policies on improving water efficiency and water productivity at the required temporal and spatial scales.

Decoupling of economic growth from water use is critical for food security in LAC as water resource restrictions is one of the most important barriers to food production. The increase in irrigation activities has contributed to the substantial growth in agricultural production, enabling humanity to feed its growing population. However, more efficient use of green water (rainwater stored in the soil as soil moisture) and blue water (surface water and groundwater) has been stressed as one of the most important factors to achieve greater agricultural productivity (Pasha, 2002; Molden et al., 2003; Rosegrant et al., 2003).

Although improved methods and technologies have produced efficiency gains in all economic sectors, in some regions the need and potential exists for further improvements to ensure food security for a growing world population while minimizing the impacts on ecosystems and their goods and services.

Yet, the region is moving towards meeting the Millennium Development Goals (MDGs), but poor farming practices, unregulated human activity (or poorly implemented or -monitored existing regulations), including industrial development and urban poverty, have negatively affected LAC's water resources (UNEP, 2013). Additionally, given the region's rate of population growth, rapid urbanization and current patterns of water use, sustaining an adequate water supply for future generations is an increasingly important issue. There are many opportunities to enhance water efficiency and management in the region.

This chapter reviews the efficiency of water resources use in LAC. For this purpose, first of all, it provides the concepts and definitions together with the drivers for water efficiency. Second, it analyses the efficiency of water resources use in Latin America, looking at the water users in different sectors: urban and industry, agriculture, energy and the environment. Finally, it provides a summary of challenges and opportunities for enhanced water efficiency and management across the region.

10.1.2 Definitions and approaches

Achieving an efficient use of natural resources and other factors of production is a common goal of many current policies towards sustainability. Efficiency can be defined in general terms as the ratio between a desired output and an input, that is, the quantity of resource consumed in the process. Improving efficiency means creating more value with less resource consumption. However, depending on the scale and the disciplinary approach, the formulation of this indicator and the possibility of increasing it, imply different approaches (Jollands, 2006). Particularly in the case of water, three main interpretations for efficiency are usually recognized: technical efficiency, water productivity and economic or allocation efficiency (GWP, 2006).

- Technical efficiency considers the rate of physical application of water to its desired purpose. This factor can be defined for all the water uses in every sector. In agriculture, the value depends mainly on the technique (e.g. surface, drip irrigation) but also on the management system such as the mode of application of water linked to this technique (turns, on demand) and other factors (maintainability) allowing for the correct use of technology. Thus, factors other than the change of technique can lead to efficiency improvement.
- Water productivity is defined as the ratio between an output linked to a use and a water volume input. It provides a description of how well water resources are made productive (i.e. generating value) in their different uses.
- Economic or allocation efficiency deals with the objective of allocating the resource in order to maximize the net social benefits for society. It represents a general criterion characterizing the distribution of water between users (not a technical ratio attached to a specific use) (Wichelns, 2002). Possibilities to improve efficiency are linked to the economic instruments and governance arrangements, such as water markets, water rights reallocation, or the virtual water trade, leading to a higher benefit from the use of the available resources.

A comprehensive assessment of the relationship between green water (rainwater stored in the soil as soil moisture), blue water (surface water and groundwater) and grey water (volume of freshwater polluted) and economic efficiency should also consider the efficiency of the use of other resources, such as financial capital, labour or energy, in obtaining water services. Indeed, not only obtaining more benefit per unit of water is important but also more water per unit of other resources (GWP, 2006). For instance, this is also relevant in the debate on the efficiency between private and public sectors (Pierce, 2012).

It is also important to remember that these definitions are only valid within the broader economic context and other social objectives in order for efficiency not to be considered the final objective (Adger et al., 2003). For the most part, a higher efficiency does not mean that total consumption will be reduced as other incentives may govern resource use. Moreover, efficiency can make a resource cheaper, or increase its availability, incentivizing new uses (Pfeiffer and Lin, 2010; Dumont et al., 2013).

10.1.3 Determinants of the adoption of water conservation technologies

Theoretically speaking, the scarcer a resource becomes, the more likely it is that technologies will be adopted to save this resource. Empirical studies demonstrate that the scarcity of water resources is an important driver of water-saving technology adoption (see e.g. Schuck et al., 2005). However, water-saving technology adoption will increase in response to augmented water shortage only if users perceive that adoption will lead to water savings or generate other benefits.

In agriculture, the most important determinant of technology adoption is ultimately the farmer's perception of the incremental benefits and costs to his own farm income (Sharma and Sharma, 2004; Blanke et al., 2007). Hence farm-level perceptions of the water-saving properties and the impacts on income of each water-saving technology are critical determinants of the successful adoption of water conservation technologies.

Perry et al. (2009) state that farmers invest in improved irrigation technology for a variety of reasons, including increased income, risk aversion/food security, convenience and reduced costs. Varying prices for market goods, land, labour, water, electricity, energy, inputs, technology and soil management change farmers' perceptions on the value of water relative to these inputs. The farmers respond to market rules searching for the highest return per unit of land or water, depending on the relative scarcity of both resources (Ali and Talukder, 2008).

Studies demonstrate that public, government-supported extension of water-saving technologies has a positive effect on adoption of water conservation technologies (Schuck et al., 2005). Generally speaking, government policies promote the adoption of water-saving to incentivize water users to increase their technical and economic efficiency (Sharma and Sharma, 2004; Blanke et al., 2007).

Dagnino and Ward (2012) found that water conservation subsidies that promote a change from surface to drip irrigation can increase the demand for water despite the absence of new depletable supplies. Findings show that where water rights exist, water rights administrators will need to safeguard against increased depletion of the water source with increased subsidies that reward reduced water applications. There is a need for good water accounting as discussed by Molden et al. (2010), to take into account these environmental impacts of the adoption of water conservation technologies.

10.2 Methodology and data for evaluating water use efficiency and its socio-economic implications

The methodology follows the different approaches to efficiency as presented in the introduction: technical efficiency, water productivity, economic efficiency and efficiency in the provision of water.

10.2.1 Methodology and data to evaluate technical efficiency

10.2.1.1 Specific uses/local scale

Technical efficiency considers the rate of physical application of water to its desired purpose (eq. 1). Therefore, it is a percentage indicating how well a technique or mode of distribution delivers water.

Thus, the technical efficiency (eff) can be defined as:

$$eff = \frac{\text{water delivered for the intented use}}{\text{water withdrawals}} \tag{1}$$

This expression is valid for all water uses. For instance, in the urban sector, efficiency of water delivery characterizes how much water is lost during its distribution to the final user. However, a priori this ratio must be considered as a partial indicator only. Particularly low efficiencies calculated according to this indicator do not mean that excess water is wasted or lost as return flows can generate value once they go back to the river basin.

A more detailed characterization of water use and reuse potentialities can be obtained based on the quantification of fractions (Perry, 2007). Water use is divided into:

- Consumed fraction (evaporation and transpiration) comprising beneficial consumption (water evaporated or transpired for the intended purpose) and non-beneficial consumption (water evaporated or transpired for purposes other than the intended use);
- Non-consumed fraction, comprising the recoverable fraction (water that can be captured and reused) and non-recoverable fraction (water that is lost to further use).

This allows for the differentiation between uses that remove the water from further use (evaporation, transpiration, flows to sinks) and those uses that have little quantitative impact on water availability (e.g. navigation, most domestic uses).

An alternative expression of efficiency could be the ratio between water delivered for the intended use and total water evaporated (eq. 2). It is particularly meaningful in the case of irrigation, as it would indicate the distribution between evaporation and plant transpiration.

$$\text{eff} = \frac{\text{water delivered for the intended use}}{\text{total evaporated water}} \tag{2}$$

10.2.1.2 At the basin scale

The principal consequence of not identifying the potential reusability of return flows is that an increase in efficiency (first definition) may lead to downstream users being deprived from resources they were receiving. Other unintended effects should also be taken into account. For instance, switching from surface irrigation to sprinkler or drip irrigation implies that farmers will potentially have greater flexibility in their water use (on demand instead of turns), allowing the improvement of yields or growing crops that are more sensitive to water shortage (Dumont et al., 2013). This will increase water productivity but also water consumption.

The traditional approach of technical efficiency applied at the catchment or river basin level implies the consideration of the ratio between water consumption (total evapotranspiration, ET) and the basin's total resources. For a closed basin (i.e. where all the resources are allocated) this ratio is close to 100%. This result has sometimes been interpreted as efficiency and cannot be improved in this situation. However, this refers to technical efficiency and not economic efficiency (see section 10.2.3). In a closed basin, therefore, there exists the possibility of improving the total value of water use, even though 100% technical efficiency is achieved.

10.2.2 Methodology and data to evaluate water productivity of specific uses

Water productivity (WP), defined as WP = product/water consumed [mass/volume] (i.e. the inverse of the sum of the green and blue water footprint), is used at plant, field and farm scale. Many times total withdrawal is considered in the expression of WP. It should be observed, however, that this would lead to technical efficiency ratios (as described in the previous section).

Looking at the biophysical level first, WP is an efficiency parameter of the crop production process, where water (as well as other inputs) is subject to a transformation process of crop or biomass production, owned and managed by the farmer. We define green water productivity $WP_{green}=yield/ET_{green}$ as the water productivity in rain-fed agriculture. For irrigated agriculture, blue water productivity is the difference between total water productivity and green water productivity ($WP_{blue}=WP_{total}-WP_{green}$).

In the industrial sector, water use efficiency is commonly determined as the ratio of production and water withdrawal. Here we use consumption in the denominator, not withdrawal.

The notion of WP can also be applied in a wider sense, by attributing different values to the numerator. This is commonly done in water valuation approaches, where economic attributes can be given in monetary terms (e.g. US$), social attributes (e.g. jobs, food security), or environmental attributes (e.g. carbon sequestration, biodiversity).

Pollution is not formally included in water efficiency or productivity measures, yet polluted water may reduce yield and hence enters the equation for crop WP indirectly. However, it ought not to be neglected, especially when considering urban environments, industry and other sectors. In the end, water pollution is also a form of water use that subtracts from other uses (e.g. due to pollution of return flows or salinization). It is therefore worth pursuing efficiency increases in those areas, which means: lowering the pollution per unit of production.

10.2.3 Economic efficiency: characterizing the allocation of water resources at the basin scale or amongst other geographical areas

Indeed, at this scale allocative efficiency considers re-allocating and co-managing water among uses by re-allocating water from lower value to higher value uses within and between sectors, thereby mitigating adverse impacts (Wichelns, 2002; Molden et al., 2003). At the same time environmental flow requirements need to be identified and managed (Richter et al., 2011). The total amount of water allocated in a river basin needs to be based on the maximum sustainable water footprint level of that basin (Hoekstra, 2013).

Value can be expressed in monetary terms (e.g. $/litre), food calorie terms (e.g. kcal/litre), energy terms (e.g. MJ/litre). Evaluation of water productivity should be carried out both in a physical sense (more crop per drop), and in an economic sense (more value per drop), in order to obtain the greatest benefit.

Economic water productivity (as defined in the previous section) provides a tool to attribute value and productivity to all water uses and users within a hydrological domain, and not only those pertaining to irrigated agriculture. When based on hydrological accounting of actual water consumption, a value (whether economic, social, ecological or agronomic) can be attributed to all water uses and reuses, including those that tend to be left unaccounted for in irrigation efficiency approaches as 'wasted fractions' non-utilized by irrigation (van Halsema and Vincent, 2012).

The water available within a catchment or river basin for allocation purposes is determined by the water balance equation:

$$P=ET+R+D\pm\Delta S \tag{3}$$

where P is precipitation, ET is evapotranspiration, (evaporation, E and transpiration T), D is drainage and ΔS is the change in soil moisture. In order to assess whether or not a new technology that is available to farmers is beneficial to society, one needs to calculate net social returns instead of net private returns. The two concepts are identical, except that net social returns value all inputs and outputs at social prices, not market prices. Social prices are identical to market prices when well-functioning markets exist. When well-functioning markets do not exist, as is almost always the case with water, then one must attach a social value to water, which is defined as the value of the water in the best alternative use (at the margin) (Barker et al., 2003).

10.3 Technical efficiency in the use of water resources in Latin America from the production perspective

10.3.1 Urban and industrial uses

According to UN data for the year 2011, 78% of the population in the LAC region is concentrated in cities and this figure is increasing. Indeed it is expected to reach 86% by 2050. This trend carries with it the difficult task of satisfying the needs of existing megacities and balancing the environmental impacts that derive from them such as increased direct and indirect water consumption. Efficiency increases in the use of water in this context represents a way of limiting water stress and thus reducing the impacts of population growth and urbanization.

In LAC, technical efficiency in urban water supply is rather low. In Brazil, 37.57% of the water is lost (ANA, 2013). In Nicaragua, this figure reaches 25% in urban areas (GWP, 2011), and in the case of Colombia it was 20.5% in 2004 (ICC, 2007). The Inter-American Development Bank reported that 56% and 60% of the water was either lost or irregularly consumed in the water sector in Ecuador and Venezuela respectively (CAF, 2013). Approximately 36% of the water is lost in Mexico (Aguilar and Castro, 2010).

The Americas Association of water regulators surveyed water utilities in 2011. They obtained responses from twenty-three utilities in seven countries. On average, water conveyance efficiency is reported to be 58.81%, but ranges from 30.88% in Paraíba to 92.5% in Ceará, both states of Brazil. However, only ten out of twenty-three companies reported any data.

Nonetheless these figures do not reflect the complete picture. The quality of the services needs to be improved (CEPAL, 2010), not only the quality of the service as such (pressure, hours of service, reliance) but also the quality of the water for consumption (GWP, 2011).

Many problems for urban water management are rooted outside the urban scope. A recent report (GWP, 2011) mentions that unsustainable land management (soil and forest management), as well as industrial and agricultural pollution affect urban water availability and quality. Solutions for water provision and degradation are more feasible if a more systemic view of water resources, considering ecosystem services, is taken, which would require the adoption of integrated water management (GWP, 2012). Under this framework, water planners link basin level water management to the cities' water management and also consider the combined management of surface and groundwater resources (GWP, 2012).

10.3.2 Agricultural use

In LAC, as in many other regions of the world, agriculture is the main user of freshwater. However, the large and growing proportion of the population living in urban areas as well as the increased water demand from a growing industry and mining sector in LAC, in addition to reduced water supplies due to increased water pollution and climate change will put considerable pressure for continued transfers of water away from agriculture.

Trends in individual country's economies in LAC, the contribution and importance of agriculture to each of these national economies, trends in agricultural exports and the share of people employed in agriculture are all important factors underlying the development of irrigation and other water uses in the region. Since LAC's GDP growth for 2012 is projected to be 3.2% and 4.0% in 2013, compared to 1.6% and 2.2% in the OECD countries (see Chapter 4) and given that the decoupling of economic growth from water uses and water pollution is not yet generalized in the LAC region, increasing water efficiency in agriculture is a major challenge. However, there has been a decrease in the investment in irrigation in LAC in the last years (Molden, 1997; CAWMA, 2007; Ringler et al., 2010).

Focusing on South America and the Caribbean, the total irrigated area is around 18.6 millions of hectares; corresponding to only 7% of the world's total estimated irrigated area (CAWMA, 2007). Brazil has 3.5 million irrigated hectares, followed by Chile, Argentina and Bolivia. In general, irrigation in South American countries has been inefficient; a major weakness is the failure to provide adequately for the operation and maintenance of irrigation systems once construction or installation is completed (Garces-Restrepo et al., 2007).

Thus, in general irrigation efficiency of the existing systems in LAC countries falls below expectations. However, some efficient irrigation systems exist in the region, such as the case of banana production in Ecuador and fruit and vineyards in Chile (Ringler et al., 2010). With few exceptions, agricultural water use in general has been inefficient in LAC due to the predominance of traditional surface irrigation technologies; FAO (2003) reports that 95.6% of irrigated lands in LAC are surface irrigated; 2.7% use sprinklers and just 1.7% use localized irrigation (drip and micro-sprinkler). These percentages indicate that there is considerable potential to increase water productivity in the region by switching to more efficient water application methods (de Oliveira et al., 2009).

In the LAC region, the levels of technical irrigation efficiency are medium to low, in the range between 30% and 40% (Figure 10.2). In its country database, FAO (2013) includes average irrigation efficiencies for LAC countries (referred to as water requirement ratios) ranging from 18% (Costa Rica) up to 48%, 51% and 65% (Brazil, Paraguay and Puerto Rico respectively). The average for the region is reported at 39%, whereas the world average is 56%. Field estimates in various irrigation projects in Brazil, for example, resulted in average actual and potential water application efficiencies of 40 % and 60%, respectively, for conventional and improved irrigation systems (Ringler et al., 2010). The introduction of efficient irrigation systems in Chile during the past fifteen years has led to a significant increase in the proportion of irrigated land with efficient irrigation technology; at present, 30% of Chile's total irrigated surface is equipped with efficient irrigation technologies such as drip and sprinkler systems. This trend has led to an overall irrigation efficiency of 58% in the last ten years. Brazil shows progress towards a better application of water with 59% of irrigated lands being under surface irrigation, 35% with sprinkler irrigation, and 6% with localized irrigation; here water scarcity and farm characteristics have encouraged the use of more efficient irrigation methods. Thus, in order to ensure water and food security in LAC, there is a need to improve water efficiency, both in humid and arid regions.

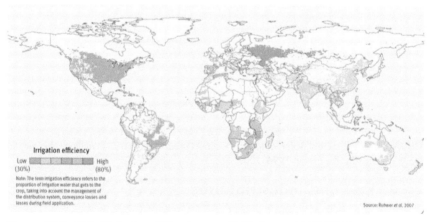

Figure 10.2 Global irrigation efficiencies, year 2000. *Source: UNEP (2012).*

In a comprehensive evaluation of 144 projects that adopted sustainable agricultural technologies and practices, including several studies in LAC, Pretty et al. (2006) demonstrate that the greatest opportunity for improvement in water productivity, i.e. marketable yield divided by crop water consumption, is in rain-fed agriculture. Water-related risks due to high rainfall variability can successfully be reduced by improved farm management, thereby avoiding low productivity or crop failure. Adequate measures include (supple-mental) irrigation, soil, and nutrient and crop management.

However, inadequate agricultural water use in LAC is salinizing, waterlogging, and eroding agricultural lands and polluting water for agricultural use. Most salinization problems originate from the inefficient use of water. Argentina and Chile have about 35% of their irrigated lands affected by salinity whereas 30%, equivalent to 250,000ha, of the coastal region of Peru under irrigation is also impacted by this problem. In Brazil 40% of the irrigated land in the northeast is affected by salinity as a result of improper irrigation (Ringler et al., 2010).

10.4 Water productivity in the use of water resources in Latin America from the production perspective

In the LAC region as a whole, the largest water user is the agricultural sector, amounting to 99% of the green and blue water consumption and 46% of the nitrogen-related pollution (Figure 10.3). Urban water supply represents as much as 0.5% of the total water consumed and 37% of the total nitrogen pollution. Meanwhile the industrial sector represents just 0.1% of the total water consumed and 17% of the total nitrogen pollution.

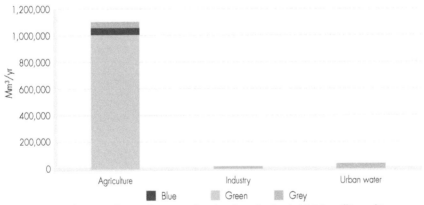

Figure 10.3 The annual water footprint of national production in LAC (in million cubic metres, Mm³), average for the period 1996–2005. *Source: Mekonnen and Hoekstra (2011)*

10.4.1 Urban and industrial uses

Figures 10.4 and 10.5 show the water footprint for domestic water supply and for industrial production for several countries of the LAC region. These values are inversely related to water productivity as was defined in section 10.2.2.

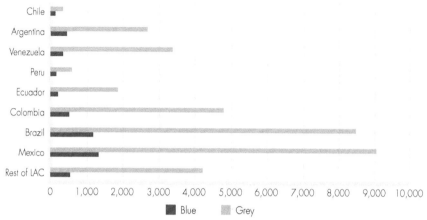

Figure 10.4 Annual water footprint of domestic water supply (in million cubic metres, Mm³), average for the period 1996–2005. *Source: Mekonnen and Hoekstra (2011)*

The water footprint of domestic water supply is determined by the grey water footprint; the grey footprint represents close to 88% of LAC's total water footprint for domestic water supply. Mexico has the highest value for its grey water footprint of domestic water supply and sanitation, followed by Brazil, Colombia, Venezuela and Argentina, in decreasing order. These five countries represent approximately 80% of LAC's domestic water supply grey water footprint. On the other hand, Chile has one of the lowest grey water footprints for domestic water supply for the southern sub-region. This is a reflection of the significant increase in the coverage of water treatment in the past decade, which has changed from 10% in 1990 to 80% in 2010.

Chile and Peru have the lowest blue water footprints for domestic water supply, accounting for 6% of LAC's total domestic supply blue water footprint. In contrast, Brazil and Mexico have the highest blue water footprints; theirs being eight times that of Chile and Peru.

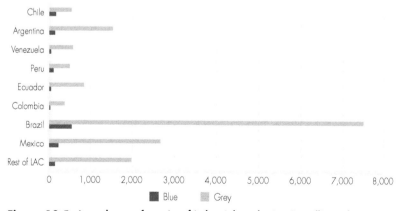

Figure 10.5 Annual water footprint of industrial production (in million cubic metres, Mm³), average for the period 1996–2005. *Source: Mekonnen and Hoekstra (2011)*

As was the case with the domestic water supply water footprint, the industrial production water footprint is mainly composed of the grey water footprint. The industrial production grey water footprint accounts for over 90% of its total water footprint. However, it is important to note that the industrial grey water footprint is less than half the value of the domestic water supply grey footprint. Brazil is by far the country with the highest industrial production grey water footprint. Mexico, the country with the second highest grey water footprint related to industrial production has a grey water footprint 65% lower than that of Brazil. Chile and Peru have the lowest figures, while Argentina has a medium-level industrial production grey footprint.

Mexico and Brazil also have the highest industrial production blue water footprint, thus these countries have the lowest industrial water productivities. The highest blue water productivities for industrial production are found in Colombia, Ecuador, and Venezuela; their industrial blue water footprints range from 20 to 45Mm³/yr. Medium industrial blue water productivity countries are Argentina, Chile, and Peru, with industrial blue water footprints from 102 to 158Mm³/yr.

10.4.2 Agricultural use

It is evident from Figure 10.6 that the LAC region relies extensively on rain-fed production systems, as the green water footprint is the most important component of the total crop production water footprint in LAC. Crop production in Argentina and Brazil has the highest crop production water footprint (Figure 10.6) whilst Mexico has a crop water footprint close to the average for the rest of LAC.

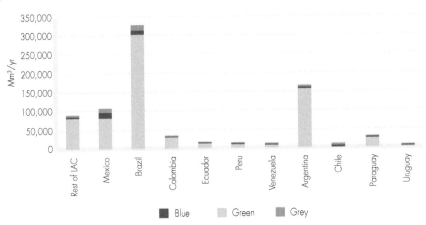

Figure 10.6 Total water footprint of agricultural crop production for the LAC region (average 1996–2005). *Source: Mekonnen and Hoekstra (2011).*

Mexico and Brazil have the highest blue water footprint for crop production, ranging from 9,000 to 14,000Mm³/yr. Medium-range crop production blue water productivities can be found in Peru, Argentina, Chile and Colombia; the average blue water footprint of these countries ranges between 2,500 and 4,000Mm³/yr.

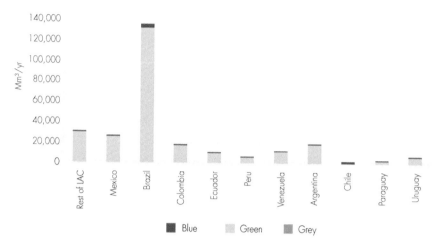

Figure 10.7 Water footprint of livestock production (Mm³/yr), period 1996–2005. *Source: Mekonnen and Hoekstra (2011).*

As in the case of crop production, the most significant component of livestock's water footprint in LAC is the green water footprint (Figure 10.7). Brazil stands out as the country with the highest water footprint of LAC countries although livestock's blue water footprint only represents 4% of the total water footprint in this country.

10.4.3 Energy production

The relationship between water and energy is mainly characterized by hydropower generation. The main hydropower producing countries in the world belong to the OECD and are responsible for 42% of the entire hydroelectric output. Asian countries are responsible for 26%, where China is the main contributor. LAC has a hydropower production share of 20%, mostly contributed by Brazil, which produces almost 12% of the world's total. In Brazil, 75% of the electric power is provided by hydropower.

Hydropower generation is generally associated with a reservoir, which accumulates water in order to maintain a regular flow regime. The evaporation rates in these reservoirs drive water losses in watersheds with hydroelectric dams or reservoirs. This factor gives hydropower dams a consumptive profile in terms of water use, an important fact which is in general overlooked in national or regional water plans.

An interesting indicator of water efficiency in the case of hydropower reservoirs is the ratio between the amount of water evaporated and the capacity for electricity generation. Mekonnen and Hoekstra (2011), exploring this indicator, have presented a preliminary study on hydroelectricity water efficiency. The authors used an evaporation database of thirty-five hydropower reservoirs throughout the world, eight of them in Brazil. The authors' results indicate that hydropower's blue water footprint averages from 140 and 244L/kWh for potential capacity and real charges, respectively.

In Brazil there are more than a hundred hydropower reservoirs with nominal capacities over 30MW. Sousa and Reid (2010) presented a blue water footprint assessment of the main Brazilian hydropower reservoirs, based on their estimated evaporation rates. They have found values ranging from 0.47 to 399.84L/kWh, for sixty-six studied reservoirs. The average blue water footprint was 35.46 L/kWh. The results are not directly comparable to Mekonnen and Hoekstra's range due to methodological differences with respect to real evaporation estimates. For the case of Chile, the average blue water footprint of hydroelectric reservoirs was 45L/kWh.

A similar study conducted by Torcellini et al. (2003) has estimated an average blue water footprint of 68L/kWh for the US's hydropower reservoirs. Blue water footprints of hydropower reservoirs are generally much higher than those of other energy sources. For example, Torcellini's values for hydropower reservoirs are thirty times higher than those found for thermoelectric plants.

10.5 Economic efficiency in the use of water resources in Latin America from the production perspective

As mentioned in section 10.4, in the LAC region as a whole, the largest water user is the agricultural sector, accounting for 99% of the green and blue water consumption and 46% of the nitrogen-related pollution (Figure 10.3), while it accounts for between 1 and 23% of the total Gross Domestic Product (GDP) and employs from 1 to 36% of the economically active population. Urban water supply represents as much as 0.5% of the total water consumed and 37% of the total nitrogen pollution. Meanwhile the industrial sector represents just 0.1% of the total water consumed and 17% of the total nitrogen pollution, while it contributes from 15 to 68% to the GDP that it generates and employs from 13 to 32% of the economically active population.

Economic efficiency of water use for the industrial sector in LAC is on average US$ 155/m³ (see Figure 10.8). Agriculture's water efficiency in LAC is significantly lower, with an average value of US$ 5/m³.

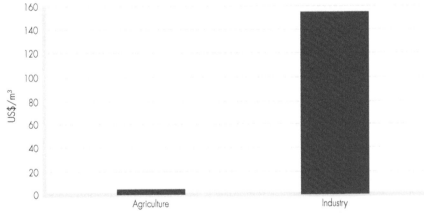

Figure 10.8 Economic water productivity (US$/m³) in agriculture and industry in LAC countries (2011). *Source: Mekonnen and Hoekstra (2011).*

10.5.1 Industrial use

As Figure 10.9 indicates, Colombia, Venezuela, and Uruguay are LAC countries with the highest economic water efficiencies in the industrial sector. These countries present economic water efficiencies from US$ 280/m³ to US$ 300/m³.

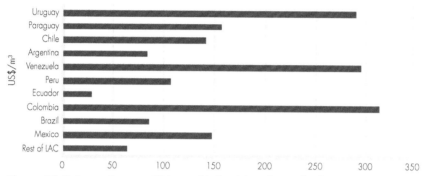

Figure 10.9 Economic water efficiency of industrial production for the LAC region (average 1996-2005) (US$/m³). *Source: Mekonnen and Hoekstra (2011).*

Paraguay, Mexico, Chile and Peru show medium figures for the economic water efficiency indicator for their industrial sector (US$ 140/m³ to US$ 155/m³). The countries with the lowest economic water efficiency in their industrial sectors are Ecuador, Argentina, and Brazil, with an economic efficiency indicator which varies between US$ 27/m³ and US$ 80/m³.

10.5.2 Agricultural use

As pointed out previously, agriculture is the productive sector with the lowest economic water efficiency, with values between US$ 0.15/m³ and US$ 35/m³ (see Figure 10.10). The highest economic water efficiencies can be found in Venezuela and Uruguay. All other LAC countries have low economic water efficiencies for their agricultural sectors which are all less than US$ 1/m³.

Figure 10.10 Economic water efficiency of agricultural production for the LAC region (average 1996–2005) (US$/m³). *Source: Mekonnen and Hoekstra (2011).*

10.6 Environmental impacts of increased water efficiency

With reference to the environment, environmental efficiency is defined as the ratio of the minimum feasible use of an environmentally detrimental input to the observed use of said input, given the technology and the observed levels of outputs and conventional inputs (Reinhard et al., 2002). Whilst resilience is defined as the ability of a system to withstand perturbations or shocks (Gunderson and Light, 2006). In the case of water ecosystems, these perturbations could come from droughts or floods for example, or could be related to changes in water availability and water quality. Thus, improved efficiency in water use in an economic sector such as agriculture or urban water demand could increase the occurrence of environmental impacts in other ways (Box 10.1). However, it could also improve water availability in terms of both the quantity and the quality of the water. For example, the modernization of irrigation infrastructures is likely to increase energy demand, which in turn could increase water requirements to produce this energy. Thus, it is important to consider these environmental effects when projects that increase water efficiency are evaluated. The current challenge is to improve water efficiency whilst maintaining environmental sustainability (Ulanowicz et al., 2009).

Box 10.1 Environmental implications of irrigation modernization

[Adapted from Eugenio Gómez Reyes 'Inventario de recursos hídricos e implicaciones de la modernización del riego' in LA-Mexico (2012)]

Modernization of irrigation is widely viewed as a water-conservation strategy by policy makers who wish to increase water availability for human consumption and the environment. However, the adoption of more efficient irrigation technologies has not always achieved this desired result. There may be a rebound effect; water efficiency means the same production can be delivered with less water, but in fact more can be produced with the same amount of water. Furthermore, irrigation modernization reduces return flows, decreasing available water resources downstream. Additionally, reducing return flows leads to less leaching of pollutants; however, water available to absorb the contamination is also reduced. Ward and Pulido-Velázquez (2008) developed an integrated basin-scale analysis in the Upper Río Grande basin of North America (New Mexico) in order to study the effects of several water conservation policies on irrigation use and on water saved. They observed that incentive-based water conservation tools promote a change in the crop mix with more productive and water-intensive crops thus increasing the net farm income but also increasing the total water depleted. Subsequently, the adoption of water conservation technologies leads, in several cases, to an expansion of irrigated acreage (Pfeiffer and Lin, 2010).

During the last few decades, several irrigation programmes have been developed by Mexico's government in an attempt to improve water efficiency in irrigated agriculture. These water conservation programmes have often been developed without considering important factors in decision-making such as an integrated basin-scale analysis. Despite the large financial resources allocated in these projects, the main objective of water saving has not been achieved. Similarly, an analysis of Chile's agricultural census data of 1997 and 2007 indicates that irrigation efficiencies have increased significantly reaching 58% in 2007. However, during the same period, agriculture's water footprint increased. Thus it can be seen that policies focused on reducing water application do not necessarily always lead to water conservation.

As such Li and Yang (2011), conclude that a system's network must maintain a balance be-tween two essential but complementary attributes: efficiency and resilience. This is demonstrated in the current renewed interest of environmental water flows. In general it is an 'abstract water use' representing water quantities that ought to be maintained in streams and underground in order to sustain the system's functionality. According to Holling and Meffe (1997) the pathology of natural resource management arises when the range of natural variation in a system is reduced thereby producing resilience losses. In short, the balance to be struck be-tween the efficiency in the system (performance) and its resilience (reserve capacity) means en-suring more resource efficient systems: which use less land, water and inputs in order to pro-duce more food sustainably, while at the same time maintaining resilience to changes and shocks. Thus, this section introduces a certain note of caution in the pursuit of efficiency.

10.7 Conclusions and recommendations

The LAC region is fortunate enough to be endowed with an abundance of freshwater, possessing 31% of the world's freshwater resources (UNEP, 2010). This has contributed to the general perception that water is an abundant resource that is always available. This culture of abundance combined with a low educational level of farmers has resulted in the inefficient use of water. Moreover, several factors such as population growth, rapid urbanization, and increased water demands due to increased economic growth are putting considerable pressure on available water resources. Decoupling economic growth from water use is at the core of innovation strategies for sustainable consumption, production and ultimately resource efficiency.

In LAC, as in other regions of the world, agriculture is the main user of freshwater and more than 90% of the water consumed by this sector is green water. The greatest opportunity for an improvement in water productivity and efficiency is in rain-fed agriculture through enhanced and known water management practices. In general, irrigation efficiency of the existing systems in LAC countries also falls below expectations, due to the predominance of traditional surface irrigation technologies. In this region, irrigation effi-ciency ranges between 30 and 40% with the average reported at 39%; whereas the world average is 56%. These percentages indicate that there is a great potential of increasing water productivity in the region by switching to more efficient water application methods. However, future increases in irrigation efficiency in LAC countries must minimize unwanted consequences such as salinization, waterlogging, and increases in total water consumption (rebound effect).

Urban water use in LAC also has low technical water efficiency figures relative to developed countries; on average, water conveyance efficiency is reported to be 58.81%. Therefore increasing water demands due to a growing population and rapid urbanization requires increased technical efficiencies in the urban sector. There is also room for improvement with regard to the water efficiency in the electric sector.

Thus LAC countries must improve their water use efficiencies by addressing the following three major challenges (UNEP, 2011; 2013). First, the development of a water accounting system that considers the environment. This is essential in order to achieve the goals of increased water efficiency in a sustainable manner. That is, minimizing undesired environmental impacts such as salinization, decreased water availability for downstream users and increased total water consumption. Second, the implementation of transparent and comprehensive accounting systems will serve as an incentive to adopt best water management practices in agriculture so as to reduce environmental impacts. Third, the development of effective coordination mechanisms between authorities from different sectors and policies, at both national and river basin level, could ensure that their policies and objectives are mutually consistent and do not undermine each other.

In addition, appropriate policy instruments must be considered that provide adequate incentives to use water resources efficiently and ultimately achieve a more sustainable use of water in all sectors. This means that water users must consider water as a valuable resource; that is, water should be considered an economic good, as was originally recognized at the Dublin conference on Water and the Environment (ICWE, 1996). There are several policy instruments available that internalize the value of water resources when making water-use decisions; examples of these are water tariffs, water pricing, and water rights markets, among others. Chapter 13 gives an in-depth analysis of the use of these policy instruments in LAC countries.

References

Adger, W.N., Brown, K., Fairbrass, J., Jordan, A., Paavola, J., Rosendo, S. & Seyfang, G. (2003). Governance for sustainability: towards a 'thick' analysis of environmental decision making. *Environment and Planning A*, 35(6): 1095–1110.

Aguilar, B.I. & Castro, J.L. (2010). *Principios Gerenciales y Eficiencia Operativa de los Servicios del Agua: Los casos de Saltillo, Tijuana y Monterrey*, Banco de Desarrollo de América del Norte (BDAN).

Ali, M.H. & Talukder, M.S.U. (2008). Increasing water productivity in crop production – A synthesis. *Agricultural Water Management*, 95(11): 1201–1213.

ANA (2013). Agência Nacional de Águas (Brazil). [Online] Available from: www.ana.gov.br. [Accessed May, 2013].

Barker, R., Dawe, D. & Inocencio, A. (2003). Economics of water productivity in managing water for agriculture. In: Kijne, J.W., Barker, R. & Molden, D. (eds). *Water Productivity in Agriculture: Limits and Opportunities for Improvement*, Wallingford, UK, CABI Publishing, pp.19–35.

Blanke A., Rozelle S., Lohmar B., Wangc, J. & Huang, J. (2007). Water saving technology and saving water in China. *Agricultural Water Management*, 87: 139–150.

CAF (2013). Corporación Andina de Fomento. [Online] Available from: www.caf.com/es. [Accessed May, 2013].

CAWMA (2007). Comprehensive Assessment of Water Management in Agriculture. *Water for Food, Water for Life: A Comprehensive Assessment of Water Management in Agriculture*, International Water Management Institute, London, Earthscan.

CEPAL (2010). Economic Commission for Latin America. *Objetivos de Desarrollo del Milenio. Avances en la sostenibilidad ambiental del desarrollo en America Latina y el Caribe.* [Online] Available from: www.eclac.cl/cgi-bin/getprod.asp?xml=/publicaciones/xml/6/38496/P38496.xml&xsl=/dmaah/. [Accessed May, 2013]. Online report number LC/G.2428-P.

Dagnino, M. & Ward, F.A. (2012). Economics of agricultural water conservation: empirical analysis and policy implications. *International Journal of Water Resources Development*, 28 (4): 577-600.

Dumont, A., Mayor, B. & López-Gunn, E. (2013). Is the rebound effect or Jevons paradox a useful concept for better management of water resources? Insights from the irrigation modernisation process in Spain. At the water confluence. *Aquatic procedia*, 1: 64–76.

FAO (2003). Food and Agricultural Organization of the United Nations. *Unlocking the Water Potential of Agriculture.* [Online] Available from: www.fao.org/docrep/006/y4525e/y4525e00.htm#Contents. [Accessed May, 2013].

FAO (2013). Food and Agricultural Organization of the United Nations. *Irrigation water requirement and water withdrawal by country.* [Online] Available from: www.fao.org/nr/water/aquastat/water_use_agr/index7.stm. [Accessed May, 2013]. Online report.

Garces-Restrepo, C., Vermillion, D. & Muñoz, G. (2007). *Irrigation Management Transfer*, Rome, FAO. Water Reports number 32.

Gilmont, M. (2013). Decoupling dependence on natural water: reflexivity in the regulation and allocation of water in Israel, *Water Policy*. In press.

GWP (2006). Global Water Partnership. *Taking an integrated approach to improving water efficiency*, Technical Committee (TEC). Technical Brief No. 4.

GWP (2011). Global Water Partnership. *Situación de los recursos hídricos en Centroamérica: hacia una gestión integrada:* Publicaciones de la oficina regional para Centroamérica. [Online] Available from: www.gwp.org/Global/GWP-CAm_Files/SituaciondelosRecursosHidricos.pdf. [Accessed May, 2013]. Online report.

GWP (2012). Global Water Partnership. *Towards Integrated Urban Water Management.* Perspectives papers.

Gunderson, L. & Light, S.S. (2006). Adaptive management and adaptive governance in the Everglades ecosystem. *Policy Science*, 39: 323–334.

Hoekstra, A.Y. (2013). *The water footprint of modern consumer society.* Routledge, London.

Holling, C.S. & Meffe, G. (1996). Command and control and the pathology of natural resource management. *Conservation Biology*, 10: 10328-10337.

ICC (2007). International Consulting Corporation. [Online] www.icc-us.com. [Accessed May, 2013].

ICWE (1996). International Conference on Water and the Environment. *The Dublin Statement and Report of the Conference.* [Online] Available from. docs.watsan.net/Scanned_PDF_Files/Class_Code_7_Conference/71-ICWE92-9739.pdf. [Accessed May, 2013].

Jollands, N. (2006). Concepts of efficiency in ecological economics: Sisyphus and the decision maker. *Ecological Economics*, 56(3): 359–372.

LA–Mexico (2012). *Report of Mexico.* Contribution to the Water and Food Security in Latin America Project. Water Observatory Project, Madrid, Fundacion Botín.

Li, Y & Yang, Z.F. (2011). Quantifying the sustainability of water use systems: calculating the balance between network efficiency and resilience. *Ecological Modelling*, 222: 1771-1780.

Mekonnen, M.M. & Hoekstra, A.Y. (2011). *National water footprint accounts: the green, blue and grey water footprint of production and consumption,* UNESCO-IHE. Value of Water Research Report Series No. 50.

Molden, D. (1997). *Accounting for water use and productivity.* International Irrigation Management Institu-te, Colombo, Sri-Lanka. SWIM Paper No. 1.

Molden, D., Murray-Rust, H., Sakthivadivel, R., & Makin, I. (2003). A water-productivity framework for understanding and action. In: Kijne, J.W., Barker R., & Molden D. (eds). *Water Productivity in Agriculture: Limits and Opportunities for Improvement,* Wallingford, UK, CABI Publishing, pp. 1-18.

Molden, D., Oweis, T., Steduto, P., Bindraban, P., Hanjra, M. & Kijine, J. (2010). Improving agricultural water productivity: Between optimism and caution. *Agricultural Water Management*, 97: 528-535.

de Oliveira, A.S., Trezza, R., Holzapfel, E.A., Lorite, I. & Paz, V.P.S. (2009). Irrigation water management In Latin America. *Chilean Journal of Agricultural Research*, 69 (Suppl. 1): 7–16.

Pasha, M.R. (2002). Inefficient water management: its impact on economic growth. *The Journal*, 7(4): 87–98.

Perry, C. (2007). Efficient rrigation: inefficient communication; flawed recommendations. *Irrigation and Drainage*, 56: 367–78.

Perry, C., Steduto, P., Allen, R. & Burt, C. (2009). Increasing productivity in irrigated agriculture: Agronomic constraints and hydrological realities. *Agricultural Water Management*, 98: 1517–1524.

Pfeiffer, L. & Lin, C.Y. (2010). The effect of irrigation technology in groundwater use. *Choices* 25(3).

Pierce, G. (2012). The political economy of water service privatization in Mexico City, 1994–2011. *International Journal of Water Resources Development*, 28(4): 675–691.

Pretty, J., Noble, A., Bossio, D., Dixon, J., Hine, R.E., Penning de Vries, P. & Morison, J.I.L. (2006). Resource conserving agriculture increases yields in developing countries. *Environmental Science and Technology*, 40(4): 1114–1119.

Reinhard, S., Lovell, C.K., & Thijssen, G. (2002). Analysis of environmental efficiency variation. *American Journal of Agricultural Economics*, 84(4): 1054–1065.

Richter, B.D., Davis, M.M., Apse, C. & Konrad, C. (2011). A presumptive standard for environmental flow protection. *River Research and Applications*, 28(8): 1312–1321.

Ringler, C., Biswas, A. & Cline, S.A. (eds). (2010). *Global Change: Impact on Water and Food Security,* Berlin-Heidelberg, Springer-Verlag.

Rosegrant, M., Cai, X., & Cline, S. (2003). Will the world run dry? global water and food security. *Environment 45(7)*: 24–36.

Sharma, P. & Sharma, R. (2004). Groundwater markets across climatic zones: a comparative study of arid and semiarid zones of Rajasthan. *India Journal of Agricultural Economics*, 59: 138–150.

Schuck, E., Frasier, W.M., Webb, R.S., Ellingson, L.J. & Umberger, W.J. (2005). Adoption of more technically efficient irrigation systems as a drought response. *Water Resource Development*, 21: 651–662.

Sousa Júnior, W.C. & Reid, J. (2010). Uncertainties in Amazon hydropower development: Risk scenarios and environmental issues around the Belo Monte dam. *Water Alternatives*, 3(2): 249–268.

Torcellini, P., Long, N. & Judkoff, R. (2003). *Consumptive Water Use for US Power Production.* Con-ference Paper presented at the ASHRAE Winter Meeting Anaheim, California, January 24–28, 2004. Colorado, National Renewable Energy Laboratory.

UNEP (2010). United Nations Environment Programme. International Payments for Ecosystem Services. UNEP Division of Technology, Industry and Economics. Economics and Trade Branch. [Online] Available from: www.unep.ch/etb/events/IPES%20Side%20Event%20Bonn/IPES%20SUM%20FINAL.pdf. [Accessed May, 2013].

UNEP (2011). United Nations Environment Programme. *Resource Efficiency: Economics and Outlook for Latin American Case Studies: MERCOSUR, Chile and Mexico.* [Online] Available from: www.pnuma.org/reeo. [Accessed June, 2013].

UNEP (2012). United Nations Environment Programme. *Global Environment Outlook 5* (GEO-5), Nairobi, Kenya.

UNEP (2013). United Nations Environment Programme. *The State of Natural Resource Manage-ment in Latin America and the Caribbean: Opportunities for Sustainable Practices and Prioriti-sation of Resources:* UNEP Division of Technology, Industry and Economics (UNEP DTIE) and UNEP Regional Office for Latin America and the Caribbean (PNUMA ROLAC).

Ulanowicz, R.E., Goerner, S.J., Lietaer, B. & Gomez, R. (2009). Quantifying sustainability: Resi-lience, efficiency and the return of information theory. *Ecological Complexity*, 2009: 27–36.

van Halsema, G.E. & Vincent, L. (2012). Efficiency and productivity terms for water management: A matter of contextual relativism versus general absolutism. *Agricultural Water Management*, 108: 9–15.

Ward, F.A. & Pulido-Velázquez, M. (2008). Water conservation in irrigation can increase water use. *Proceedings of the National Academy of Sciences*, 105(47): 18215–18220.

Wichelns, D. (2002). An economic perspective on the potential gains from improvements in irriga-tion water management. *Agricultural Water Management*, 52 (3): 233–248.

World Bank (2013). *World Development Indicators Database.* Washington DC, The World Bank.

REFORMING WATER GOVERNANCE STRUCTURES

Coordinator:
Lucia De Stefano, Water Observatory – Botín Foundation, and Universidad Complutense de Madrid, Spain

Authors:
Pedro Roberto Jacobi, PROCAM /IEE Universidade de São Paulo, Brazil
Lucia De Stefano, Water Observatory – Botín Foundation, and Universidad Complutense de Madrid, Spain
Elena López-Gunn, I-Catalist, Complutense University of Madrid, and Water Observatory – Botín Foundation, Spain
Miguel Solanes, IMDEA Agua – Madrid Institute for Advanced Studies, Spain
Gonzalo Delacámara, IMDEA Agua – Madrid Institute for Advanced Studies, Spain
Gonzalo Marín, Fundación Canal de Isabel II, Spain
Antonio Embid, Departamento de Derecho Público, Universidad de Zaragoza, Spain
Vanessa Empinotti, PROCAM /IEE Universidade de São Paulo, Brazil
Elisa Blanco, Departamento de economía agraria – Pontificia Universidad Católica de Chile, Santiago, Chile
Guillermo Donoso, Pontificia Universidad Católica de Chile, Santiago, Chile
Marta Rica, Water Observatory – Botín Foundation, and Universidad Complutense de Madrid, Spain
Natalia Uribe, WaterLex, Switzerland
Alejandro Jiménez, Stockholm International Water Institute, Sweden

Contributors:
Luis F. Castro, School of Civil Engineering, Universidad Nacional de Ingenieria, Peru
Thalia Hernández Amezcua, Universidad Autónoma de Mexico, Mexico
Julio M. Kuroiwa, Laboratorio Nacional de Hidráulica – Universidad Nacional de Ingeniería, Lima, Peru
Marielena N. Lucen, Ministry of Energy and Mines, Peru
Julio I. Montenegro, School of Civil Engineering, Universidad Nacional de Ingenieria, Peru
Rosario Pérez Espejo, Universidad Autónoma de Mexico, Mexico
Patricia Phumpiu Chang, Centro del Agua para América Latina y el Caribe – ITESM, Monterrey, Mexico
Pedro Zorrilla-Miras, Cooperativa Terrativa, Madrid, Spain

Highlights

- Achieving long-lasting water and food security needs to be based on a solid foundation, represented by governance institutions that are able to ensure a fair framework for development. During the past three decades Latin America and Caribbean (LAC) has undergone significant institutional water reforms triggered by a number of factors, among which are the demands from civil society for more inclusive, sustainable, efficient and effective water governance, as well as the influence of international organizations promoting the introduction of Integrated Water Resources (IWRM) and other paradigms in LAC water governance structures.

- Some common trends in those reforms include: a shift towards decentralization, often complemented with the creation of coordination and supervising bodies at a higher level; the formulation of new water laws and policies that include a number of IWRM principles (environmental sustainability, integration, participation, accountability, transparency, cost recovery, etc.); the legal support of the right to water and sanitation; and the creation of water use levies and tariffs for cost recovery.

- In some countries the focus is now on adjusting and implementing those reforms, while others are still in the process of debating and formulating them. The main challenges for the implementation of ongoing reforms are related to the lack of integrated planning of water use, the poor coordination of the main stakeholders (both governmental and non-governmental), and the need for management instruments that may fit local conditions better.

- In its search for improved water security, LAC has pioneered the recognition of the access to safe water and sanitation as a human right. The countries' attention is now on the implementation of that right. The inclusion of the right to water and sanitation in most of the constitutional texts or laws is a first important step, which, however, has to be followed by clear financial and regulatory efforts.

- During the past three decades, private and public domestic operators have participated in the provision of water and sanitation. The analysis of past experiences suggests that the focus of reforms should be on creating favourable conditions for a quality and equitable service, which can be achieved only through ensuring strong governance, in general and specific for water.

- Funding of the water sector remains a challenge; governments struggle and usually fail to meet financial requirements. Despite the gradual introduction of tariffs and charges, revenues from the water sector are still insufficient to cover its financial needs. International public and private investors play a key role in filling that gap, with a clear emphasis on the development of infrastructure for domestic supply provision.

11.1 Introduction

A constant challenge worldwide is set by the need to count on adaptive institutions that strengthen democracy and promote growth and social development. In Latin American and Caribbean (LAC) countries there is a clear need to improve access to water, guarantee the quality of water for all uses, and enhance ecosystem services (Akhmouch, 2012). This makes the challenge of improved water governance particularly present and pressing in LAC countries, which often lack adequate institutional water systems (Crase and Gandhi, 2009; Akhmouch, 2012; Jiménez-Cisneros & Galizia-Tundisi, 2012). This chapter focuses on 'blue' water governance, which is a key instrument to achieving water security, while it does not deal *explicitly* with food security. Indeed, although well-performing water institutions do contribute to water security and therefore to food security (Chapter 1), the governance structures framing food security lie outside the water sector. As for green water, in other chapters it is pointed out that key inputs to agriculture and food production are water (blue and green) and land, whose use and management are strongly intertwined in practice but normally managed by different institutions. While this chapter focuses on the governance of the blue part of the land-water system, the institutional framework dealing with land and ecosystem management is discussed in Chapter 14.

Water governance can be defined as a system that makes water management more effective, accountable and participatory, thus strengthening the role of multiple stakeholders in institutional capacity building, improving coordination, broadening participation and consolidating partnerships (Jacobi, 2009). Water governance structures in some LAC have undergone reforms that implied not only re-orientation of policy priorities and approaches, but also the restructuring of institutional frameworks. This has led to the need for new intermediate institutions that enable a negotiated approach to water governance. Two issues hamper the capacity of institutions to improve and adjust to constantly changing conditions: the lack of proper evaluation of the quality of policies – often a consequence of lack of transparency and accountability that may favour some actors and their private interests over others; and the lack of adequate control over bureaucratic systems. Institutional reforms involved changes in the 'rules of the game', expressed by the coexistence of formal laws, informal norms and practices, and organizational structures, as well as strengthening institutional capacity.

The analysis of institutional experiences in the past two decades indicates a wide range of water governance approaches in LAC, which is telling that water management is a social and political issue as well as a technical one. The need to reform institutions has been mainly driven by the fact that the State had to respond to growing demands from civil society and, in particular, from economic sectors to improve its actions. Institutions are also reformed in order to respond to the need to improve their transparency, stimulate social capital, strengthen accountability, promote public interest, reduce institutional obstacles, and improve policy implementation and performance of the public and private sectors.

This chapter deals with water governance and its institutional reaches in LAC, with a special focus on Brazil, Chile, Costa Rica, Mexico and Peru. It first revisits the circumstances that triggered reforms undertaken in the different countries, and presents some reflections about their implementation currently and in the future. Then, the chapter analyses some of the elements that characterize institutional changes promoted by those reforms, while it leaves to other chapters of this book the in-depth description of other aspects (e.g. participation, transparency and accountability, economic instruments, etc.). With that perspective in mind, the role and characteristics of the legal systems for water use that frame and enable water governance, the recognition of the right to water and sanitation as a human right and the conditions needed to ensure its implementation are analysed. Finally, the chapter deals with the challenge of funding reforms and with how countries tap into national and international sources in order to address this issue.

11.2 Institutional setup: past, present, future

In this section, the main characteristics and challenges of reforming water governance structures are considered. The legal and organizational systems presented here constitute the framework within which four different types of actors operate: the state (public) institutions; market (private sector) institutions; activist (NGO) institutions; and civil society in a broad sense (Allan, 2013). Most of the water is used by the private sector (farmers, agribusiness, mining companies, etc.) as one input to their production activity. For these actors the market is the main driver determining production choices and the associated water uses (*ibid.*). One of the main tasks of the water institutional setup presented in this chapter is framing the use of water as a production input and ensuring that it is compatible with long-term water security.

11.2.1 Water reforms in LAC: triggers and trends

Since the 1980s, virtually all countries in the LAC region underwent institutional reforms of their water sector (Jacobi et al., 2009; Hernández et al., 2012) or at least have engaged in a lively debate on how to adjust their water institutions to new challenges posed by the need to address water and food security both as a country and at the scale of urban and rural communities. These reforming processes have been triggered by a number of factors. First, countries need to adjust to new and unseen socio-economic dynamics and the alteration environmental processes brought about by globalization and a strong economic development largely based on the exploitation of natural resources (see Chapters 3 and 4). For instance, in Peru water policy reform was driven by the need to update the 1969 General Water Law, which presented limited cohesion between water quantity, water quality and environmental considerations and did not recognize the economic value of the resource (MINAG, 2009). Second, processes of democratization have spurred demands from society for more inclusive, effective and environmentally sustainable water governance, which had to be reflected in an upgrade of water institutions. Thus, in Brazil the main driver for reforms was the need to approach water management from a regional standpoint and the need to consider the multiple uses of water, as well as the effects of their interrelations (Jacobi et al, 2009). Third, in some cases, major political changes have triggered water reforms. For instance, in Chile the major Water Code reform was driven by the shift towards a more decentralized political context. Economic liberalization enacted during the military regime of 1973–1989 included the 1981 National Water Code, which established transferable water use rights and facilitated water markets (Hearne and Donoso, 2005). Last but not the least, multilateral players – mainly the World Bank and the Inter-American Development Bank – and different international cooperation agencies are often perceived as important drivers of reform and as providers of comprehensive technical and financial support, as well as pro-reform decision-makers (Castro, 2007; Wilder, 2010).

Reforms have taken place mainly through the modification of the legal system and often with the approval of a new Water Act (see Section 11.3); the definition of water resources policies and guiding principles for water management; and in some cases even through bottom-up, informal reforms that have tried to anticipate or adjust top-down mandates to the local contexts (Kauffman, 2011). As a result, LAC countries exhibit coexistence of different approaches to the right to water and water services (as a human right, as a commodity, as a public service); coexistence of a set of formal and informal rules and standards that define different institutional models of water management; and coexistence of multiple state, private and social actors involved in decision-making processes (Hernández et al., 2012). Indeed, different political systems, political-administrative structures and institutional arrangements for water governance define the dynamics of public, private and public capacities for management with different performance results, according to the history and background of each country.

Being aware of the difficulties of generalizing when considering a diverse region such as LAC, it is useful to point out some features of the institutional setting that can be observed in some of the countries. Several LAC countries have decentralized at least some water functions (Table 11.1). In those decentralized models, domestic water supply and sanitation is usually transferred to the local level, while higher-level sub-national governments are responsible for water resources management (Akhmouch, 2012). The decentralization process often has gone hand in hand with the definition of the river basin as a water management unit (see Chapter 2), and in Peru specifically the 2009 Water Act reinforces the need to decentralize water management (participation of users, national regional and local government in the decisions process). In Colombia, the reform of the constitution in

1991 and the subsequent approval of the 1994 water legislation aimed to strengthen private water management institutions, increase private participation in the operation and redefine the role of government in providing public services. In that context, the state's main role is to regulate, support, plan and control the provision of these services, thus driving a process of decentralization and privatization in water management, transferring the operation of water services to the private sector (Hernández et al., 2012).

Table 11.1 Allocation of responsibilities in water governance at sub-national level and the role of the central government in selected LAC countries

COUNTRY	ROLE OF CENTRAL GOVERNMENT (dominant actor or joint role with sub-national governments)	ALLOCATION OF ROLES AND RESPONSIBILITIES IN WATER POLICY DESIGN AND IMPLEMENTATION
ARGENTINA	Joint	Municipalities, inter-municipal bodies, Provinces, River basin organizations
BRAZIL	Joint	Municipalities, Water-specific bodies, States
CHILE	Dominant	Municipalities
COSTA RICA	Dominant	Municipalities, Inter-municipal bodies, Regions, River basin organizations
MEXICO	Dominant	Municipalities, Regions, Water-specific bodies, River basin organizations
CUBA	Dominant	Regions, Municipalities, River basin organizations
DOMINICAN R	Dominant	River basin organizations
EL SALVADOR	Dominant	Municipalities, Inter-municipal bodies, Water-specific bodies, River basin organizations
GUATEMALA	Joint	River basin organizations, Municipalities.
HONDURAS	Joint	Municipalities, Inter-municipal bodies, Water-specific bodies
NICARAGUA	Joint	Regions, Municipalities, Inter-municipal bodies, Water-specific bodies, River basin organizations.
PANAMA	Dominant	Municipalities, others (water committees)
PERU	Joint	Regions, Municipalities, Water-specific bodies, River basin organizations

Source: own elaboration based on Akhmouch (2012).

A second feature common to several LAC countries is the increase of participation of stakeholders in decision-making processes (see Chapter 12), with special emphasis on the role of water users, which in some cases have acquired large control over water use through their associations. For instance, in Mexico the 1992 National Water Law, modified in 2004, created watershed councils to promote and facilitate – at least on paper – the participation of civil society organizations in planning, decision-making, implementation and monitoring of the national water policy at a basin level (Wilder, 2010). In the new institutional design, however, the federal water management agency CONAGUA assumed a policy making and overseeing role and retained key strategic functions (*ibid.*). In Chile, the 1981 Water Code significantly reduced the State's intervention in water resources management to a minimum and increased the management powers of water use right holders, organized into water user associations (Hearne and Donoso, 2005). However, multiple central authorities (ministries, departments, public agencies) continue to be involved in water policy making and regulation at central government level (Donoso, 2014).

While decentralization of water management and participation of water user organizations have been common features in some countries (e.g. Brazil, Chile, Mexico, Peru and Costa Rica), differences arise when taking these guidelines into practice. Brazil and Mexico, for example, implemented decentralized management and established the watershed as the management unit. In Chile, users and water users associations play a central role in the administration of water rights and there have been only timid attempts to establish river basin master plans (Hearne and Donoso, 2005). In Peru, the institutional landscape is characterized by partial decentralization to manage water at a basin level and the establishment of the National Water Authority in charge of managing water resources by basin (Kuroiwa et al., 2014).

The strong demands for democratization and for well-functioning institutions – both in general and in the water sector – has caused vigorous claims for increased accountability of all those involved in determining, influencing or implementing public policies. This has promoted important advances, at least on paper, in terms of transparency and accountability in the LAC region. These advances have often originated from outside the water sector but undoubtedly their effects can be perceived also within it (see Chapter 12).

Another feature common to several LAC countries is the definition of national or regional water policies and strategies that recall principles of IWRM such as policy integration, coordination and cooperation, integrated management of different water sources, environmental sustainability, public participation, planning at a watershed level (Regional Process of the Americas, 2012). Brazil represents a good example of this. During the 1980s, the degradation of Brazil's water resources in areas of large urban–industrial concentration led to pressure from civil society in favour of the improvement of water sources. Thereby, consensus was reached around the need for: the creation of a national water resources system considering multiple water uses, the adoption of references for regional management, decentralized and participatory management, a national water resources information system and technological and capacity development in the area (ANA, 2002; Jacobi et al., 2009). The Water Law came into force in 1997 and consisted of the basic legal text that created the Water Resources National Policy and the National Management System of Water Resources. The resulting policy is based upon four basic principles: a) adoption of the water basin as the management unit; b) the consideration of multiple uses; c) water as an economic good, with an economic value, encouraging its rational use; and d) participatory and decentralized management, providing opportunities to users and the organized civil society to participate in decision-making processes (Barth, 1999; Pagnoccheschi, 2003; Jacobi, 2004). In a similar way, Costa Rican water policy establishes among its goals the achievement of a balance between the use of water resources for human development and the sustainability of ecosystems. The guiding principles for accomplishing this are: integrated water resources management, establishing the human right of access to drinking water and basic sanitation, considering water a public-domain good, using a comprehensive ecosystem approach, encouraging the participation of all stakeholders, and the polluter pays principle.

Other common features that can be identified in the evolution of water institutions in the region are discussed in other sections of this chapter: the legal recognition of the right to water and sanitation and its implications in terms of implementation (Section 11.4) and the early stages of the reinforcement of water tariffs and charges as a means to increase revenues for the water sector and to improve water use efficiency (Section 11.5).

11.2.2 Implementing water reforms: the way forward

In the LAC countries there are both external and internal variables that cause water institutions to operate below par despite the formulation of water reforms. External factors are related to the overall trends in governance and levels of economic and human development already analysed in other parts of this book (Chapters 4 and 6), which constitute crucial enabling conditions for the success of any substantial improvement of water governance. When looking specifically at the water sector, the as yet limited citizen participation, the mismatch between hydrological and administrative boundaries and the insufficient capacity of local and regional governments in relation to their responsibilities have been identified among the most important challenges when designing water policy in several LAC countries (Akhmouch, 2012; Table 11.2).

Moreover, the lack of coordination across administrative levels and sectors creates a duplication of some functions and activities, inefficiencies in the allocation of resources, insufficient and partial performance of certain functions, overlap between institutions, and conflicts of power between them. In this context, institutional problems have led to excessive delays in processing and management decisions; technical shortfalls in the implementation of tasks; and lack of the necessary financial and human resources to carry out the assigned functions (Hernández et al., 2012).

Mexico and Brazil represent two of the most advanced and modern water governance systems in Latin America due to the legislation and institutional reforms focused on watershed management and societal participation, but the implementation of their institutional reform is still under way. For instance, in Brazil there are significant differences between states and also between Water Basin Committees in relation to the consolidation of the current decentralized institutional model (Bechara Elabra and Magrini, 2013), which points to the complexity of the ongoing institutional restructuring. To complete institutional reforms, this restructuring needs to be fully implemented and the National Water Plan be approved. In addition to the modification of the territorial model, major changes are linked to an increased process of privatization of services through public–private partnerships so as to ensure investments that governments are not able to afford. Meanwhile in Mexico there is a need to coordinate the decision-making process and improve communication between different sectors, so as to reach agreement and allow for different stakeholders to participate in decisions. According to Serrano (2007), the consolidation of the reform is incomplete, and the lack of regulations is causing a bottleneck situation within the process.

Table 11.2 Main challenges in water policy making and their relative importance in selected LAC countries

MAIN CHALLENGES IN WATER POLICY MAKING	VERY IMPORTANT	SOMEHOW IMPORTANT	NOT IMPORTANT
Limited citizen participation	Argentina, Chile, Costa Rica, Guatemala, Mexico, Nicaragua, Panama	Brazil, Dominican Republic, Honduras, Peru	
Horizontal coordination across ministries	Argentina, Brazil, Costa Rica, Dominican Republic, Honduras, Nicaragua, Panama	Chile, Guatemala, Mexico, Peru	
Mismatch between hydrological and administrative boundaries	Brazil, Costa Rica, Dominican Republic, Guatemala, Nicaragua, Panama, Peru		Argentina, Honduras
Local and regional government capacity	Chile, Guatemala, Honduras, Mexico, Nicaragua, Panama	Argentina, Brazil, Costa Rica, Peru	
Vertical coordination between levels of government	Brazil, Dominican Republic, Guatemala, Honduras, Panama	Argentina, Chile, Mexico, Nicaragua, Peru	
Economic regulation	Chile, Guatemala, Mexico, Panama, Peru	Argentina, Costa Rica, Dominican Republic, Honduras, Nicaragua	
Managing geographically specific areas	Argentina, Chile, Costa Rica, Panama	Honduras, Nicaragua	Brazil, Dominican Republic, Guatemala, Peru
Allocation of water resources	Guatemala, Mexico, Nicaragua, Panama	Chile, Dominican Republic, Honduras	Argentina, Brazil, Costa Rica
Horizontal coordination among sub-national actors	Costa Rica, Honduras, Panama, Peru	Brazil, Chile, Dominican Republic, Mexico, Nicaragua	Guatemala
Managing the specificities of rural areas	Chile, Costa Rica, Panama	Argentina, Dominican Republic, Honduras, Mexico, Nicaragua, Peru	Guatemala
Managing the specificities of urban/ metropolitan areas	Argentina, Chile, Panama	Brazil, Costa Rica, Honduras, Mexico, Nicaragua, Peru	Dominican Republic, Guatemala
Enforcement of environmental norms	Costa Rica, Mexico, Panama	Chile, Dominican Republic, Honduras, Nicaragua, Peru	Argentina, Brazil, Guatemala

Source: own elaboration based on Akhmouch (2012).

Although operational principles (e.g. accountability, transparency, equity) are established, there are still complications related to the definition of responsibilities and functions.

In Chile, among the internal problems, the principal one is quite possibly the lack of a superior public authority that effectively coordinates all functions performed by public and private institutions in relation to water, supported by the enforcement of water user organizations (Hearne and Donoso, 2005).

In Costa Rica the approach to water resources management has been expressed through a Water Policy and a National Plan of Integrated Water Resources Management. However, these policy instruments are still not fully effective in changing water management practices, since administrative, operational and regulatory roles between government agencies and other water users have not yet been well defined (Astorga, 2010).

11.3 Legal nature of water and water rights

Whereas the general organizational setting and overall principles define the actual (or target) framework for water governance, the legal nature of water (who owns it, who can use it and how) represents the basic 'bricks' or, more precisely, the 'foundations' of the 'institutional building' in each country. Any change in the organizational system and any attempt to change the water policy orientation will have to take into account the water rights system and decide whether to adjust to it, make little amendments or engage in a far-reaching (and far more challenging) reform of those legal foundations.

When talking about water rights in a given country, as a starting point one ought to consider whether it has a Water Act or not. Most of the LAC countries do have one, which for the most part was passed or amended during the past decade. In many cases, the Water Act is complemented with legislation specific for domestic supply and in other cases there is only domestic water supply legislation (Figure 11.1). Having a Water Law, however, does not necessarily imply that this includes all the elements that are widely accepted to be considered good water management principles, especially in the case of Water Acts prior to the 1990s. Additionally, even in the most modern Water Acts, where these issues are included, their formulation or degree of implementation is often lacking (e.g. see Chapter 12 for public participation provisions; Chapter 15 for management at a river basin level).

11.3.1 Ownership of water resources

Unique features distinguish water from other natural resources: mobility, variability and uncertainty in supply, bulkiness, indivisibility, diversity of social, cultural and environmental functions, sequential and multiple use, interdependency among uses and users within a given river basin system, and conflicting cultural and social values. These characteristics can lead to multiple market failures, such as vulnerability to monopoly control and natural monopolies, imperfect competition, externalities, sub-optimal allocation of public-good attributes, risk, uncertainty, imperfect information, and potential for social and environmental inefficiencies and inequity. Institutions must address these failures in order to ensure efficient resource use and allocation. Thus, water is different from an ordinary commodity, although it can be traded using due caution. It is a free access and sometimes a common good, which, in absence of regulation is characterized by non-exclusion and rivalry and thus is prone to free riders. The characteristics of water have important consequences concerning its ownership, water rights systems, management institutions,

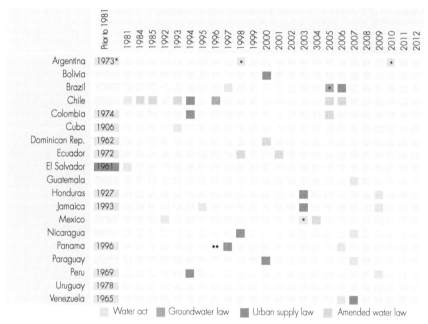

Figure 11.1 Timeline of the approval of the Water Act, domestic supply legislation and specific groundwater law in selected LAC countries. One asterisk indicates laws that apply only to part of the country's territory (province or state). Two asterisks: it is a law on natural resources with a specific section on water. *Source: own elaboration based on data from WaterLex and FAO Legal Office WaterLex.*

and conflict-solving mechanisms (Hanemann, 2006). Thus most regulatory schemes consider the establishment of exclusive access through the definitions of water use rights.

In most legal systems, water belongs to the public domain of the State. The principle of public ownership and control is a feature of both Western and Eastern water law (Bonfante, 1929; Wohlwend, 1975; Caponera, 1992; Ke, 1993). In general, legislation in the LAC region defines water as a 'public domain', 'national waters', 'national goods of public domain', 'property of the Nation' etc. Public ownership of water resources is the principle in force e.g. in Argentina, Brazil, Chile, Ecuador and Mexico, along with other LAC countries (see Table 11.3). However, similar terms do not mean the same thing in different countries. For example, the concept of public property in Chile has little to do with the features found in other countries.

11.3.2 Water rights

Although water belongs to the public domain, water use rights granted to economic agents are protected as private property. A system of secure and stable water rights is an incentive for investments in the development and conservation of water resources, and prevents the social unrest that would result from ignoring existing uses at times of change

Table 11.3 Ownership of water in selected LAC countries

COUNTRY	OWNERSHIP OF WATER [1]
ARGENTINA	The provinces have the original dominion over the natural resources existing in their territory.
BRAZIL	Ownership of water resources rests with the Union and, in some cases, with the states.
CHILE	With few exceptions, water is national property.
CUBA	Ownership of water resources is vested originally in the State.
DOMINICAN REP.	All waters in the country, without any exception, are the property of the State.
ECUADOR	Surface and underground water resources [...], including those which were previously privately owned, are deemed to be national property and for public use.
GUYANA	The State is the owner of all waters of the country and its rights of use.
MEXICO	The ownership of land and waters within the boundaries of the national territory corresponds to the Nation.
PANAMA	All waters within the national territory are public domain goods belonging to the State and belong to it.
PARAGUAY	Surface and ground waters are public domain property of the State.
PERU	Natural resources, renewable and non renewable, are patrimony of the Nation.
URUGUAY	Surface waters as well as subterranean waters, except for rainwater, integrated into the hydrological cycle constitute a unitary resource of public interest, which, as the public hydraulic domain, constitutes part of the public domain of the State.
VENEZUELA	All the waters are goods of public domain belonging to the Nation.

Source: own elaboration based on data from FAO Legal Office WaterLex, WaterLex Legal Database on the Human Right to Water and Sanitation, www.senado.gov.ar, www.congreso.gob.pe, and www.tsj.gov.ve.

in water legislation (Conac, 1991). A water right is usually a right to use (i.e. withdraw water or dispose polluting effluents). Ownership normally means a usufructuary power, and not ownership of the body of water itself (Getches, 1990; Tarlock et al., 2002). However, property rights to water use are conditioned.

11.3.3 Conditions on water rights

In most countries water rights are complemented by a requirement of effective and beneficial use. In virtually all jurisdictions, the allocation and permanency of water rights are contingent upon allocating them to a socially recognized beneficial use (CEPAL, 1995). When water rights are not utilized they are lost under the forfeiture and abandonment provisions of water legislation. Other conditionalities on water rights include provisions concerning no harm to third parties and the environment. Furthermore, in some countries water rights have been adjusted as new knowledge developed or conditions change, since the government has a permanent duty to monitor the use of water, under public trust obligations. Rights not subject to conditionalities of effective and beneficial use facilitate monopolization and have other negative features in cases of water trade: they can be traded according to their nominal entitlements, and not on the basis of effectively consumed

---

1 Non-official translations.

water. Chile allows the trading of nominal water entitlements, just as Australia does. As a result, trade deprives the environment and users of export areas of water, available so far. Negative externalities to the environment and third parties are thus difficult to control (Young, 2010, 2011, 2012; Donoso, 2011).

11.3.4 Theory versus practice

It is worth mentioning the difference between written water law and its implementation in practice. It is possible to find Water Acts that are very elaborated and complete, but this does not necessarily mean that they are fully implemented and enforced on the ground. Shortcomings in this sense can be observed in the management of water resources by river basin, the limited role of water tariffs, the difficulties associated with the protection of water and water ecosystems or the achievement of true public participation. Pitfalls in the design and reliability of water rights registers are also common even in countries with a well-developed legal water system as is the case of Chile. This is particularly important in the case of groundwater, where the establishment and continuous updating of registers of water use rights is considered to be crucial in laying the foundations of groundwater management (GEF, 2012).

Even if in Argentina, Brazil, Colombia or Mexico the situation is notably better than in the remainder of the region, LAC still faces challenges in terms of designing and enforcing more advanced legal water systems. For instance, the poor application of environmental laws to protect water quality is a clear shortcoming in the region, where mining, industry and even urban areas can be non-compliant with the law without serious legal or economic consequences (see for instance Chapter 9). This also applies to the non-compliance in other sectors, as is the case of the Madre de Dios river (Peru). Here there is illegal exploitation of gold following intense deforestation and large amounts of mercury are used to separate gold from the metal ore. There is no control of the effluents, which are left untreated and cause severe water pollution (Kuroiwa et al., 2014). This suggests that water protection cannot be achieved only with water-related laws and that, in any case, their effectiveness is linked to a global improvement of the rule of law, poverty reduction and the building capacity of the local population.

Another notable gap – which is not unique to the region (De Stefano & Lopez-Gunn, 2012) – is the enforcement of groundwater water rights (GEF, 2012). Groundwater is a classic example of common pool resource and for this reason it is prone to overuse in the absence of sound management practices. An example of poor enforcement of legal regulation can be found in the Guanajuato State, where the economy and a fast-growing population have led to the drilling of around 17,000 wells since the early 1970s. Those wells ten years ago were abstracting approximately 4,000 Million m^3/yr (about 1,200 Million m^3/yr more than the renewable resource). Aquifer depletion was occurring at rates of 2–3m/yr, and had important effects on water security in the area (Foster et al., 2004).

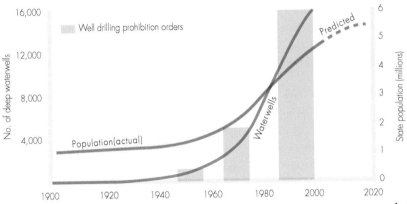

Figure 11.2 Growth of population and water well drilling in Guanajuato State, even during well drilling prohibition orders. *Source: Foster et al. (2004)*

In the 1990s the Mexican federal government made major efforts to register and control groundwater abstraction, including the issuing of three well-drilling bans, but the number of deep wells experienced a sharp increase despite the bans (Figure 11.2). Thus, the lack of capacity for field implementation and the clash of interests between the law and socio-economic trends favoured by groundwater use caused lack of consistent enforcement of the bans and pointed to the need for finding solutions to aquifer depletion not only based on command-and-control approaches (Foster et al., 2004).

11.4 The recognition of the human right to water and sanitation and the MDGs

In LAC the access to adequate water and sanitation is still a major challenge, both in terms of the share of population served and in terms of the need to address large spatial and social disparities in the service coverage (Chapter 6). There is no doubt that addressing this challenge is not just a matter of building water infrastructure but also a matter of counting on institutions that are able to create favourable conditions (regulatory, financial, social) that allow infrastructures to meet the goal they were designed for. For instance, if institutions fail in preserving the ecosystems that actually provide water, it will be increasingly more difficult (and expensive) to actually supply the pipeline network with good quality water. If institutions fail in setting up a sound and long-lasting system to finance the operation and maintenance of existing water distribution and sanitation systems, the quality and equity of the service will inevitably suffer. Thus, the broad recognition in LAC of the right to safe and clean drinking water and sanitation as a human right could act as a starter or a catalyst for institutional reforms.

In July 2010, the United Nations General Assembly (UNGA) formally recognized the right to water and sanitation as a human right (HRWS), essential for the full enjoyment of life and all human rights (UNGA 64/292). The human right to water and sanitation entitles everyone to sufficient, safe, acceptable, accessible, and affordable water and sanitation services for personal and domestic uses, which are delivered in a participatory, accountable and non-discriminatory manner (WASH, 2012). Two months later the Human Rights Council affirmed by consensus that access to water and sanitation was a legally binding human right (HRC 15/9)[2] (Figure 11.3). During the last decades, claims and international pressure mounted for the recognition of the HRWS, with a parallel claim, particularly rooted and strong in LAC, of a series of environmental rights (Chapter 14). The UNGA resolution has now shifted attention towards the implementation of the human right to water, towards adequate financing, 'capacity building' and technology transfer, as well as adequately allocating responsibilities at international and national levels.

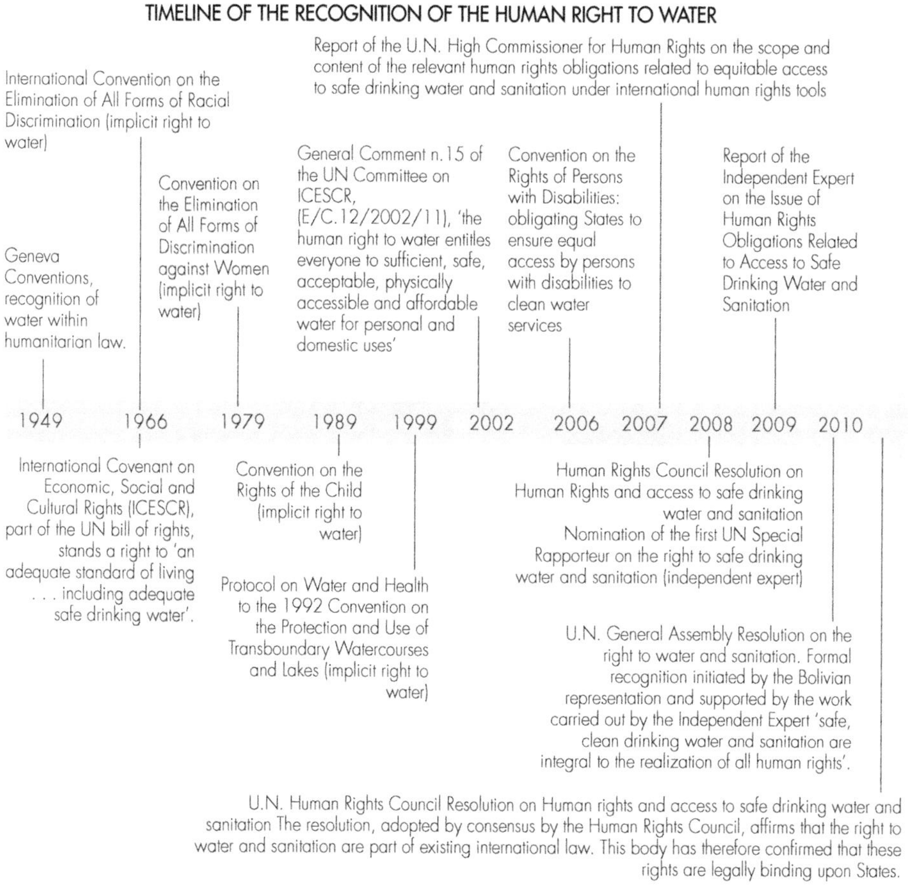

TIMELINE OF THE RECOGNITION OF THE HUMAN RIGHT TO WATER

Report of the U.N. High Commissioner for Human Rights on the scope and content of the relevant human rights obligations related to equitable access to safe drinking water and sanitation under international human rights tools

International Convention on the Elimination of All Forms of Racial Discrimination (implicit right to water)

Convention on the Elimination of All Forms of Discrimination against Women (implicit right to water)

Geneva Conventions, recognition of water within humanitarian law.

General Comment n.15 of the UN Committee on ICESCR, (E/C.12/2002/11), 'the human right to water entitles everyone to sufficient, safe, acceptable, physically accessible and affordable water for personal and domestic uses'

Convention on the Rights of Persons with Disabilities: obligating States to ensure equal access by persons with disabilities to clean water services

Report of the Independent Expert on the Issue of Human Rights Obligations Related to Access to Safe Drinking Water and Sanitation

1949   1966   1979   1989   1999   2002   2006   2007   2008   2009   2010

International Covenant on Economic, Social and Cultural Rights (ICESCR), part of the UN bill of rights, stands a right to 'an adequate standard of living . . . including adequate safe drinking water'.

Convention on the Rights of the Child (implicit right to water)

Protocol on Water and Health to the 1992 Convention on the Protection and Use of Transboundary Watercourses and Lakes (implicit right to water)

Human Rights Council Resolution on Human Rights and access to safe drinking water and sanitation
Nomination of the first UN Special Rapporteur on the right to safe drinking water and sanitation (independent expert)

U.N. General Assembly Resolution on the right to water and sanitation. Formal recognition initiated by the Bolivian representation and supported by the work carried out by the Independent Expert 'safe, clean drinking water and sanitation are integral to the realization of all human rights'.

U.N. Human Rights Council Resolution on Human rights and access to safe drinking water and sanitation The resolution, adopted by consensus by the Human Rights Council, affirms that the right to water and sanitation are part of existing international law. This body has therefore confirmed that these rights are legally binding upon States.

Figure 11.3 Timeline: international legal and political recognition of the human right to safe water and sanitation. *Source: modified and updated from Maganda (2011).*

2 The Human Rights Council confirmed that the human right to water and sanitation is derived from Articles 11 and 12 of the International Covenant on Economic, Social and Cultural Rights and is therefore legally binding on the 160 countries which have ratified the Treaty (status as of 18-02-2013).

The large majority of LAC countries voted in favour of the above-mentioned UN General Assembly resolution (Figure 11.4), reinforcing a new generation of solidarity and collective rights such as the right to environment. However, as often happens, the main stumbling block is in their implementation. At the interim evaluation of the Millennium Development Goals (MDGs) presented at Rio+20 in June 2012, statistics looked promising. According to the Joint Monitoring Programme[3] (WHO-UNICEF, 2012), 94% of the population have secure water access and 80% have access to sanitation, although these measures have been questioned by newer indicators (Flores et al., 2013). However, statistics hide great interregional disparity, differences between urban and rural, a marked diversity in the quality, sustainability and efficiency of water services, as well as notable differences between wealthy and poor areas in the same country (Chapters 4 and 6). As LAC is a region characterized by great income distribution inequality, it is essential to look beyond national coverage rates to understand the challenges ahead.

No data   Absent   Abstain   In favour

Figure 11.4 **Map on voting for UN General Assembly resolution recognizing the human right to safe drinking water and sanitation.** *Source: own elaboration*

States' international human rights obligations require them to go well beyond the targets set in the MDGs (for a methodological discussion see: Easterly, 2007; Albuquerque, 2012), whose indicators do not include or account for basic components of the human

---

3 The Joint Monitoring Programme of World Health Organization (WHO) and UNICEF measures the progress in meeting the MDG targets on water and sanitation to 'halve, by 2015, the proportion of people without sustainable access to safe drinking-water and basic sanitation'. It establishes categories of what are 'improved' and 'unimproved' sources of drinking water and sanitation facilities (WHO-UNICEF, 2012, p. 33), based in estimations about types of facilities used.

right to water and sanitation.[4] Thus, the right to water and sanitation must inform a state's design and implementation of its MDG policies (see Albuquerque, 2012) including the need to go beyond averages towards targeting groups that face discrimination and systemic exclusion.

Legal and institutional frameworks for water and sanitation often support the sustainability of interventions by creating a legal reference point for actors seeking to hold states accountable for their efforts (*ibid.*). Since the late 1960s and early 1970s, a series of pioneer LAC countries like Bolivia, Costa Rica, Uruguay and Venezuela started to include in their constitutional frameworks the implicit or explicit right to water. In the 1990s and early 2000s, more countries had enshrined this right into their constitution (Table 11.4, Figure 11.5), and HRWS now is present in the legislation of fifteen countries covering more than 75% of the population in LAC (Maganda, 2011; Waterlex, 2013).

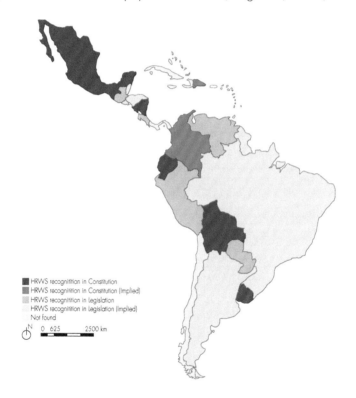

Figure 11.5 **Map on inclusion of Human Right to safe drinking water and sanitation (HRWS) in constitutions.** *Source: own elaboration*

---

4 Availability, quality, acceptability, accessibility, affordability, non-discrimination, access to information and participation, accountability and sustainability.

Table 11.4 Table summarizing State recognition of the human right to safe drinking water and sanitation (HRWS) in national constitutions, laws and policies in selected LAC countries. Sentences by the Constitutional Courts, which can represent very relevant advances in the field, are not included in this table.

COUNTRIES	HRWS recognition	SUMMARY [5]
ARGENTINA	In Legislation (Implied)	Every person may make use of public water free of charge (...) to satisfy domestic needs of drinking and hygiene (...) It is prohibited, however, to contaminate the environment.Art.25. Water Code of the Province of Buenos Aires, Law 12.257 of 9 December 1998.
BOLIVIA	In Constitution	I. Everyone has the right to water and food. Art. 16. New Constitution of Bolivia, 2009.
BRAZIL	In Legislation (Implied)	[Basic] public sanitation services shall be delivered in accordance with the following fundamental principles: I-universal access [...]. Art. 2. Law on Basic Sanitation, 2007.
CHILE	In Legislation (Implied)	Not in constitution but included in the legislation
COLOMBIA	In Constitution (Implied)	It will be a fundamental objective of state activity to address the unmet needs regarding health, education, environmental sanitation and drinking water. [...]. Art. 366. Constitution of Colombia, 1991, as last amended April 1, 2005.
COSTA RICA	In Legislation	Access to drinking water is an inalienable human right and must be guaranteed constitutionally. Art. 1.1. Executive Decree No. 30480-MINAE of 5 June 2002.
DOMINICAN R.	In Constitution (Implied)	The state shall ensure the improvement of nutrition, sanitation services and hygienic conditions, [....]. Art. 8. Constitution of the Dominican Republic, 2002.
ECUADOR	In Constitution	The human right to water is essential and cannot be waived. Art. 12. Constitution of the Republic of Ecuador, 2008.
EL SALVADOR	In Legislation (Implied)	The cities and urban populations shall be provided with services for the supply of drinking water (...). Art. 61. Health Code, Decree No. 955 of 1988, as last amended 2008.
GUATEMALA	In Legislation	a) Principle of Equality: Access to water for satisfaction of the vital and essential needs of the population and the improvement of these is a fundamental biological and social right of every human being. Article 2: Principles. General Water Law, Law No. 3702 of 26 September 2007.
GUYANA	In Legislation (Implied)	Subject to subsection (2), every public utility (...) shall make every reasonable effort to provide service to the public in all respects safe, adequate, efficient, reasonable and non-discriminatory. Section 25: Duty to provide adequate service. Public Utilities Commission Act, Act No. 10 of 1999.
HONDURAS	In Legislation (Implied)	The present law establishes the norms applicable to drinking water and sanitation services (...) as a basic instrument for the promotion of the quality of life of the population and for securing of sustainable development as an intergenerational legacy. Art 1. Decree No. 118-2003, Framework Law for the Drinking Water and Sanitation Sector.
MEXICO	In Constitution	Every person has the right to access, safe disposal and sanitation of water for personal and domestic use in sufficient quantity and quality. Article 4. Constitution of the United States of Mexico (1917, as last amended in 2011).
NICARAGUA	In Constitution	It is the obligation of the state to promote, facilitate and regulate the provision of (...), water, (...) and the population has an inalienable right to have access to these services. Art. 105. Constitution of the Rep. of Nicaragua. 1987, as of Sept. 2010.
PARAGUAY	In Legislation	b) Access to water for the satisfaction of basic needs is a human right and shall be guaranteed by the state in adequate quantity and quality. Art 3. Law on Water Resources, Law 3239 of 10 July 2007.
PERU	In Legislation	Access to water for the satisfaction of the primary needs of the human person has priority, even in times of scarcity, because it is a fundamental human right. Article III: Principles. Water Resources Act, June 2009.
URUGUAY	In Constitution	Access to drinking water and access to sanitation constitute basic human rights. Art. 47. Constitution of the Republic of Uruguay, 1967, as last amended 31 October 2004.
VENEZUELA	In Legislation	The principles governing the integrated management of water resources (...) are the following: Access to water is a fundamental human right. [...]. Art. 5. Water Law, 2 January 2007.

Source: own elaboration based on information from WaterLex Legal Database on the Human Right to Water and Sanitation (www.waterlex.org/waterlex-legal-database/index.php).

5 Non-official translations. Direct access to official documents through the WaterLex Legal Database.

11.4.1 Initiatives for implementation

The recognition of the HRWS and its consideration at a constitutional level is undoubtedly a milestone in the movement for universal access to these basic services. The HRWS framework applies to all stakeholders regardless of their nature: from states and citizens to public and private operators, who are involved in realizing its implementation and operationalization (Regional Process of the Americas, 2012), though the responsibilities differ among all stakeholders.

In Brazil, as of 2011 the federal government has put in place the programme 'Water for All', focused on the provision of water for poor rural communities of the semi-arid region of Brazil, and the main actors have been community organizations, NGOs and national and state governments in partnership with municipalities (see Figure 11.6). The provision of water cisterns has been promoted by a coalition of NGOs with the collaboration of households of all municipalities involved in the programme (Agua para Todos, 2013).

Similarly, in Chile, the national programme for public water supply in rural areas ('Programa Nacional de Agua Potable Rural') has been in place since 1994 and has increased water coverage in concentrated and semi-concentrated rural localities by over 95%. In this regard Uruguay can be taken as a model for extending the access to water, now with 100% coverage throughout the country. In addition, many countries are receiving support from the Spanish Fund for Water and Sanitation in Latin America initiated in 2007, which, with an estimated budget of US$1,500 million, aims to support the achievement of the human right to water in nineteen countries of the region.

In a region with a long history of inequality there are important citizen initiatives and social movements that contribute to monitoring governmental actions, and ultimately contribute to the achievement of the right to water. As an example, in 1998 the Central American Water Tribunal (CAWT) was set up for conflicts related to water ecosystems in Central America, creating a public space for democratic participation in water debates. In 2000 the CAWT became the Latin American Water Tribunal (LAWT) in order to increase the impact of this body throughout the region (Ávila, 2010). Similarly, rural water committees of the different regions have created associations at different levels (national, regional and continental) to share their concerns and raise the political profile of rural water in their countries (e.g. Confederación Latinoamericana de Organizaciones Comunitarias de Servicios de Agua y Saneamiento).

11.4.2 Public and private domestic supply service

The discussion about the recognition and adoption of the HRWS often goes hand in hand with the debate about the pros and cons of the privatization of the supply of domestic water service.[6] In this context, LAC represents a formidable 'laboratory' of

---

6 The term 'privatization' is used to describe different types of participation by private or government companies, with a range of contracts in which the government can transfer responsibilities related to a series of aspects such as water services, maintenance, investment, expansion, etc. (Budds and McGranahan, 2003).

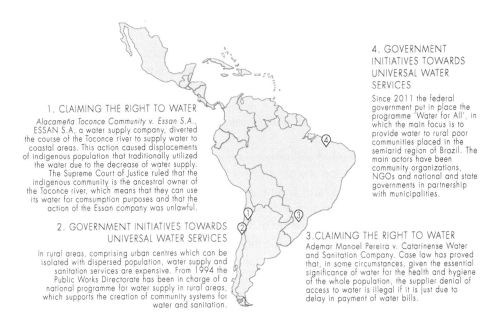

1. CLAIMING THE RIGHT TO WATER

Alacameña Toconce Community v. Essan S.A., ESSAN S.A, a water supply company, diverted the course of the Toconce river to supply water to coastal areas. This action caused displacements of indigenous population that traditionally utilized the water due to the decrease of water supply. The Supreme Court of Justice ruled that the indigenous community is the ancestral owner of the Toconce river, which means that they can use its water for comsumption purposes and that the action of the Essan company was unlawful.

2. GOVERNMENT INITIATIVES TOWARDS UNIVERSAL WATER SERVICES

In rural areas, comprising urban centres which can be isolated with dispersed population, water supply and sanitation services are expensive. From 1994 the Public Works Directorate has been in charge of a national programme for water supply in rural areas, which supports the creation of community systems for water and sanitation.

4. GOVERNMENT INITIATIVES TOWARDS UNIVERSAL WATER SERVICES

Since 2011 the federal government put in place the programme 'Water for All', in which the main focus is to provide water to rural poor communities placed in the semiarid region of Brazil. The main actors have been community organizations, NGOs and national and state governments in partnership with municipalities.

3.CLAIMING THE RIGHT TO WATER

Ademar Manoel Pereira v. Catarinense Water and Sanitation Company. Case law has proved that, in some circumstances, given the essential significance of water for the health and hygiene of the whole population, the supplier denial of access to water is illegal if it is just due to delay in payment of water bills.

Figure 11.6 Map with examples of the implementation of the Human Right to Water and Sanitation. *Source: own elaboration.*

different approaches to water services provision. As a matter of fact, during the past three decades LAC governments have explored (and moved back and forth between) different paths to address the pressing challenge of providing adequate water and sanitation to their citizens.

Institutional reforms aimed at diminishing the role of the State in the provision of various services – including water – have been the key for many LAC countries since the 1980s (ECLAC, 2012a). These processes have included the privatization of water services and sanitation in many cities, due to what were considered favourable conditions for privatization, namely: cities with a relatively large middle class, poor financial conditions of public operators, and the momentum of neoliberal policies pushed by international organizations such as the World Bank or the International Monetary Fund (Budds and McGranahan, 2003). However, the reality was that many privatization processes did not always flourish. While concession contracts in Argentina and Bolivia were not successful (see Chapter 13), in others like Mexico these contracts have now taken root. The main aspect linked to failures in the implementation of water management programmes is related to weak or absent regulatory frameworks. This has led to problems such as unjustified asset and income transfers, and failure to ensure efficiency and new investment after privatization (Hantke-Domas and Jouravlev, 2011). Among the causes of this failure Castro (2007) points to corruption, lack of adequate or strong government regulation, lack of private investment, inadequate consideration of inclusive policies designed to reduce inequality, and as a result, resistance movements by civil society. However, the analysis of experiences worldwide and in the region suggests that the debate should not be focused on the 'dilemma' private vs. public service but rather on creating a legal and financial framework suitable to ensure an adequate service provision.

The analysis of water and sanitation service provision shows that the macro-economic context and the value of water as a key element in the economy, as well as sound governance (both of context and sectoral variables) are critical to the sustainable development of water services. Moreover, the design of the industrial structure of water supply and sanitation impinges on the ability to deliver services to the population. Assets are long-lived, allowing investments to be delayed and quasi-rents to be captured once initial investments have been made (Massarutto, 2007; Guasch et al., 2008). Fragmented services lose economies of scale, increase transaction costs, make services more expensive and may facilitate the capture by vested interests (Foster, 2005; ADB, 2009). Water supply and sanitation services have decreasing average costs (Krause, 2009). Therefore, both efficiency and equity are achieved by selecting optimal size in terms of economies of scale. At the same time, they require important investments, especially when new sectors of the population have to be served. This entails having guarantees of continuity of ownership in order to recover investments through tariffs. Adequate regulation of a natural monopoly, strategic planning of public policies, prioritization of water in public budgets and decisions with adequate subsidies for lower-income citizens are requisites for the institutional design of water and sanitation systems.

While each contract will have its own singularities, countries will need to consider the contractual and regulatory duties of contractors. In terms of implementing regulation, there are differences between, on the one hand, contracts and, on the other hand, comprehensive general regulation, franchizing and concessions. Almost 90% of water supply and sanitation privatizations in LAC during the 1990s were concessions, i.e. contracts (Estache et al., 2003). After a first wave of privatization of water supply and sanitation in the 1980–1990s mainly by international operators, during the 2000s there has been a radical reduction of their presence. Ducci (2007) identifies four main reasons for this decrease: a change in the overall strategy of the operator, e.g. in search of new business opportunities in other regions; re-orientation of the national policy in relation to water supply and sanitation; collapse of the financial and economic balance of existing water provision contracts; and social and political conflicts. As a consequence, it is clear that state-owned water companies will continue being the backbone of water supply and sanitation in Latin America (*ibid.*). Nonetheless, it should be noted that, for the characteristics of the service provided, there are incentives for members of the public sector (politicians, managers and employees of the utility itself) to capture quasi-rents (Wallsten and Kosec, 2008). It seems therefore important to identify alternatives for their control and regulation in order to ensure their accountability, e.g. through the establishment of clear service standards (in terms of quality, service reliability, tariffs affordability, etc.) and their strict enforcement by an independent supervising body.

11.5 Financing of the water sector

No institutional or legislative reforms can take place without solid financial backing. Thus there is little doubt that each country must address the permanent challenge of ensuring sufficient funds to sustain and further develop its water sector and the institutions that enable its functioning.

11.5.1 What needs to be financed?

Financing needs for water policy are contingent upon economic development levels. Some of the countries in LAC are currently going through a very incipient stage of water resource exploitation; and water policy within that context is very much a question of building canals to take runoff resources to where they are needed or, alternatively, boreholes to withdraw groundwater, where available. In these countries (or at a given stage for almost every country), water policy has focused on fostering irrigation and urban development, requiring substantial financing for capital investment (OECD, 2009). In some of the countries in the region, however, more and more often society's demands for participation, equity and environmental protection add new layers to water policy and create new funding needs.

Essentially, there are three major items to be financed (Figure 11.7): water resource management, including water use (both withdrawal and wastewater disposal) through charges or fees, plus forfeiture for non-use of water use rights; water service provision through public works (infrastructures), via water tariffs; and where a sui-generis or effective IWRM approach is in place, river basin management (i.e. joint water and land use management), conceivably through the use of payment for environmental services schemes or compensatory measures or levies.

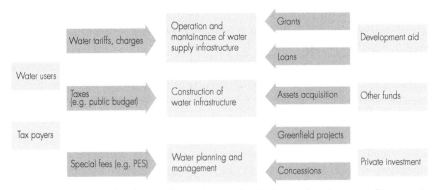

Figure 11.7 Water-related expenditures that need to be financed and sources of incomes in LAC countries. *Source: own elaboration*

Some of the countries in the region have faced severe foreign exchange shortages in the past due to sub-optimal saving rates or current account deficits. Over time, this has led to high levels of indebtedness (Adler and Iakova, 2013) or even a debt crisis (Reinhart and Rogoff, 2011). That debt burden for decades represented a significant restriction for economic development (Rodrik, 2011). It greatly hindered any possibility to harness the necessary resources in order to finance water policies, which in turn has a twofold impact: on the one hand, the financing gap impedes water policy as such; on the other hand, the need to repay an ever-increasing foreign debt led some countries to turn to their comparative advantage in terms of natural capital endowment, both increasing their exports of natural resources – including water-intensive goods – and also enduring lower levels of environmental quality overall (ECLAC, 2012a; OECD/UN-ECLAC, 2013).

11.5.2 Where and how to lever funds?

11.5.2.1 National financing

Not many countries in the region rely on their own (national) resources to finance water policy and, if they can, it is usually just for some water services (i.e. Chile and its sanitation service). Their funding gap (which is mainly a fiscal one, in those countries with no public budget surplus) refers to insufficient or unstable revenues to implement water policies at different levels of government (Hernández et al., 2012). However, a sustained growth pattern over the past few years in some countries (namely Brazil, Colombia, Chile, Peru, and Uruguay, amongst others, or Paraguay and Panamá very recently) should lead to improved financial self-sufficiency[7] (ECLAC, 2012a).

In this context, each country has to take its own decision on how to finance its water needs. The advantage of water tariffs is that they lighten the burden over national budget, which allows the diversion of revenues to sectors that are more difficult to finance on the basis of direct charges. These tariffs generate incentives for higher water use efficiency in business (control of revenues and costs), through the consolidation of a direct relationship between revenues and services provided (served clients and supplied volumes, recollected and treated). In addition, a clear signal is provided to consumers of the real cost of services, therefore fostering a more rational use. Further, tariffs make service provision less vulnerable to macro-economic fluctuations.

To date, the use of tariffs levied on the use of natural resources is not widespread in the region (see Chapter 13) but in those countries where tariff schemes have been implemented, this has meant a sort of self-funding source as well as a partial cost-recovery mechanism. As with taxes and charges, they tend to feed into the public budget at different government levels. Revenues from these taxes and charges are very unlikely earmarked for water policy purposes. However, as social efforts, be they user contributions or public investment, are often if not always insufficient, credit or private investment may also be required, either from domestic or foreign sources.

While multilateral development banks have been a traditional and important source of financial resources for the water sector in LAC (see over), private banks have not represented

---

7 For the decade 2000–2010, per capita GDP in the LAC region grew by an average of 1.9% per annum, as compared with 0.3% for 1980 to 2000, and 3.3% for 1960–1980.

such a reliable funding source: any water project has the potential to generate sufficient cash-flow to pay for the loan; though there are some risks associated with exchange rate fluctuations (this led to the failure of different Build, Operate and Transfer projects in Mexico in 1995). Capital markets, in turn, are well developed in countries such as Brazil or Chile, but have not played a major role elsewhere.

In LAC, despite funding flows from international sources, governments struggle and usually fail to meet financial requirements. This has led, amongst other things, to an increasing interest in water use charges or fees (both for water abstraction and wastewater disposal; see Chapter 13 for specific examples). This interest has a number of common features in the region:

- There is a search for new approaches since traditional ones, due to the lack of operational capability, have not been effective in most cases (see Easter and Liu, 2005, for cost recovery in irrigation and drainage projects; Ferro and Lentini, 2013, for water and sanitation).
- Many of the approaches to water use charging are deemed on the basis of ideology (rather than technology). Furthermore, there are double-dividend aspirations (Fullerton et al., 2008) and, occasionally, rent-seeking behaviour (Delacámara and Solanes, 2012).
- Within a context of increasing water scarcity, the public sector's attention shifts away from supply to combined supply and demand management, thus requiring further use of financial and economic policy instruments.

Despite the existence of such charges or fees, in almost all cases levies are not actually paid, but are paid just by a minority or are negligible for water users. However, this does not mean that water use charging is an easy endeavour. There are major obstacles: the lack of proper definitions of water use rights, including a pre-condition of payment for right purchase and holding; the level of information required (who uses water, how, how much, where, what actual revenue might be actually obtained, etc.); the weakness of procedures for the operational effectiveness of charging schemes; and the social and political acceptability of these levies, among others.

11.5.2.2 International financing

In LAC, national funds needed for developing and operating the water sector are complemented by public and private international sources. According to two major public databases of OECD and the World Bank,[8] during the period 2000–2011 the international

---

8 The contribution of international sources to the financing of part of water-related investments can be assessed through two major public databases: the Creditor Reporting System (CRS) of the Development Assistance Committee (DAC) of the Organization for Economic Cooperation and Development (OECD) for public funds (/stats.oecd.org/index.aspx?DataSetCode=CRS) and the Private Participation in Infrastructure Database (PPI) of the World Bank (http:77ppi.worldbank.org). From these databases it is possible to extract data about water and sanitation projects, hydropower and irrigation projects. Data correspond to investments committed on an annual basis and expressed in current US dollars. In CRS, the analysis presented in this chapter considers sectors with codes 14000, 23065 and 31140; in PPI infrastructure associated with water domestic supply and sanitation is considered.

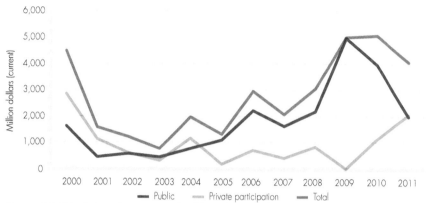

Figure 11.8 Evolution of international public and private funding to the Latin American water sector over the period 2001–2011 *Source: own elaboration based on data from CRS (2013) and PPI (2013).*

overall (public and private) investment commitment in LAC amounted to 33,238 million current US$, being the public investment about 66% of the total amount (21,877 million US$) (Figure 11.8).

Public grants and loans include both the Official Development Aid (ODA)[9] and Other Official Flows (OOF) that cannot be included in the ODA category. Between 2000 and 2011 the OOF to LAC amounted to over US$14,701 million, more than twice ODA flows in the region, which were US$7,170 million. Almost all of the OOF (99%) were loans, while ODA consisted of loans and grants in similar shares (50% and 48%, respectively) (CRS, 2013). Overall, since 2001 there is a clear positive trend in public investment, reaching its maximum in 2009 (US$4,972 million), which marked a tipping point towards a decline (Figure 11.9). The peak during the period 2008–2011 is due to the activation of Spain's cooperation fund for water and sanitation in the LAC region (US$1,500 million over a four-year period), whose investment commitments amount to 53% of the ODA of the period 2001–2011 and made Spain the main donor to the region in 2008 and 2009.

Over the 2000–2011 period, Japan was the main contributor to the ODA (35.24%), followed by Spain (24.07%) and Germany (11.93%). The main recipients were Peru (16.38%), Brazil (13.24%) and Bolivia (10.01%). As for the OOF, most of the funds were allocated to Brazil (30.05%), Argentina (17.09%) and Colombia (13.62%), while the main funding providers were the Inter-American Development Bank (54.1% of the OOF)

---

9 ODA is defined as 'flows of official financing administered with the promotion of the economic development and welfare of developing countries as the main objective, and which are concessional in character with a grant element of at least 25 percent (using a fixed 10 percent discount rate). By convention, ODA flows comprise contributions of donor government agencies, at all levels, to developing countries (bilateral ODA) and to multilateral institutions. ODA receipts comprise disbursements by bilateral donors and multilateral institutions. Lending by export credit agencies with the pure purpose of export promotion is excluded' (IMF, 2003)

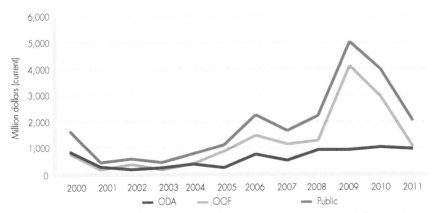

Figure 11.9 Evolution of international public investment during the period 2001–2011.
Source: own elaboration based on data from CRS (2013).

and the World Bank (45.01%). Projects associated to large urban water supply and sanitation received 44% of the ODA, while small systems (rural and peri-urban), hydropower and agriculture received 28%, 10% and 3%, respectively (CRS, 2013).

In terms of private participation in investments in the water sector, during 2000–2011 LAC received 32% of the world's investment in the above-mentioned water-related sectors with private participation (Figure 11.10) being especially significant in 2001 (60% of the global investments) and in 2011 (78%) (PPI, 2013).

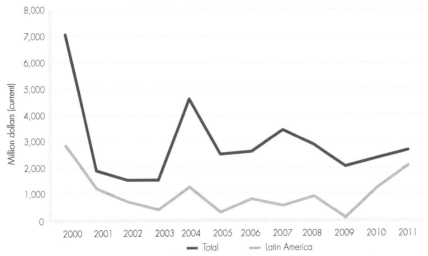

Figure 11.10 Global and regional private investment in the water sector. *Source: own elaboration based on data from PPI (2013).*

Between 2001 and 2011 almost 70% of private investment occurred in Chile, Brazil and Mexico (Figure 11.11), principally due to the support of big companies. The participation of private operators was noticed in the agricultural, industrial and sanitation sectors, characterized by the concessions of important systems, which represented 53% of the overall investment. By far, water supply and sanitation was the main recipient of private funds: about 77% of the total investments, mainly through contract for the construction or the rehabilitation of water supply systems, operation and transfer. Water purification and wastewater treatment plants received only 21% of the total investment, mainly through Build-Operate-Transfer (BOT) projects (PPI, 2013).

No data
<400 million US$
400–1,000 million US$
1,000–2,000 million US$
>2,000 million US$
N   0   625   2500 km

Figure 11.11 Geographical distribution of investments with private participation in the water sector during the period 2001–2011. *Source: own elaboration based on data from PPI (2013).*

From these figures it can be concluded that during the past decade international investors and organizations have played a significant role in funding the water sector, with special emphasis – for both public and private funds – in the development of infrastructure to provide water and sanitation to the population of the LAC region.

References

ADB (2009). Asian Development Bank. *Urban Water Supply Sector Risk Assessment: Guidance Note*, Manila, Philipines.

Adler, G. & Iakova, D. (2013). *External Conditions and Debt Sustainability in Latin America*, Western Hemisphere Department. IMF Working Paper No. WP/13/27.

Agua para Todos. (2013) [Online] Available from: www.integracao.gov.br/pt/web/guest/resultado-da-busca1?p_auth=MV6uc2sr&p_p_auth=MnjxzO1R&p_p_id=20&p_p_lifecycle=1&p_p_state=exclusive&p_p_mode=view&p_p_col_id=column-2&p_p_col_pos=1&p_p_col_count=2&_20_struts_action=%2Fdocument_library%2Fget_file&_20_groupId=10157&_20_folderId=167983&_20_name=11701. [Accessed April, 2013].

Akhmouch, A. (2012). *Water Governance in Latin America and the Caribbean: A Multi-Level Approach:* OECD Publishing. OECD Regional Development Working Paper No. 2012/04.

Albuquerque, C. (2012). *The future is now – Eliminating inequalities in sanitation, water and hygiene.* [Online] Available from: www.ohchr.org/Documents/Issues/Water/SRWatsanInequalitiesConsultation.pdf [Accessed July, 2013].

Allan, J.A. (2013). Food and water security: beyond water resources and the water sector. In: Lankford, B., Bakker, K., Zeitoun, M. and Conway, D. (eds). *Water Security: Principles, Perspectives, Practices*, London, Earthscan.

ANA (2002). Autoridad Nacional del Agua. *Relatório Conjuntura de Recursos Hídricos.* [Online] www.ana.gov.br. [Accessed July, 2013].

Astorga, Y. (2010). *Estado y gestión del recurso hídrico en Costa Rica* (informe final), Duodécimo informe sobre el estado de la nación en desarrollo humano sostenible.

Ávila, P. (2010). The role of indigenous claims to water and indigenous knowledge. Water conflicts and human rights in indigenous territories of Latin America. In: Rosenberg International Forum on Water Policy *Water For The Americas: Challenges & Opportunities*, Buenos Aires, Argentina.

Barth, F.T. (1999). Aspectos institucionais do gerenciamento de recursos hídricos. In: Rebouças, A. da Cunha, Braga, B. and Tundisi, J. G (eds). *Águas Doces no Brasil: capital ecológico, uso e conservação*, São Paulo, Brazil, Escrituras Editora. pp. 563–577.

Bechara Elabra, L. & Magrini, A. (2013). The Brazilian Water Resources Management Policy: fifteen years of success and challenges. *Water Resources Management*, 27: 2287–2302.

Bonfante, P. (1929). *Instituciones de Derecho Romano.*,Trad. de la 3a ed. ital. Luis de Bacci y Andres Larrosa, Madrid, Campuzano Horma Fernando.

Budds, J. & McGranahan, G. (2003). Are debates on water privatization missing the point? Experiences from Africa, Asia and Latin America. *Environment and Urbanization*, 15 (2): 87–113.

Caponera, D. (1992). *Principles of Water Law and Administration*, The Netherlands, Balkema.

Castro, J. E. (2007). Water governance in the 21st century. *Ambiente e Sociedade*, 2 (10).

CEPAL (1995). Comisión Económica Para América Latina y Caribe. *Mercados de derechos de agua: entorno legal.* [Online] Available from: www.eclac.cl/publicaciones/xml/0/5670/Lcr1485e.pdf. [Accessed August, 2013]. Online report No. LC/R.1485.

Conac, F. (1991). Land and water rights issues in irrigated schemes in sub-Saharan Africa: conflicts to be avoided. *DVWK Bulletin*, 16/ Beck, 366 (1).

Crase, L. & Gandhi, V. (2009). *Reforming Institutions in Water Resource Management. Policy and Performance for Sustainable Development*, London, Routledge.

CRS (2013). Creditor Reporting System of the Development Assistance Committee of the Organization for Economic Cooperation and Development for public funds.

Delacámara, G. & Solanes, M. (2012). Policies, structure and regulation in water and sanitation services. In: Heller, L. (ed.). *Basic Sanitation, Environmental Health and Public Policies: new paradigms for Latin America and the Caribbean*, Pan-American Health Organization.

De Stefano, L. & Lopez-Gunn, E. (2012). Unauthorized groundwater use: institutional, social and ethical considerations. *Water Policy*, 14 (1): 147–160.

Donoso, G. (2011). *Case Studies. The Chilean Water Allocation Mechanism*, established in its Water Code of 1981 Deliverable number: D6.1.

Donoso, G. (2014). Integrated water management in Chile. In: Martínez-Santos, P., Aldaya, M.M. & Llamas, M.R. (eds). *Integrated Water Resources Management in the 21st Century: Revisiting the paradigm*. Boca Raton, FL, CRC-Press. Forthcoming.

Ducci, J. (2007). *Salida de operadores privados internacionales de agua en América*, Inter-American Development Bank.

ECLAC (2012a). United Nations Economic Commission for Latin America and the Caribbean. *Economic Survey of Latin America and the Caribbean. Policies for an adverse international economy*, Santiago de Chile. Report No. LC/G.2546-P.

ECLAC (2012b). United Nations Economic Commission for Latin America and the Caribbean. *Social Panorama of Latin America*, Santiago de Chile. Report No. LC/G.2557-P.

Easter, K.W. & Liu, Y. (2005). *Cost Recovery and Water Pricing for Irrigation and Drainage Projects*. The International Bank for Reconstruction and Development, Washington DC, The World Bank. Agriculture and Rural Development Discussion Paper No. 26.

Easterly, W. (2007). *How the Millennium Development Goals are Unfair to Africa*, Global Development and Economy, Brookings Global Economy and Development. Working Paper No. 14.

Estache, A., Guasch, J.L. & Trujillo, L. (2003). *Price Caps, Efficiency Payoffs and Infrastructure Contract Renegotiation in Latin America*, Washington DC, World Bank. Policy Research Working Paper No. 3129.

Flores, Ó., Jiménez, A. & Pérez-Foguet, A. (2013). Monitoring access to water in rural areas based on the human right to water framework: a local level case study in Nicaragua. *International Journal of Water Resources Development*, 29(4): 605–621.

Ferro G., & Lentini E. (2013). *Políticas tarifaria para el logro de los Objetivos de Desarrollo del Milenio (ODM): situación actual y tendencias regionales recientes*, Colección de Documentos de Proyectos, Santiago, Chile, CEPAL. Report No. LC/W.519.

Foster, S., Garduño, H. & Kemper, K. (2004). *Mexico. The 'Cotas': Progress with Stakeholder Participation in Groundwater Management in Guanajuato*, Sustainable Groundwater Management, Lessons from Practice, The World Bank. Case Profile Collection No. 19.

Foster, V. (2005). *Ten Years of Water Service Reform in Latin America: Towards an Anglo-French Model*, The International Bank for Reconstruction and Development, Washington DC, The World Bank, Water Supply and Sanitation Sector Board Discussion Paper Series. Paper No. 3.

Fullerton, D., Leicester, A. & Smith, S. (2008). *Environmental Taxes*. Prepared for the Report of a Commission on Reforming the Tax System for the 21st Century, Chaired by Sir James Mirrlees, The Institute for Fiscal Studies. Oxford University Press.

Getches, D. (1990). *Water Law in a Nutshell*, St. Paul, Minessotta, USA, West Publishing Company.

GEF (2012). Global Environmental Facility. *Groundwater Governance: A Global Framework for Country Action*, Montevideo, Uruguay, Latin America and the Caribbean Regional Consultation. Final Report. 18–20 April 2012.

Guasch, J.L, Laffont, J.J. & Straub, S. (2008). Renegotiation of concessions contracts in Latin America. Evidence from the water and transport sectors. *International Journal of Industrial Organization*, 26(2): 421–42.

Hanemann, W.M. (2006). The economic conception of water. In: Rogers, P.P., Llamas, M.R. & Martinez-Cortina, L. (eds). *Water Crisis: Myth or Reality?*, London, Taylor & Francis. pp. 61–91.

Hantke-Domas, M. & Jouravlev, A. (2011). *Lineamientos de política pública para el sector de agua potable y saneamiento*. Documento de proyecto, CEPAL.

Hearne, R., & Donoso, G. (2005). Water institutional reforms in Chile. *Water Policy*, 7: 53-69.

Hernández, A., Flórez, J. & Hincapié, S. (2012). *Impact of Accountability in Water Governance and Management Regional Analysis of Four Case Studies in Latin America*, Universidad de los Andes/UNDP Virtual School. Commissioned Study Draft Version.

IMF (2003). External Debt Statistics: *Guide for Compilers and Users – Appendix III, Glossary*, IMF, Washington DC.

Jacobi, P.R. (2004). A gestão participativa de bacias hidrográficas no Brasil e os desafios do fortalecimento de espaços públicos colegiados. In: Coelho, V. & Nobre, M. (orgs.), *'Participação e Deliberação*, São Paulo, Brazil, Editora 34. pp. 270–289.

Jacobi, P.R. (2009). Governança da água no Brasil. In: Ribeiro, W. (org.), *Governança da água no Brasil: uma visão interdisciplinar.*, São Paulo, Annablume.

Jacobi, P.R., de Almeida Sinisgalli, P.A., Medeiros Y. & Ribeiro Romeiro, A. (2009). Governança da água no Brasil: dinâmica da política nacional e desafios para o futuro. In: Jacobi, P.R. and Sinisgalli, P.A. (orgs.), *Dimensões político institucionais da governança da água na América Latina e Europa*, São Paulo, Annablume. pp. 227–243.

Jiménez-Cisneros, B. & Galizia-Tundisi, J. (Coord.) (2012). Diagnóstico del Agua en las Américas. Mexico, Red Interamericana de Academias de Ciencias (IANAS) and Foro Consultivo Científico y Tecnológico, AC. Mexico.

Kauffman, C.M. (2011). Transnational Actors and the Power of Weak Laws: Decentralizing Watershed Management Without the State, Paper Presented at the 2011 *Annual Meeting of the American Political Science Association*, Seattle, Washington.

Ke, L. (1993). *Water Resources Administration in China. Law Proposal No 2249-A. 199*, Brazil, June 1993.

Krause, M. (2009). *The Political Economy of Water and Sanitation*, New York, Routledge.

Kuroiwa, J.M, Castro, L.F., Lucen, M.N. & Montenegro, J.I. (2014). Integrated water resources management in Peru – the long road ahead. In: Martínez-Santos, P., Aldaya, M.M. & Llamas, M.R. (eds). *Integrated Water Resources Management in the 21st Century: Revisiting the paradigm*. Boca Raton, FL, CRC-Press. Forthcoming.

Maganda, C. (2011). *Water: a Resource in Europe, a Right in Latin America?*, XIVth World Water Congress, International Water Resources Association (IWRA), Porto de Galinhas, Pernambuco, Brazil.

Massarutto, A. (2007). *Liberalization and private sector involvement in the water industry: a review of the economic literature*, Milan, IEFE (Centre for Research on Energy and Environmental Economics and Policy), Bocconi University. Working Paper No 6.

MINAG (2009). Ministerio de Agricultura. *Policy and National Strategy of the Water Resources in Peru*, Lima, Ministry of Agriculture.

OECD (2009). Organization for Economic Co-operation and Development. *OECD Glossary of Statistical Terms*, OECD.

OECD/UN-ECLAC (2013). Organization for Economic Co-operation and Development. *Latin American Economic Outlook 2013*, SME policies for structural change, Report No. LC/G.2545.

Pagnocceschi, B. (2003). Política Nacional de Recursos Hídricos. In: Paul, E. (org.). *Políticas ambientais no Brasil: análises, instrumentos e experiências*, São Paulo, LITTLE. pp. 241–258.

PPI (2013). *Private Participation in Infrastructure Database*, The World Bank. [Online] www.ppi. worldbank.org. [Accessed March, 2013].

Regional Process of the Americas (2012) *Americas' Water Agenda: Targets, Solutions and the Paths to Improving Water Resources Management*. [Online] Technical Document prepared at 6th World Water Forum, Marseille. February 2012. Available at: www.gwp.org/Global/ GWP-CAm_Files/Americas'%20Water%20Agenda.pdf [Accessed October, 2013]

Reinhart, C.M. & Rogoff, K.S. (2011). From financial crash to debt crisis. *American Economic Review*, 10: 1676–1706.

Rodrik, D. (2011). *The Future of Convergence*, HKS Faculty. Research Working Paper Series No RWP11-033.

Serrano, J.D. (2007). *La Gobernanza del agua en México y el reto de la adaptación en zonas urbanas: el caso de la Ciudad de México*, Anuario de Estudios Urbanos, UAM-Azcapotzalco.

Tarlock A., Corbridge, D., James, N. & Getches, D. (2002). *Water Resources Management: A Casebook on Law and Water Policy*, 5th edition, University Casebook Series, New York, Fundation Press.

Wallsten, S. & Kosec, K. (2008). The effects of ownership and benchmark competition: An empirical analysis of U.S. water systems. *International Journal of Industrial Organization*, 26(1): 186–20.

WASH (2012). United Freshwater Action Network (FAN Global). WaterLex. *The Human Right to Safe Drinking Water and Sanitation in Law and Policy - A Sourcebook*. [Online] Available from: www.freshwateraction.net/sites/freshwateraction.net/files/RTWS-sourcebook.pdf. [Accessed March, 2013]. Online report

WATERLEX (2013). *Waterlex database*. [Online] Available from: www.waterlex.org/waterlex-legal-database/index.php. [Accessed March, 2013].

WHO-UNICEF (2012). *Progress on Drinking Water and Sanitation*. [Online] Available from: www. unicef.org/media/files/JMPreport2012.pdf [Accessed July, 2013]. Online report.

Wilder, M. (2010). Water governance in Mexico: political and economic apertures and a shifting state-citizen relationship. *Ecology and Society*, 15(2): 22.

Wohlwend, B.J. (1975). Hindu Water and Administration in Bali, Proceedings of the Conference on *Global Water Law Systems*, Valencia, Spain.

Young, M. (2010). *Environmental Effectiveness and Economic Efficiency of Water Use in Agriculture: The Experience of and Lessons from the Australian Water Reform Programme*, OECD.

Young, M. (2011). *The Role of the Unbundling Water Rights in Australia's Southern Connected Murray Darling Basin*, WP 6 IBE EX-POST Case Studies, Adelaide, Australia, University of Adelaide. IBE-Review Report No. D-6.1.

Young, M. (2012). Australia's rivers traded into trouble. Opinion. *Australian Geographic*, MAY-9-2012.

12

THE ROLE OF STAKEHOLDERS IN WATER MANAGEMENT

Coordinator:
Lucia De Stefano. Water Observatory – Botín Foundation, and Universidad Complutense de Madrid, Spain

Authors:
Vanessa Empinotti, PROCAM /IEE Universidade de São Paulo, Brazil
Lucia De Stefano. Water Observatory – Botín Foundation, and Universidad Complutense de Madrid, Spain
Pedro Roberto Jacobi, PROCAM /IEE Universidade de São Paulo, Brazil
Úrsula Oswald Spring, Centro Regional de Investigaciones Multidisciplinarias,Universidad Nacional Autónoma de México, Mexico
Pedro Arrojo Agudo, Departamento de Análisis Económico, Universidad de Zaragoza/Fundación Nueva Cultura del Agua, Spain
Miguel Solanes, IMDEA Agua – Madrid Institute for Advanced Studies, Spain
Guillermo Donoso, Pontificia Universidad Católica de Chile, Chile
Patricia Phumpiu Chang, Centro del Agua para América Latina y el Caribe – ITESM, Mexico

Contributors:
Julio M. Kuroiwa, Laboratorio Nacional de Hidráulica – Universidad Nacional de Ingeniería, Peru
Rosario Pérez, Universidad Autónoma de México, México

Highlights

- The abundance of still largely unexploited natural resources and the sustained growth pattern of many countries in the Latin America and Caribbean (LAC) region contribute to the creation of situations where different needs, interests and understanding of the concept of socio-economic development have led to tensions and conflicts.

- Poor legal compliance, insufficient legal instruments and lack of funds are often at the root of significant environmental damage and conflicts in the LAC countries. Disputes are mainly related to the construction and operation of water works, water diversion, industrial and mining pollution and the privatization of the water supply and sanitation coverage.

- Advocacy networks play a key role in empowering and giving national and international visibility to the local population directly affected by environmental degradation or social injustice.

- During the past few decades the demands from civil society organizations in LAC for a larger participation in decision-making processes supported the inclusion of participatory practices in the new institutional arrangement and the creation of new spaces for negotiation such as river basin committees and water councils.

- Formal participation is uneven in terms of level of involvement of stakeholders and is mainly limited to water users (usually the ones representing large scale economic activities). Other interests not associated to water rights or the views of indigenous population are often underrepresented in formal forums and social activism still prevails as the main means to voice their demands.

- In LAC, besides lobbying and direct access to the highest state leaderships, the private sector has two new strategies to influence the decision-making processes: as one of the main stakeholders in participatory formal institutions and through their partnership with international NGOs and development agencies in defining new rules for water certification and water accounting, that can lead to new water policies in the future.

- Most LAC countries have passed information transparency laws, which apply also to water-related public information. The actual implementation of the legal obligations to information disclosure is fostered by benchmarking initiatives and watchdog studies promoted by civil society and international organizations, mainly for the water and sanitation sector.

12.1 Stakeholders organizations and their spaces for negotiation

Latin America and the Caribbean (LAC) is a region well known for its social and economic contrasts. High levels of poverty and inequality coexist with high rates of growth and raw material exploitation. The lush nature, combined with large expanses of land also indicates a high concentration of natural resources. Development practices and economic growth lead to tensions between different social groups and actors about how this region should be. In this context, water and the struggles to access and control it have contributed to the construction of the political and natural landscape of the region.

Water access is disputed by different sectors of society and activities throughout the whole region. Tensions come from energy, mining, irrigation, urban demands and their impact over livelihoods of local and traditional populations as well as the environment. On the other hand, water pollution and the access to domestic water supply, particularly the privatization of water services, have become major sources of conflict in the last two decades.

Such tensions represent the challenge of promoting multiple water uses whilst guaranteeing its universal access as part of the strategy to ensure water security in this region. In order to deal with this, stakeholders[1] – ranging from economic agents to indigenous organizations – have been using and developing different strategies to express their agendas as well as to influence the decision-making processes and water governance in LAC. Such strategies include public demonstrations and campaigns, lobbying, participation on councils and committees, the proposal of new regulations, denouncing conflicts to the courts and asking for transparency on how decisions are made. These interactions are happening in different spaces of negotiation and discussion, involving different actors and networks, and in different moments.

Usually, when discussing civil society organizations and their participation in water management, the analyses focus mainly on practices of public participation that consider formal participatory institutions as the main venues for negotiation. However, the stakeholders' repertoire goes beyond such negotiation spaces and reflects the understanding of social participation as the direct involvement of an array of people in decision-making and implementation of water policy or management through the opportunity to express their voices and articulate their arguments in public forums (Berry and Mollard, 2010).

Even though nowadays many of the LAC countries have undergone water reforms (Chapter 11) in which stakeholders' participation has become part of the institutional arrangement, activism and public demonstrations still take place. Such strategy unveils

---

1 Stakeholders are understood as individuals, groups or institutions that are concerned with, or have an interest in the water resources and their management. Even though public sector agencies are also stakeholders, in this chapter, the focus will be on private sector organizations, NGOs and social actors.

how water reform, by itself, was not able to decrease water access inequalities through the creation of new spaces for negotiation. Actually, in many cases, reforms have worsened the situation as elites and corporations have taken advantage of government interventions (Boelens et al., 2011). As a consequence, some social actors believe that there are moments in which public demonstrations are more effective (relative to formal participation) in bringing a specific claim to the attention of the general public or to compel the state to include specific topics on the official agenda (Empinotti, 2007). On the other hand, in some cases civil society organizations have been withdrawing from water councils and committees, in which their representation is outnumbered and decide to focus their actions on other strategies such as direct lobbying, unilateral partnerships with the government and public demonstrations (Warner, 2005; Empinotti, 2011).

At the same time, the importance of the private sector – mainly farming, food traders and manufacturing – in the management of water has been unveiled as strategic in order to guarantee water and food security (Allan, 2013). Because of this, initiatives such as water certification and indicators of water efficiency have become new channels to promote alliances among manufactures, food producers, NGOs and development agencies that can lead to new agreements for regulated water use in the production sector (Pegram et al., 2009; Empinotti, 2012; Empinotti and Jacobi, 2013). Finally, the approval of transparency laws throughout the region, pushing for accountability and corruption control, has become an opportunity for civil society organizations to ask for information and to control the government's expenditure on infrastructure projects and plans to increase water availability in LAC.

Such considerations show that the analysis of how stakeholders influence water management should go beyond the understandings proposed by the concept of participatory citizenship and multi-stakeholders platforms, and also include other spaces impacting decision-making processes such as the courts, non-state market-driven governance systems and the increased attention to transparency and access to information.

Acknowledging the importance of different channels of expression and negotiation beside councils and committees, this chapter explores the different strategies that stakeholders apply in order to influence water governance in the LAC region, with a special focus on Brazil, Chile, Costa Rica, Mexico and Peru. The chapter starts describing the main sources of tensions regarding water in the region and the reasons behind it. Then it looks at how disputes and the disregard for traditional community-based water management practices lead to activism and advocacy that represent informal but important spaces of participation for civil society organizations, such as NGOs, social movements and networks. It also discusses whether courts are (or are not) spaces in which stakeholders can voice their claims. Following that discussion the chapter analyses how formal participation is taking place in these countries. In this case, stakeholders are members of the new spaces of negotiation such as river basin committees and water councils. Another space that has been increasing in relevance in the last few years is related to water use certification and water indicators, transforming the private sector into a key player in water management. Finally, the last section will present how accountability practices and transparency laws are becoming tools that stakeholders can use to influence how water is managed in LAC countries.

12.2 Tensions over water and social activism in LAC

In LAC, the origins of tensions over water are complex and diverse (Arrojo 2005; Arrojo, 2010; Oswald, 2011; Aguariosypueblos.org., 2013). They are generally rooted in different understandings of water allocation, national and regional priorities, contrasting views of development and environmental care, cultural and economic interests, and livelihood defence (Larrain and Schaeffer, 2010). They often originate from the development of economic activities and at times from the institutional reforms promoted to facilitate said economic development (Boelens et al., 2011). Thus, the regional growth supported mainly by commodities exports (Sinnott et al., 2011; ECLAC, 2013) is likely to accentuate tensions associated with dam construction, water diversions, urbanization and mining taking place at domestic and transboundary scales (Table 12.1).

Table 12.1 Features of main water conflicts in LAC

ISSUE	CONFLICT TRIGGER	STAKEHOLDERS INVOLVED
HYDROPOWER PLANTS AND DAMS	Loss of territory and livelihoods as a consequence of dam construction and operation	Rural works, indigenous populations, state, private sector, NGOs, social movements
WATER DIVERSION	Taking water from regions under water stress, prioritizing urban over rural areas and agribusiness activities.	Rural workers, small farmers, indigenous people, NGOs, unions, associations, church, state, agribusiness, municipalities.
IRRIGATION	Farmers do not respect previous formal and/or informal agreements with regard the amount of water they should take from the water body. Priority of agro-export activities over small farmers and indigenous farming practices.	Small farmers, state, agribusiness, indigenous populations, social movements.
MINING	Impact on water resources quality and availability for other economic activities and domestic supply; non-compliance with legislation, destruction of natural landscape (e.g. Deforestation)	Indigenous people, small farmers, fishermen, mining companies, water supply companies, NGOs, local and regional government.
URBANIZATION	Water pollution jeopardizes domestic water supply even in areas that naturally are water abundant.	State, municipalities, NGOs.
CONCESSIONS	Privatization of drinking water, wastewater treatment plants with inadequate service and high prices.	Multinational enterprises, local and national governments, international tribunals (WTOCV).

Source: own elaboration.

In LAC, approximately 60% of territory is included in transboundary basins: the Amazon basin alone includes eight countries with more than 8,000km of shared borders (Rebagliati, 2004). Since each country has sovereignty over its water bodies, yet the river basin could be shared, often water uses impact neighbouring countries. In this context, the main reasons for tension are related to flow control, overuse of water, pollution from upstream countries and the impact of water uses over traditional livelihoods and the environment. Tensions between countries over water often find a venue to be managed in treaties and international agreements. As a matter of fact, overall only around 15% of the South American transboundary population and area is not covered by at least one treaty or an international River Basin Organization (De Stefano et al., 2012), and interestingly relationships over South American shared waters are far less confrontational than in other regions of the world (Wolf et al., 2003; Yoffe et al., 2003; De Stefano et al., 2010; Biswas, 2011). In some cases, multi- and unilateral agreements and financial support have contributed to managing some of the tensions in the region as, for instance, in the Colorado and Bravo rivers (USA and Mexico). In other cases, such as in the Lempa River (Guatemala, Honduras and El Salvador) and the Orinoco basin (Colombia, Venezuela and Brazil) tensions over water eventually led to the creation of cooperation and integration plans for the shared basins.

Conflicts over water can be triggered by environmental consequences of water uses or by their social implications and, even if a certain dispute can be focused only on one of those two factors, they cannot be taken apart (Castro, 2008). For instance, the increase in agro-export activities in LAC has pushed for intensive land use and the expansion of irrigation practices which increased pressure on water availability and ecosystems (Castro, 2008; Boelens et al., 2011). At the same time, changes in farming practices have often led to the loss of traditional knowledge and the disruption of livelihoods, showing that the changes in water use contribute to displace small and indigenous farmers that are replaced by the agro-export model (Boelens et al., 2011). Similarly, the construction of water infrastructure to meet the increasing needs for energy and water in LAC impacts rivers' ecosystems and, at the same time, contributes to the loss of territories and traditional livelihoods, pushing population to urban areas and disrupting local economies (Zhouri and Oliveira, 2007; Oliver-Smith, 2009; Boelens et al., 2011). In mining, the combination of highly polluting production processes with inadequate environmental legislation (and/ or disrespect of it) has had a negative impact on soil, biodiversity, water, and aquifers in almost all countries of LAC (Flota et al., 2012) and is at the root of intense conflicts throughout the region (Figure 12.1). Destruction of upstream ecosystems providing crucial services to urban supply systems and pollution of aquifers are common in LAC metropolitan areas (e.g. Mexico City, São Paulo, Rio de Janeiro, and Lima) and directly affect the capacity of water utilities to provide safe water to households.

Water services privatization has become one of the main sources for conflict in LAC during the last two decades. The rapid increase of urban populations combined with the lack of sufficient funds for the creation and maintenance of public supply services often pushed local and federal governments to grant water supply and sewage concessions to the private sector. Private companies or concessionaries are often reluctant to expand the water supply network to poor suburbs and shanty towns, where the recovery of investments via water tariffs is unviable and governmental subsidies are required. The lack of effective supervising bodies, however, often contributes to the establishment of abusive practices, like unaffordable prices or non-compliance of water supply standards. Although these practices can occur also in case of public providers, they have been especially obvious in some private water concessions in the region, obliging some governments to cancel concessions due to public opposition.

Figure 12.1 Location and number of mining conflicts in LAC. *Source: OLCA (2013)*

In this context, during the past few decades the LAC region has witnessed several grassroots mobilizations around water, which at times have led to intense confrontations (Box 12.1; Bell et al., 2009; FNCA, 2009). Collective actions often start in communities directly affected by a certain decision, but soon they come into contact with existing networks on the frontline in question, composed by both national and international NGOs. LAC is very diverse and it is difficult to generalize about the most salient features of water-related social activism in the region, as the emergence and characteristics of social movements is heavily influenced by the national socio-political context where they emerge and to which they have to adjust (Zibechi, 2006). However, a common thread of many of these social movements is the defence of the public (community) nature of natural resources and the opposition to their transformation into mere economic goods (Seoane, 2006). Moreover, their main way of influencing decisions is outside formal participation venues described in the next sections. Demonstrations, activism actions and legal litigation become means for some civil society organizations to gain a seat at the negotiating table or, for those that are already present at the table, to increase their negotiation power in formal participation venues.

Box 12.1 Examples of grassroots movements in LAC

Beyond networks of affected people, coalitions often occur, for example in Brazil with the MAB (Movement People Affected by Dams), in strategic partnership with Via Campesina and the MST (Landless Movement), thus achieving a strong impact of their actions nationwide.

The Cochabamba conflict on water privatization known as the 'Water War' not only ignited a continental and even global revolt against the privatization of water services, but cornered the Bolivian government and strengthened the role of Evo Morales as a national opposition political leader. In this case, the regional alliance of unions and city residents with indigenous irrigation communities was essential.

The movement of people affected by toxic pollution of the Santiago River, in Mexico, became so strong that the government had difficulties to deal with it, to the extent that the outbreaks of indignation in rural communities received the support of university researchers, neighbourhood associations and unions of the city Guadalajara.

The movement of Mazahua women, also in Mexico, put the federal government on the ropes when it progressed from being a protest of a small number of communities to a revolt of the Mazahua people, to finally mobilizing tens of thousands of citizens in Mexico City, who endorsed their claim to safe drinking water in their homes as a human right. In 2011, the Mexican Congress granted water as a human right in the Constitution.

Often grassroots movements find a counterpart in organized activism networks. Today, in LAC there are strong national and international networks against open pit mining, oil exploitations, large dams and the privatization of water and sanitation services. These networks provide local communities with information and technical assistance, legal advice and media projection, often in collaboration with important sectors of the scientific community. The incorporation of local communities into these setups is one of the keys to the success of activism networks. When they manage to transform the 'indignation' of whole territories into regional or national citizen mobilization, these movements expose a social conflict difficult to ignore (See Box 12.1). From there, complex political processes are usually open, in which the governments and transnational corporations are not only challenged, but questioned and conditioned. When this occurs, a political component soon emerges that ends up having parliamentary consequences or even producing changes in government. An example is the inclusion of water as a human right in LAC countries' Constitutions such as in Mexico. Because of this Latin America became the first region in the world to institutionalize such a claim (see also Chapter 11).

Social movements and networks, which tend to be non-violent, also resort to the courts. Despite the frequent successes obtained on the legal front, in LAC these favourable rulings are rarely effective in practice, which suggests the limited strength of laws and courts in some LAC countries (Box 12.2). This is why social movements rely primarily on non-violent resistance in their territories and citizen mobilization at regional and national levels. Often, the action moves to the international arena, either through important and prestigious ethical courts, such as the Latin American Water Tribunal or the Court of the People, or taking their complaints to the United Nations or the home countries of transnational corporations that are their opponents in the conflict.

Box 12.2 The use of the justice system to influence decisions

Since the 1980s, two parallel processes have taken place worldwide. On the one hand, greater decentralization and public participation were encouraged and promoted, sometimes without adequate attention to local capabilities and resources. On the other hand, developing countries have signed international treaties for the protection of private international investments. In practice, however, countries often did not fully understand what they were agreeing to (IISD, 2006). International investment agreements signed by central governments override decisions taken at local or municipal level. Countries transfer national jurisdiction to international investment courts that can only operate at the request of investors. International arbitration is thus a market created by investors that applies principles for the protection of investors, without having responsibilities for issues of local importance.

Investment agreements are signed by central governments without community participation. In addition, communities and the public are not necessarily parties to investment litigation (although their participation may be allowed by decisions of the arbitration courts) since their participation is contingent to the sovereign will of the arbitration courts. Thus, no matter the importance that litigation may have for local communities, cases are litigated only by governments and investors. Arbitration courts have condemned countries to pay compensation for environmental measures taken by local governments in relation to water resources (Álvarez, 2004). Thus, in the context of international arbitration, local issues and community participation risk irrelevance: local public interest is of little relevance to arbitration courts, since their mandate consists principally of protecting investors' interests. In fact it can be said that investment arbitration treaties and investment arbitration often empty the public participation processes of their original meaning and power.

Even though demonstrations and activism are important vehicles for civil society organizations to express their demands and points of view about water management decisions to authorities and production sectors, the reform of water institutions is increasingly creating new spaces for negotiation. Because of that, civil society organizations are becoming relevant stakeholders in decision-making processes related to water also through formal participation as discussed below.

12.3 Formal participation as a space for negotiation

Water reforms that have taken place in LAC since the 1980s[2] have restructured the institutional arrangements and introduced or officiated councils, committees and forums in which stakeholders are recognized as members (for more see Chapter 11). Even though participation in most of the cases is already part of the institutional engineering, its understanding and level of implementation vary from country to country. Indeed, institutional arrangements such as the main unit of water management, the scale at which participatory decision-making processes happen and the types of spaces for negotiation are intertwined factors that shape public participation in each country (Table 12.2).

Table 12.2 Comparative overview of participatory levels in selected LAC countries

	LAW	TYPE OF SPACES FOR NEGOTIATION	SCALE OF ACTION	MEMBERS
BRAZIL	Law 9433/1997	Watershed committees	State and Federal	Federal, state and local representatives, users, civil society organizations
		National Water Council	Federal	Federal, state and local representatives, users, civil society organizations
		State Water Councils	State	State and local representatives, users, civil society organizations
CHILE	Water Code 1981	Water Users Associations, Water communities, Water Channel Associations, Monitoring Communities	Local	Users, NGOs, social movements
COSTA RICA	Water Law 1942	Supplying Water and Sanitation Systems Association (ASADAS), Public consultations	Local and national	Users and civil society organizations
MEXICO	Water Law (Ley Aguas Nacionales - LAN) 1992/2004	River basin committees	Regional (watershed)	State, users associations, NGOs, enterprises, Academia
PERU	Law 23899/2009 (Ley de Recursos Hídricos)	Basin council / Water users organizations	Local and regional	Users, universities, associations, campesinos and natives communities, state, local and regional representatives

Source: own elaboration.

In the countries presented on Table 12.2, formal participation is understood as part of a strategy that will lead the competent water authorities to share the decision-making processes with different stakeholders. Despite that, the State continues to be the main and ultimate decision maker. Additionally, water authorities are responsible for influencing

---

2 The institutional water reforms started to take place in 1981, through the Chilean Código de Agua, followed by the Ley Aguas Nacionales in Mexico in 1992, the Lei das Águas in 1997 in Brazil and, recently, in 2009, the Ley de Aguas de Peru. However, countries such as Costa Rica still have not undergone institutional water reforms and water is still managed by institutions placed under different ministries that barely interact with each other (Table 12.2). Such dynamics were common in other Latin American countries such as Peru until 2008.

the speed in which participatory spaces are created, as well as defining and enforcing the rules to make them active (Scott and Banister, 2008). Users and other civil society organizations can participate in the control and maintenance of the system at local level, or make suggestions when water management plans are elaborated. The only exception is Chile, where the legislation identifies the market (instead of the State) as the main force influencing water rights allocation (Bauer, 1998).

This reflects how social participation takes place and its impact on water governance. For instance, Chile focuses on water management at the level of the water bodies and therefore its institutional arrangements establish that participation should happen mainly at local level through Water Users Associations, Water Communities, Water Channel Associations and Monitoring Communities. These are spaces where water is managed and controlled on a daily basis and conflicts among different water users should be negotiated (Bauer, 1997, 1998). The Peruvian system also allows for this type of participation through the Juntas de Usuarios y Comités responsible for operating and distributing water locally as well as for collecting water taxes and tariffs at local level. This type of participation is known as activity-specific participation in which stakeholders are asked to undertake specific tasks, working as executors instead of planners, defining how water should be allocated and who should have access to it (Pretty, 1995; Agarwal, 2001; Chambers, 2005; Empinotti, 2007). Participation at the local level is instrumental.

On the other hand, the Mexican, Brazilian and Peruvian systems assume river basins as the unit of water management and concentrate stakeholders' participation at the river basin and regional level. Participation takes place in the form of stakeholders input into planning, coordination and implementation of river basin plans as well as to build consensus among the members of these councils. In these cases, participation is basically a consultation since stakeholders are asked for opinions and suggestions during the elaboration of water management plans, although in Peru and Mexico their impact over the final decisions is still quite limited (Wester et al., 2005; Jiménez-Cisneros and Galizia-Tundisi, 2012). On the other hand, in the Brazilian context, stakeholders are able to influence decisions made in the river basin councils. Indeed, in those councils the number of seats for users and civil society organizations combined can outnumber those of the State, thus providing them with decisional power if their interests converge on a specific issue, while not one of the sectors alone can approve a proposal without the support of others. However, the impact of these negotiated river basin plans is void at the moment that the government disregards them as a tool to support its decisions in the construction of water infrastructure and water allocation, consequently weakening the water institutions and contributing to the understanding that the State still holds the main stake over water management in the country (Empinotti, 2011). Besides, in the Brazilian institutional structure, water governance also takes place at national and state level through the National Water Council and the State Water Councils respectively. In these councils, stakeholders and the State are responsible for defining the main guidelines for water management and for regulating water legislation. Nevertheless, the State has the majority of seats at the national and state councils, thus reducing the role of stakeholders and transforming participation into a consultative practice, hence maintaining the State as the main decision maker (Jacobi, 2009).

In Costa Rica, even though the Water Law does not create spaces for participation, other laws such as the Association Law and Law 8660/2008 allow associations that regulate the water distribution at the local level (Supplying Water and Sanitation Systems Association – Asociaciones Administradoras de Sistemas de Acueductos y Alcantarillados Sanitarios, ASADAS) and a national agency to promote participation of civil society organizations (Regulatory Authority for Public Services – Autoridad Reguladora de los Servicios Públicos, ARESEP). Users participate in different moments at the local level, and at the national level the participation of civil society organizations takes place while public consultation meetings are promoted by ARESEP. Recently the Ministry for Environment, Energy and Seas has reinforced water policy by creating the Vice Ministry for Water and Seas. However, it is early to see the results of these organizational changes.

Notwithstanding that the State maintains control over water institutions, it is worth emphasizing that water reforms have reinforced the participation of the private sector, as water users, within the decision-making processes, empowering this sector in comparison with other social actors. One of the reasons for that is how legislation defines stakeholders. For instance, in the Chilean, Mexican and Peruvian cases, stakeholders' participation occurs mainly through users associations and state agencies, thus allowing the private sector to become a main actor in the process with access to negotiation spaces that were not in place before. For this reason, stakeholders' participation is constrained to the scope and interests of each users association, including mining and electricity companies. This has led to uncoordinated actions, specifically related to bodies of water that, in the long run, can affect the sustainability of the river basin (IIC, 2011). These characteristics reflect the bias towards a technocratic and utilitarian perspective of water since the institutional arrangements consider that only sectors such as agriculture, industry, fishery or the mining industry should be involved in decision-making processes. From this perspective, water management should be restricted to direct users, the State, or the market as in the case of Chile, with little consideration of other perceptions such as those of NGOs, social movements or even unions, leaving social actors marginalized in the water governance processes.

The Brazilian, Mexican and Peruvian models, however, allow other organizations, besides users, to participate in water-related advisory or decision-making bodies. In the Brazilian context civil society organizations are represented by NGOs, communitarian and professional associations, unions, universities, research institutes and indigenous communities (Lei das Águas n. 9433, 1997). The Peruvian legislation reserves seats for natives and traditional communities in the river basin committees along with users (Ley de Recursos Hídricos n. 29338, 2009). In Mexico, rural groups, small businesses, environmental organizations and social platforms should be part of the river basin committees but they are systematically excluded from the councils (Boelens et al., 2011).

It is important to point out that natives and traditional populations are also underrepresented sectors in formal participatory forums, which exemplifies that there is a distance between having a seat, being allowed to negotiate and the ability to have your claims transformed into practices (Agarwal, 2001). Indeed, even though the Brazilian and Peruvian legislation recognizes and enables seats for these groups, they usually represent around 2% of the total council, thus barely having any power during the voting processes. However, their presence in participatory institutions at least allows for the introduction of their own agenda into the discussion, even if they have little guarantee that their claims will be addressed.

As a consequence of the persistent control of governments at different organizational levels over the participatory forums and their recommendations, civil society organizations such as NGOs, research institutes and social movements given visibility to their claims through activism and advocacy, while the private sector intensifies its influence through parallel forums and alliances with some civil society organizations in defining parameters for water certification and water efficiency indicators. The use of spaces to influence decision-making processes that go beyond the formal participatory institutions reflects the logic of the system's characteristics. First of all, multi-stakeholder platforms, such as water councils and committees, focus on consensus-building by providing a conductive space for mutual understandings. This is a recommended practice where a single actor does not dominate the field and there is a basic willingness to communicate (Warner, 2005). One of the main purposes should be to forestall conflict situations by discussing the water management practices and interventions among different stakeholders (ibid.). In this context, conciliation techniques help building a positive relation between the parties of a given dispute (Sgubini et al., 2004). The success of conciliation over environmental conflicts resides in strengthening collective imaginaries on the importance of rights and duties involved in the protection of the environment (Velásquez Muñoz, 2004).

Nonetheless, there are at least two problems that go against the multi-stakeholder platform assumptions described above. First, water issues are complex problems in which different actors have antagonist views on how to solve them, considering how water should be allocated and by whom (Warner, 2005; Jacobi, 2006). Second, decisions over water allocation and the construction of water infrastructure take place at government level, and then they are brought to councils and committees. Stakeholders' discussions concentrate mainly on decisions previously made, which leads some actors to believe that their participation is only to legitimize the government's decisions. As a consequence, frustrated civil society organizations withdraw from councils and committees since they perceive their participation as inefficient in promoting their own agenda or in changing government's plans. Thus, the government and private sector's agenda are the ones prevailing and influencing the water management in the region (Boelens et al., 2011; Empinotti, 2011).

In general, most participatory processes in LAC remain at the information and consultation stages. A meaningful and interactive participation would require devolving mandates down to the lowest practicable level and giving people the right to say 'no' to interventions proposed by the government. Nonetheless and even though in many cases formal participation is still incipient and does not meet initial expectations, it should be acknowledged that it is contributing to share water governance decisions and to expose problems and conflicts about how water is allocated in different regions and contexts.

12.4 Water certification as a new space for negotiation

Another space for negotiation that is effective but usually not recognized takes place when civil society organizations and private sector organizations discuss and propose new approaches to future public policies regarding water management. The private sector is one of the main water users and consumers. Industrial and agricultural practices together correspond to more than 90% of water consumption in the world (World Bank, 2010; Hoekstra and Mekonnen, 2012), which makes the private sector the main water user. Consequently, this sector's interest over water issues focuses on guaranteeing its access to water resources as a means to reducing water-related risks for its business activities. In order to achieve this goal, lobbying practices and the proposal for new market mechanisms become strategies to shape future water policies.

While green NGOs lobby the State and legislative bodies for an environmental agenda, or professional associations push for a technical approach to manage water, organizations representing the private sector's interests have been focusing mostly on securing regulations that do not constrain business and on ensuring that the regulatory environment is consistent across government departments, predictable and stable over time and applied to all the companies in a similar way (Pegram et al., 2009). In this context, the water law reforms that have occurred in LAC during the past three decades were an opportunity for the private sector to influence the process. In the Brazilian case, the industrial sector was one of the most active groups in Congress during the negotiation of the 1997 Water Law, advising their representatives and influencing the final text. In Peru, during the debates over and the formulation of the 2009 Water Resources Law in Congress, the private sector – mainly the National Mining, Energy and Petroleum Association – was able to actively influence the final text. Some of their agenda was translated into law through the authorization of economic activities in headwaters, and the introduction of concepts such as efficiency in water use and equity in access (Budds and Hinojosa-Valencia, 2012).

At the same time, multinational and international industries direct their attention to the discussion of water use indicators and future certifications related to water use in the production process. The strategy is to discuss and elaborate rules among companies and civil society organizations that could become the reference for future public policies. Researchers identified such practices as private governance and non-state market-driven governance systems that allow the private sector to influence the rules that will impact their production practices in the future (Cashore, 2002; Smith and Fischlein, 2010). One of the consequences of this strategy was the inclusion of water indicators into the corporations' social environmental responsibility portfolio. Such a trend has developed during the last six years, when the water footprint method, combined with the ISO initiative to create a protocol on water use, attracted the corporations' attention (Daniel and Sojamo, 2012; Empinotti, 2012). These initiatives of multinational organizations were triggered by their interest in assessing the water-related risks for their business – from both a regulatory and a physical point of view – as well as the need to address the consumers' expectation for environmental commitment (Hepworth, 2012; Larson et al., 2012).

The debate over the water footprint and other initiatives captured the attention of transnational corporations such as Coca-Cola, SABMiller and Nestlé, which compete in the international scale and have their production chain spread all over the globe. Moreover, LAC industries discussing such issues are usually large exporters of raw materials and have their main consumer markets abroad (see Table 12.3).

Interestingly, these types of initiatives have low participation rates of LAC industries in comparison to other regions of the world. For instance, only 36% of the invited Latin American corporations adhered to the Carbon Disclosure Project's Water Initiative, compared to 62% in North America, 80% in Europe, 80% in Africa, 51% in East Asia and 62% in Southeast Asia and Oceania (Deloitte, 2012). In the CEO Mandate, only two out of a total of ninety-one endorsing companies are from LAC. This could be partially explained by the geographical distribution of the corporations' headquarters but there could be also other more substantive reasons that ought to be explored.

While many of the Latin American companies listed in Table 12.3 are still in the process of calculating their water footprint, those that already have their results, in general, treat them confidentially and are discussing them internally. Initiatives for water accounting, however, did not promote changes in water governance practices nor did they trigger the discussion of new public policies, following an international trend (Hepworth, 2012, Sojamo and Larson, 2012). An interesting exception occurred in Brazil, where the industrial sector and international environmental NGOs engaged in a lively discussion over indicators for water efficiency and regulation (Empinotti, 2012). While supporting NGOs' initiative by creating a broad water indicator, the industrial sector was concerned with the possibility of having a public policy defining the acceptable amount of water that each sector should use. From the industrial perspective, such a reference would increase State control over water rights and distribution. Such concern led the industrial sector to redirect the discussion initially driven by governmental agencies towards the use of certifications that acknowledge the industries initiatives in reducing their water use, instead of establishing rules that could define and limit the average amount of water that can be allocated to each industrial sector (Empinotti, 2012).

During the past decades, the interaction between the private sector and civil society organizations has shaped environmental discussions and new public policies. However, it is understandable that the private sector participation in water-related spaces for negotiation serves the ultimate goal of ensuring water access for its production processes. Thus, there is no guarantee that water will be better or more equally distributed among the different society sectors or that water use will be more sustainable, if this does not revert positively in business activity. For this reason, the participation and contribution of the private sector to water governance should be adjusted and constantly evaluated, to push the private sector to understand water as a common good and human right, following the principles defined by most of LAC water legislations.

Table 12.3 Latin American companies involved in water networks and initiatives on water accounting tools

	SECTOR	COUNTRY	INITIATIVE
Natura	Cosmetics	Brazil	WFN [1] / WBCSD [2]
FIBRIA	Pulp	Brazil	WFN/WBCSD
Cimentos Liz	Cement	Brazil	WBCSD
Abril Group	Media	Brazil	WBCSD
Petrobrás	Oil	Brazil	WBCSD
Suzano Papel e Celulose	Pulp and paper	Brazil	WBCSD
Votorantim	Cement, metals, energy, steel, agribusiness	Brazil	WBCSD
Grupo Orsa	Pulp and paper	Brazil	WBCSD
Banco do Brasil	Banking	Brazil	The CEO Water Mandate [3]
Vale	Mining	Brazil	CDP – Water Initiative - WBCSD [4]
Cia. Siderurgica Nacional – CSN	Steel	Brazil	CDP – Water Initiative
ABInBev	Beverage	Belgian-Brazilian	BIER – water stewardship [5]
Quimico del Campo	Metals	Chile	WFN
Vinã Concha y Toro	Wine	Chile	WFN
Vinã del Martino	Wine	Chile	WFN
Vinã Errazuriz	Wine	Chile	WFN
Codelco	Mining	Chile	WBCSD
Masisa	Timber	Chile	WBCSD
Empresas CMPC	Pulp and paper	Chile	WBCSD
Cementos Argo	Cement	Colombia	WBCSD
EPM Group	Energy and water	Colombia	WBCSD
Grupo Nutresa	Food	Colombia	CEO Mandate
Ecopetrol	Oil	Colombia	CDP – Water Initiative
CEMEX	Cement	Mexico	WBCSD
Wal Mart de Mexico	Retail	Mexico	CDP – Water Initiative
Fresnillo	Mining	Mexico	CDP – Water Initiative

Source: own elaboration based on data from WBCSD, WFN, CDP, CEO Water Mandate
1 Water Footprint Network Initiative, 2 World Business Council for Sustainable Development
3 UN Global Compact's CEO Water Mandate, 4 Carbon Disclosure Project (CDP) Driving
Sustainable Economies, 5 Beverage Industry Environmental Roundtable – Water Stewardship.

12.5 Accountability and information transparency: two faces of the same coin

Accountability and transparency are often pointed out as 'silver bullets' against corruption and bad governance in the water sector (Stalgren, 2006; Transparency International, 2008; Asís et al., 2009; UNDP, 2011; Regional Process of the Americas, 2012), which, in turn, are considered to have a key role in poor service provision, environmental degradation, society inequity and other important failings of the water sector.

Accountability implies being held responsible for one's actions: from the approval of e.g. a new water infrastructure, down to the decision of turning a tap on and off to provide water in a specific location or for a specific use. Thus, it is a relationship between those that are held accountable (e.g. politicians, government officials, private companies or individual citizens) and those entitled to demand accountability (e.g. social and State actors) and to apply sanctions in cases of poor performance or abuses (Hernández et al., 2013). Accountability entails answerability, i.e. the existence of formal processes where actions are judged according to specific criteria. Answerability, in turn, requires access to information by those who demand accountability and the obligation to justify one's actions and decisions if required to do so. For many, information transparency and justification alone, however, do not guarantee accountability, as it is necessary to have in place mechanisms and bodies with enforcement capacity, i.e. to apply sanctions for not meeting the established standards or not playing by the rules (Schedler, 1999; Schedler, 2004; Fox and Haight, 2007; Peruzzotti, 2008).

This section focuses on societal accountability[3] of the public authorities or companies that manage water resources or provide water services. The 'right to know' for constituents, customers or civil society organizations in general is usually pursued through two different strategies: the first is top-down, which means that public institutions proactively provide information to the public on issues relevant to water management (Fox and Haight, 2007). Typically this strategy is implemented through the internet, as proactive information. This method is relatively easy and inexpensive and it contributes to reducing the number of requests for information (Mendel, 2009). The second strategy (bottom-up) implies that the public files information requests normally following well-established procedures. However, even if these strategies are in place, information provision can be 'opaque', since the material is only nominally available given that it is often presented in a way that is difficult to understand/use or, more importantly, because it is not reliable (Fox and Haight, 2007).

12.5.1 Legal provisions to foster access to information

The analysis of existing initiatives to ensure information transparency suggests that in LAC there is a keen perception of this issue and a large body of legal provisions to pursue it. The legal basis comes from specific articles in the Constitution (e.g. in Colombia, Ecuador or Mexico) and/or from specific laws that deal with the issue of access to information. Most of the laws address the freedom of information in general, but in several countries there are also laws that regulate the access to environmental information (e.g. in Argentina), which is particularly relevant to water management. In some cases, sector laws like the Brazilian water law also establish the creation of an information system that should contain information to support decision-making processes related to water governance and management.

---

3 Downward or societal accountability means being answerable to a constituency (users, customers or society in general).

Since the end of the 1990s in the LAC region there has been a surge of information transparency laws (Figure 12.2), and currently about two-thirds of the countries in the region already have a specific law for access to information in place. Moreover, transparency portals are becoming a common way of conveying information to the citizens in a centralized way and examples of it can be found in Peru, Guatemala, Bolivia, Colombia, Chile, Brazil and Mexico. Usually they are websites managed by the government devoted to publishing public financial information regarding public companies, municipalities and government procurement (Solana, 2004) and can be a tool to empower civil society organizations. These portals, however, rarely have specific information about water.

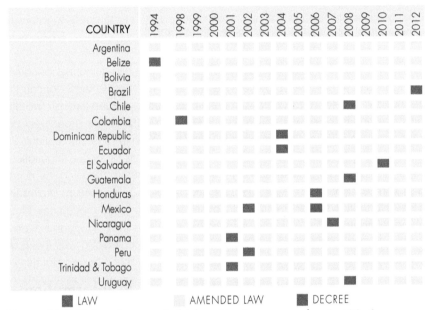

Figure 12.2 Timing of approval of information transparency law in LAC. *Source: own elaboration with data from Mendel (2009), Vleugels (2009) and Michener (2010).*

A comparative analysis of key elements of transparency laws in twelve LAC countries (Michener, 2010) suggests that the weakest points of the existing legal provisions for information transparency are related to the regulation of exceptions and of the appeals in cases of information denial, while the scope of the law and the duty to publish are quite well developed (Table 12.4).

LAC laws, and in particular the Peruvian one, have progressive rules in relation to the duty of public bodies to publish information in a proactive manner. Several countries have, at least on paper, well-developed systems to foster agile access to information. For example, in Mexico, Nicaragua and Ecuador there are specific rules on how to make information, that is subject to proactive publication, easy accessible (e.g. information index, list of classified information) (Mendel, 2009). The dominant trend in all countries in the region is to make increasingly more information available on a proactive basis, particularly via the internet, even in cases when it is not required under an information transparency law (*ibid.*).

Table 12.4 Strengths and weaknesses of the existing transparency laws in several LAC countries. Scores are from 0 to 3. Colours are for interpretation only. Criteria for the scoring can be found in Michener (2010).

COUNTRY	SCOPE OF THE LAW	PROCEDURAL GUARANTEES	DUTY TO PUBLISH	EXCEPTIONS	APPEALS	SANCTIONS
Brazil	1.5	2.1	1.5	1.9	1.5	2.2
Colombia	1.5	1.7	1.3	1.1	1.2	1.3
Chile	2.2	2.7	3.0	1.9	2.5	2.1
Dominican Republic	1.4	1.7	1.3	1.3	2.0	1.0
Ecuador	2.1	1.4	2.1	1.4	1.3	2.3
Guatemala	2.3	2.9	2.5	2.4	2.3	1.3
Honduras	2.1	1.9	2.8	1.4	1.7	2.3
Mexico	2.6	2.9	3.0	2.3	2.8	2.3
Nicaragua	2.8	2.9	2.3	1.9	1.7	1.7
Panama	2.3	1.9	1.5	1.6	1	2.0
Peru	2.0	1.7	2.3	2.3	1.7	1.7
Uruguay	1.6	2.0	2.0	1.9	1.5	1.7

Source: own elaboration based on Michener (2010).

Many LAC transparency laws, but not all of them, impose the duty to publish not only to public corporations but also to private bodies, which receive funding through public contracts. In Peru, the obligation is even extended to all bodies exercising a public power or performing a public function (Mendel, 2009). In some countries, like Chile and Colombia, only corporations with 50% public ownership are covered, although a large block of State involvement ought to adhere to the principle of openness, since significant involvement of the State in a corporation normally signals a public interest in its operations (*ibid.*).

The two most common options for appeal in case of refusal of information are internal complaints or complaints to an independent oversight body and/or the courts. Many laws – e.g. in the Dominican Republic, Guatemala and Peru – include legal provisions for an internal appeal, usually to a higher authority within the same body which originally refused the request. Chile, Honduras and Mexico appoint an independent administrative oversight body for the review of denials of information (*ibid.*). Most laws in Latin America, as well as globally, include a regime of sanctions for individuals who obstruct access to information, and some also provide for the direct responsibility of public bodies. In some countries – like the Dominican Republic and Peru – it is a criminal offence to obstruct access to information, while in other countries – like Chile, Honduras and Mexico – the law provides for administrative liability (*ibid.*).

12.5.2 Implementing legal provisions: are they enough to have transparent water sector?

As for any legislation, the mere existence of a legal framework is no guarantee of achieving satisfactory access to information, either because of flaws in the design of the law or, more often than not, due to difficulties in its implementation. In LAC, most of the existing comparative studies focus on the strength on paper of legal provisions, but several watchdog initiatives and academic studies also point out gaps in the implementation of the law (Alianza Regional, 2009; Fraga, 2012; IDB, 2012; Soto and Rojas, 2012; Torres, 2012). But how does this apply to water? Do legal provisions for information transparency manage to make the water sector truly transparent? In the water sector, benchmarking exercises typically assess the technical performance of water utility companies (Table 12.5). In some cases, they also include criteria related to governance, financial performance, or customer service. Thus, they do not assess information transparency but do contribute to making water and sanitation companies more transparent. Transparency benchmarking as such is rare. An ongoing initiative to improve Brazil's water agencies' transparency is based on a methodology first applied in Spain and has been adapted to Brazil (Empinotti and Jacobi, forthcoming). It represents an exception to the rule in that it looks at water management as a whole and not only at a specific sector (e.g. water utilities).

Table 12.5 Examples of benchmarking initiatives of water and sanitation utilities companies

EVALUATING AGENCY	COUNTRIES	PERIODICITY	INDICATORS
Interamerican Development Bank	LAC countries	On request	Service quality, business management efficiency, operating efficiency, access to service, investment planning and execution, financial sustainability, environmental sustainability, corporate government and accountability
Fitchratings	Mexico, Colombia, Panama	Yearly	Control, coverage, charges, cash, capital, capacity, legal compliance, community and clients
Grupo Regional de Trabajo de 'Benchmarking' de la Asociación de Entes reguladores de agua y saneamiento	Several LAC countries	Yearly	Performance indicators: service structure, operational structure and service quality, economic indicators
Superintendencia Nacional de Servicios de Saneamiento	Peru	Quarterly/ yearly	Access and quality of the service, billing, economic and financial sustainability, management efficiency, governance, customer service and eco-efficiency
Superintendencia de Servicios Públicos Domiciliarios	Colombia	Monthly/ quarterly	Registration of property, control of assets, fixed assets insured, compliance of contractual agreements, settlement of contracts
Superintendencia de Servicios Sanitarios	Chile	Yearly	Water treatment, drinking water quality, water continuity, accuracy in billing, complaints
Benchmarking Central American Water Utilities	Cosra Rica, El Salvador, Guatemala, Honduras, Nicaragua, Panama	Yearly	Water quality, water standards, leaks, operation costs, water consumption, connections networking, service coverage, metering, water cost
International Benchmarking Network International Bank	International	Yearly	Service coverage, production, non-revenue water, metering practices, network performance, cost and staffing, quality of service, billings and collection, financial performance, assets, affordability of services, process indicators

Source: own elaboration.

As mentioned previously, the internet is a powerful tool for conveying relevant information about water management to society. An online search of some key information in some countries in LAC provides insights into areas where there is a very progressive and proactive information provision and issues that still have a poor coverage (Table 12.6). Interestingly, online consultation seems to be a real possibility in the considered countries. Water authorities use the internet to make water rights registers accessible or publish documents for public consultation, but they rarely use it to record and make public the received comments. Significant gaps in information provision are related to the application of water law infractions, sanctions and the follow-up of the execution of public works.

Table 12.6 Online availability of information about selected issues in five LAC countries. The table reflects the information available online on February 2013 and that which could be found by consulting the websites of public organizations in charge of managing water resources in each country.

COUNTRY	BRAZIL	CHILE	PERU	COSTA RICA
Comments received to water-related documents issued for public consultation	Usually yes	Yes, for EIA studies	No	No
Registers of water right	Yes, in most of the Brazilian states	Yes	Yes	Yes
Statistics about water law infractions and sanctions	Usually not	No	No	No
Background studies supporting the planning process	Yes, in most of the cases	Yes	No	No
Data about the incomes from water tariffs	Yes, where water tariffs are in place	Yes (non-use tariff)	No	No
Data on the process for granting water-related contracts and tenders	Yes but not always easily accessible	Yes, in most of the cases	No	No
Data on follow-up and control of public works execution (duration and cost)	Not in a consolidated way	Yes	Not in a consolidated way	No

Source: own elaboration.

From the above, it can be concluded that most LAC countries have well-developed and in some cases very progressive information transparency laws, which can contribute to the transparencys of water-related public bodies. The actual implementation of the legal obligations to information disclosure is ongoing and it surely fostered by benchmarking initiatives and watchdog studies promoted by civil society and international organizations, mainly for the water and sanitation sector. To provide water for human uses in a sustainable way, however, it is key to have a holistic approach and consider the system that provides those resources – rivers, aquifers watersheds, wetlands. Thus, the next step is to assess and seek information transparency in the management of the whole system, and not only at the end of the pipe, where water is supplied.

References

Agarwal, B. (2001). Participatory Exclusion, Community Forestry, and Gender: An Analysis for South Asis and a Conceptual Framework. *World Development*, 29: 1623–1648.

Aguariosypueblos.org. (2013). *Agua rios y pueblos*. [Online] www.aguariosypueblos.org/. [Accessed July, 2013].

Alianza Regional (2009). *Saber Más I: Informe regional sobre la situación actual del acceso a la información pública*. Alianza Regional para la Libre Información y Expresión. 143° Período de Sesiones de la Comisión Interamericana de Derechos Humanos. [Online] Available from: www.redrta.org/Publicaciones/Alianza%20Regional.pdf. [Accessed July, 2013].

Allan, J.A. (2013). Introduction: Can improving returns to food-water in Africa meet African food needs and the needs of other consumers?. In: Allan, J.A., Keulertz, M., Sojamo, S. & Warner, J. (eds). *Handbook of Land and Water Grab in Africa*. pp 1–8. Routledge, Abingdon.

Álvarez, G. (2004). Técnicas Medioambientales TECMED, S.A. v. United Mexican States: Introductory Note. ICSID Case number ARB(AF)/00/2. *ICSID Review*, 19: 154–157.

Arrojo, P. (2005). *El reto ético de la nueva cultura del agua: funciones, valores y derechos en juego*. Paidós, Barcelona.

Arrojo, P. (2010). *The Global water Crisis*. Barcelona, Justícia, C. I.

Asís, M.G.d., O'Leary, D., Ljung, P. & Butterworth, J. (2009). *Improving Transparency, Integrity, and Accountability in Water Supply and Sanitation: Action, Learning, Experiences*. The World Bank, Washington.

Bauer, C. (1997). Bringing Water Markets Down to Earth: The Political Economy of Water Rights in Chile, 1976–1995. *World Development*, 25: 639–656.

Bauer, C. (1998). Slippery Property Rights: Multiple Water Uses and the Neoliberal Model in Chile, 1981–1995. *Natural Resources Journal*, 38: 109–155.

Bell, B., Conant, J., Olivera, M., Pinkstaffm C. & Terhorst, P. (2009). *Changing the Flow: Water Movements in Latin America*, Food and Water Watch, Other Worlds, Reclaiming Public Water, Red VIDA Transnational Institute.

Berry, K. & Mollard, E. (2010). Introduction: Socila Participation in Water Governance and Management. In: Berry, K. & Mollard, E. (eds). *Social Participation in Water Governance and Management*. pp.20–28. Earthscan, London.

Biswas, A.K. (2011). Transboundary Water Management in Latin America: Personal Reflections, *International Journal of Water Resources Development*, 27(3): 423–429.

Boelens, R., Mesquita, M.B.d., Gaybo, A. & Peña, F. (2011). Threats to Sustainable Future: Water Accumulation and Conflict in Latin America. *Sustainable Development Law and Policy*, 12: 41–45, 67–69.

Budds, J. & Hinojosa-Valencia, L. (2012). Restructuring and Rescaling Water Governance in Mining Contexts: the co-production of waterscapes in Peru. *Water Alternatives*, 5: 119–137.

Cashore, B. (2002). Legitimacy and the Privatization of Environmental Governance: How non-state market-driven (NSMD) governance systems gain rule-making authority. *Governance: An International Journal of Policy, Administration, and Institutions*, 15: 503–529.

Castro, J.E. (2008). Water Struggles, Citizenship and Governance in Latin America. *Development*, 51: 72–76.

Chambers, R. (2005). *Participation: review, reflections and future. Ideas for Development*, Earthscan, London.

De Stefano L., Edwards P., de Silva L. & Wolf, A.T. (2010). Tracking Cooperation and Conflict in International Basins: Historic and Recent Trends. *Water Policy*, 12(6): 871–884.

De Stefano, L., Duncan, J., Dinar, S., Stahl, K., Strzepek, K.M.,& Wolf, A.T. (2012). Climate Change and the Institutional Resilience of International River Basins. *Journal of Peace Research*, 49 (1): 193–209.

Daniel, M. & Sojamo, S. (2012). From Risk to Shared Value? Corporate strategies in building a global water accounting and disclousure regime. *Water Alternatives*, 5: 636–657.

Deloitte (2012). *Collective Response to Rising Water Challenges*. CDP Global Water, London. Report 2012.

ECLAC (2013). Economic Commission for Latin America and the Caribbean. *Preliminary Overview of the Economies of Latin America and the Caribbean*. ECLAC, Santiago, Chile.

Empinotti, V.L. (2007). *Re-framing Participation: the Political Ecology of Water Management in the Lower São Francisco River Basin - Brazil*. University of Colorado. Ph.D.

Empinotti, V.L. (2011). E se eu não quiser participar? O caso da não participação nas eleições do comitê de bacia do rio São Francisco. *Ambiente e Sociedade*, 14: 195–211.

Empinotti, V.L. (2012). O setor privado e a Pegada Hídrica: risco, oportunidade e vulnerabilidade. In: Empinotti, V. & Jacobi, P. R. (eds). *Pegada Hídrica: inovação, co-responsabilização e os desafios de sua aplicação*. pp. 119–135. Annablume, São Paulo.

Empinotti, V.L. & Jacobi, P.R. (2013). Novas práticas de governança da água? O uso da pegada hídrica e a transformação das relações entre o setor privado, organizações ambinetais e agências internacionais de desenvolvimento. Desenvolvimento e Meio Ambiente, 27: 23–36.

Empinotti, V.L. & Jacobi, P.R. (forthcoming). *Transparência na Gestão dos Recursos Hídricos no Brasil*. Universidade de São Paulo, São Paulo.

Flota, E.d.R.P., Villela, M.d.J.E. & Hernández, Á.G. (2012). Concesiones Mineras en Tierras Ejidales: Detrimento de la Propiedad Social. *Revista Iberoamericana para la Investigación y el Desarrollo Educativo*, 9.

Fox, J. & Haight, L. (2007). Las reformas a favor de la transparencia: teoría y práctica'. In: Fox, J., Haight, L., Hofbauer, H. & Sánchez, T. (eds). *'Derecho a saber. Balance y perspectivas cívicas'*. Fundar-Centro de Análisis y Investigatición, Cidade do México.

FNCA (2009). Fundación Nueva Cultura del Agua. *Agua, Ríos y Pueblos. Homenaje a quienes luchan por defender los ríos y sus gentes*. Pedro Arrojo Agudo (coord.): Fundación Nueva Cultura del Agua. [Online] Available from: www.aguariosypueblos.org/en/ [Accessed July, 2013].

Fraga, I. (2012). *Brazil's Access Law Active, But Problems Still Remain*. Washington DC. [Online] Available from: www.freedominfo.org/2012/07/brazils-access-law-active-but-problems-still-remain/ 2013. [Accessed May, 2013].

Hepworth, N. (2012). Open for Business or Opening Pandora's Box? A Constructive Critique of Corporate Engagement in Water Policy: An Introduction. *Water Alternatives*, 5: 543–562.

Hernández, A., Flórez, J. & Hincapié, S. (2013). *Impact of Accountability in Water Governance and Management. Regional Analyis of Four Case Studies in Latin America*, Universidad de los Andes and UNDP Virtual School. Draft report No. I.

Hoekstra, A.Y. & Mekonnen, M.M. (2012). The water footprint of humanity. *Proceedings of the National Academy of Sciences*, 109: 3232–3237.

IDB (2012). Interamerican Development Bank. *Open Government and Targeted Transperancy: Trends and challenges for Latin America and the Caribbean*. Interamerican Development Bank, Washington DC.

IIC (2011). Instituto Ingenieros de Chile. *Temas Prioritarios para una Politica Nacional de Recursos Hídricos*. Santiago, Chile. [Online] Available from: documentos.dga.cl/REH5332.pdf. [Accessed July, 2013].

IISD (2006). International Institute for Sustainable Development of Canada. *Pakistan attorney general advises states to scrutinize investment treaties carefully*. Investment Treaty News: International Institute for Sustainable Development. [Online] Available from: www.iisd.org. [Accessed May, 2013].

Jacobi, P.R. (2006). Participação da Gestão Ambiental no Brasil: Os Comitês de Bacias hidrográficas e o desafio do Fortalecimento de Espaços Públicos Colegiados. In: Alimonda, H. (ed.). *Los Tormentos de la Materia*. pp. 169–194. CLACSO Libros, Buenos Aires.

Jacobi, P.R. (2009). Governança da Água no Brasil. In: Ribeiro, W. (ed.). *Governança da Água no Brasil: uma visão interdisciplinar*. pp 35 -60. Annablume, São Paulo, Brazil.

Jiménez-Cisneros, B. & Galizia-Tundisi, J. (Coord.) (2012). Diagnóstico del Agua en las Américas. Mexico, Red Interamericana de Academias de Ciencias (IANAS) and Foro Consultivo.

Larrain, S. & Schaeffer, C. (2010). *Conflicts over water in Chile: Between human rights and market rules*. Heinrich Böll Foundation, Santiago, Chile.

Larson, W., Freedman, P., Passinsky, V., Gubb, E. & Adriaens, P. (2012). Mitigating Corporate Water Risk: Financial Market Tools and Supply Management Strategies. *Water Alternatives*, 5: 582–602.

Mendel, T. (2009). *The Right to information in Latin America: a comparative legal survey*. Quito, Ecuador.

Michener, R.G. (2010). *The Surrender of Secrecy: Explaining the Emergence of Strong Access to Information Laws in Latin America*. University of Texas at Austin. Ph.D.

Oliver-Smith, A. (2009). *Development and Dispossession. The Crisis of Forced Displacement and Resettlement*. School for Advanced Research Press, Santa Fé, Argentina.

OLCA. (2013) Observatorio Latinoamericano de Conflcitos Ambientales. Sistema de Información para la gestión comunitaria de Conflictos Socio-ambientales mineros en Latinoamérica [Online] Available from: basedatos.conflictosmineros.net/ocmal_db/ [Accessed October, 2013].

Oswald Spring, Ú. (2011). *Water Resources in Mexico. Scarcity, Degradation, Stress, Conflicts, Management, and Policy*, Springer Verlag, Berlin.

Pegram, G., Orr, S. & Williams, C. (2009). *Investigating Shared Risk in Water: corporate engagement with the public policy process*. [Online] Available from: assets.wwf.org.uk/downloads/investigating_shared_risk.pdf. [Accessed August, 2013]. Online report.

Peruzzotti, E. (2008). *Marco Conceptual de la Rendición de Cuentas*. [Online] Available from: www.agn.gov.ar/ctrc/documentos%20ctrc/peruzzotti%20ii.pdf. [Accessed August, 2013].

Pretty, J.N. (1995). Participatory Learning of Sustainable Agriculture. *World Development*, 23: 1247–1263.

Rebagliati, E.B. (2004). *Avanzando la Agenda del Agua: aspectos a considerar en América Latina*. En Series sobre Elementos de Políticas, Fascículo 2. Organización de Estados Americanos/Unidad de Desarrollo Sostenible y Medio Ambiente.

Regional Process of the Americas. (2012). Americas' Water Agenda: Targets, Solutions and the Paths to Improving Water Resources Management. [Online] Technical Document prepared at

6th World Water Forum, Marseille. February 2012. Available at: www.gwp.org/Global/GWP-CAm_Files/Americas'%20Water%20Agenda.pdf [Accessed October, 2013].

Schedler, A. (1999). Conceptualizing Accountability. In: Schedler, A., Diamond, L. & Plattner, M. F. (eds). *The Self-Restraining State: Power and Accountability in New Democracies.* Lynne Rienner Publishers, Colorado.

Schedler, A. (2004). *¿Qué es la rendición de cuentas?.* [Online] Available from: works.bepress.com/andreas_schedler/6/. [Accessed July, 2013].

Seoane, J. (2006). Movimientos sociales y recursos naturales en América Latina: resistencia al neoliberalismo, configuración de alternativas. *Sociedade e Estado,* 21(1): 85–107.

Scott, C.A. & Banister, J.M. (2008). The Dilema of Water Management 'Regionalization' in Mexico under Centralized Resource Allocation. *Water Resources Development,* 24: 61–74.

Sgubini, A., Prieditis, M. & Marighetto, M. (2004). *Arbitration, Mediation and Conciliation: differences and similarities from an International and Italian business perspective.* [Online] Available from: www.mediate.com/articles/sgubinia2.cfm. [Accessed August, 2013].

Sinnott, E., Nash, J. & Torre, A.d.l. (2011). *Natural Resources in Latin America and the Caribbean: Beyond Booms and Busts?.* [Online] Available from: openknowledge.worldbank.org/bitstream/handle/10986/2482/555500PUB0Natu1EPI1991501801PUBLIC1.pdf?sequence=1. [Accessed July, 2013].

Smith, T. & Fischlein, M. (2010). Rival Private Governance Networks: Competing to define the rules of sustainability performance. *Global Environmental Change,* 20: 511–522.

Sojamo, S. & Larson, E.A. (2012). Investigating Food and Agribusiness Corporations as Global Water Security, Management and Governance Agents: The Case of Nestlé, Bunge and Cargill. *Water Alternatives,* 5: 619–635.

Solana, M. (2004). *Delivering Public Financial Information to Citizens in Latin America.* Civil Society Team. The World Bank, Washington DC.

Soto, M. & Rojas, C. (2012). *Resultados Índice Latinoamericano de Trasnparencia Legislativa Argentina-Chile-Colombia-México-Perú.* Mexico DF, Mexico.

Stalgren, P. (2006). *Corruption in the Water Sector: Causes, Consequences and Potential Reform.* Swedish International Water Insitute.

Transparency International (2008). *Corruption in the Water Sector.* Global Corruption Report 2008. Cambridge University Press, New York.

Torres, N. (2012). *Acceso a la información en Colombia: 124 anos después.* Washington DC, USA. [Online] Available from: www.freedominfo.org/2012/07/acceso-a-la-informacion-en-colombia-124-anos-despues/2013. [Accessed May, 2013].

UNDP (2011). United Nations Development Programme. *Fighting Corruption in the Water Sector.* UNDP, New York.

Velásquez Muñoz, C.J. (2004). Conciliación de conflictos ambientales. División de Ciencias Jurídicas de la Universidad del Norte .Revista de derecho, 21: 106–130. [Online] Available from: hdialnet.unirioja.es/servlet/articulo?codigo=2347489. [Accessed July, 2013].

Vleugels, R. (2009). Overview of all 90 FOIAs . September 2009. [Online] Available from: www.access-info.org/documents/Fringe_Special_-_90_FOIAs_-_sep_7_20091_resource.pdf. [Accessed October, 2013].

Warner, J. (2005). Multi-Stakeholder Platform: integrating society in water resources management?. *Ambiente e Sociedade,* 8: 1–19.

Wester, P., Scott, C.A .& Burton, M. (2005). River basin closure and institutional change in Mexico's Lerma Chapala Basin. In: Svendsen, M. (ed.). *Irrigation and River Basin Management: Options for Governance and Institutions.* CABI Publishing, Wallingford.

Wolf, A.T., Yoffe, S. & Giordano, M. (2003). International Waters: identifying basins at risk. *Water Policy,* 5 (1): 29–60.

World Bank (2010). *Development and Climate Change.* The World Bank, Washington DC. World Development Report 2010.

Yoffe, S., Wolf, A.T. & Giordano, M. (2003). Conflict and Cooperation Over International Freshwater Resources: indicators of basins at risk. *Journal of the American Water Resources Association,* 39(5): 1109–1126.

Zhouri, A. & Oliveira, R. (2007). Desenvolvimento, conflitos sociais e viol ência no Brasil rural: o caso das usinas hidroelétricas. *Ambiente e Sociedade,* 10: 119–135.

Zibechi, R. (2006). *Movimientos sociales: nuevos escenarios y desafíos inéditos.* OSAL, Observatorio Social de América Latina (Año VII no. 21). Buenos Aires, Argentina: CLACSO (Consejo Latinoamericano de Ciencias Sociales).

13

ECONOMIC INSTRUMENTS FOR ALLOCATING WATER AND FINANCING SERVICES

Authors:
Alberto Garrido, Water Observatory – Botín Foundation, and CEIGRAM, Technical University of Madrid, Spain
Guillermo Donoso, Pontificia Universidad Católica de Chile, Santiago, Chile
Oscar Melo, Pontificia Universidad Católica de Chile, Santiago, Chile
Miguel Solanes, IMDEA Agua – Madrid Institute for Advanced Studies, Spain

Highlights

- All countries with post-Dublin (1992) new water legislation have implemented more or less sophisticated economic instruments and financial mechanisms to treat water and supply service as an economic good.

- The notion of water as a public resource domain, coupled with the need to increase cost recovery rates, is at the root of the legislative foundations of all economic instruments applied to water.

- The 'polluter-pays-principle' is also a designing principle in all modern legislation, but in practical terms there are numerous difficulties that hinder its application. Environmental taxation has been implemented in some countries, but the revenue collected is still low, and does not act as a true deterrent for polluters.

- There are several examples of advanced water charging in agriculture, which differ amongst crops, irrigation technology and areas. After decades of little or no cost recovery rates in irrigating schemes, some countries, such as Argentina, Mexico, Peru and Brazil, have taken significant steps to make farmers pay for operation and maintenance costs of the infrastructure supplying their water.

- Chile is the sole Latin America and Caribbean (LAC) country with decades of experience in water trade mechanisms. It seems that recently passed laws in other countries have not been developed nor have they enabled trading mechanisms, whereas the 1981 Chilean Water Code and its subsequent amendments had specific provisions defining water rights as tradable. Market prices for water rights are quite high, with the mining sector being one of the major purchasers.

- Payments for ecosystem services (PES), and in particular Payments for Watershed Services (PWS), have seen an important growth in the past years, bringing renewed hopes for a conservation approach that could succeed where other approaches have failed. LAC has led this development and is continuing to develop new initiatives, although strong growth is observed in other parts of the world.

- To be more efficient and effective, PES should be applied according to size, service per unit of land and type of watershed. Most large (national) schemes are government funded through special taxes, and receive funding from multilateral/aid organizations or governments thus threatening the scheme's sustainability.

13.1 Introduction

Principle 4 of the Dublin Statement[1] reads that 'Water has an economic value in all its competing uses and should be recognized as an economic good.' The Dublin Statement also claimed that '[The] Application of the "polluter pays" principle and realistic water pricing will encourage conservation and reuse.'

Economic instruments are used to allocate water resources, manage demand, reduce pollution discharges, finance water service costs and incentivize environmentally positive actions (positive externalities). Water and food security demands that scarce resources should be properly managed and services sufficiently financed. This chapter reviews four kinds of economic instruments, namely, (a) tariffs, levies and charges, (b) environmental taxes (c) water markets and (d) payments for ecosystem services.

As will be reviewed in this chapter, the urban supply sector is undergoing a second round of reforms, after the feverish privatization processes of the late 1990s (see Chapters 8 and 11). The challenges have been well diagnosed: how to expand the networks in order to reach the continuously growing population of cities; to bring drinking water and sanitation services to all neighbourhoods and households whilst at the same time keeping water prices at reasonable levels. Improvements and innovations are abundant, and the LAC region is clearly on track towards improving most indicators (see Chapter 6).

Ferro and Lentini (2013) reported that evaluations of the Interamerican Development Bank (IDB) indicated that to meet the water-related Millennium Development Goals (MDG) in LAC investments amounting in 2003 to US$16.5 billion in drinking water services, 22 billion in sanitation, and 17.7 billion in treatment of serviced waters, totalling 56.2 billion (approximately US$ 200 per person) were necessary.

In the field of irrigation, tariffs always face opposition and have been questioned as effective mechanisms to allocate scarce resources (Molle and Berkoff, 2007). And yet, around the region we have seen numerous initiatives on cost-recovery objectives, which have then evolved towards demand-management instruments. Ensuring adequate and self-sustained operation and maintenance is the main target.

Environmental taxes have been implemented in some countries, and this is one of the policy areas that will require longer implementation processes. Also the region has seen tremendous growth in the use of payments for ecosystem services (PES, see Chapter 14 for an assessment of LAC's ecosystem services).

The chapter also looks at water trading mechanisms. Little or nothing has been truly implemented in the region except in Chile, where trading occurs regularly in many regions and prices vary according to changes in the supply and demand.

---

1 The Dublin Statement On Water and Sustainable Development (1992).

13.2 Water tariffs

13.2.1 Fees or charges for the use of water resources

Many countries consider that the use of natural resources imposes costs on society and requires conservation and management activities. In order to reimburse the state for these costs of conserving the natural sources, many countries have established charges or fees that all users must pay.

In Mexico, for instance, at least eight categories are defined, whose rates increase when water is scarcer in the region (see Table 13.1). Note that the rate for irrigation is zero in all the regions and that for hydropower or exceeding the concession for irrigation do not vary with the regions' availability.

Table 13.1 Levies for water use for different zones in Mexico, 2010 (US$ cents per m³, exchange rate Mexican peso/US$ of 2010)

	AVAILABILITY AREAS								
TYPE OF USE	1	2	3	4	5	6	7	8	9
General regime	143.15	114.57	95.46	78.70	62.02	56.07	42.21	11.20	0.00
Drinking water, for consumers greater than 300/l per person	5.67	5.67	5.67	5.67	5.67	5.67	2.64	1.32	0.66
Drinking water, for consumers less than 300/l per person	2.84	2.84	2.84	2.84	2.84	2.84	1.32	0.66	0.33
Agricultural, within the concession	0.00	0.00	0.00	0.00	0.00	0.00	0.00	0.00	0.00
Agricultural, for units beyond the concession	1.01	1.01	1.01	1.01	1.01	1.01	1.01	1.01	1.01
Spas and Recreational centres	0.08	0.08	0.08	0.08	0.08	0.08	0.04	0.02	0.01
Hydropower	0.03	0.03	0.03	0.03	0.03	0.03	0.03	0.03	0.03
Fish farms	0.02	0.02	0.02	0.02	0.02	0.02	0.02	0.02	0.02

Source: CONAGUA (2012)

In Costa Rica, different conceptualizations of water charges have evolved since the enactment of the 1942 Water Law. Presently, users must pay a charge called the 'environmentally adjusted water use charge' (*Canon ambientalmente ajustado por aprovechamiento de aguas*), which has two components: (a) an aggregate value which differs on the type of use (hydropower, agricultural, household consumption or industrial), and takes into account cost estimates and marginal valuations; and (b) a payment for the water environmental service. In general, the water charge (canon) in Costa Rica is considered a success story, with benefits identified in (a) the more efficient water allocation mechanism and reduced pressures; (b) the revalorization of the water resources; and (c) stakeholder's participation in designing the instrument (LA–Costa Rica, 2012).

In Chile, since 2005 there has been a 'non-use' fee (*patente de no uso*) that is charged to the users with surplus water rights who do not have the infrastructure required to make effective use of the water. It is calculated differently for consumptive and non-consumptive uses and varies from region to region. The elevation difference between the abstraction point, the return flow point and the length of the non-use period are also taken into account. The main objective of this fee is to 'correct' the distortions that were generated by the initial allocations (Melo et al., 2004)

Brazil's 1997 water law establishes that water be considered an economic good and introduces water fees with the triple objective of communicating the value of water, rationalizing its use and generating revenue for the further development of water resources. The model for setting water tariffs (*cobrança*) has been followed by a somewhat flexible and adaptive methodology (Formiga-Johnson et al., 2007). See Box 13.1 devoted to the basin of Paraíba du Sol in Brazil.

Peru passed the Law of Water Resources in 2009, which was later developed into a detailed regulation including a financial and economic regime. It defines a fee for using water resources, in lieu of the fact that they belong to the Nation's domain. Fees are differentiated by users (Article 177) and then collected revenue is used to fund basins' planning, administration and environmental protection among other goals. Interestingly, users that obtain individual or collective certificates of 'efficient use' can obtain fee rebates and also access water preferentially.

Box 13.1 An integrated approach in the Brazilian Paraíba do Sul Basin (PDSB)

PDSB covers 5.5 million hectares, located in Brazil's economic epicentre, covering the states of Sao Paulo, Rio de Janeiro and Minas Gerais, servicing 180 cities totalling 5.6 million people (8.7 million in the metropolitan area of Rio de Janeiro are outside the basin but are served through an inter-basin transfer). Four elements are identified in order to enable the implementation of a bulk pricing reform in the PDSB: (a) an inclusive and bottom-up negotiation; (b) collected fees would be invested in the basin; (c) a paradigm shift accepting the notion of water as an economic good was to be embraced by key actors in the basin; (d) advanced technical knowledge dating back several decades, so that committee members agreed on the primary problems and the role that bulk pricing would play in solving them.

The approved formula includes three components: a withdrawal component, a consumption component and an effluent dilution component. Upon the first implementation period it was found that the system had some flaws and was due for revision in 2006. There were several drawbacks that were corrected: (a) coping with illegal users; (b) taking the treatment of non-paying users more seriously; (c) solving the asymmetric status of users in different States, given that they were subject to different jurisdictions.

Some lessons can be drawn from this example. First, the formula was simple and had low implementation risks; second, the system had a hybrid approach with market-inspired schemes that preserved the role of the state (ANA and CEIVAP[2]); third, the idea of water being an economic good was deeply ingrained among users and the professional circles in the CEIVAP; fourth, the problems were well-diagnosed, with pollution being the direct one, and a consensus around the most practical means to face them was easily built among users and agencies; fifth, cross-cutting three important states, a federal component was required and essential; sixth and lastly, there were attractive incentives for implementation, including matching funds from the national programme to combat pollution, and revenues were earmarked for specific and visible basin projects. And yet, loris (2010) found some weaknesses and reported that, between 2003 and 2006, the charging scheme was responsible for collecting a total of 25.4 million Brazilian reals (US$10.85 million at the exchange rate of 2 July, 2005), which is considerably less than the budget required to restore the environmental quality of the basin.

Source. Formiga-Jhonson et al. (2001) and loris (2010)

13.2.2 Irrigation charges and fees

Irrigation schemes charge farmers fees to meet the operations and maintenance (O&M) costs. IWMI, USAID and FAO agreed that attention should be paid to five items (Molle and Berkoff, 2007). First, rational water use should be achieved by careful control of distribution and by allocating water to broadly meet crop requirements, with fees having little or no impact on irrigation performance. Second, the presumable efficiency gains from irrigation tariffs would most probably be realized by the control of supply or some kind or quotas. Third, the most critical financial factor is the level of fiscal autonomy of the irrigation agency, providing an incentive for cost-effective performance. Fourth, cost recovery should be contextualized to factor in irrigators' ability to pay, and O&M activities should be prioritized for cost recovery strategies. Fifth, subsidized users should repay some of the investments, but should not be expected to pay the extra-costs imposed by inefficient or miscalculated investments or overstaffed organizations.

Despite these caveats, it is also true that irrigation water given free of charge would also generate welfare losses, in the form of opportunity cost and externalities. Furthermore, many large countries like Mexico or small countries like Suriname have suffered the abandonment of irrigation infrastructures because of insufficient fees collection and poor cost-recovery rates.

Consider the case of Mexico. Irrigated agriculture is extremely important in terms of both irrigated acreage (more than 5.5 million hectares) and total water use. Since the passing of the Water Law in 1992 and the creation of the National Water Commission, Mexico

---

2 National Water Agency of Brazil (Agência Nacional de Águas) and Integrated Comittee of the Hydrographic Basin of Rio Paraíba do Sul (Comitê de Integração da Bacia Hidrográfica do Rio Paraíba do Sul).

embarked on a massive policy reform to allocate the water management of its large water districts to the recently created users associations (WUAs). This involved setting up new institutions such as basin agencies, giving WUAs managing capacity to administer both capital assets and water resources, and transferring the financial responsibility of running districts and collecting charges to the WUAs. During the devolution process, water prices increased by 45–180% and government O&M subsidies were removed. Molle and Berkoff (2007), citing other sources, claimed that O&M charges have been quite low (equivalent to 2–7% of the gross product), and that maintenance may be suboptimal in many cases. Garrido and Calatrava (2009) reported significant increases in irrigation water charges upon the implementation of the devolution process.

There are about 3.5Mha under irrigation in Brazil, although 29Mha are estimated to be suitable for irrigation by the National Water Agency (ANA). The Irrigation Law, enacted in 1979, and its regulations provide for the cost recovery of investment and O&M costs of government-supported irrigation projects through water charges to beneficiaries.

There is an interesting case of volumetric control and two-part charging mechanism in the Chancay-Lambayeque in Peru (Vos and Vincent, 2011). The Chancay-Lambayeque irrigation system achieved high performance with on-demand delivery to some 22,000 smallholders in a command area of some 100,000ha. Full cost recovery rates, accompanied by the requirement to pay in advance, reinforced the management and ensured the control of water use and cropping operations. Rates were US$0.003 per m^3 (four soles, the Peruvian currency, for a service module of 576m^3) in 1995, and were adjusted with inflation reaching US$0.005 per m^3 in 2010.

13.2.3 Charges for urban consumers

13.2.3.1 Regulatory frameworks

The design of the industrial structure for water supply and sanitation impinges on the ability to deliver services to the population. Assets are long-lived, allowing investments to be delayed and quasi-rents to be captured once initial investments have been made (Guasch et al., 2008). Fragmented services lose economies of scale, increase transaction costs, make services more expensive, and may facilitate capture by vested interests (Foster, 2005; ADB, 2009). Water supply and sanitation services have decreasing average costs (Krause, 2009) and therefore both efficiency and equity are achieved by selecting optimal size in terms of economies of scale.

Economies of scale lead to natural monopolies that must be regulated to ensure that the market operates as if it were a competitive market in order to achieve the maximum social welfare. Regulation should guarantee that the service is safe, sufficient, regular, physically accessible, convenient, and affordable. In terms of implementing regulation there are differences between, on the one hand, specific contracts, and on the other, comprehensive general, regulation, franchizing and concessions. Almost 90% of water supply and sanitation privatizations in LAC during the 1990s were made through concessions (Estache et al., 2003). More developed countries prefer to grant licences controlled by general regulations of compulsory application, approved by law, and enforced by fully empowered, permanent, professional regulators (Jouravlev, 2005). Chile has embarked on a process of privatization of water supply and sanitation that has been considered a success (see Box 13.2).

Efficiency covers costs while considering equity by facilitating improvements in the quality of services and their expansion to the poor. The Brazilian case has its particularities, as the private sector represents presently around 10% of the total concessions in the country, it has been constantly growing and changes are underway through the growth in the implementation of concessions, and the private utilities association expects to reach 40% by 2023. One of the causes is the lack of investment capacity of municipalities and state-owned companies to maintain and renew equipment. While there are diverse regulation frameworks in LAC, state-owned companies continue to be very relevant, but their main challenges are lack of accountability and regulation (see Chapter 11).

Box 13.2 Privatization of water services in Chile

In the system of water supply and sanitation in Chile, there is a tariff law according to which the Superintendence of Sanitary Services (SISS) periodically conducts studies to set the maximum prices that are authorized to sanitation concessionaires. These rates are set so as to allow each company to cover investment and operating costs and to obtain an agreed return on the investment required to provide the service of production and distribution of drinking water, collection, wastewater disposal and treatment. In order to establish efficiency incentives, water rates are set based on an efficient firm model, so that the values and parameters entering the formulas are not the actual company's, but of a fictitious company called 'business model'. The business model has been a useful tool in regulating utilities in Chile in recent decades. Currently, however, it has shown some problems. One the one hand, the current rates are not a real incentive to reduce water consumption. On the other hand, rates are set for the next five years, independently of potential water shortages, or water abundance which occur with much greater frequency, a variability that is not being captured by the price. Regarding the operation of private water companies, they operate through concessions, which may be overthrown if these do not meet quality standards, flow, or tariff standards.

13.2.3.2 Tariff levels and structures

In a very recent study of 308 large cities around the world, Zetland and Gasson (2012) evaluated the differences in water charges and researched reasons behind these differences. They found that the average water tariff for urban consumers was US$1.21 per m^3 (σ=1.13; max=7.54), whereas the wastewater tariff was US$1.02 per m^3 (σ=1.07; max=5.68). The following factors are identified by Zetland and Gasson (2012) to explain water and wastewater tariffs around the world: (a) labour costs; (b) regulatory price control which aim at minimizing tariffs; (c) public vs. private organizations; (d) water scarcity; (e) the age and condition of infrastructure; (f) subsidization schemes and the type of socially targeted policies.

Ferro and Lentini (2013) recently assessed the pricing policies in the LAC region. They assembled data from fifteen major utilities from Argentina, Brazil, Chile, Colombia, Costa Rica, Ecuador, Panamá, Paraguay, Perú and Uruguay, with a total population served of approximately 100 million people (see Table 13.2).

Table 13.2 Average monthly bill and average price in the main fourteen water utilities in LA

UTILITY	AREA OF SERVICE	YEAR	AVERAGE BILL (US$)	AVERAGE PRICE (US$/m³)
AySA	Buenos Aires + 17 municipalities	2011	4.82	0.17
ASSA	Province of Santa Fe, Argentina	2011	5.87	-
SABESP	Estado de São Paulo (Brazil)	2010	48.43	2.63
COPASA	Belo Horizonte, Minas Gerais (Brazil)	2010	36.09	2.94
Aguas Andinas	Metro Santiago de Chile	2011	38.98	1.77
Aguas de Antofagasta	Antofagasta, Chile	2011	67.69	3.54
EAAB	Bogotá, Colombia	2011	31.82	2.64
ACUAPAR	Cartagena S.A. (ACUACAR) Distrito de Cartagena de Indias, Colombia	2011	32.51	1.98
SEDAPAL	Lima, Peru	2011	27.74	1.01
SEDACAJ	Cajamarca, Peru	2011	12.93	0.94
AyA	Supplier of drinking water and sanitation, Costa Rica	2010	22.72	1.07
EMAAPQ	Quito, Ecuador	2011	19.76	0.72
IDAAN	Supplier of drinking water and sanitation, Panamá	2011	15.54	0.3
OSE	Supplier of drinking water and sanitation, Uruguay	2009	26.27	1.8

Source: Ferro and Lentini (2013)

Increasing block tariffs (IBT) have become commonly used because they fulfill three goals (Olivier, 2010): (a) affordability and fairness, with a highly subsidized first block (subsistence first block); (b) resource conservation (higher consumption is charged at a higher price); (c) economic efficiency, with the higher block corresponding to short-term marginal cost of provision.

Small private operators are often in the business of supplying the poor, using tankers and informal companies selling water to the poor, usually at many times the price of tap water (see Box 13.3). Nauges and Strand (2007) found that average tap water price (PPP corrected) in three Salvadorean cities is about US$0.25 per m³, and in the marginal quarters in Tegucigalpa, about $US0.4 per m³. The average non-tap price in Tegucigalpa is US$8.43 per m³.

The history of the Buenos Aires water concession is now a classical example of mismanagement and poor regulatory practice. The domestic supply service was awarded to the Aguas Argentinas Consortium in 1993, when only 70% of the metropolitan area population was connected to the water system and 58% to the sewerage system. In the suburban areas, these percentages were even lower, 55 and 36%, respectively, but almost 100% in the Capital District. Coverage targets specified expansions to the benefit of the poorest households in marginal areas. In 2003, the coverage and sewerage rates lagged behind targets by 47 and 70%. To compensate for the increasing investment costs of servicing new customers, initially estimated at US$1,120, totally out of the price range for the poorest consumers, the regulator approved increasing the rates of existing consumers by 93% from US$17.57 per month in May 1993 to US$33.88 in 2002. Casarin et al. (2007) observed that the concession left 1 million people unserved, and only 50% and 25% of the expansion targets with water connection and sewerage services.

Lima's water system was on the verge of collapse at the end of the 1980s. Severe under-financing, under-maintenance and little or no expansion were all parts of a vicious cycle facing rapidly growing cities in developing countries (Fernández-Maldonado, 2008). A new law to regulate sanitation services opened the door to private capital and created the SUNASS,[3] the regulatory body. In 2006, 3.9 million new customers were added on top of the 3.1 existing ones in 1980, and still 1 million Limeños were left served with trucks selling water at US$2.2 or US$3 per m³, which was in 2006 nine times more than the socially regulated SEDAPAL's tariff ($0.33 per m³). After 2006, revenue collected through tariffs was 90% of the costs, and because of the cross-subsidies only 11% of the customers paid more than the cost of provision. Presently, Lima's water problems are still unsolved: more than one-third of the serviced water is not billed, and in 2007 only 13% of its wastewater was treated.

Manaus, capital of the Amazonas state of Brazil, has 1.7 million inhabitants, in addition to another half million in the suburban areas. Drinking water reached 80% of the people in 2004; although access to sanitary networks reached only 7% of the households (Olivier, 2010). An attempt was made to embed a cross-subsidy mechanism so that the wealthiest and industrial consumers would subsidize socially targeted consumers, but failed because not enough revenue was generated in the former two groups. As a result, tariffs for the poorest consumers had to be raised by 31% to ensure that the company would not lose money. Furthermore, the largest consumers had the option to disconnect from the network, taking advantage of loopholes in groundwater regulations. In the Metropolitan Region of São Paulo similar difficulties were found when readjusting the tariffs for the poorest customers, who paid in the early 2000s slightly higher average prices than richer households, and in terms of percentage of disposable income ten times more (Ruijs et al., 2008). According to Ioiris and Costa (2009) the minimal payment for water services (the so-called 'social tariff') was significantly higher in Rio de Janeiro than in other parts of Brazil, which certainly contributed to the high rate of unpaid debt: in CEDAE (Rio de Janeiro) it was R$30 for 15m³/month; DMAE (Porto Alegre), R$7.5 for 10m³/month; and SABESP (São Paulo): R$4.42 for 10m³/month (all 2008 data).

---

3 SUNASS: Superintendencia Nacional de Servicios de Saneamiento, Peru (www.sunass.gob.pe).

Box 13.3 Social equity: social tariffs

Most large LAC cities have been growing rapidly in the last decades, requiring continuous expansion of drinking water and sanitary networks. Charging the expansionary costs on new customers, generally in marginal areas, would be unaffordable for the poorest households. One difficulty of socially targeted policies is that if social rates are not sufficiently compensated by the revenue collected from regular customers, the water operator may be dissuaded to expand the network to add more marginal consumers.

Most pro-poor policies and arrangements involve one or a combination of the following features:

- A minimum volume free of charge, which in LAC ranges between 4 and 15m³ per month and per household. The first priced block, that varies between 18 and 25m³ per month and household, is set at an affordable cost. In Chile, 15m³ per month is the maximum serviced at subsidized price; in Colombia 20m³ per month is offered at subsidzed rates; in São Paulo paying the flat rate gives a rate to 10m³ per month free of charge.

- Consideration of affordable tariffs. Capacity to pay or affordability are dubious concepts for which there is no clear theoretical foundation. Various authors and organizations have defined various thresholds in percentage terms of the household's income (5%, by The World Bank; Vergès, 1%; PNUD, 3%; IAD, 5% for the poorest households). The findings are that in Campinas, Brazil, charges are below 2%; about 5% in LAC cities with no pro-poor provisions; 1.8% in Arequipa, Perú; 9.8% in Cost Rica, whereas in cities with pro-poor provisions, it ranges from 0.9% in Ceará, Brasil, and Trujillo, Perú to 8.4% in Bogotá, Colombia. In Chile the goal is to keep the water and sanitation bill below 3%.

- An increase in the flat rate accompanied with a reduction of the volumetric rate, increasing the billing frequency, reduced or limited service as opposed to disconnection for non-paying customers, and a control over sumptuary consumption (car washes, swimming pool).

Source: Ferro and Lentini (2013)

13.3 Economic instruments applied to water quality management

In addition to command-and-control (CAC) instruments, two types of economic instruments (EI) have received the most recent attention: discharge fee programmes, which charge plants for each unit of pollution emitted, and marketable permit programmes, which assign plant emissions allowances that they may trade with other plants. Caffera (2010) claims that the experience in the region with economic instruments in pollution control is limited to three programmes: Santiago de Chile's Total Suspended Particles' Emissions Compensation Programme (ECP) of 1992 and its extensions to industry emissions of

Nitrogen Oxides and Particulate Matter in 2004; Colombia's 1997 Discharge Fee for Water Effluents' contents of Biochemical Oxygen Demand and Total Suspended Solids; and Costa Rica's 2009 Environmental Fee for Water Discharges of Chemical Oxygen Demand and Total Suspended Solids.

Colombian Law 99 of 1993 established the legal foundation for a national discharge fee programme. While the programme was plagued with difficulties and serious non-compliance in the first five years after 1997, BOD and TSS discharges dropped significantly following the initiation of the program in 1997 (Caffera, 2010). This could have resulted from the economic incentive and efficiency properties of the new discharge fee programme or because of the improved permitting, monitoring, and enforcement of both the new discharge fees and existing emissions standards.

In reviewing, the Colombian discharge fee, Caffera (2010) indicated that its main problem was the broad non-compliance by municipal sewerage companies. Because emissions of these sources did not decrease, the environmental quality targets were not met, and the fees never stopped increasing. In view of this, a new decree (Decree #3100), was enacted (later modified by Decree #3440 of 2004), which introduced the following changes: (1) it mandated the regional and municipal authorities to establish (a) individual targets of pollution reduction for municipal sewage companies and sources whose loads are more than a fifth of the total loads received by the water body, and (b) group targets for the rest of the sources, according to the group's type (industrial branch, etc.); (2) it mandated the regional and municipal authorities to ask the municipal sewage companies to present a Plan for Pollution Management in accordance with the pollution reduction target; (3) it changed the method by which the fee is adjusted. However, Caffera (2010) wrote 'it is obvious that the changes sought to leave the municipal sewage companies and large polluters outside the fees' program, changing a monetary incentive to invest in pollution abatement by a prescriptive-type pollution abatement plan' (p.13).

Inspired by the Colombian programme, Costa Rica implemented an Environmental Fee for Discharges which puts a price on each kilogramme of COD and TSS discharged. The Costa Rican programme also faced implementation difficulties. It was challenged in court by the sugar cane industrial-agricultural union, on the basis that the fee was a tax, something that could only be decreed by the congress, the appeal was ruled against by the Supreme Court. The Ministry of the Environment approved a new decree (#34431) in 2008, which changed the amount and structure of the fee. Other implementation difficulties were related to the lack of trained personnel, of databases, and of monitoring equipment. The collection of fees was estimated to be only 80% of the total potential and as such prevented the purchasing and installation of treatment plants and monitoring equipment. Costa Rican regulators found that the most difficult sources of pollution originate from public utilities providing water services such as sanitation, drinking water, and irrigation.

In Chile the Decree #70 of the Ministry of Public Works established in 1988 that water utilities can charge for water provision but also water collection and disposal services. At the time very few cities had isolated collection and treatment services. But since the investments required to provide these services can be included in water tariffs once they are operational, water utilities now collect and treat almost all urban water. Water discharges from other sources are still regulated through traditional command and control methods (Donoso and Melo, 2006).

13.4 Payments for environmental services

Some of the goods and services provided by ecosystems are traded in markets, but others are not. In the latter case some or all of the costs of providing, and the benefits of using, these goods and services are not transmitted through prices, what economists call an externality. The main idea behind payments for environmental services (PES) is to establish the incentives lacking due to the existence of an externality, by putting in place a mechanism that compensates suppliers/producers and charges beneficiaries of the ecosystem service. This section introduces the concept of PES, which is further developed and expanded upon in the next chapter (14).

While different approaches that use market-based mechanisms have been labelled as PES, more recently the concept has been narrowed down. For example Wunder (2005) defines PES as '(1) a voluntary transaction in which (2) a well defined environmental service (or a land use likely to secure that service) (3) is "bought" by a (minimum of one) buyer (4) from a (minimum of one) provider (5) if and only if the provider continuously secures the provision of the service (conditionality)' (p. 3). An alternative and less restrictive definition is proposed by Porras et al. (2008), and considers only three criteria: that an environmental externality; is addressed with a payment, is voluntary in the supply side, and has conditionality.

As Chapter 14 explains, several payment mechanisms can be used including in cash or in kind transfers between governments and landowners, tradable development rights, voluntary contractual arrangements, and product certification and labelling (MEA 2005). The former ones are the most common in schemes that conform to the current PES definition. PES could deliver environmental and social co-benefits. The payment component of PES schemes, on the other hand, could have a relevant role in poverty alleviation (Pagiola et al., 2002).

Landell-Mills and Porras (2002) identified sixty-one watershed initiatives, twenty-two of them in LAC, but only eleven where in a pilot or mature stage of development and were still ongoing by 2006 (Porras et al., 2008). Of these projects, six are implemented at a national level in Colombia, Costa Rica, El Salvador, Guatemala and Mexico. There are also some regional initiatives that are replicated in several countries, like the Regional Integrated Silvopastoral Ecosystem Management Project (RISEMP) in Colombia, Costa Rica and Nicaragua, funded by the Global Environmental Fund (GEF) and the World Bank (WB), and the Programme for Sustainable Agriculture on the Hillsides of Central America (PASOLAC) in El Salvador, Honduras and Nicaragua, funded by the Swiss Agency for Development and Cooperation (SDC). More recently, Bennet et al. (2013) identified 205 active programmes in 2011 worldwide, twenty-eight of them in LAC (Ecuador, Colombia, Brazil, Mexico, Costa Rica and Bolivia). These authors also report that initiatives in this region are putting more emphasis on building social capital and more frequently use payments in-kind. Table 13.3 presents a summary of some of the most significant Payment for Watershed and Water-related services (PWS) initiatives in Latin America.

Table 13.3 PES schemes for watershed protection and water-related ecosystem services in LAC

NAME	COUNTRY	ACTIVITY PAID FOR	SERVICE	SELLER	SCALE	SPATIAL EXTENT (hectares)	YEARS	AMOUNT TRANSACTED IN 2011 (million USD)
RISEMP	Colombia, Costa Rica, Nicaragua	Biodiversity, carbon, watershed	Restoration (silvopasture)	NGOs, Intern. Org., States	International (3 countries)	3,500	2002–2008	393.8
Pimampiro	Ecuador	Watershed	Conservation/minor restoration	Municipal government	Local	496	2000–present	4.6
PSA program	Costa rica	Carbon, watersheds, biodiversity, landscape	Conservation/minor restoration	Public sector, Intern. Org.	National	270,000	1996–present	340
PSA PROGRAM	Mexico	Watershed	Conservation and restoration	Private and communities	National	600,000	2002–present	82.5
Los Negros	Bolivia	Watershed, biodiversity	Forest and paramo conservation	Farmers	Local	2,774	2003–present	8.0

Source: adapted from Wunder et al. (2008). Notes: RISEMP ended in 2008, and the amount transacted is estimated from Pagiola et al. (2004) an is an average for the duration of the programme. For Pimampiro, the amount transacted is calculated from Patanayak et al. (2010). For the rest of the programmes the amount transacted from Watershed Connect website.

Martín-Ortega et al. (2012) reviewed thirty-nine PES programmes in LAC, which have been summarized in Table 13.4. There is a great variety of approaches and partnerships, but most focus on forests' and land conservation to protect watersheds.

The next chapter (14) will also review PES, jointly with biodiversity markets, REDDs and CDMs and other instruments.

13.5 Water markets as a water allocation mechanism: the case of Chile

With increasing water scarcity and decreasing supply augmentation options, water managers and policy makers see interest in implementing market allocation systems (Rosegrant and Gazmuri, 1995; Easter et al., 1999; Saleth and Dinar, 2004). Efficiency and activity of water markets (WM) are intrinsically linked to the design of institutional and physical water systems (Bjornlund and McKay, 2002).

With WM the price of water rights (WR) reveals the opportunity cost of water, creating incentives to use water efficiently and employ it in its most productive use. WM are expected to lead to a socially optimal and efficient allocation by inducing two key changes. First, water is transferred from low-value users to high-value users. Second, WM

Table 13.4 Main characteristics of water-related PES programs in LAC

ASPECT	DEFINITION
COUNTRIES	10 in Costa Rica; 6 in Ecuador; 4 in Bolivia, Brazil, Colombia and Mexico; 2 in El Salvador and Nicaragua; 1 in Guatemala and Honduras
CONTEXT	Local specific component, 92.1% National components only, 26.3%
ENVIRONMENTAL ASPECT	42.1% Undefined Among the remaining 57.9%, 77.3% targeted deforestation and land cover; 31.8% water pollution; 22.7% water overuse
STAKEHOLDERS	40% a leading national NGO 23.7% Municipality 18.3% Governmental 16% Semi-autonomous agencies
INTERMEDIATION	78.9% use intermediary 21.1% direct transaction between buyers and sellers.
AGENTS INVOLVED	96.4% landowners and farmers
PAID ACTIONS	73.7% have more than one action, with a majority focusing on forest conservation and reforestation for water catchments 23.7% forest management
TARGETS	91.3% Aim at improving water supply 53.3% Aim at improving in-stream supply (water flow regulation for hydropower)
DIFFERENTATION	42% include some kind of differentiation (from 2 to 12, average 2.14), according to: 74.8% type of activity 23.9% type of forest or land feature
EVOLUTION OF SCHEMES	42.1% include several transformation stages

Source: Martín-Ortega et al. (2012)

generate greater investments in water conservation technologies due to the trade induced price increase (Chong and Sunding, 2006).

However, WMs can also result in third-party effects, speculative behaviour in water trade, social and environmental externalities. The Chilean government introduced a tax for holding unused water rights as a reaction to speculative behaviour and WR hoarding, which did not inhibit but did distort the market (World Bank, 2011). In Chile, trades need to be registered and approved by Water User Associations (WUA) so as to reduce negative third-party effects caused by return flows (Donoso, 2006).

Once initiated, markets ideally evolve towards maturity. In a mature market, allocation and productive efficiency of water are maximized (Bjornlund, 2002). Researchers assess market maturity in different ways, such as by the number of transfers or by price dispersion. Frequent transfers and small price dispersions indicate mature markets. However, if water rights are initially allocated to high value uses, few transactions are required for a mature market (Easter et al., 1999). Price dispersions can also be caused by geographical flexibility and reliability of infrastructure of irrigation canals (Hadjigeorgalis, 2004; Donoso et al., 2012), commodity prices (Challen, 2000) and quantities traded (Bjornlund and McKay, 2002), leaving both measures open to improvement.

Since the establishment of the water allocation mechanism based on a market of WR in Chile, a series of empirical and theoretical studies have been carried out to determine: the existence of a WM, the market activity measured through the number of transactions; WM efficiency; bargaining, cooperation, and strategic behaviours of market participants; and marginal gains from trade.

Several authors (Cristi and Trapp, 2003; Quentin et al., 2012) find evidence that markets are more active in those areas where water is a scarce resource with a high economic value. These studies indicate that the market mechanism has, in general, represented an efficient water allocation system. This is the case of the Limarí Valley, where water is scarce with high economic value, especially for the emerging agricultural sector. Inter-sectoral trading has transferred water to growing urban areas in the Elqui Valley and the upper Mapocho watershed, where water companies and real estate developers are continuously buying water and account for 76% of the rights traded (Donoso et al., 2012).

Table 13.5 presents WR transaction data based on data of the Dirección General de Aguas (DGA), for the period 2005–2008. The results for this four-year period show that there were 24,177 transactions of which 92.3% were independent of other property transactions, such as land. The value of transactions independent of other property transactions is US$4.8 billion, which on average is US$1.2 billion per year. The average WR price is US$215,623. WR prices in the north of the country are greater than in the south, which indicates that the market at least in part reflects the relative scarcity of water. WR prices present a high coefficient of variation of 465. However, price dispersion is lower in the more active markets.

A key conclusion of these studies is that WM are driven by demand from relatively high-valued water uses and facilitated by low transactions costs in those valleys where WUAs and infrastructure assist the transfer of water. Market functioning differences are explained by scarcity, the distribution infrastructure and water storage capacity, and the proper functioning of WUAs. More frequent transactions in the 21st century than in 1980s and 1990s indicate a degree of maturity in the public's knowledge concerning the new legislation and possibly a growing demand for water.

Analysing WM in Chile, Jouravlev (2005) concluded that they (i) facilitate the reallocation of water use from lower to higher value users, (ii) mitigate the impact of droughts by allowing for temporal transfers from lower value annual crops to higher valued perennial fruit and other tree crops, and (iii) provide lower cost access to water resources than alternative sources such as desalination.

By analysing the effect of WM, it can be seen that numerous problems have been resolved through their implementation. The use of such an allocation mechanism has allowed users to consider water as an economic good hence internalizing its scarcity value; constitutes an efficient reallocation mechanism which has facilitated the redistribution of rights already granted; has permitted the development of mining in areas in the semi-arid

Table 13.5 Water rights (WR) transactions and prices for the period 2005-2008.

	TOTAL TRANSACTIONS of WR	TRANSACTIONS OF WR INDEPENDENT OF OTHER GOODS SUCH AS LAND	WR TRANSACTION VALUES (ONLY WR TRANSAC- TIONS INDEPENDENT OF OTHER GOODS) (10^6 US$)	AVERAGE WR TRANSACTION PRICE (US$)
I	568	564	20	36,121
II	153	131	216	1,652,519
III	16	15	8	530,933
IV	3,489	3,448	550	159,615
V	3,191	2,839	517	182,029
RM	4,804	4,226	2,312	547,095
VI	2,315	2,010	509	253,367
VII	6,518	6,159	622	101,059
VIII	2,330	2,162	29	13,432
IX	494	487	8	16,805
X	225	223	23	103,390
XI	68	68	0	2,588
XII	6	6	0	20,200
Total	24,177	22,338	4,817	215,623

Source: World Bank (2011)

northern region of Chile by buying water rights from agriculture; has resolved problems associated to water deficits derived from a significant increase of water demand caused by the significant population growth in the central region of Chile and additionally has helped to solve water scarcity problems above all in instances when a rapid response has been required (Donoso, 2006; World Bank, 2011).

The problems that WM have not been able to resolve are water use inefficiency in all sectors, not only in the agricultural sector, environmental problems, and the maintenance of ecological water flows. A major challenge of WM in Chile is how to ensure optimal water use without compromising the sustainability of rivers and aquifers. The sustainability of northern rivers and aquifers is at present jeopardized due to the over-allowance of WRs by the DGA. On the other hand, increased consumptive WR market activity has generated increased conflicts with downstream users due the existence of WR-defined over return flows.

Research in Chile on the impact of water markets on small farmers has been limited and no reliable conclusions have been reached to date. Some critics contend that small farmers have not regularized their rights, risking losing them, and in other cases have sold their water rights thus losing their means of subsistence. But Hadjigeorgalis (2008) shows that the WM in the Limarí basin has been successful in moving water and water rights from low- to high-valued uses and that resource-constrained farmers use temporary WM as a safety net. She did not find inequity with respect to offer prices; resource-constrained farmers receive the same offer prices as wealthier ones. The Limarí watershed has the most complex irrigation reservoirs system in the country, which has allowed the spontaneous development of a spot market for water volumes. Although this market has represented an important 'pressure valve' to withstand dry years, it also faces efficiency challenges that need to be addressed (Alevy et al., 2011).

While the institutional replication of the Chilean WM may seem like an option for LAC countries faced with increasing water scarcity and decreasing supply augmentation options, the contextual uniqueness of each WM makes the establishment of universal rules for replication difficult (Shah, 2005).

13.6 Implications for improving water and food security

The implementation of economic instruments provides revenue to finance water services and should provide incentives to agents to act more responsibly. Urban tariffs are the fundamental source of revenue to expand coverage of drinking water and sanitation. It seems that a significant part of the investment costs that are required to meet the water MDGs in the region cannot be funded by the targeted households. And yet, it is clear that implementing adequate tariff structures is essential to make progress and bridge the gaps reported in Chapter 6. Improved sanitation, will not only reduce the prevalence of many water-borne diseases, but also improve the ecological status of numerous important rivers and waterways.

As discussed in this chapter, pollution charges are meant to deter contaminants' discharges and generate revenue to fund monitoring and mitigation actions. The cases of Colombia and Costa Rica show that large and medium-size cities are among the heaviest pollutants. A vicious cycle commonly prevails not only in LAC but in virtually all countries where urban tariffs are below US$1.5 per m³. Below this level proper urban water treatment and secure drinking water supply in adequate conditions are barely possible. Improving water security indicators has a large cost (aproximately US$100 per person and year, including drinking water supply and sanitation).

Therefore, three aspects converge and have implications for improving water security indicators: (a) a balanced and efficient tariff regime for urban water, accompanied by pro-social provisions; (b) better implementation of pollution charges and the polluter-pays-principle; and (c) payments for ecosystem services and watershed conservation. All three of them complement each other, but it seems that in LAC there is a long way to go in terms of sanitation and urban and industrial wastewater treatment. For the moment, PES are very limited to avert the consequences of urban and industrial growth, and are focused only on areas of high ecological value or headwaters of specific rivers.

Irrigation water prices are essential to increase food production sustainably. Billions of dollars of investment in irrigation have been wasted as a result of insufficient and poor maintenance of infrastructure. Adequate pricing of irrigation water is also essential to ensure that water resources are not wasted or assigned to low-value crops. Investment in irrigation, new and that in need of rehabilitation or technical improvements have been estimated at US$95 billion cumulatively up to 2050 (Schmidhuber et al., 2009), or US$ 7 billion up to 2030 (Faurès, 2007). These represent huge investments that may need proper tariff mechanisms and financial structuring. More stable food production will surely result from it, improving also food security indicators.

References

ADB (2009). Asian Development Bank. *Urban Water Supply Sector Risk Assessment: Guidance Note, Manila, Philipines*, Asian Development Bank.

Alevy, J., Cristi, O. & Melo, O. (2011). *Proyecto Mercado Electrónico del Agua en Chile*, Santiago, Chile, Universidad del Desarrollo y Pontificia Universidad Católica de Chile.

Bennett, G., Carroll, N., & Hamilton, K. (2013). *Charting New Waters: State of Watershed Payments 2012*, Washington, DC, Forest Trends. [Online] Available from: www.ecosystemmarketplace. com/reports/sowp2012. [Accessed May, 2013].

Bjornlund, H. (2002). Water exchanges: Australian experiences, Proceedings from the conference *Allocating and Managing Water for a Sustainable Future: Lessons from Around the World*, 11–14 June 2002, Natural Resources Law Center, Boulder, Colorado, University of Colorado.

Bjornlund, H. & McKay, J. (2002). Aspects of water markets for developing countries: experiences from Australia, Chile, and the US. *Environment and Development Economics*, 7 (4): 769–795.

Caffera, M. (2010). The use of economic instruments for pollution control in Latin America: lessons for future policy design. *Environment and Development Economics*, 16 (3): 247–273.

Casarin, A.A., Delfino, J.A. & Delfino, M.E. (2007). Failures in water reforms: Lessons from Buenos Aires's concession. *Utilities Policy*, 15: 234–247.

Challen, R. (2000). *Institutions, Transaction Costs, and Environmental Policy: institutional reform for water resources*, Cheltenham, Edward Elgar Publishing.

Chong, H. & Sunding, D. (2006). Water markets and trading. *Annual Review of Environment and Resources*, 31(1): 239–264.

CONAGUA (2012). Comisión Nacional del Agua. *Zonas de disponibilidad para el cobro de derechos*. [Online] Available from: www.conagua.gob.mx/atlas/usosdelagua37.html. [Accessed March, 2013].

Cristi, O. & Trapp, A. (2003). *Mercado de Agua para Irrigación y Uso Urbano: una Aplicación a la Cuenca del Río Elqui, Chile*, Water Partnership Program (BNWPP), Water Rights System Window, July 2003, World Bank.

Donoso, G. (2006). Water markets: case study of Chile's 1981 Water Code. *Ciencia e Investigación Agraria*, 33 (2): 157–171.

Donoso, G. & Melo, O. (2006). Water quality management in Chile: use of economic instruments. In: Biswas, A.K., Tortajada, C., Braga, B. and Rodriguez, D.J. (eds). *Water Quality Management in The Americas*. Berlin, Springer-Verlag.

Donoso, G., Melo, O. &. Jordan, C. (2012). *Estimating Water Rights Demand and Supply: Are buyer and seller characteristics important?*, Departamento de Economía Agraria, Facultad de Agronomía, Pontificia Universidad Católica de Chile.

Easter, K.W., Rosegrant, M.W. & Dinar, A. (1999). Formal and informal markets for water: Institutions, performance, and constraints. *The World Bank Research Observer*, 14 (1): 99–116.

Estache, A., Guasch, J.L., Trujillo, L. (2003). *Price Caps, Efficiency Payoffs and Infrastructure Contract Renegotiation in Latin America*, Washington, DC, World Bank. Policy Research Working Paper No. 3129.

Faurès, J.M. (2007). Reinventing irrigation. In: Molden, D. (ed.). *Water for Food, Water for Life: A Comprehensive Assessment of Water Management in Agriculture*, Water Management Institute, London, Earthscan/Colombo International.

Fernández-Maldonado, A.M. (2008). Expanding networks for the urban poor: Water and telecommunications services in Lima, Peru. *Geoforum*, 39: 1884–1896.

Ferro, G. & Lentini, E. (2013). *Políticas tarifarias para el logro de los Objetivos de Desarrollo del Milenio (ODM):situación actual y tendencias regionales recientes*, Santiago de Chile, Chile, CEPAL.

Formiga-Jonhson, R.M. , Kumler, L. & Lemos, M.C. (2007). The politics of bulk water pricing in Brazil: lessons from the Paraíba do Sul basin. *Water Policy*, 9: 87–104.

Foster, V. (2005). *Ten Years of Water Service Reform in Latin America: Toward an Anglo-French Mode*, Washington DC, The International Bank for Reconstruction and Development, The World Bank. Water Supply and Sanitation Sector Board Discussion Paper Series, Paper No. 3.

Garrido, A. & Calatrava, J. (2009). *Agricultural Water Pricing: EU and Mexico*, Paris, OECD. [Online] Available from: www.oecd.org/dataoecd/25/38/45015101.pdf [Accessed May, 2013]. Online report.

Guasch, J.L, Laffont, J.J. & Straub, S. (2008). Renegotiation of concessions contracts in Latin America. Evidence from the water and transport sectors. *International Journal of Industrial Organization*, 26(2): 421-42.

Hadjigeorgalis, E. (2004). Comerciando con Incertidumbre: Los Mercados de Agua en la Agricultura Chilena. *Cuadernos de Economía*, 40 (122):3–34.

Hadjigeorgalis, E. (2008). Distributional impacts of water markets on small farmers: Is there a safety net?. *Water Resources Research*, 44(10).

Ioiris, A. & Costa, M.A.M. (2009). The challenge to revert unsustainable trends: uneven development and water degradation in the Rio de Janeiro metropolitan area. *Sustainability*, 1 (2009): 133–160.

Ioris, A. (2010). The political nexus between water and economics in Brazil: a critique of recent policy reforms. *Review of Radical Political Economics*, 42: 231–250.

Jouravlev, A.S. (2005). Integrating Equity, Efficiency and Environment in the Water Allocation Reform, Presentation at the *International Seminar Water Rights Development*, Beijing, China, 6–7 December 2005.

Krause, M. (2009). *The Political Economy of Water and Sanitation*, New York, Routledge.

LA–Costa Rica (2012). Report of Costa Rica. Contribution to the Water and Food Security in Latin America Project. Water Observatory Project. Madrid, Fundación Botín.

Landell-Mills, N &. Porras, T.I. (2002). *Silver Bullet or Fools' Gold? A global review of markets for forest environmental services and their impact on the poor*. Instruments for sustainable private sector forestry series. London, UK, International Institute for Environment and Development.

Martín-Ortega, J., Ojea, E. & Roux, C. (2012). *Payments for Water Ecosystem Services in Latin America: Evidence from Reported Experience*, Bilbao, Spain, Basque Centre for Climate Change (BC3). BC3 Working Paper Series No. 2012–14.

MEA. (2005). Millennium Ecosystem Assessment. *Ecosystems and Human Wellbeing*, Washington DC, Synthesis/Island Press.

Melo, O., Donoso, G. & Jara, E. (2004). Profundidad de Mercado, Asignación Inicial y Alternativa a la Patente de No Uso. *Revista de Derecho Administrativo Económico*, 13: 171–180.

Molle, F. & Berkoff, J. (2007). *The Lifetime of an Idea. In Irrigation Water Pricing. The Gap between Theory and Practice*, Wallingford, CAB International.

Muller, J. & Albers, H.J. (2004). Enforcement, payments, and development projects near protected areas: How the market setting determines what works where. *Resource and Energy Economics*, 26: 185–204.

Nauges, C. & Strand, J. (2007). Estimation of non-tap water demand in Central American cities. *Resource and Energy Economics*, 29 (2007): 165–182.

Olivier, A. (2010). Water tariffs and consumption drop: an evaluation of households' response to a water tariff increase in Manaus, Brazil. *Water Policy*, 12 (2010): 564–588.

Pagiola, S., Agostini, P., Gobbi, J., de Haan, C., Ibrahim, M., Murgueitio, E., Ramírez, E., Rosales, M. & Ruíz, J.P. (2004). *Paying for Biodiversity Conservation Services in Agricultural Landscapes*, Washington, DC, The World Bank. Environment Department Paper No. 96.

Pagiola, S., Landell-Mills, N. & Bishop, J. (2002). Making market-based mechanisms work for forests and people. In: Pagiola, S., Bishop, J. & Landell-Mills, N. (eds). *Selling Forest Environmental Services: Market-based Mechanisms for Conservation and Development*, London, Earthscan.

Pattanayak, S.K., Wunder, S. & Ferraro, P.J. (2010). Show me the money: do payments supply environmental services in developing countries?. *Review of Environmental Economics and Policy*, 4(2): 254–274.

Porras, I., Grieg-Gran,M. & Neves, N. (2008). *All that glitters: A review of payments for watershed services in developing countries*, London, International Institute for Environment and Development. Natural Resource Issues No. 11.

Quentin Grafton, R., Libecap, G., McGlennon, S., Landryj, C. & O'Brien, B. (2012). An integrated assessment of water markets: a cross-country comparison. *Review of Environmental Economics and Policy*, 5 (2): 219–239.

Rosegrant, W.M. & Gazmuri, R. (1995). *Reforming water allocation policy through markets in tradable water rights: Lessons from Chile, Mexico and California*, IFPRI. EPTD Discussion Paper No. 6.

Ruijs, A., Zimmermann, A. & van den Berg, M. (2008). Demand and distributional effects of water pricing policies. *Ecological Economics*, 66: 506–516.

Saleth, R.M. & Dinar, A. (2004). *The Institutional Economics of Water: A Cross-Country Analysis of Institutions and Performance*, Cheltenham, UK, Edward Elgar.

Schmidhuber, J.J., Bruinsma, J. & Boedker, G. (2009). Capital requirements for agriculture in developing countries to 2050. *Expert Meeting on How to Feed the World in 2050*, Rome, FAO.

Shah, T. (2005). The new institutional economics of India's water policy. In: van Koppen, B., Butterworth, J. & Juma, I. (eds). *African Water Laws: Plural legislative frameworks for rural water management in Africa*, Proceedings of a workshop held in Johannesburg, South Africa, 26–28 January 2005. pp.1–25.

Vos, J. & Vincent, L. (2011). Volumetric water control in a large-scale open canal irrigation system with many small holders: The case of Chancay-Lambayeque in Peru. *Agricultural Water Management*, 98: 705–714.

World Bank (2011). *Chile: Diagnóstico de la Gestión de los Recursos Hídricos*. Washington, DC, USA, The World Bank. [Online] Available from: www-wds.worldbank.org/external/default/WDSContentServer/WDSP/IB/2011/07/21/000020953_20110721091658/Rendered/PDF/633920ESW0SPAN0le0GRH0final0DROREV.0doc.pdf [Accessed July, 201]. Online report.

Wunder, S. (2005). *Payments for Environmental Services: some nuts and bolts*, Bogor, Indonesia, CIFOR. Occasional Paper No. 42.

Wunder, S., Engel, S. & Pagiola, S. (2008). Taking stock: a comparative analysis of payments for environmental services programs in developed and developing countries. *Ecological Economics*, 65(4): 834–852.

Zetland, D. & Gasson, D. (2012). A global survey of urban water tariffs: are they sustainable, efficientand fair?. *International Journal of Water Resources Development*, First article: 1–16.

14

LEGAL FRAMEWORK AND ECONOMIC INCENTIVES FOR MANAGING ECOSYSTEM SERVICES

Authors:

Bárbara A. Willaarts, Water Observatory – Botín Foundation, and CEIGRAM, Technical University of Madrid, Spain

Olga Fedorova, CEIGRAM, Technical University of Madrid, Spain

Diego Arévalo Uribe, Water Management and Footprint. CTA – Centro de Ciencia y Tecnología de Antioquia, Colombia

Gabriela de la Mora, Instituto de Investigaciones Sociales, Monterrey, Mexico

Marta Echavarría, Fundadora y Directora de Ecodecisión, Colombia

Elena López-Gunn, I-Catalist, Complutense University of Madrid, and Water Observatory – Botín Foundation, Spain

Patricia Phumpiu Chang, Centro del Agua para América Latina y el Caribe – ITESM, Monterrey, Mexico

Contributors:

Marta Rica, Water Observatory – Botín Foundation, and Complutense University of Madrid, Spain.

Highlights

- Payment for Ecosystem Services (PES) are rapidly emerging in LAC as complementary conservation measures to classic command and control policies. The majority of the PES programmes being implemented so far are focused on protecting headwaters in order to ensure the provision of water related services for urban areas. Carbon and biodiversity markets are less developed yet.

- Water-related PES schemes tend to be set up within a well-defined geographical setting, i.e. the watershed, which makes relatively easier to identify service sellers and buyers. Carbon and biodiversity services deliver benefits on much broader scales (often global), making difficult the identification of those service 'buyers' that should pay for supporting ecosystem services.

- The existing PES schemes in LAC have arisen from specific social and political arrangements between public and private actors involved in conservation rather than through legal mechanisms that have fostered these schemes. Legal frameworks explicitly supporting PES or PES like schemes are emerging across many countries, particularly in Peru, Brazil, Colombia and Mexico.

- The great majority of the PES initiatives in LAC have been developed at a local scale. Nevertheless, the development of institutional and legal frameworks related to ecosystem services management occurs on at least two political-geographical scales: national and sub-national (provincial governments and municipalities).

- PES could benefit from a legal framework but it is not a requirement for implementation. However, stable and enforceable contractual law and clear and secure land tenure along with property rights are necessary conditions for successful implementation.

14.1 Introduction

Latin America and the Caribbean (LAC) region has an outstanding natural capital and contributes to the provision of multiple ecosystem services (ES) at a wide range of scales. ES are here understood as 'all those benefits, material and in-material provided by nature, which contribute to human wellbeing' (adapted from MA, 2005). They include productive services like food, drinking water, fibre or minerals, but also all those other benefits derived from the well functioning of ecological processes (e.g. clean water, climate regulation, soil formation) and biodiversity conservation (e.g. eco-tourism, pollination, natural medicines).

The importance of LAC's natural capital is evidenced by the fact that this region holds approximately 70% of the world's vertebrates biodiversity (IUCN, 2013), 40% of the global aboveground carbon stocks (FAO, 2010a), 30% of total blue freshwater resources (FAO, 2013) and 13% of world heritage sites (UNESCO, 2013). Yet, the fast pace of development taking place in the region is generating a large pressure on LAC's natural capital, causing important environmental impacts and the loss of multiple ES, particularly regulating services (see Chapter 3). Two important factors explain current pressure on LAC's natural capital: 1) the prevailing economic model, which is natural resource use-intense and highly coupled yet; and 2) the large and often poorly urbanization process, which has large impacts on freshwater ecosystems and constitutes the most important driver of point water pollution. High commodity prices have stimulated the rapid growth of the primary sector in LAC (mostly of agriculture and mining), generating large negative environmental externalities (e.g. deforestation, diffuse pollution, soils degradation, etc.) and low interest in internalizing these costs to remain competitive, i.e. maintain its comparative advantage and support the prevailing *cheap* food policies. Similarly, urban growth encompasses a growing water demand to meet citizens needs, i.e. infrastructure development and water transfers, and yet investments in wastewater treatment plants are scarce, exacerbating the water pollution problem.

During the last decade different initiatives are emerging to incentivize the conservation and sound management of critical ES. Among all the different initiatives, economic incentives and payment for ecosystem services (PES) schemes are emerging as complementary strategies to traditional command and control environmental regulations, in an attempt to internalize the cost of non-market ES and deter its progressive degradation. Such schemes are also surfacing in those cases where no regulatory framework exists for managing natural resources but interest in preserving ES is significant (e.g. ensuring water quality to downstream urban citizens).

LAC is currently a leading region in the implementation of PES – particularly of water-related programmes – (Martín Ortega et al., 2012; Bennet et al., 2013), although the effectiveness of these programmes remains so far unclear, due to a variety of problems, including absence of baseline conditions, lack of clearly defined land tenure and property rights, and the financial un-sustainability, in many ongoing initiatives. Given the development path this region still has ahead, this chapter aims to review the success of ongoing PES schemes in LAC, as well as their institutional setting, to assess whether these instruments are useful and can foster a more green growth in this region and what would be the challenges ahead. Accordingly, Section 2 provides a fresh and up-to-date outlook on existing PES programs across LAC; Section 3 summarizes the legal and institutional setting in place for managing ES in the region; and lastly, section 4 analyses what are the main challenges and threats of PES schemes.

14.2 Economic incentives for managing water and land sustainably: payment for ecosystem services

As discussed in Chapter 13, economic mechanisms and incentives like PES, that pursue the integration of positive environmental externalities are increasingly being proposed as a promising approach for conserving ecosystem services. Mechanism such as PES are not intended to replace traditional command and control measures but to complement them by making them more acceptable (FAO, 2010b). In fact, PES incentives can support existing regulations, reducing the expected gain from non-compliance and even define opportunity costs for PES schemes (*ibid*). Also, when command and control regulations do not exist or are ineffective, PES might provide room for inclusive solutions that involve different stakeholders, as long as a stable contractual legal environment is in place (Grieg-Gran et al., 2006).

Yet, there is no overall agreement in the literature on what are PES schemes and what are not. Wunder (2007) has defined a set of criteria a PES scheme should fulfil to be distinguished from other incentive types: (1) a voluntary transaction in which (2) a well-defined environmental service (or land likely to in which a well-defined environmental service (or a land use likely to secure that service) (3) is 'bought' by a (minimum of one) buyer (4) from a (minimum of one) provider (5) if and only if the provider continuously secures the provision of the service (conditionality). Lately, however, there has been much debate over definitions when applied in practice, since many so-called PES schemes do not fulfil all the criteria set above. Other definitions providing a more encompassing approach to PES have been provided by Sommerville et al. (2009), who considers PES as an umbrella term where different schemes can be classified to '(1) transfer positive incentives to environmental service providers that are (2) conditional on the provision of the service, where successful implementation is based on a consideration of (1) additionality and (2) varying institutional contexts'. Muradian et al. (2010) propose a different conceptual framework, in which PES should not be limited to market transactions, but regarded as 'a transfer of resources (monetary or not) between social actors, which aims to create incentives to align individual and/or collective land use decisions with the social interest in the management of natural resources'. Such a framework would help PES schemes to not be rejected in a number of communities e.g. the indigenous Andean communities, which are highly sceptical on the monetarization of nature and ES, due to its public and collective prevailing nature (Wunder, 2006; FAO, 2010b).

Since there is yet not a clear consensus on what PES encompass, we have chosen to adopt a broad definition and generally refer to PES as 'any transaction, voluntary or regulated where there is a payment or exchange of credits (not necessarily monetary) between a buyer and seller that promotes some improvement of an ecosystem service' (adapted from Stanton et al., 2010). This implies that agreements such as 'reciprocal agreements', 'benefit sharing mechanisms', 'mitigation obligations' or 'offsets' are here included under the umbrella of PES or PES-like incentives.

LAC is today one of the frontrunners in the implementation of PES worldwide (Martin-Ortega et al., 2012). The reasons are diverse but probably influenced, among other factors, by the large number of ongoing environmental problems, the importance of its vast natural capital and perhaps also the cultural values of LAC society towards nature. PES schemes in LAC took-off in the 1990s with Costa Rica taking the lead thanks to the development of the national PES programme *Pago por Servicios Ambientales* in 1997. Ongoing PES programmes in LAC can be classified into four main categories: water, biodiversity, carbon and marine programmes. Table 14.1 summarizes the main characteristics of ongoing water, carbon and biodiversity PES schemes. The countries supporting the largest and most diverse number of active PES programmes are Brazil, Mexico and Costa Rica. Ecuador is the country holding the largest number of active water-related PES. Among the different schemes, water-related PES are still the most popular initiatives (see Box 14.1) followed by carbon programmes. Biodiversity markets have not yet proved as popular in LAC. The underlying reasons for the success of water-related PES schemes could be partly attributed to the fact that these ecosystem services deliver their benefits within well-defined geographical settings (basin or a watershed), making it easier to identify service providers and beneficiaries, and facilitating the negotiation process. Conversely, actors engaged in carbon and biodiversity initiatives are harder to identify, since the benefits normally exceed the limits of a well-defined spatial unit (basin, country, continent), further complicating the negotiations and identification of services beneficiaries (and thus, buyers).

Table 14.1 Overview of PES and PES-like initiatives found across Latin America and the Caribbean

TARGET ES	PAYMENT/MARKET TYPE	FREQUENCY	NUMBER OF ACTIVE PROGRAMMES BY COUNTRY	
WATER	Bilateral agreements (voluntary)	+ +	Bolivia	(5)
			Brazil	(4)
	Beneficiaries-paid fund (~ trust funds)	+ + +	Colombia	(3)
			Costa Rica	(2)
			Ecuador	(10)
	Water quality trading & offsets (regulatory)	+	Mexico	(3)
			Peru	(1)
BIODIVERSITY	Cap and trade (mitigation & compensation)	+ + +	Argentina	(1)
			Brazil	(2)
	Voluntary provisioning	+	Colombia	(1)
			Costa Rica	(1)
	Government-mediated payments (buyers of land to preserve an area)	+	Mexico	(1)
			Paraguay	(2)
CARBON	Forestry based projects (REDD, afforestation, reforestation)	+ + +	Brazil	
			Costa Rica	
			Ecuador	
	Renewable energy investments (wind, landfill, biomass)	+ +	Peru	*
			Panama	
	Investments in energy efficiency and fuel switch	+ +	Mexico	
			Nicaragua	

Source: own elaboration based on Bennett et al. (2013); Madsen et al. (2010) and Peters-Stanley and Hamilton (2012).

* No information was found on the number of carbon related programmes, rather on the amount of offset per country. In 2011 countries who achieved emission reductions through voluntary markets were: Brazil (>5 $MtCO_2e$/year) and to a lesser extent Peru, Ecuador, Panama, Nicaragua and Mexico (< 0.5 $MtCO_2e$/year).

Box 14.1 Watershed payment for ecosystem services in Latin America

Since the early 1990s various water-related PES schemes have been developed in Latin America (LA) to achieve win–win solutions that allow both finance conservation as well as stakeholder engagement at different levels. However, only a few of these PES schemes have so far been successful. Most have not managed to consolidate a common structure, with a lack a clear policy and institutional framework. Thus there are notable threats to this type of initiative, which prevent them from being successful.

Water-related PES programmes can be classified into: payments for watershed services (PWS), water quality trading (WQT) markets, and reciprocal or in-kind agreements (Stanton et al., 2010). Globally, between 2000 and 2008 the number of water related PES programmes had grown 500%, from fifty-one to almost 288[1] (Bennett et al., 2013). Among these 288 initiatives, the majority were PWS (75%) and the remaining (25%) were WQT. By 2011 the number of programmes had slightly dropped to 205. Some 60% of these programmes (128) are being developed in China and the US, whereas LA accounts for twenty-three active initiatives, the majority of which located in the Andean countries (Figure 14.1). Between 2008 and 2011 the number of water-related PES in LA declined (-22%). This variation, however, could be due to different factors like changes in the methodology used to record water-related PES schemes since 2008.

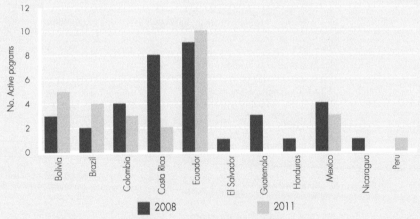

Figure 14.1 Watershed PES trends in the Latin America region. *Source: based on Stanton et al. (2010) and Bennett et al. (2013)*

Despite the negative overall trends, Peru, Bolivia, Brazil or Ecuador show an increase in water PES schemes. According to Bennett et al (2013), water PES in LAC are expected to grow in the years to come due to new water funds being created and increased funding for national programmes (e.g. in Mexico and Ecuador).

---

1 The information available is very heterogeneous in quality and quantity which makes it difficult to establish trends on the current state of PES initiatives.

Regarding the nature of the PES schemes, voluntary programmes prevail in LAC, particularly among water and carbon initiatives, while biodiversity markets are for the most part regulatory based (cap and trade schemes). Examples of voluntary agreements of water-related PES schemes include the bilateral agreements where service buyers, e.g. drinking water companies and hydroelectric generators (public or private), pay upstream service sellers, e.g. landowners, within a watershed to improve their land use management practices to ensure service provision downstream i.e. sufficient water and/ or of good quality (Table 14.2). Besides the monetary exchanges, in-kind payments, e.g. provision of agro-inputs, technical training or land tenure security are also frequent types of transactions. Monetary payments are frequently determined in two ways: either through the opportunity cost or by the estimation of willingness to pay. In some cases the price per hectare paid to landowners is estimated as an anti-poverty subsidy in order to provide a 'fair' income to poor communities.

Table 14.2 Main characteristics of water-related payments for ecosystem services programmes

WATER PES TYPES	PAYMENT MECHANISM	ACTIVITIES FUNDED	TRANSACTION TYPES	ACTORS INVOLVED
PUBLIC (GOVERNMENT)	Direct Subsidies	Improving land management activities (Best Agricultural Practices; Ecological restoration)	Monetary or in-kind payments (Agro-inputs, technical training, or tenure security)	**Buyers/Funders:** Governments, NGOs, private companies **Sellers/Beneficiaries:** Land owners, informal stewards, government nature reserves, NGOs with title and management responsibility of protected areas
	Land Purchase			
	Transfer of development rights	Forest Management Practices (Afforestation/ Reforestation)		**Administrators:** those establishing the specifics of the transaction and facilitating any negotiation between the buyers and sellers
PRIVATE (COMPANIES, NGOS)	Subsidies from private sources rights	Protection measures aiming at promoting economic activities alternative to those driving land degradation and deforestation.		**Intermediaries:** facilitators of the transaction or implementation of the project
	Fees for watershed protection			**Funders:** Governments and donors (multilateral banks, NGOs, private interests) financing part of the project in addition to the buyers

Source: own elaboration based on Stanton et al. (2010)

Other voluntary water-related PES schemes rapidly emerging in LAC are the 'water trust funds'. Such water funds are normally created by different public and private partners, including agent donors, who create a long-term financial mechanism or a trust fund as defined by local financial regulations. The returns of this fund and sometimes some portion of the principal investment are directed towards watershed actions e.g. restoring degraded lands, adopting sustainable farming practices, reforestation and educating children about sustainable water management (Bennett et al., 2013). This fund is managed by a stakeholder board also called 'administrators' who are different from the service buyers; they make joint decisions about best investments across the watershed. These types of water funds are currently the dominant form of active watershed PES schemes in LAC. In fact, a new public–private initiative, the so-called 'Latin American Water Funds Partnership', a joint initiative supported by the Nature Conservancy, the FEMSA Foundation, the Inter-American Development Bank (IDB) and the Global Environment Facility (GEF) has committed $27 million to develop and spread water funds across LAC. The Partnership plans to support at least thirty-two funds in total, protecting more than 2.8 million hectares in the coming years.

Regarding carbon payments, they can be either regulated or voluntary. Regulated markets, specifically designed for developed countries or Annex I parties of the United Nations Framework Convention on Climate Change (UNFCCC), are bound to certain emission reduction targets. Options to meet this commitment include different mechanisms where the project-based Clean Development Mechanism (CDM) is among the most common. The project-based CDM refers mostly to projects funded by developed countries or companies which contribute to reducing emissions in developing countries and make progress towards sustainable development. In doing so, developed countries offset part of their emissions and at the same time promote low carbon economies in developing countries. Renewable energy projects (e.g. wind, biomass, and landfill) and forestry projects (afforestation and reforestation) are the most common initiatives currently active in LAC. Since LAC countries are non-Annex I, they are not bound to any emission reduction commitment and therefore carbon market initiatives are all voluntary. Voluntary markets include emission reduction projects like REDD (Reduced Emissions from Deforestation and Forest Degradation), a source of funds still under negotiation from which LAC countries could benefit by receiving compensation payments for maintaining forests and preventing deforestation. The designers of REDD hope to deliver additional sustainable development benefits beyond simple carbon sequestration, creating a triple 'win' for climate change mitigation, biodiversity conservation, and poverty alleviation (Johns, 2012). However, UNFCCC negotiations on REDD are ongoing and significant aspects of the final design remain unresolved (see Box 3.2, Chapter 3).

Payments specifically designed for biodiversity conservation are very limited in LAC. Biodiversity markets are normally established with the purpose of creating a payment that can help to protect or restore habitats and species. Principally there are three types: regulatory compliance, government-mediated payments, and voluntary provisioning. The prevalence of the regulatory type in LAC is probably related to the fact that biodiversity markets are not yet well developed. Such regulatory transactions occur in countries like Brazil or Paraguay, where companies and developers are enforced by the national law or by the constitution to mitigate and compensate the environmental impacts of their activities. For instance, the Brazilian Forestry Code (Codigo Florestal, enacted 1965) stipulated that landowners must keep a certain percentage of natural vegetation on their land. In those cases where deforestation and vegetation clearance will exceed the legal quota, compliance with the law can still partly be met through off-site conservation, e.g. compensating other landowners within the same watershed to retain more than the minimum percentage of native vegetation cover. These Forest Code offsets have the potential to evolve into a formal bank, which is still under discussion at the state level. However, their success requires strict law enforcement to avoid uneven ecological compensations (e.g. destruction of a high-quality habitat by purchasing a low quality one). Some voluntary biodiversity markets also exist in LAC. For instance, government-mediated payments have, for a long time, been the most frequent mechanism to achieve conservation goals. Such types of payment involve governments and in some cases non-profit organizations purchasing land or creating payment programmes for biodiversity stewardship in those cases where there is public demand for biodiversity goods and services. Other voluntary markets include the Conservation Trust created in Paraguay where project developers can pay into a fund to compensate for damages as required by the Paraguayan Constitution. Emerging voluntary biodiversity PES schemes will include land markets for habitats of high biodiversity value, payments for biodiversity management, payments for private access to view a species or see its habitat, tradable rights and credits and biodiversity-conserving businesses (Bishop et al., 2008).

Overall, it is important to highlight that, while many of the PES schemes are focused on a single service, the number of programmes aimed at protecting or simultaneously restoring multiple ecosystem services is growing. These combined programmes are known as either 'bundle' or 'stacked' payments (Cooley and Olander, 2011). A bundle payment is a unique payment for the conservation or restoration of an area that simultaneously delivers multiple services (e.g. payments executed through the Costa Rica PES programme *Programa de Servicios Ambientales-PSA*). In this case, landowners receive a single payment for preserving or restoring the forest with the intention of ensuring the provision of multiple services such as carbon sequestration, biodiversity conservation, maintenance of the landscape aesthetic and the provision of hydrological services. Stacked PES programmes are separate payments sold by a landowner to different buyers with the intention of securing different services within the same area.

14.3 Enabling conditions for implementing incentives supporting ecosystem services

14.3.1 Constitutional recognition and existing laws on ES and PES

The social, economic and biophysical aspects of ecosystem services have received considerable attention in the past. However, little analysis exists on the legal and institutional setting and to what extent such frameworks support or hamper the flow of ecosystem services. In LAC, most constitutions recognize the right of people to enjoy a good quality environment and the duty of the state to preserve it, although in practice this has very different conceptualizations. This recognition is due to two factors: first, to an increase in the standard of living conditions; and second, to the growing importance of post-material and ecological values in society, spurred by international summits such as Stockholm (1972), Rio (1992), Rio +10 (2002) and Rio +20 (2012). This environmental awareness links with deep, entrenched autochthonous concepts such as 'good living' (*buen vivir*). Globally, over 177 countries explicitly recognize a 'right to a clean environment' (Boyd, 2012), fifteen of which belong to LAC (see Figure 14.2). LAC as a region in fact leads both in the recognition of a right to a clean environment, guaranteeing the environment as an individual's right, and also recognizing nature in terms of rights, as is the case of Ecuador (Murcia, 2011). Yet the main criticism associated with environmental rights in LAC, and elsewhere, is the weaknesses of mechanisms that could enforce this protection, since these rights are not always fundamental rights (Olivares, 2010).

ES are not explicitly considered in any of the LAC constitutions except for Ecuador, which acknowledges in article 74 of its political constitution that '*ecosystem services will not be subject to appropriation; their production provision, use and exploitation shall be regulated by the state*'. The inclusion of ES conservation in the constitutions has a great potential to give legal standing to the value of nature and/or ecosystem services, thus creating an acquiescent regulatory frame for developing pro-conservation mechanisms. However, reality shows that explicitly recognizing ES can sometimes limit environmental conservation. In Ecuador, ownership and exploitation of ES is attributed to the government, which instead of supporting local PES schemes is working towards obtaining international financing for ES maintenance, which may undermine the establishment of locally funded PES schemes (Southgate and Wunder, 2009). The non-explicit consideration of ES in political constitutions might not pose a problem as long as it does not prevent the development of initiatives aimed at preserving and maintaining them (Greiber, 2009). As previously mentioned, PES schemes operate effectively when property rights and land tenure are defined and it is easy to sign PES contractual agreements by the different stakeholders (FAO, 2010b).

From this constitutional acknowledgement, different legal frameworks have been developed to protect LAC's natural capital, e.g. biodiversity, carbon, forests, water and protected areas (see Table 14.3). Yet, no country in LAC has passed a national law on the general regulation of ES nor on PES, although Brazil, Peru, Mexico and Colombia are pending approval of such legislation. Regional and sub-national regulations on PES have been yet established in Brazil and Mexico. As Table 14.3 shows, in most countries, ES management falls under the umbrella of a wide range of environmental laws, predominantly those of forestry and water resources. Within these different environmental laws, arrangements have been set up regarding specific regional PES programmes. For instance, in Costa Rica under the Forestry law 7575 the managing body of the national PES programme, FONAFIFO, was established; and in Mexico legal frameworks and government funding channels were set up from the outset of their respective PES programmes (PSAH and PSA-CABSA) (Hall, 2008).

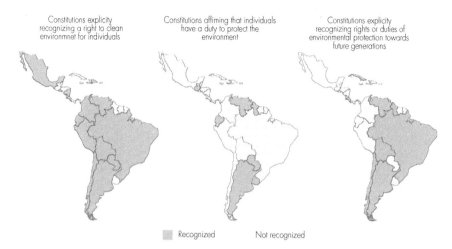

Figure 14.2 Constitutional recognition of the right to a clean environment in LA. *Source: own elaboration based on NCJM (2011) and RBA (2013)*

Most legal frameworks that refer to ES and PES foresee the creation of valuation instruments of natural resources according to their social and economic contribution and the ES these provide. Whether all these regulatory frameworks are effective or not in managing and preserving ES is hard to assess since, in many cases, no baseline exists to compare and analyse progress (See Table 14.4). Thus a more detailed assessment is needed, as well as determining to what extent such measures balance out other policies that hinder the sustainable management of ES.

Legislation at all levels – from local to national – can play an important role in the promotion and implementation of PES and PES-like schemes. Particularly, legislation is required for those PES programmes set out at a national level or those that have international financing in order to be implemented, such is the case of REDD+ programmes and cap-and-trade (Greiber, 2009). Given the limited amount of national PES programmes and associated legislation, in terms of numbers, the majority of PES schemes are local. However, this represents a good starting point since legislation improvements could benefit from practical experience, with local projects informing regional and national legislation which, in turn, provides greater legal certainty and a framework that enables, rather than restricts, regional and local initiatives.

Additionally where PES schemes are regulated, attention must be paid to its integration in the existing legal and institutional frameworks. In the case of public PES it is important not to make the process overly bureaucratic. In the case of private PES, these would benefit from specific legal frameworks that go beyond basic contract law. However, for both private or public PES schemes to be up-scaled, a robust legal framework is required to ensure both formal coherence and effectiveness.

Table 14.3 Legal frameworks supporting ecosystem services directly or indirectly

		COSTA RICA	PERU	BRAZIL	COLOMBIA	MEXICO	ARGENTINA
Laws directly supporting PES	SUB-NATIONAL			Espírito Santo: LN 9,864 PES Law. Amazonas: "Law On Climatic Change, Environmental Conservation And Sustainable Development (Bolsa Floresta)**		Sustainable Development Of Forests Law, Morelos State***; Sustainable Development Of Forests Law, Baja California State**	
	NATIONAL	Pending	N° 2386/2007-cr Ecosystem Services Law (Drafted 2009) Pending	Payment For Ecosystem Services Law (Pending) Law 5.487/09,	Draft National PES Strategy (pending)	National Development Plan 2013 – 2018*	
Laws indirectly supporting ES and/or PES	ENVIRONMENT	General Environment Law 755; Law On Soil Use, Management & Conservation 7779	General Environment Law, 28611. Law On Sustainable Use Of Natural Resources, 26821. Supreme Decree 012-2009MINAM. National environmental policy.	LN 6938 National Environmental Policy.	Renewable Natural Resources And Environmental Protection Code; General Environmental Law 99; Climate Change Convention, Law 164/94.	General Law Of Ecological Equilibrium And Environmental Protection; Climate Change General Law	General Environmental Law 25.675
	BIODIVERSITY	Biodiversity law 7788; Conservation Of Wild Life Law 7317	Conservation And Sustainable Use Of Biological Biodiversity Law 26839.		Law Decree 216 de2003 which determines the objectives, structure of Ministry of Environment, Housing and Land Development.	General Law On Wildlife	Protection And Conservation Of Wild Fauna Law 1981
	WATER	Water Law 276	Law On Water Resources No. 29338; Law Decree On Water 17752	Water Law LN 9433	Water Law 99/93	Federal law on Rights(art223)(National Programme for Hydrological Environmental Services (PSAH))**; National Water Law	Federal Agreement On Water 2003
	PROTECTED AREAS	Services Of Natural Parks 6084	Natural Protected Areas Act 26834	National Protected Areas System Law (9985/00)	Decrees 1865/94 and 708/2001 establish the development of regional policies of environmental management		
	FOREST	Forestry Law 7575** (FONAFIFO)	Forestry And Wild Fauna Act 27308	Brazilian Forest Code LN4.771; Management Of Public Forests, Implementation Of Brazilian Forest Service, (Regional)	Decree 900/97 Regulates Forestry Incentive Certificate	Development law Sustainable Forestry The Sustainable Rural Development Act; Presidential Decree 4thApril2001 creating National Forest Commission (CONAFOR)**	Forest Act No 13.273
	OTHERS	Public Services Regulating Authority Act 7593; Income Tax Act No.7092		Substituto De Proyecto De Ley 792,1,190, 1,667 ,1,920, 1,999 Y 2,364 (Programa Bolsa Floresta).**	National Development Plan Laws: 152/94 & 1151/2007*; Tax legislation: Law 223/95, Decree 1996/1999, Law788/2002, Decree 3172/2003	Agreement establishing the Rules of Operation for the granting of payments for the Programme to develop ES markets for carbon sequestration and those derived from biodiversity and to stimulate the establishment and improvement of agroforestry systems (PSA-CABSA)**	

Source: own elaboration

* No direct mention/regulation of PES as such but recognizes ES and the need for their conservation and preservation thus creating the enabling conditions.

** Legislation creates organisms and/or financial mechanisms for specific national PES programmes (programme name and/or organization set up).

Table 14.4 Summary of the advantages and disadvantages of having legal regulation for ecosystem services payment schemes

ADVANTAGES	DISADVANTAGES
PES become *legitimate* policy instruments creating **legal certainty** which enhances PES effectiveness	Possible further fragmenting of environmental legislation
Scope of PES instruments clarified	May conflict with other legal frameworks
Can streamline the process of setting out a PES programme by decreasing bureaucracy and tax incentives	May hamper implementation through increased bureaucracy and discrimination of eligibility for other financial subsidies especially for smaller PES
Only way of creating and implementing a **national** PES scheme	

Source: Greiber (2009) and FAO (2010b)

14.3.2 Land tenure and property rights in LAC

For any PES scheme, secure land tenure and clearly defined property rights are crucial for their effective implementation (Dent and Kauffman, 2005; FAO, 2010b; Contreras-Hermosilla, 2011; Larson and Petkova, 2011; Montagnini and Finney, 2011). One of the key aspects of a PES scheme is to establish a transaction whereby the service seller contracts an obligation to either stop, maintain or undertake specific land use activities and in some cases even gain rights to trade the service such as in the case of carbon sequestration credits (Muradian et al., 2010). Thus the PES contractual agreement always requires that the tenure rights of all actors are clearly defined and recognized.

Land tenure as defined by the FAO (2002) is the 'relationship, whether legally or customarily defined, among people, as individuals or groups, with respect to land'. (For convenience, 'land' is used here to include other natural resources such as water and trees). Land tenure is an institution, i.e., rules invented by societies to regulate behaviour. Rules of tenure define how property rights to land are to be allocated within societies. They define how access is granted to rights to use, control, and transfer land, as well as associated responsibilities and restraints. In simple terms, land tenure systems determine who can use what resources for how long, and under what conditions.' Property rights on the other hand, define how the land (and all natural resources present on that territory) or property can be used, controlled and transferred.[1]

---

1 FAO's simplified representation of property rights includes:

– use rights: rights to use the land for grazing, growing subsistence crops, gathering minor forestry products, etc.

– control rights: rights to make decisions how the land should be used including deciding what crops should be planted, and to benefit financially from the sale of crops, etc.

– transfer rights: right to sell or mortgage the land, to convey the land to others through intra-community reallocations, to transmit the land to heirs through inheritance, and to reallocate use and control rights.

The current situation of land tenure and land rights recognition in LAC is very heterogeneous given the extension of the region; however, there are some characteristic trends that are common for the majority of countries. Land and tenure security are still incomplete, even though most countries have established property registries with cadastres,[2] in many countries less than 50% of their national territory is covered by the cadastre (Figure 14.3). It is important to note that there are very different idiosyncrasies between continental Latin American countries and the Caribbean islands regarding land tenure. The Caribbean is characterized by the prevalence of state-owned land, which is not legitimized by its citizens, who follow alternative collective forms of land tenure. While in continental Latin America land tenure institutions are more entrenched, both formally and customarily, tenure security is still not achieved as less than half of farmers have solid title deeds over their lands (ECLAC, FAO, IICA, 2012). All over LAC several programmes of land titling are under way, which would provide a more secure environment for the widespread implementation of PES programmes. However, past experience has shown that titling programmes may bring increased disputes. Therefore there is a need to differentiate between the problems of access and distribution of land among farmers, as well as the territorial claims of indigenous populations (ECLAC, FAO, IICA, 2012; Van Dam, 2011).

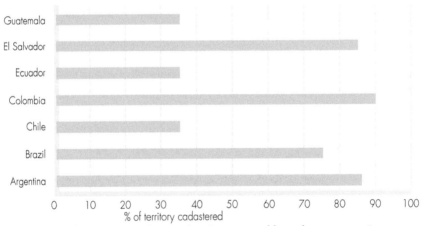

Figure 14.3 **The percentage of national territory covered by cadastre survey.** *Source: own elaboration based on data of CPCI (2011).*

To sum up, the lack of clearly defined land tenure systems and unsecure property rights undermines the possibility of effectively spreading the implementation of PES programmes across LAC regions for several reasons. First, identifying the legitimate users is complex:

---

2 FAO (2002) defines a parcel-based land information system that includes a geometric description of land parcels, usually represented on a cadastral map. In some jurisdictions it is considered separate from, but linked to, the register of land rights and holders of those rights (land register), while in other jurisdictions the cadastre and land register are fully integrated.

in many LAC countries ES and natural resources are publicly owned and managed by the state, thus a contract cannot be signed unless the state enables the de facto users and communities to use and benefit from PES schemes. In cases of private ownership the de facto users of land are sometimes not legally recognized as either owners or tenants, and thus payments cannot be made to them since they cannot contract such obligations. In other cases de jure users are not capable or willing to allow usufruct users to sign contractual agreements. Unsettled disputes of land claims also impede the effective implementation of PES schemes. Furthermore, there have been cases, such as Costa Rica, where users that had benefited from land reform or other governmental subsidies were not eligible for PES programmes and the other way around, once a user became part of PES programmes they were then excluded from receiving other financial support (Grieg-Gran et al., 2005).

Second, users that do not have secure land tenure[3] have no incentive to participate in PES schemes, or implement more sustainable land practices because they do not have any guarantee of obtaining the long-term benefits. Third, PES needs to be perceived as fair for the effective implementation of the programme. In order to achieve this, allocations need to be carried out carefully to not provide benefits to the large estate owners (*latifundios*) rather than those most in need. Such distribution would add an element of income redistribution and social equity to PES.

14.3.3 Institutional arrangements

There is no blueprint for an ideal institutional set up. Instead, institutions should be adjusted to national and local circumstances, in particular the prevailing governing structure. Overall the basic requirements for a PES scheme among private users, as stated in FAO (2010b), are:

* the absence of any legal provision that outlaws PES schemes
* basic contractual law: '*pacta sund servanda*' (contracts need to be fulfilled)
* civil law to enforce contract rights in case of non-compliance.

Public PES schemes require more regulation, since a public entity needs to be created or enabled in order to implement the scheme. However, public institutions at all levels fulfil important PES-related functions. Local institutions connect PES to the reality on the ground, regional institutions help to overcome administrative boundaries and national institutions can introduce PES visions and coordinate related policies. Private institutions may complement public institutions in the development and implementation of PES schemes. They can bring more flexibility and independence, which are important external capacities, as well as additional financial resources.

---

3 FAO (2002) defines tenure security as: the certainty that a person's rights to land will be protected. People with insecure tenure face the risk that their rights to land will be threatened by competing claims, and even lost as a result of eviction. The attributes of security of tenure may change from context to context: investments that require a long time before benefits are realized require secure tenure for a commensurately long time.

An appropriate institutional framework for PES needs to consider three financial dimensions: increasing available funds through specialized fundraising and fund-managing institutions; limiting institutional transaction costs; and providing sufficient financial means to ensure institutional performance.

As far as the management and administration of PES schemes are concerned, national institutions should perform only those activities which cannot be performed effectively at a more immediate or local level. Trust is fundamental to the long-term success and sustainability of PES programmes. Good governance – in particular public participation, transparency and access to information, as well as accountability and the rule of law – helps to build trust and is therefore key in the context of managing ES correctly.

14.4 Challenges and threats in PES implementation for LAC

Some key challenges regarding PES initiatives in LAC have been identified, particularly those related to water. These are four key challenges that can help explain why a large number of the initiatives, despite their potential, have had a low level of implementation.

First, *low stakeholder engagement and high dependence on foreign capital*. PES schemes are multi-stakeholder initiatives. One of the key added value is that PES are powerful tools for raising awareness and actively involving important stakeholders from different sectors of society (public sector, private sector, civil society). To ensure the uninterrupted provision of ecosystem services, funding conservation should be based on co-responsibility linked to risk management through the identification of specific threats and vulnerabilities. Most existing PES schemes in LAC are promoted by international institutions, using in many cases imported models that include foreign funding, which should in theory work as seed funding. However, this seed funding may finally create a dependency for local stakeholders and make the initiative unsustainable over time. If the institutional structure is weak and there is high dependency on external resources, financial sustainability could become uncertain. One of the conclusions of the latest Ecosystem Market Place report (Bennet et al., 2013) is the limited participation of the private sector in PES initiatives. Most initiatives so far have been promoted by NGOs or public sector entities, which may indicate the limited knowledge the private sector has about environmental risks.

Second, *lack of stability and clarity in the institutional and legal framework*. The prevailing sectorial approach to manage natural resources, the existence of weak public institutions and the unconsolidated regulatory framework, are some of the major constraints for a PES initiative to be effectively and efficiently integrated into water management in the region. In relation to the legislative framework: first, the lack of stability in the legislative, regulatory and institutional contexts make PES initiatives highly vulnerable due to their long implementation timelines (more than four to five years). Second, it is necessary to fulfil two basic conditions for the creation of a sustainable market scheme: a) define land property rights and b) establish trust between the supplier and the buyer of the environmental service (the role of the intermediary). The lack of viable and sustainable land titling programmes is a problem which affects particularly native communities, attempting to respect their traditional rights over land. Third, the use of public funds in PES schemes is essential for the active involvement of the public sector. However, these have to be accompanied by means to charge fees, and use and manage public funds, facilitating involvement in PES schemes.

Third, *lack of government coordination of isolated, small and disjointed initiatives.* A large number of local initiatives started a few years ago and the few schemes that are still active demonstrate a lack of coordination between initiatives. Government involvement as a regulator and coordinator could facilitate this process, particularly given the widespread lack of regulatory frameworks for PES scheme implementation. At present there is some evidence of significant progress in the implementation of regulatory frameworks and national programmes, which allows for the coordination of existing isolated initiatives. This is the case in Peru where a bill-regulating environmental services is being discussed. The case of Colombia that has recently approved methodological guidelines for watershed payments,[4] through which the government can create incentives to promote the investment in ecosystem services. Article 111 of the Colombian national environmental law (law 0953, published May 2013) establishes legal and institutional frameworks for purchasing strategic areas for water supplies. This purchase requires a minimum investment of 1% of municipal budgets in Colombia in order to establish hydrological PES.

Fourth, *vision of PES from a local, social and cultural perspective in the LAC region.* The PES approach raises a wide number of questions on the risks and opportunities posed by these schemes due to the ideological opposition to commoditizing nature or economic valuation of ES. The multiple values of water frameworks not only include the value of water for production, but also include water as a fundamental human right, water as the provider of ecological sustainability for the environment and biodiversity, water as the source of cultural sustenance of people and as a natural provider of social relations. This concept goes against the conceptualization of water as a tradable product involved in a market scheme, where ecosystem services can be bought and sold. In the Latin American region, particularly Bolivia and Ecuador, indigenous communities have been opposed to PES schemes based on the ancient Andean worldview of the relationship between people, earth and water. This situation has forced several ventures to change their names from 'payments' to 'compensation', so as to transform the relationships from a purely economic transaction. Landowners can receive cash, as well as in-kind payments, which can include income-generating activities as well as education and health benefits to communities.

To conclude, PES and PES-like mechanisms are only some of the possible solutions to the dichotomy of, on the one hand, increased production and consumption, and, on

---

4 'Methodological Guide for the Design and Implementation of Economic Incentive Payment for Environmental Services – PES'.

the other, the conservation and preservation of natural resources. Nevertheless, they are interesting instruments, combining environmental conservation and economic development, which is especially important for rural communities. The extent to which PES programmes can be financially sustainable in the long term and whether good practices and behaviour promoted by PES can become entrenched and adopted even after the programme has finished remains still unclear. One thing is for sure, any strategy aiming at preventing the degradation of LAC's natural capital, will require instruments (being PES just one among many other options) which can provide development alternatives for those who steward ecosystem services. And this will require as well the compromise of the international community since LAC's capital delivers benefits far beyond it borders.

References

Bennett, G., Carroll, N., & Hamilton, K. (2013). *Charting New Waters: State of Watershed Payments*. Washington, DC, Forest Trends. [Online] Available from: www.ecosystemmarketplace.com/reports/sowp2012. [Accessed March, 2013].

Bishop, J., Kapila, S., Hicks, F., Mitchell, P. & Vorhies, F. (2008). *Building Biodiversity Business*. Shell International Limited and the International Union for Conservation of Nature, London, and Gland, Switzerland. 164 pp.

Boyd, D.R (2012). The Constitutional Right to a Healthy Environment. *Environment* [Online]. Issue July–August 2012. Available from: www.environmentmagazine.org/Archives/Back%20Issues/2012/July-August%202012/constitutional-rights-full.html [Accessed July, 2013]

Cooley, D. & Olander, L. (2011). *Stacking ecosystem services payments: risks and solutions*. Duke Nicholas Institute for Environmental Policy Solutions, Washington, DC, Working Paper NI WP, 11-04.

Contreras-Hermosilla, A. (2011). People, governance and forests—the stumbling blocks in forest governance reform in Latin America. *Forests*, 2(1): 168–199.

CPCI (2011) Comité Permanente sobre el Catastro en Iberoamérica. *DATA CATASTRO* edición No.3 Abril. [Online] Available from: www.catastrolatino.org/documentos/Datacatastro_edicion_3.pdf [Accessed August, 2013].

Dent, D. & Kauffman, S. (2005). Green water credits. In: *Presentation at the FAO/Netherlands Conference on Water for Food and Ecosystems*, The Hague, Netherlands.

ECLAC, FAO, IICA. (2012). Economic Commission for Latin America and the Caribbean, Food and Agriculture Organization & Instituto Interamericano de Cooperación para la Agricultura. *The Outlook for Agriculture and Rural Development in the Americas: A Perspective on Latin America and the Caribbean*. Santiago, Chile, FAO.

FAO (2002). *Land Tenure and Rural Development*. Rome, FAO Land Tenure Studies.

FAO (2010a). *Global Forest Resource Assessment. Main Report*. [Online] Available from: www.fao.org/docrep/013/i1757e/i1757e.pdf. [Accessed March, 2013].

FAO (2010b). Payments for environmental services within the context of the Green Economy. In: *Stakeholder Consultation: From environmental externalities to remuneration of positive externalities in the Agricultural Food Sector*. Rome 27–28 September, FAO. [Online] Available from: www.fao.org/docrep/013/al922e/al922e00.pdf [Accessed August, 2013].

FAO (2013). AQUASTAT database. [Online]. Available from: www.fao.org/nr/water/aquastat/main/index.stm [Accessed June, 2013].

Greiber, T. (ed.). (2009). *Payments for Ecosystem Services: Legal and Institutional Approach.* Gland, Switzerland, IUCN. Report No. 78.

Grieg-Gran, M., Porras, I. & Wunder, S. (2005). How can market mechanisms for forest environmental services help the poor? Preliminary lessons from Latin America. *World development,* 33(9): 1511–1527.

Grieg-Gran, M., Noel S. & Porras, I. (2006). *Lessons learned from payments for environmental services.* Report 2006/05 (also issued as GWC Report 2), ISRIC – World Soil Information, Wageningen, Netherlands [Online] Available from: www.isric.org/isric/webdocs/docs/GWC2_Lessons%20learned%20(July%202006).pdf [Accessed August, 2013].

Hall, A. (2008). Paying for environmental services: the case of Brazilian Amazonia. *Journal of International Development,* 20(7): 965–981.

IUCN (2013). International Union for Conservation of the Nature. *The IUCN Red List of Threatened Species.* Version 2013.1 [Online] Available from: www.iucnredlist.org. [Downloaded on March, 2013].

Johns, B. (2012). *PES and REDD+: The Case of Costa Rica.* American University, U.N.-Mandated University for Peace. MSc Thesis.

Larson, A. M. & Petkova, E. (2011). An introduction to forest governance, people and REDD+ in Latin America: obstacles and opportunities. *Forests,* 2(1): 86–111.

MA (2005). Millennium Ecosystem Assessment. *Ecosystems and Human Well-being: Synthesis.* Washington, DC, Island Press.

Madsen, B., Carroll, N. & Moore Brands, K. (2010). *State of Biodiversity Markets Report: Offset and Compensation Programs Worldwide.* [Online] Available from: www.ecosystemmarketplace.com/documents/acrobat/sbdmr.pdf [Accessed August, 2013].

Martín-Ortega, J., Ojea, E. & Roux, C. (2012). *Payments for Water Ecosystem Services in Latin America: Evidence from Reported Experience,* Bilbao, Spain, Basque Centre for Climate Change (BC3). BC3 Working Paper Series No. 2012–14.

Montagnini, F., & Finney, C. (2011). Payments for environmental services in Latin America as a tool for restoration and rural development. *Ambio,* 40 (3): 285–297.

Muradian, R., Corbera, E., Pascual, U., Kosoy, N. & May, P. H. (2010). Reconciling theory and practice: An alternative conceptual framework for understanding payments for environmental services. *Ecological Economics,* 69 (6): 1202–1208.

Murcia, D. (2011). El Sujeto Naturaleza: elementos para su comprensión, In: Acosta A. and Martínez E., (eds). *La Naturaleza con derechos: de la filosofía a la política,* Quito, Ediciones Abya Yala. pp. 287–316.

NCJM (2011). Dutch Section of the International Comission of Jurists. *Human Rights and Environment.* Stakeholder input report for the Office of the High Comissioner of Human Rights. Available from: www.njcm.nl/site/uploads/download/433 [Accessed June, 2013].

Olivares, A. (2010). El nuevo marco institucional ambiental en Chile. *Revista Catalana del Derecho Ambiental,* 1(1): 1–23.

Peters-Stanley, M. & Hamilton, K. (2012). *Developing Dimension: State of the Voluntary Carbon Markets 2012.* Ecosystem Marketplace [Online] Available from: www.forest-trends.org/documents/files/doc_3164.pdfwww.forest-trends.org/documents/files/doc_3164.pdf [Accessed June, 2013].

RBA (20013). Rights-based Approach to Conservation. *The Human Right to a clean environment.* [Online] initiative supported by the Environmental Law Centre (ELC) of the International Union for the Conservation of Nature (IUCN). Available at: community.iucn.org/rba1/Pages/The%20 Human%20Right%20to%20a%20clean%20environment.aspx [Accessed February, 2013].

Sommerville, M. M., Jones, J. P. & Milner-Gulland, E. J. (2009). A revised conceptual framework for payments for environmental services. *Ecology and Society,* 14 (2): 34.

Southgate, D. & Wunder, S. (2009). Paying for watershed services in Latin America: a review of current initiatives. *Journal of Sustainable Forestry,* 28(3–5): 497–524.

Stanton, Y., Echavarria, M., Hamilton, K. & Ott, C. (2010). State of Watershed Payments: An Emerging Marketplace. Ecosystem Marketplace. [Online] Available from: www.foresttrends. org/documents/files/doc_2438.pdf [Accessed June, 2013].

Van Dam, C. (2011). Indigenous territories and REDD in Latin America: Opportunity or threat?. *Forests,* 2 (1): 394–414.

UNESCO (2013). United Nations Educational, Scientific and Cultural Organization. *World Heritage Sites and Biosphere Reserve Database* [Online] Available from: www.en.unesco. org/. [Accessed June, 2013].

Wunder, S. (2006). Between purity and reality: taking stock of PES schemes in the Andes. *Ecosystem Market Place,* 1 (4), The Katoomba Group.

Wunder, S. (2007). The efficiency of payments for environmental services in tropical conservation. *Conservation Biology,* 21(1): 48–58.

15

RETHINKING INTEGRATED WATER RESOURCES MANAGEMENT: TOWARDS WATER AND FOOD SECURITY THROUGH ADAPTIVE MANAGEMENT

Coordinator:
Elena López-Gunn, I-Catalist, Complutense University of Madrid, and Water Observatory – Botín Foundation, Spain

Authors:
Elena López-Gunn, I-Catalist, Complutense University of Madrid, and Water Observatory – Botín Foundation, Spain
Aziza Akhmouch, Organisation for Economic Co-operation and Development (OECD), France
Maite M. Aldaya, Water Observatory – Botín Foundation, and Complutense University of Madrid, Spain
Virginia Alonso de Linaje, Universidad Complutense de Madrid, Spain
Maureen Ballestero, Global Water Partnership, Costa Rica
Manuel Bea, Geosys S.L.,Spain
Ricardo Hirata, Universidad de Sao Paulo, Brazil
Julio M. Kuroiwa, Laboratorio Nacional de Hidráulica – Universidad Nacional de Ingeniería, Lima, Peru
Beatriz Mayor, Universidad Complutense de Madrid, Spain
Lorena Perez, I-CATALIST, Universidad Complutense de Madrid, Spain
Patricia Phumpiu Chang, Centro del Agua para América Latina y el Caribe – ITESM, Monterrey, Mexico
Christopher Scott, University of Arizona, Tucson, USA
Fermín Villarroya, Universidad Complutense de Madrid, Spain
Pedro Zorrilla- Miras, Cooperativa Terrativa, Madrid, Spain

Contributors:
Pedro Roberto Jacobi, PROCAM /IEE Universidade de São Paulo, Brazil
Andrea Suarez, HIDROCEC – Universidad Nacional de Costa Rica, San Jose, Costa Rica
Roberto Constantino Toto, Universidad Autónoma de México, México

Highlights

- Integrated Water Resources Management increasingly means looking at the anthropo-hydrogeological cycle, thus considering a range of conventional and non-conventional resources which are part of water resources management, such as conjunctive use, the potential of rainwater harvesting, water reuse and virtual water trade.

- Virtual water is an important component of integrated strategies in redistributing water resources. On the whole, in terms of agricultural products, the Latin America and Caribbean (LAC) region was a net exporter of green virtual water (141.5km^3/yr) especially from Argentina and Brazil, and a net importer of blue virtual water (16.1km^3/yr) especially Mexico, during the period 1996–2005.

- There are many opportunities for LAC to achieve more sustainable, equitable, and efficient use of their resources thus facilitating a transition towards a green economy, already present in numerous successful cases. Although many challenges still need to be faced; in many cases economic growth in LAC has been achieved through intensive use of natural resources like land and water – coupled with an increase in the levels of pollution and the loss of ecosystems and biodiversity. Collectively, these represent a serious challenge to water-security.

- In the LAC countries water governance occurs at very different levels – from the international political sphere down to the irrigation district level. Despite the progress made during the past decade, coordination of all these levels, i.e. achieving integrated water resources management, and strengthening stakeholders' involvement are fundamental to ensuring the legitimacy of the process and thus achieving clearly stated policy goals.

- The LAC region is in active pursuit of water security through IWRM with a clear focus on social equity and environmental quality and the way forward is clear, with a well-defined pathway. However, it will require institutional communication, political will and a strong dose of civil-society engagement in the planning process; the building blocks required for a resilient, robust future.

15.1 Introduction

IWRM is coordination (process), water security is the goal (result, status). IWRM is a process of change, which takes place continuously and dynamically. Water Security is a development objective. (Christopher Scott)

The Integrated Water Resources Management (IWRM) paradigm has just celebrated its twenty-first birthday in 2013, a period over which it has become dominant in both the water sector and sustainable development circles. It was born in 1992 as a result of the International Conference on Water and the Environment in Dublin and at Rio de Janeiro with Agenda 21 (Ait-Kadi, 2013). Its conceptual and implementation framework was developed by the Global Water Partnership, under the auspices of the World Water Council (GWP/TAC, 2000; GWP, 2004). IWRM is defined as '*a process which promotes the coordinated development and management of water, land and related resources in order to maximize economic and social welfare in an equitable manner, without compromising the sustainability of vital ecosystems and the environment*' (GWP/TAC, 2000).

Yet, due to the rapidly changing times we are currently immersed in, the lifespan of concepts and paradigms is also put to the test more quickly. According to Kuhn (1962), scientific progress is the result of 'development by accumulation', i.e. when normal science is interrupted by periods of revolutionary science. The IWRM paradigm is therefore in a state of flux (GWP, 2012; López-Gunn et al., 2013). This chapter aims to identify new trends and directions, as well as potential changes in its conceptual basis, particularly from fast-emerging complementary concepts such as water security (GWP/TAC, 2000; Grey and Sadoff, 2007; Pochat, 2008; GWP, 2010; Cook and Bakker, 2012; UN Water, 2013) analysed in Chapter 6. Along these lines, are there enough anomalies in the IWRM paradigm to warrant major changes? This chapter will argue that in order to 'speed up' the implementation of IWRM it is fundamental to ask new questions about its main tenets. The chapter analyses and evaluates the main ingredients of the IWRM paradigm, looking at a) the integration of resources, b) of sectors and c) across organizations. IWRM acquires real added value once a series of clear and specific policy goals are set, e.g. those provided by water security or the upcoming Sustainable Development Goals (SDGs) on water (Sachs, 2012) that in 2015 will effectively replace the merely target-oriented Millennium Development Goals (MDGs).

15.2 'W and R' in IWRM

This chapter will first revisit the resource base and consider how to re-think the hydrological cycle by adopting an 'anthropo-hydrogeological' cycle, i.e. a cycle in the context of the new era of the Anthropocene (Steffen et al., 2011). Building on Chapter 2, it also considers interactions within the unitary water cycle affected and modified by human use, and also innovative ways of thinking about water such as the concept of virtual water.

15.2.1 Water resources: the 'anthropo-hygeodrological cycle'

As highlighted in Chapter 2, the Latin America and Caribbean (LAC) region has great wealth in terms of water resources and presents a resource intensive development pattern, where much of the population lives in cities and human activities deeply and radically alter the water cycle in terms of its quantity and quality in time and space (Figure 15.1). The increasing demand for water on the one hand, and supply constraints on the other, implies a need to rethink the hydrological cycle in order to increase water security for both urban and rural areas, but also from a sectorial point of view (agriculture, mining or energy). The understanding and correct quantification of water in its different forms (atmosphere, surface and underground) are fundamental for the proper management of water resources and this also includes the need of breaking down any false paradigms about sustainability. Thus a first step for IWRM is proper water accounting, where the concept of 'water savings' does not necessarily detract from other uses (see Chapter 10 on water efficiency).

From an IWRM perspective, it is therefore necessary to characterize each source of water available in the water-cycle and their interdependencies. The opportunities offered by both conventional and non-conventional resources add increasing complexity to water management, which will require a new matrix-based approach considering an anthropo-hygeodrological cycle (Galbraith, 1971; Barlett and Goshal, 1990). In modern societies, there are six main sources of water: surface water (lakes, rivers and reservoirs), groundwater (aquifers), soil water (edaphic), precipitation water (rain harvesting), water reuse (treated or untreated), and desalinated water, to which a seventh – 'virtual water' – should be added (as will be discussed below). The first two are the most commonly used for the large water supply systems of cities and agricultural areas. In LAC this represents more than 90% for the cities water supply. The fourth (rain harvesting) has been used for a long time by families in poor regions (in semi-arid zones of Brazil, for example) as an adaptation mechanism and it is starting to be used more widely as an additional source of water in some cities. Desalination and water reclamation are also being implemented in LAC countries due to the increasing costs of obtaining water from conventional sources. In specific locations these new resources can represent a key strategic option for addressing local problems. For example, desalination for mining or for public water supply in Chile and northeastern Brazil respectively is an emergent trend.

The coordination and integration of both conventional and non-conventional sources is likely to be fundamental for specific locations in order to reduce water risks and pressures. Groundwater and surface water feature a clear complementarity in many aspects, which is crucial in order to increase water security for societal needs, e.g. public water supply and economic activities. In many cases, the problem of water supply in cities or for crops production is related to seasonal rain variation (periods of drought) and also to a lack of water infrastructure. Aquifers can store large amounts of water, as available 'natural (green) infrastructure', though there are few cases of planned joint management of surface and groundwater in LAC countries. Some positive examples are in Lima (Peru) and some

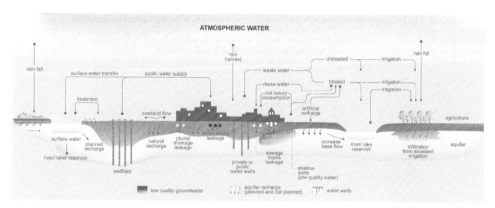

Figure 15.1 The 'anthropo-hygeodrogeological' cycle. *Source: expanded from Foster et al. (2011)*

cities in Mexico, but these are the exception rather than the rule due to the high level of technical knowledge and institutional coordination required. As a consequence, the high-quality, drought-resilient capacity of groundwater resources tends to be underestimated (Garduño et al., 2006). It is also necessary to recognize that there are more cases of spontaneous (or unplanned) conjunctive use than a planned conjunctive management of groundwater and surface water (López-Gunn et al., 2011). This is the case in the State of São Paulo, Brazil, where 15% of cities are supplied by both surface (main source) and groundwater resources (complementary source, i.e. 12,000 wells in the metropolitan area of São Paulo) (Hirata et al., 2006). Although surface and groundwater represent the 'bulk' of apparent resources, a wider perspective should also consider the opportunities of non-conventional resources and the largely unseen or 'forgotten' water resources of virtual water flows and green water. Thus for IWRM, particular attention should be paid to the range of resources and the advantages and disadvantages of each.

From the perspective of IWRM, it is also important to go beyond the evaluation of supply and demand interventions to a more systemic perspective. In this sense, the diversification of resources means a probable reduction of risk (Table 15.1), which allows for the re-visiting of supply side engineering measures, in order to consider alternatives such as rainwater harvesting, aquifer recharge enhancement (with an excess of surface runoff or reclaimed wastewater), desalination, and urban wastewater reuse. Likewise, examples of demand side measures are water conservation, promoting crop changes, improving irrigation efficiency (e.g. irrigation water use quotas, covering open canals, economic incentives to use high-pressure systems or the use of low-pressure water distribution pipes in agricultural areas) or measures that incorporate seasonal and spatial aspects.

One important issue for the integrated management of this resource portfolio refers to the allocation of responsibilities and information. With regard to this, sound information on resource use, accurate water accounting and extended participation would make integrated water resources management more likely (see section 15.4.2). For example,

Table 15.1 Comparative features of different components of water resource portfolios

FEATURE	GROUND-WATER	SURFACE WATER	DESALINATED WATER	RAIN HARVEST WATER	RECLAIMED WASTE WATER
		HYDROLOGICAL CHARACTERISTICS			
STORAGE VOLUMES	Very large	Small to moderate			
RESOURCE AREAS	Relatively available	Restricted to water bodies	Restricted to saline water location	Relatively unrestricted	Restricted to water availability
FLOW RATES	Very low	Moderate to high	Depends on the infrastructure	Depends on the infrastructure	Depends on the infrastructure
RESIDENCE TIMES	Generally decades/centuries	Mainly weeks/months	Centuries	Hours/days	Months/years
DROUGHT PROPENSITY	Generally low	Generally high	Low	High	Low
EVAPORATION LOSSES	Low and localized	High for reservoirs	Low	High	Low
RESOURCE EVALUATION	High cost and significant uncertainty	Low cost and often less uncertainty	High and often less uncertainty	Low and often less uncertainty	High and often less uncertainty
ABSTRACTION IMPACTS	Delayed and dispersed	Immediate	Low to moderate	Low	Low to moderate
NATURAL QUALITY	Generally (but not always) high	Variable (but generally needs treatment)	(-)	Generally high to moderate	(-)
POLLUTION VULNERABILITY	Variable natural protection	Largely unprotected	(-)	Associated to atmospheric contamination	(-)
POLLUTION PERSISTENCE	Often persistent in the short to medium term	Mainly transitory	(-)	(-)	(-)
		SOCIO-ECONOMIC FACTORS			
PUBLIC PERCEPTION	Not well known by the public	Aesthetic, predictable	Moderate	Moderate	Low
DEVELOPMENT COST	Generally modest	Often high	High	High or modest (depending on technology used)	Low
DEVELOPMENT RISK	Less than often perceived	More than often assumed	Less than often perceived	Less than often perceived	Less than often perceived
STYLE OF DEVELOPMENT	Mixed public and private	Largely public	Mixed public and private	Mixed public and private	Largely public

Source: expanded from Tuinhof et al. (2006)

in Costa Rica the regulatory framework does not allow for the use of groundwater, which makes joint management almost impossible. This links up with transparency on resource use (see Chapter 12), adequate data gathering and the availability of good water registers. For example, in the case of Mexico the Registro Público de Derechos de Agua (REPDA), the main approximation tool for federal water use is incomplete and its validity rather poor. In the case of Costa Rica the water information system (SINIGIRH) compiles information on river basins from different data sources (universities, AyA, ICE, IMN, SENARA, MINAE) into a single database and aspires to improve the hydrologic and hydrogeological information by strengthening the network of metering stations in order to support decision making.

Box 15.1 Extreme water security? Floods, droughts, population growth and migration in the Andes

One of the main functions of water management is dealing with water availability and in particular with climate variability which includes extreme events such as floods, droughts and general climatic changes. Water management when there is too much or too little water, and under a new scenario where underlying baseline resource conditions are subject to change due to climate change, are real stress tests for IWRM. Focusing on the Andean region, composed of Colombia, Ecuador, Peru and Bolivia, we briefly discuss issues related to extreme events, IWRM and water security. In the case of floods, there is a large portion of the population exposed to floods (approximately 15% of the population; see Table 15.2) (General Secretariat of the Andean Community, 2009). As can be seen in Figure 15.2 and Table 15.2 the areas most affected by droughts are in southeastern Peru and southwestern Bolivia. The population that has the potential for being affected by droughts reaches 19% of the total. An extreme drought can cause the total loss of work and capital for a small community. In addition and less well known, the absence of humidity can cause the presence of pests. The areas more prone to droughts have the lowest population growth rates (see Figure 15.2). This indicates that climate variability affects people significantly, forcing them to move to areas in which jobs may be more secure (Figures 15.3). Knowledge and data on climate variability and change can facilitate improved water resource management to reduce the vulnerability of people and areas most exposed, thus increasing system resilience. This is especially if information is produced on how this variability and change affects other systems e.g. economic system (losses), and impact on social system (e.g. migration).

Table 15.2 Population prone to suffering droughts and floods in the Andean Community countries

		UNITS	BOLIVIA	COLOMBIA	ECUADOR	PERU	ANDEAN COMMUNITY
POPULATION	TOTAL	Million	9,427	48,889	13,215	27,254	**92,785**
	EXPOSURE TO FLOODS	Million	600	5,232	2,428	8,459	**13,710**
		%	6%	12%	18%	20%	**15%**
	EXPOSURE TO DROUGHTS	Million	1,819	8,235	4,547	2,616	**17,217**
		%	19%	19%	34%	10%	**19%**
AGRICULTURAL AREA	TOTAL	Km²	268,954	533,431	115,342	256,118	**1,173,845**
	EXPOSURE TO FLOODS	Km²	57,000	120,000	14,000	34,000	**225,000**
		%	21%	22%	12%	13%	**19%**
	EXPOSURE TO DROUGHTS	Km²	88,000	59,000	24,000	120,000	**291,000**
		%	33%	11%	21%	47%	**25%**

Source: own elaboration based on data from the General Secretariat of the Andean Community (2009).

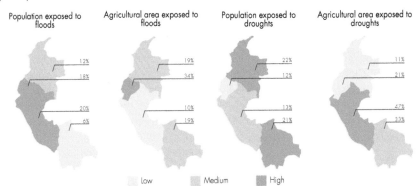

Figure 15.2 Population and areas most affected by droughts and floods in the Andean **Community.** *Source: own elaboration based on data from the General Secretariat of the Andean Community (2009).*

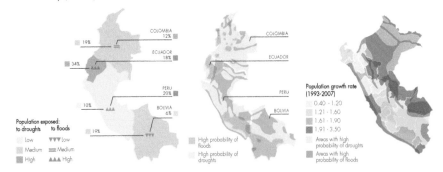

Figure 15.3 Population and areas most affected by droughts and floods in the Andean **Community and Peru.** *Source: own elaboration based on data from the General Secretariat of the Andean Community (2009).*

15.2.2 Innovations in resource 'thinking': virtual water in IWRM

The virtual water concept represents an important dimension of IWRM, particularly because it links water to use. However, it alone cannot determine optimal water resource allocation in importing and exporting to and from LAC countries and regions since water resources management requires consideration of multiple objectives and trade-offs from different options (Allan, 2011; Yang et al., 2013). The problem of water scarcity can be addressed by different means, i.e., improving water use efficiency locally, transferring water from outside, and transferring virtual water into the region in order to reduce local water demand. These measures are not mutually exclusive and can be combined to form an integrated approach in addressing water security problems. Thus, the trade of virtual water is one important component of integrated strategies in tackling water security (Guodong, 2003). The essence is that countries/regions can undertake economic activities (including agriculture) in which they have a comparative advantage. Virtual water strategies could potentially improve overall water use efficiencies in agriculture by adjusting crop structure and importing most water-intensive crops, thereby easing the level of water stress in specific regions, particularly in arid areas or areas with high population growth (Yang et al., 2013). However, it is fundamental to take the local context into account and to consider whether the local economy can import virtual water in exchange for other value added exports. With regard to agricultural products, during the period 1996–2005 the LAC region was a net exporter of green virtual water ($141.5km^3/yr$) and a net importer of blue virtual water ($16.1km^3/yr$), as concluded by Mekonnen and Hoekstra (2011), exporting through agricultural products three times more virtual water than it consumed. Thus when considering water security for countries with lower water availability, virtual water is a key element.

The water footprint indicator provides additional information for policy makers that can complement the classical measure of water withdrawals. Traditional national water use accounts only refer to the direct blue water withdrawal within a country. Beyond this, the water footprint assessment provides additional information on green and blue water consumption and pollution (grey water) including data on direct and indirect water use (virtual water flows), which makes the water footprint very different from other IWRM indicators (Table 15.3). By just looking at water use within its own country, most governments have a partial view of the sustainability of national consumption. In order to support a broader analysis and better informed decision making, national water use accounts could be extended to national water accounting on the basis of the water footprint methodology or other similar water accounting methods (Molden, 1997; Molden and Shakzivadivel, 1999; Molden et al., 2007; Perry, 2012). The specification on whether water resources are being used or consumed, and also whether they refer to blue (surface or groundwater) or green water (soil water) would provide a stronger information base from which to formulate national water plans and specific river basin plans, which are coherent, well aligned and integrated with national policies in relation, for example, to the environment, agriculture, energy, trade, foreign affairs and development cooperation (Hoekstra et al., 2011). Ideally, economic values and also energy implications would also be taken into consideration, as discussed in the next section.

Table 15.3 Total water footprint and total virtual water flows in Latin American countries

| | TOTAL WATER FOOTPRINT | | | | TOTAL VIRTUAL WATER FLOWS | | | | | | | | | | | |
| | | | | | VIRTUAL WATER IMPORT | | | | VIRTUAL WATER EXPORT | | | | NET VIRTUAL WATER IMPORT | | | |
	Green Water Mm³/yr	Blue Water Mm³/yr	Grey Water Mm³/yr	Total Wf Mm³/yr	Green Water Mm³/yr	Blue Water Mm³/yr	Grey Water Mm³/yr	Total Import VW Mm³/yr	Green Water Mm³/yr	Blue Water Mm³/yr	Grey Water Mm³/yr	Total Export VW Mm³/yr	Green Water Mm³/yr	Blue Water Mm³/yr	Grey Water Mm³/yr	Total Net VW Mm³/yr
ARGENTINA	176,194	5,708	9,189	191,091	4,285	266	1,116	5,667	93,307	1,863	2,875	98,045	-89,022	-1,597	-1,758	-92,377
BELIZE	677	14	177	868	70	14	22	105	445	19	67	530	-375	-5	45	-425
BOLIVIA	31,559	601	284	32,444	1,212	134	182	1,528	3,551	23	38	3,613	-2,339	111	144	-2,084
BRAZIL	135,966	13,826	31,930	481,722	29,475	2,368	3,701	35,544	105,427	1,944	5,121	112,492	-75,952	424	-1,421	-76,949
CHILE	9,143	2,797	3,888	15,828	4,720	369	987	6,076	1539	600	851	2,990	3,180	-230	135	3,085
COLOMBIA	50,173	2,384	7,210	59,767	8,065	742	1,714	10,521	11,778	190	1,286	13,254	-3,713	552	428	-2,733
COSTA RICA	5,412	428	1,437	7,277	2,240	386	488	3,114	4,638	246	606	5,489	-2,398	141	-117	-2,374
ECUADOR	26,444	2,443	3,366	32,253	2,192	207	410	2,809	6,362	399	396	7,156	-4,170	-191	14	-4,347
EL SALVADOR	5,202	134	879	6,215	2,025	346	335	2,705	3,305	101	252	3,657	-1,280	245	82	-953
GUATEMALA	13,248	378	1,030	14,656	2,359	424	481	3,264	6,964	177	1,133	8,273	-4,605	247	-652	-5,010
GUYANA	1,632	257	135	2,024	251	58	48	357	719	135	34	888	467	-77	14	-530
HONDURAS	7,573	182	600	8,355	870	195	223	1,288	6,723	108	284	7,115	-5,853	87	61	-5,827
MEXICO	109,021	16,453	23,053	148,527	65,407	14,169	12,724	92,299	13,128	8,870	4,107	26,105	52,279	5,298	8,617	66,194
NICARAGUA	5,877	230	333	6,440	751	113	158	1,022	2,389	69	68	2,525	-1,638	44	90	-1,504
PANAMA	2,556	141	478	3,175	805	106	246	1156	1,137	248	164	1,549	-332	-143	81	-394
PARAGUAY	32,845	323	655	33,823	437	44	134	616	12,520	56	153	12,729	-12,082	-12	-19	-12,113
PERU	18,040	4,553	3,022	25,615	6,983	526	971	8,479	2,499	559	474	3,532	4,484	-34	497	4,947
SURINAME	290	80	74	444	96	12	24	132	92	31	27	151	4	-19	4	-19
URUGUAY	11,504	888	344	12,736	975	61	193	1,230	6,652	732	203	7,587	-5,676	-671	-10	-6,357
VENEZUELA	23,341	1,926	4,844	30,111	6,832	607	1,102	8,542	1,425	128	497	2,051	5,407	479	605	6,491
TOTAL	965,697	53,746	92,928	1,113,371	140,048	21,148	25,256	284,597	284,597	16,498	18,635	319,730	-144,548	4,649	6,620	-133,279

Source: own elaboration based on data from Mekkonen and Hoekstra (2011)

15.3 The 'I' in IWRM

This section discusses issues linked to sectorial integration – or rather coordination – and the future challenges and trade-offs. It thus looks first at the nexus between food–water– energy and new concepts such as ecological boundaries and environmental security by looking at the human footprint (ecological, carbon and water) and how it fares when compared with the human development index. Both the nexus and the green economy offer important emergent sectorial themes for IWRM.

15.3.1 The water–food–energy nexus

The need for integration is particularly relevant in relation to the water, food and energy nexus to ensure water, food and energy security in the LAC region. This is because energy, food and water security partly pivot around successfully managing the interactions and potential trade-offs in the nexus. For example, the interconnections as discussed in detail in Chapter 9 are evident: the use of dams and waterfalls for hydroelectricity produc-tion and storage (water-energy); the need for energy to pump water for irrigation (Scott, 2013); the use of food crops or crop residues to obtain biofuels (food-energy); or the high water consumption required by food production (water-food) (Lundqvist et al., 2008; Hoff, 2011) (see Figure 15.4).

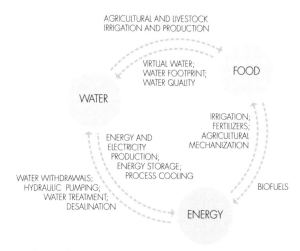

Figure 15.4 Understanding the nexus. The water, energy and food nexus. *Source: own elabo-ration.*

Within the energy–food–water nexus context, LAC is a region with abundant resources yet with important choices in terms of the prioritization of resource use. As Allan (2011) shows, this is particularly important in the case of Brazil. For example, in relation to the food/energy nexus, i.e. biofuels/soybean crops will have consequences not only for Brazil or the region but probably also impact other regions in the world. In terms of energy, in the Andean, Amazonian and Southern Cone regions, the sector is dominated by hydropower (see Box 15.2), which accounts for 60% of the total energy mix (Meisen and Krumpel,

2009). Meanwhile, Brazil is the world's second biggest producer and exporter of ethanol fuel (see Chapter 9). These energy sources are strongly dependent upon water and land availability, making these regions vulnerable to climate variations (extreme events, severe droughts, rainfall and temperature oscillations) and climate change, thus it will be necessary to consider in more depth the implications of different development models on local energy security, economic development, and food security.

Most relevant for policy makers is to make the potential synergies and trade-offs in these inter-linkages as explicit as possible. These can provide water and energy managers with new tools and cleaner paths towards sustainability and efficiency (solar decontamination, application of renewables for irrigation, dry cooling, energy production from water treatment plants, etc.).

It has been estimated that in LAC water for energy will increase by 50% in 2050 (WEC, 2010), although it should be noted that there is a high level of uncertainty around the water consumption data of primary energies (Figure 15.5). The high unitary water footprint of biofuels and their share in some of LAC's countries energy mix (especially relevant in the case of Brazil), allows bioenergy to be identified as by far the highest water consumer within the primary energy matrix, and thus highlights the importance of starting to produce some approximate numbers on this variable.

From the perspective of the nexus it is important to increase knowledge on how to achieve the balance between development, environmental sustainability and social equity. For the primary energy matrix, an IWRM 'nexus thinking' would look at synergies and trade-offs in the soybean dichotomy in terms of energy/food for countries like Argentina and Brazil who are global world producers. Furthermore, the nexus, under green growth and geographical constraints, would look in much greater depth at a gradual move to a low carbon economy, renewable energy (Meisen and Krumpel, 2009) and energy options that have a low water footprint (in terms of consumption). Costa Rica is spearheading this approach after deciding to stop the exploration and exploitation of oil and start the development of an energy matrix with 92% of the production based on renewable resources.

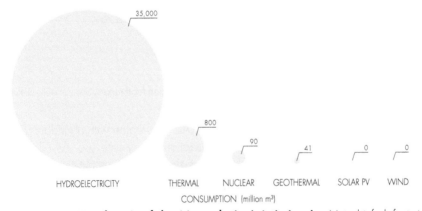

Figure 15.5 Water footprint of electricity production in Latin America. Note: biofuels footprint is not considered here as it is part of the primary energy mix. *Source: own elaboration.*

Box 15.2a Brazil: an example of energy–food nexus or trade-offs?

Brazil is the best example of the need to integrate water, energy and food trade-offs caused by the country's elevated production of biofuels. It has the greatest quantity of accessible blue and green water resources in the world and has enough technology to compensate for its lack of arable land. Moreover, it is the leading producer of sugar, second largest producer of soybean and the third largest producer of maize (Allan, 2011). Therefore, Brazil is likely to become a main exporter of virtual water embedded in food commodities globally, as well as in the raw materials of first-generation biofuels.

However, especially in the last decade, there have been side effects to this policy. Since the oil price rise in 1975, Brazil has opted for the development of nationalized biofuels production as a means to secure energy independence and give a boost to the country's economy. This process was conceived at the outset, considering land use, energy and food issues together and culminated in 2007 with the launch of the 'economic-environmental zoning' plan for the state of Minas Gerais (Coehlo et al., 2012). It consisted in the elaboration of studies about the social, economic and physical conditions (type of soil, climate, water availability, ecological values, etc.) of geographical regions in order to determine the most suitable areas to grow sugar cane with maximum yields and minimum impacts and then limiting the activity to those areas. First-generation biofuels are options for Brazil at least in the medium-term, due to its considerable availability of land and water resources. (Allan, 2011). How much this shift from food commodities exporter to biofuels exporter will impact on global food security, especially in those countries which depend on Brazil's food imports for national supply, is yet to be seen.

Box 15.2b The water–electricity (energy) nexus: what is the water footprint of electricity production in LAC?

The main sources of electricity generation in LAC are hydropower and thermoelectric power, together with biofuel production for transportation, heat and cooling. The key issue for the water–energy nexus is to determine whether increasing energy use affects water use or water consumption. For example, cooling from thermoelectric energy refers mainly to use while bioethanol refers more to consumptive use. In most of LAC, hydroelectric production plays a major role in the electric mix (see Figures 15.6 and 15.7), reaching some 100% in Paraguay, 83% in Brazil, 77.8% in Venezuela or 71.7% in Colombia (IEA, 2013). Those countries are therefore especially vulnerable to rainfall variability, such as the El Niño and La Niña phenomena and to climate change predictions reported by the IPCC's Climate and Water report (Bates et al., 2008). This variability should therefore also be taken into account for future management of the electricity sector. For the whole of LAC the total water footprint or consumptive use, estimated on the basis of IEA (2013) for the different energy technologies, is around 35,000Mm³ per year, from which almost 97% of consumptive use comes from hydroelectricity. Meanwhile, thermoelectricity and nuclear energy, the other main contributors to the electricity mix in the Andean and Amazonian regions, account for only 0–3% of the total water consumption from electricity. Coincidentally, water use for the whole of LAC accounts for 35,800 million m³, almost the same as water consumption. However, there are some aspects that must be taken into consideration. First, water use for both thermal and nuclear energy vary considerably depending on the type of cooling system used – i.e. the average value of water use can range from 68,000 million m³/ yr with once-through cooling down to 1,160 million m³/yr for closed loop systems. As cooling processes are the main water requirements for nuclear and thermal energy, clear data in this respect would be crucial for accurate water use estimations, especially within the Mesoamerican region (Mexico, 82.9%; Nicaragua, 79.6% or Guatemala, 76.7%). Along with thermal power, some other sources of renewable energy are emerging in the Mesoamerican region, such as geothermal in El Salvador (26.3%), Costa Rica (12.8%) or Nicaragua (8.6%), which for LAC in general only represents some 3% of total generation. Wind and solar photovoltaic, which have a low water footprint, are barely developed, despite their potential to decouple the water–energy nexus.

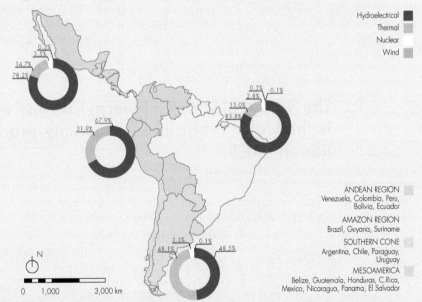

Figure 15.6 Electricity generation by source and per sub-region (Southern Cone, Mesoamerican, Amazon and Andean) in Latin America. *Source: own elaboration based on electricity data from 2009 in IEA (2013).*

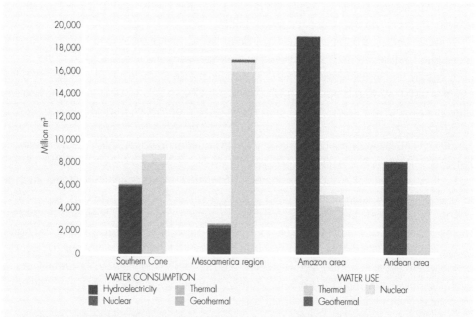

Figure 15.7 Water consumption and water use for electric generation per sub-region (Southern Cone, Mesoamerican, Amazon and Andean) in Latin America. *Source: own elaboration based on electricity data from 2009 in IEA (2013).*

15.3.2 Green growth and green economy in LAC

IWRM includes guaranteeing environmental sustainability as one of its three targets, together with efficiency and equity. At the Rio +20 meeting in June 2012, one of the main issues centred on water and green growth. In this context, the idea is to create a virtuous circle of economic incentives, able to generate the funds necessary for good water management. For example, where water is scarce – like in large parts of Mexico or Chile – incentives could focus on the rational use by agriculture as the dominant sector, via economic tools that support innovation in the use of water and force the internalization of external costs – i.e. valuation of water under realistic water prices. Environmental policy in countries such as Brazil is fairly advanced but its implementation is very slow while degradation continues in terms of deforestation (see Chapter 3) or water pollution increases. Meanwhile Costa Rica has adopted a green growth state policy, resulting in 26% of its territory being designated as areas for nature conservation and the implementation of a ban on open cast mining for heavy metals. In order to provide (financial) sustainability to these political measures, a series of economic instruments have been generated, such as a tax on fuel which is paid to environmental services producers in exchange for carbon. Meanwhile, 25% of the water tax (see Chapter 14 for more detail) is dedicated to the protection of public protected areas, and 25% for a payment of water environmental services on private lands.

In recent years, economic growth has been linked in many ways to high commodity prices (see Chapters 4 and 5), achieved at the expense of the intensification in the use of land, energy and water resources, leading to an increase in the levels of pollution and the loss of ecosystems and biodiversity (UNEP, 2009; UNEP, 2011; UN-Water, 2012a). A different development model based on a green growth approach ought to rely on a more efficient use of resources that decouples GDP growth from environmental degradation (UNEP, 2011). In LAC there has been an effort to transition towards IWRM as a framework that could help overcome this challenge (UN-Water, 2006; UN-Water, 2008; UNEP, 2012a). More generally, and as explained in Khan (2010), as countries shift to a greener set of economic arrangements, the costs of more traditional hard engineering approaches to water management become less profitable. In contrast, the cost of operating ecosystem payment schemes are much less likely to increase, providing that property and use rights and governance arrangements can ensure water-supply utilities whilst maintaining access to ecosystem services (Khan 2010; UNEP, 2011; UN-Water, 2012a). Clearly, some level of relative decoupling levels is already happening, meaning less environmental impact per unit of production (UNEP, 2011).

However, there are still challenges to achieving a 'greener' IWRM in the region (Scott and de Gouvello, 2013) (see Figure 15.8 and Table 15.4.). There is no blueprint: for countries with similar Human Development Indices (HDI), some have higher footprints than others. For instance, the three footprints of Brazil are higher than those of Peru while having the same HDI. Some countries have comparatively higher ecological footprint than others in relation to their HDI, like Costa Rica, Mexico, Argentina or Chile. On the other hand, other countries have a higher water footprint like Colombia and Peru, and Brazil has the highest carbon footprint in relation to its HDI and of all the other countries. As discussed in Chapter 3 this could be explained by changes in land use. Agriculture tends to represent 2/3 of the total water footprint (e.g. see Chapter 7), so it is key for decoupling human footprints (carbon, water and ecological), HDI and IWRM. Galli et al., (2012) propose a combined used of the three footprints in what is called the 'footprint' family, arguing that it shows a more rounded vision on all three aspects. Footprint HDI monitoring could provide a preliminary diagnosis or early indicator of the achievement of the three key elements – economics, social equity and sustainability – which can help flag up areas where further analysis is needed. In many cases, countries with a high HDI have a high ecological footprint, yet this is not the same for the carbon or water footprints.

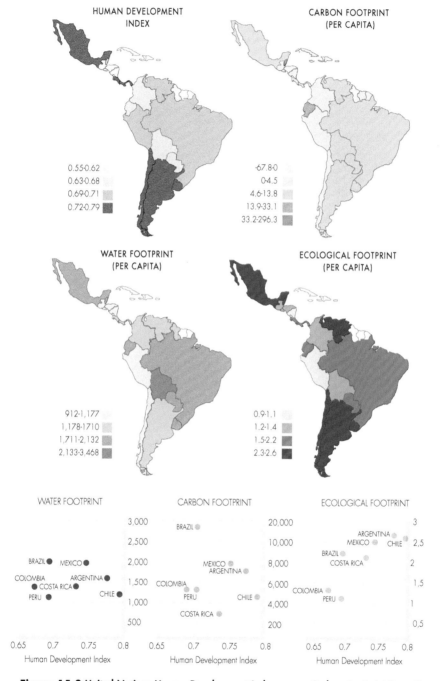

Figure 15.8 United Nations Human Development Index versus Carbon Footprint (tons C per capita per year), Water Footprint (cubic metres per capita per year) and Ecological Footprint (global hectares per capita per year). *Source: UNDP (2005), Mekonnen and Hoekstra (2011) and Ecological Footprint (2004).*

Table 15.4 United Nations Human Development Index versus Carbon Footprint (CF), Water Footprint (WF) and Ecological Footprint (EF)

COUNTRY	HUMAN DEVELOPMENT INDEX	CARBON FOOTPRINT (tC/cap/yr)	WATER FOOTPRINT (m³/cap/yr)	TOTAL ECOLOGICAL FOOTPRINT (global ha/person)
	2005	2000–2010	1996–2005	2004
ARGENTINA	0.77	6,438	1,607	2.6
BOLIVIA	0.65	5,612	3,468	1.2
BRAZIL	0.70	10,628	2,027	2.2
CHILE	0.79	3,852	1,155	2.6
COLOMBIA	0.68	4,595	1,375	1.3
COSTA RICA	0.73	2,175	1,490	2.1
ECUADOR	0.68	33,151	2,007	1.8
EL SALVADOR	0.66	1,884	1,032	1.2
GUATEMALA	0.55	-1,704	983	1.2
HONDURAS	0.58	1,990	1,177	1.4
MEXICO	0.75	7,135	1,978	2.5
NICARAGUA	0.57	-0.814	912	1.1
PANAMA	0.75	3,249	1,364	1.8
PARAGUAY	0.64	13,864	1,954	2.2
PERU	0.70	4,588	1,088	0.9
SURINAME	0.67	10,479	1,347	-
URUGUAY	0.74	7,884	2,133	2.6
VENEZUELA	0.69	7,214	1,710	2.4

Source: UNDP (2005), Mekonnen and Hoekstra (2011) and Ecological Footprint (2004)

15.4 The 'M' in IWRM

15.4.1 Integration and institutional coordination: allocation of tasks and responsibilities

This final section will look at integration in organizational terms. It draws on a recent study published by the OECD (2011) on multi-level water governance and a brief review of the main tenets of the IWRM paradigm. With a population of 596 million and growing faster than the world average, LAC countries are experiencing increasing pressure on their natural resources due to population growth, intensification of land use, increasing urbanization, climate change and natural disasters. The OECD (2012) argues that achieving water security in the LAC region is not only a question of hydrology and financing, but also equally a matter of good governance. In that framework, institutions and their coordination are essential to designing and implementing efficient, fair and sustainable water policies in the region.

Analyses on water governance are not new to LAC. The first studies on the topic date back to the end of the twentieth century. They highlighted the lack of governance strategy in the LAC water sector and revealed why most LAC countries lag behind in sustainable water management. Such reasons included the lack of political leadership, inadequate legal frameworks, poor utilities management structures, insufficient stakeholder involvement and limited financial resources. In most LAC countries, decentralization of water policies has resulted in a dynamic and complex relationship between public actors across all levels of government. To varying degrees, LAC countries have allocated increasingly complex and resource-intensive functions to lower levels of government, often in a context of economic crisis and fiscal consolidation. Yet, despite these greater responsibilities, sub-national actors were not given the financial resources to carry out their duties properly and hence coordination failures between sub-national and national governments and sub-national budgetary constraints have led to policy obstruction in several countries of LAC.

In 2011–2012, using the Multi-level Governance Framework 'Mind the Gaps: Bridge the Gaps' (OECD, 2011), the OECD carried out a survey on water governance across thirteen LAC countries (Argentina, Brazil, Chile, Costa Rica, Cuba, Dominican Republic, El Salvador, Guatemala, Honduras, Mexico, Nicaragua, Panama and Peru) in order to identify key governance obstacles to effective water management, as well as good practices for managing vertical and horizontal coordination of water policy (see Box 15.3). These countries cover a wide spectrum of options in terms of institutional settings (federal, unitary), the organization of the water sector (centralized, decentralized), water availability (water-rich and water scarce countries) and economic development (least advanced, developing and emerging countries). The survey had a particular emphasis on multi-level governance in order to analyse how public actors articulate their concerns, decisions are taken and policy makers are held accountable. The OECD defines multi-level governance as the explicit or implicit sharing of policy-making authority, responsibility, development and implementation at different administrative and territorial levels, i.e. i) across different ministries and/or public agencies at central government levels (upper horizontally); ii) between different layers of government at local, regional and provincial/state, national and supranational levels (vertically); and iii) across different actors at sub-national level (lower horizontally).

Box 15.3 Gaps to achieving effective water governance based on OECD multi-level governance challenges

Key findings were published in the report 'Water Governance in Latin America and the Caribbean: A multi-level approach' (OECD, 2012) which shows that despite a variety of situations, LAC countries share common governance and institutional challenges:

1. Sectorial fragmentation of water-related tasks across ministries and between levels of government is considered a policy gap, an important challenge to integrated water policy in 92% of countries surveyed;
2. The lack of public participation and limited involvement of water users' associations in water policy generates an accountability gap in 90% of the countries surveyed;
3. The funding gap remains a significant challenge in ten of the thirteen countries surveyed, due to unstable and/or insufficient revenues of sub-national actors in order to build, operate and maintain infrastructure;
4. In two-thirds of LAC countries surveyed, the capacity gap is a major obstacle for effective implementation of water policy at central and sub-national levels, which refers not only to the technical knowledge and expertise, but also to the lack of staff and obsolete infrastructure;
5. The information gap remains a prominent obstacle to effective water policy implementation in two-thirds of the countries, in particular regarding inadequate information generation and sharing amongst actors, as well as scattered water and environmental data;
6. Half of the countries surveyed see the mismatch between the administrative and hydrological boundaries (administrative gap) as a significant challenge to effective water management, despite the existence of river basin organizations in some of them;
7. Several LAC countries struggle to strike a balance between the often conflicting financial, economic, social and environmental agendas for the collective enforcement of water policy (objective gap).

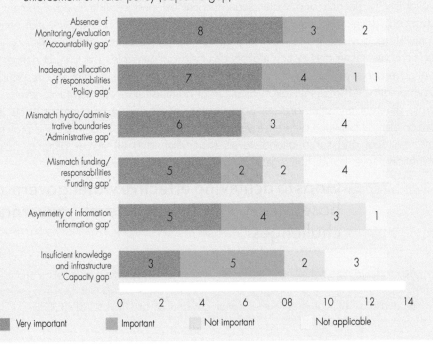

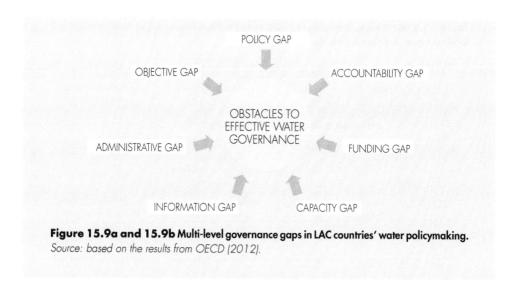

Figure 15.9a and 15.9b Multi-level governance gaps in LAC countries' water policymaking.
Source: based on the results from OECD (2012).

LAC countries have a set of policy instruments for addressing coordination and capacity challenges, but progress remains to be made in order to achieve IWRM. Meeting water governance challenges calls for more synergies to mutually reinforce actions across government, departments and agencies, as well as between researchers and decision-makers to forge science-policy dialogues (Regional Process of the Americas, 2012; Scott et al., 2012). An overview of LAC countries' experiences shows that there is a wide variety of mechanisms and instruments for integrating water policy. All LAC countries surveyed had adopted institutional mechanisms for upper horizontal coordination of water. These tools mainly consist of ministries (e.g. the Ministry of Environment in Brazil, the Ministry of Public Works in Argentina, etc.), inter-ministerial bodies or mechanisms, or specific coordinating bodies. Most countries have also engaged in efforts to coordinate water with other policy areas including regional development, agriculture and energy (see Table 15.5).

In recent years, river basin organizations have also been proposed in LAC countries as tools for effective governance, though their missions, constituencies and financing methods vary across LAC countries. While all LAC river basin organizations have functions related to planning, data collection, harmonization of water policies and monitoring, none have regulatory powers, contrary to OECD ones. The maturity of river basin organizations also varies across LAC countries especially in terms of managing competing water uses, which requires conflict resolution mechanisms in the political and legal arenas. In Brazil, the 1997 National Water Resource Strategy established river basin committees to promote multi-actor dialogues on water and arbitrate conflicts of use and implement basin management plans. In 2010, Peru started to conduct pilot exercises in six river basins. Two river basin councils have been implemented thus far and the National Water Authority (ANA) is carrying out programmes to stimulate the creation of ten additional ones, while tackling remaining challenges such as financial sustainability, capacity building, civil

Table 15.5 Ministries and institutions responsible for the management of water, energy and food resources in different Latin American countries

COUNTRY	WATER	AGRICULTURE	ENERGY
MEXICO	National Water Commission (CONAGUA) Department of Environment and Natural Resources.	Department of Agriculture, Livestock, Rural Development, Fishing and Feeding	Department of Energy Secretary of Energy
BRAZIL	National Water Agency	Ministry of Agrarian Development Ministry of Agriculture, Fishing and Supplying	Ministry of Mining and Energy
ARGENTINA	Department of Public Works Sub-department of water resources	Ministry of Agriculture, Lvestock and Fishing	Ministry of Federal Planning, Public Investment and Services Department of Energy
COSTA RICA	Ministry of Environment and Energy Water Direction	Ministry of Agriculture and Livestock	Ministry of Environment and Energy
PERU	Ministry of Agriculture. National Water Authority	Ministry of Agriculture	Ministry of Energy and Mining
CHILE	Ministry of Public Works Water Department	Ministry of Agriculture	Ministry of Energy

Source: own elaboration.

society representation and the long-term contribution of the river basin councils to national development.

LAC countries employ a wide range of mechanisms to manage the interface between actors at the sub-national level and to build capacity. Public participation is also used as a tool to increase transparency and citizen compliance in order to influence environmental protection. In Chile, when several citizens share the same groundwater drilling infrastructure, they can form associations (Asociación de Canalistas) to communally build, operate and maintain aqueducts as well as to fairly distribute water among members. A bi-national management committee was established in the Goascorán river basin between Honduras and El Salvador to engage stakeholders in the development of a basin management plan. Other tools for coordination across sub-national actors include inter-municipal collaboration, metropolitan or regional water districts, specific incentives from central and regional governments, joint financing between local actors, as well as ancestral rules.

By comparing the allocation of roles and responsibilities at the central and sub-national level in the LAC countries surveyed, the OECD has defined three models of water policy organization (Figure 15.10). These categories highlight the different coordination challenges raised by a given institutional organization, related to the frequent trade-off of decentralization; customization of water policy according to territorial specificities; and policy coherence. Within each category, the degree to which governance challenges have an impact on the performance of water policy may vary from one country to another.

In most cases, countries have developed a series of mechanisms to address the institutional challenges in their water sectors, but when other dimensions are added (e.g. capacity gaps, variety of tools in use, etc.) it would be helpful to link each model with policy objectives and desired outcomes.

Category 1	Category 2	Category 3
multi-level governance instruments need to provide an integrated and place-based approach at the territorial level	multi-level governance instruments need to integrate the involvement of different actors at central and sub-national level	multi-level governance instruments need to integrate multi-sectoral and territorial specificities in strategic planning and design at central level

Central government actors

Key challenges

| Coordination across ministries and between levels of government | Coordination across ministries, between levels of government and across actors | Coordination across sub-national actors and between levels of government |

Sub-national actors

| Examples:
Chile, Costa rica, El Salvador, Cuba, Dominican Republic | Examples:
Brazil, Peru | Examples:
Argentina, Mexico, Panama |

Figure 15.10 Preliminary categories of LAC countries. *Source: based on results from OECD (2011).*

While many technical, financial and institutional solutions to LAC water challenges exist and are relatively well known, the rate of uptake of these solutions by government has been uneven. No governance tool can offer a panacea or a one-size-fits-all response to water governance challenges in the LAC region, and local policies that take territorial specificities into account can help in many cases. Even if an optimal level of governance cannot be defined, peer dialogue and bench-learning across LAC countries facing similar challenges and with equivalent institutional organizations can help to bridge governance gaps (see Box 15.4).

Box 15.4 IWRM: information flow amongst actors and the influence of their decision-making in Costa Rica's in water policy

One of the characteristics of water management in Costa Rica is the presence of both the public sector and civil society organizations as dominant actors, e.g. the Ministry of Environment and Energy (MINAE), Regulatory Authority for Public Services (ARESEP), the Costa Rican Institute of Aqueducts and Sewage (ICAA), which supplies fresh water for 50% of the population and the presence of approximately 1,542 Associations for Administration of Rural Aqueducts (ASADAS), which are distributed throughout the country and provide drinking water to 26% of Costa Rican people, in areas where the

ICAA cannot provide that service. The diagram in Figure 15.11 displays the analysis for official functions in strategic actors. The characterization of dominance is given by the presence of Power, Interest and the Legitimization (Chevalier, 2006).

Figure 15.11 Venn Diagram of dominant, outcast and respected actors in Costa Rica's water management. *Source: LA-Costa Rica (2012).*

The decision-making in Costa Rican water management is strongly related to the official information flow amongst actors and thus the influence of these actors in the IWRM process. The result has been a convergence map (see Figure 15.12) with levels of power (high, medium and low). The upper red polygon contains academic institutions. The upper right green polygon contains civil society organizations such as NGOs supervising and executing management plans, i.e. actors with medium power, no actual vote in the decision-making process, but their opinion is taken in account. The purple polygon contains actors that regulate the availability of water for agriculture; and the brown polygon contains a critical mass of decision-making actors at the three levels of power: operators of domestic usage, hydroelectric and other productive activities.

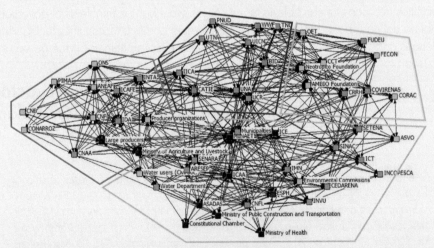

Figure 15.12 Social networks of actors in Costa Rica: connections, level of centrality and ease of access. *Source: Costa Rica FB National Report.*

15.4.2 Information technology for integrated management

Improved and more integrated water management should rely on the collection, provision and dissemination of more reliable and accurate data, to be transformed into better information, which in turn will yield better and more comprehensive water-related decisions. The key constraints and barriers to this approach are: first, the unavailability of systematic and consistent raw data compiled on adequate temporal and spatial scales; second, the lack of transparency of public bodies and private companies for sharing and allowing the open use of water data and finally, a lack of standardized methods for the audit and integration of water data; into more general accounting and decision systems. As a result, this absence of 'transparent' and assessed water information, in most countries, impedes regular reporting and evaluation of water resources and water-use trends (UN-Water, 2012b). The lack of water data and accounting and the asymmetry for different stakeholders remain pivotal issues to be tackled in IWRM. Regular demands for information come from institutions and regulators in the socio-economic, environmental or energy sectors looking for more effective and integrated data flows about water in order to monitor whether related policies are achieving the pursued goals in various dimensions. Further, there is also increasing pressure from private investors and businesses for clear and well-structured information in order to avoid or mitigate risks related to water services and water resources (see UN Global Compact CEO Water Mandate, 2007).

In LAC's least developed countries, the available funds for water activities are usually devoted to basic supply and the costs of data acquisition through conventional techniques are difficult to be met. In these cases, and also for richer countries, technological advancements could help to fill the gaps in water information via an improved cost-utility ratio. The growing availability of low-cost metering devices, the improvement in coverage and affordability of mobile handsets and the development of remote sensing (both in methodologies for generating specific data and an increased number of operating satellites) can help to monitor and record the status and dynamics of water and the environment. The potential applications of these technologies for IWRM include: the estimation of water use (especially for agriculture), the definition of water balances over large basins, the characterization of floods and other natural disasters, the analysis of water bodies' variability, the compiling of supporting information about soil moisture and groundwater levels and the monitoring of water quality (ESA, 2012; SELPER, 2012). Technology – from ground accurate sampling and conventional networks of remote sensing to ICT tools – is making the cost of water information more affordable and is becoming the key driver for a broader integration of water data and the transformation of a monopolistic, business-oriented system into a more transparent, open access and integrated vision of water information, thus benefiting IWRM.

15.5 Conclusion: IWRM as a means to a water security end?

IWRM runs the risk of being perceived as an elusive process – a nirvana (Molle, 2008) – unless the goals and targets are clearly established. Thus it could be useful to link IWRM as a process to the end goal of water security, defined as 'the sustainable availability of adequate quantities and qualities of water for resilient societies and ecosystems in the face of uncertain global change' (Scott et al., 2013). According to Allan (2003) the river basin became the central organizing unit in late modernity, even when there was evidence that global food trading processes were just as important as local hydrology in facing serious local water challenges. Yet IWRM can only be deployed if one aspect is recognized, i.e. that IWRM is seen primarily as a political process to forge and implement effective water sharing. To succeed, IWRM has to engage with what is politically feasible, thinking beyond the watershed and out of the water box, which in fact opens the realms of possibilities beyond the basin to address problems across many scales. This final concluding section will thus look at the six key policy and political ingredients for the IWRM process to succeed in the pursuit of water security.

First, one of the aspects relates to integrated planning and in particular to coordination with land use and urban planning. This was discussed in relation to footprints and HDI. For example, in the case of Brazil there is no forum for discussion of land use planning at the local level which generates serious problems with water quality, erosion and flooding. Here for example river basin authorities could provide a framework for management and planning. There is a similar case in Peru, where water councils formed on basin lines could become a permanent mechanism for coordination and dialogue between the different actors and stakeholders involved in the planning processes.

Second, from a more technical and functional perspective, a clear allocation of roles and responsibilities is very important. This must be accompanied by having the right means – financial and human – to implement policies and by fostering stable jobs, less exposed to political changes. In Peru, for example, the national and local water authorities at the moment have a lack of sufficient qualified personnel to deal with both technical and administrative issues. Brazil is similar: there is scope for additional training and institutional strengthening at all levels.

Third, in terms of economic and financial means, the case of Mexico shows there is scope for the introduction of incentives for the modulation of consumption patterns for all sectors (primary, secondary and tertiary). Furthermore, there is a need to think more deeply about the anthropo-hydrogeological cycle and the potential cost savings from internalizing ecosystem services such as storage provided by aquifers. Thus the logical sequence for IWRM could be based on strengthening the knowledge and capacity to fully record and monitor water uses, as well as to develop a holistic set of incentives targeted at the different uses.

Fourth, it is essential to play on one of the strengths of Latin America: its civil society, which at present might not be fulfilling its full potential and yet it is the key piece in the puzzle for strong political will. In a deepening of democratization processes, civil society is the cornerstone to strengthening the local population and giving a voice to local actors in shared management. Yet this also means looking at who are the main policy beneficiaries, as highlighted by levels of vulnerability to extreme events or political decisions when there are potential trade-offs e.g. in the case of food/energy. In Brazil, for example, as discussed in Chapter 14, a greater presence of local actors means a deeper questioning of inertias. Equally in Costa Rica the participation of different actors is low since there are no adequate or clear mechanisms that favour effective public participation. Oftentimes the public is informed but do not actually partake in decision making. Meanwhile in Peru the increased level of awareness about water scarcity – on the Peruvian Coast where most of the population lives – combined with clear signals of global warming, have contributed to strengthening conscience that freshwater is a scarce resource that has to be protected.

Fifth, a deeper level of institutionalization implies a modern water law, which includes key areas like the human right to water (see Chapter 11), economic instruments for a green economy and its full implementation (thus again political will). Political will could be reflected, for example, in a clear and explicitly stated water policy that identifies financial resources to be allocated (e.g. to water infrastructure) and presents clear policy and political goals at national level in order to incorporate other elements, beyond a purely technological paradigm, thereby acknowledging the resource base and its environmental functions as discussed in the section on green growth. Inevitably this will mean, on occasions, confronting vested interests, like for example in Mexico, where discussions with big users like livestock and industry need to occur in order to negotiate a reduction in their privileged incumbent position in terms of water consumption, towards more equitable use. In other cases, such as in Costa Rica would imply greater transparency, improved governability and further involvement of users in the decision on the balance of allocations, through the elimination of *Juntas Directivas* – made up by businesses to be replaced by a competition commission.

Finally, when IWRM is seen as a process it is fundamental to identify clear goals or targets as well as the sequencing or prioritization of reform (see Box 15.5). Along the lines of 'good enough governance' (Grindle, 2007; López-Gunn et al., 2012), it is about setting priorities with a clear commitment to follow through, with political priorities based on real problems with clear sequencing (Saleth and Dinar, 2004). For example, water quality and sanitation, in Brazil 21% of the population does not have access to basic sanitation (see Chapter 6). Meanwhile in Costa Rica only 4% of wastewater receives treatment. Yet the implementation of a legal decree on wastewater discharges could generate the resources needed to increase the level of treatment; an example of a virtuous circle mentioned above which relies on political will and the approval of a National Policy on Wastewater and Sanitation. Equally in Mexico a major step forward would be to expand the coverage for drinking water and sewerage. Therefore the anticipated SDGs (Sachs, 2012) in relation to water offer a golden opportunity for clear political goals and prioritization.

Political will, which comes from healthy public participation from the base of civil society and a broad civic culture, supported by outside pressure from multilateral organizations are two fundamental elements needed for IWRM to be fully implemented. It is about taking action in areas that have already had their problems diagnosed and which centre on three axes: issues of governability (institutionality, coordination, laws), infrastructure (both hard and soft), and sustainable and equitable use.

For IWRM to succeed in achieving the multiple goals of water security there must be a political will to take strong decisions that may upset the status quo and 'break away' from the traditional instated ways, facing obstacles from sectors and interests which are currently benefiting at the expense of society at large. The way forward is clear: water security through IWRM with a particular focus on social equity and environmental quality – the two pillars required for a resilient, robust future.

Box 15.5 Reflections on IWRM and water security

'IWRM in Costa Rica is understood as: comprehensiveness in resource management, economic value of water, equity in the distribution and sustainability in the use that does not compromise the future for Costa Ricans. IWRM would strengthen institutionality since it clarifies and defines a single institution as a front-runner thus defining leadership and policies. It also raises the different roles of other institutions (SENARA, ICAA, Ministry of Health, etc.) whilst additionally establishing legal, economic instruments (water charges) for resource management, monitoring, protection. Furthermore it also takes into account other areas such as capacity building, research, monitoring and the control of pollution. IWRM is a process by which ecosystems are administered, assigned, and protected and all sectors are integrated into coordinated management, from the local to the national level, from the business to the community level and from the public to the private sector, so as to ensure that every drop of water can be maximized and generate the greatest economic, social and environmental benefits. IWRM is a means to achieve water security. It is likely that there are other water management schemes that also target water security, but they will take more time, more resources, and more effort. Moving towards water security also means directing our steps towards food security, energy security, a reduction in poverty and ensuring growth with environmental sustainability, all of which are fundamental aspects of IWRM. IWRM and water security share the principles of efficient, sustainable and equitable water, thus fostering development, the eradication of poverty and the quality and quantity of the resource.' (Maureen Ballestero, Costa Rica)

'One of the major issues to be resolved is the quality of river water; the other is the need to generate resources for the management and the strengthening of local actors. It is also a strategic issue considering the key elements for water management: water as a human right, the importance to legislate on groundwater, economic instruments towards a green economy and so on. Water security is about meeting basic needs, ensuring food supply and protecting ecosystems. There is a great crossover between policies on water and sanitation, land use and urban planning and so in order to complete the planned cycle for water policy in terms of institutional and management aspects, such as quantity and quality, an essentially political solution is required alongside the political will to enforce it.' (Pedro Jacobi, Brazil)

'IWRM is not a specific action but a public administrative will for a better use of water resources, with or without considering other contexts. They are two different things: water security is a social concept with implications for the overall economy and the rights of citizens. IWRM is a set of rules and techniques for certain objectives, one of which may be water security, but water security, for what? With what priorities? To what degree? At what cost?' (Emilio Custodio, Spain)

'Water security is part of integrated water resources management. Water Security tries to establish a correct balance in the use of resources in terms of quality and quantity for the future, in a way that does not endanger sustainability. IWRM would also seek to maximize economic and social benefits to water users in harmony with the environment.' (Julio Kuroiwa, Peru)

'IWRM is a methodology and water security is a human need. Water security can be seen as an indicator for IWRM. Water security is a specific application that requires appropriate information.' (Maria Josefa Fioriti, Argentina)

'IWRM is a broader concept that, in a way, includes water security. In principle, IWRM must include issues related to water security. Water security traditionally has been treated without regard to the possibilities currently offered by virtual water trade, especially in the food sector. Both IWRM and water security should have many points in common. However, nowadays almost all water security plans only take into account the resources of the region in question, forgetting the great effect that virtual water import could have.' (Ramon Llamas, Spain)

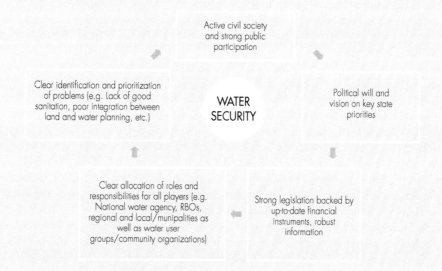

Figure 15.13 The WRM cycle to achieve water security. *Source: own elaboration.*

References

Ait-Kadi, M. (2013). IWRM: A concept fit for purpose whose time has come. In: Martínez-Santos, P., Aldaya, M.M. and Llamas, R. (eds). *Integrated Water Resources Management in the 21st Century: Revisiting the paradigm*, Abingdon, Taylor and Francis.

Allan, T. (2003). *IWRM/IWRAM: a new sanctioned discourse?*, Water Issues Study Group, School of Oriental and African Studies/King's College, London, University of London; SOAS/KCL Occasional Paper No. 50.

Allan, T. (2011). *Virtual Water: Tackling the threat to our Planet's most precious resource*, London, I.B. Tauris. pp. 384.

Barlett, C.A. & Ghoshal, S. (1990). Matrix management: not a structure, a frame of mind. *Harvard Business Review*, 68 (4):138–145.

Bates, B.C., Zundzewicz, Z.W., Wu, S. & Palutikof, (eds) (2008). *Climate Change and Water*. Technical Paper of the Intergovernmental Panel on Climate Change, IPCC Secretariat, Geneva, Switzerland. pp. 210.

Chevalier, J.M. (2006). *Sistemas de Análisis Social: Técnica Análisis social (colaboración/conflicto, legitimidad, interés, poder). Conceptos y herramientas para la investigación colaborativa y la acción social*. [Online] Available from: www.idl-bnc.idrc.ca/dspace/bitstream/10625/39895/1/128644.pdf [Accessed July, 2013].

Coelho, S., Agbenyega, O., Agostini, A., Erb, K., Haberl, H., Hoogwijk, M., & Moreira, J.R. (2012). Land and water: linkages to bioenergy. *Global Energy Assessment*: 1459–1525.

Cook, C. & Bakker, K. (2012). Water security: debating an emerging paradigm. *Global Environmental Change*, 22: 94–102.

Ecological Footprint (2004). *Living Planet Report 2004*. Available from: wwf.panda.org/about_our_earth/all_publications/living_planet_report/ [Accessed March, 2013].

ESA (2012). European Space Agency. *Looking for water in Africa: the TIGER initiative – 2009–2012 report*. [Online] Available from: earth.esa.int/web/guest/document-library/browse-document-library/-/article/tiger-initiative-looking-for-water-in-africa-2009-2012-report. [Accessed August, 2013]. Online report.

Foster, S., Hirata, R. & Howard, K. (2011). Groundwater use in developing cities – Policy issues arising from current trends. *Hydrogeology Journal*, 19 (2): 271–274.

Galbraith, J.R. (1971). *Designing Matrix Organizations That Actually Work: How IBM, Procter & Gamble and Others Design for Success*, New York, Jossey-Bass Business & Management.

Galli, A., Wiedmann, T., Ercin, E., Knoblauch, D., Ewing, B. & Giljum, S. (2012). Integrating ecological, carbon and water footprint into a 'Footprint Family' of indicators: definition and role in tracking human pressure on the planet. *Ecological Indicators*, 16 (2012): 100–112.

Garduño, H., Foster, S., Nanni, M., Kemper, K., Tuinhof, A. & Koundouri, P. (2006). *Groundwater Dimensions of National Water Resource and River Basin Planning: promoting and integrated strategy*. Washington, DC, World Bank. Briefing Note Series No10.

General Secretariat of the Andean Community (2009). *Atlas de las Dinámicas del Territorio Andino: Población y bienes expuestos a amenazas naturales*. Cali, Colombia, Corporación OSSO.

Grey, D. & Sadoff, C. (2007). Sink or Swim? Water security for growth and development. *Water Policy*, 9 (6): 545–571.

Grindle, M. (2007). Good Enough Governance Revisited. *Development Policy Review*, 25(5): 553–574.

Guodong, C. (2003). *Virtual Water – A Strategic Instrument to Achieve Water Security*, Bulletin of the Chinese Academy of Sciences. Issue 4, Denmark, Global Water Partnership/Technical Advisory Committee. [Online] Available from: www.en.cnki.com.cn/Article_en/CJFDTOTAL-KYYX200304005.htm [Accessed July, 2013].

GWP/TAC (2000). Global Water Partnership. *Integrated Water Resources Management*. TAC background papers No 4.

GWP (2004). Global Water Partnership. *Integrated Water Resources Management (IWRM) and Water Efficiency Plans by 2005. Why, What and How?*, GWP.

GWP (2010). Global Water Partnership. *What is water security?*. [Online] Available from: www.gwp.org/The-Challenge/What-is-water-security/ [Accessed August, 2013].

GWP (2012). Global Water Partnership. *IWRM pillars*. [Online] Available from: www.gwp.org/en/The-Challenge/What-is-IWRM/IWRM-pillars/. [Accessed August, 2013].

Hirata, R., Zobbi, J., Fernandes, A., Bertolo, R. (2006). Hidrogeología del Brasil: Una breve crónica de las potencialidades, problemática y perspectivas. *Boletín Geológico y Minero*, 217 (1): 25–36.

Hoekstra, A.Y., Chapagain, A.K., Aldaya, M.M. & Mekonnen, M.M. (2011). *The Water Footprint Assessment Manual: Setting the global standard*, London, Earthscan.

Hoff, H. (2011). *Understanding the Nexus.: The Water, Energy and Food Security Nexus*, Stockholm, Stockholm Environment Institute. Background Paper for the Bonn 2011 Conference.

IEA (2013). Internacional Energy Agency .*World Energy Outlook 2012*, París.

Khan, S. (2010). *The Costs and Benefits of Investing in Ecosystem Services for Water Supply and Flood Protection*. Background paper to underpin development of UNEP, 2011.

Kuhn, T. S. (1962). *The Structure of Scientific Revolutions* (1st edn), Chicago, University of Chicago Press. pp. 172.

LA–Costa Rica (2012). *Report of Costa Rica*. Contribution to the Water and Food Security in Latin America Project. Water Observatory Project. Madrid, Fundación Botín

López-Gunn, E, Llamas, M.R., Garrido, A. & Sanz, D. (2011). Groundwater management. In: Peter Rogers (ed.), *'Treatise in Water Science'*, vol. 1, Oxford, Academic Press. pp. 97–127.

López-Gunn, E., Willaarts, B., Rica, M., Corominas, J. & Llamas, R. (2012). The Spanish water 'pressure cooker': Threading the interplay between resource resilient water governance outcomes by strengthening the robustness of water governance processes. *International Journal of Water Governance*, 0: 1–28.

López-Gunn, E., Huelva, G., De Stefano, L. & Villarroya, F. (2013). Future Institutions? On the evolution in Spanish institutions: from policy takers to policy makers. In: In: Martínez-Santos P., Aldaya M.M. & Llamas MR (eds). *Integrated Water Resources Management in the 21st Century: Revisting the paradigm*. Boca Raton, FL, CRC-Press. Forthcoming.

Lundqvist, J., Fraiture, C.D., Molden, D. (2008). *Saving Water: From Field to Fork—Curbing Losses and Wastage in the Food Chain*, Water Institute, Stockholm, Stockholm International. SIWI Policy Brief.

Meisen, P. & Krumpel, S. (2009). *El potencial de América Latina con referencia a la energía renovable*, Global Energy Network Institute.

Mekonnen, M. & Hoekstra, A. (2011). *National Water Footprint Accounts: The Green, Blue and Grey Water Footprint of production and consumption*. Volume 2: Appendices, Value of water, Delft, The Netherlands, UNESCO-IHE/Institute for water education. Research Report Series No50.

Molden, D. (1997). *Accounting for Water Use and Productivity*, Colombo, International Irrigation Management Institute. SWIM Paper No 1.

Molden, D. & Sakthivadivel, R. (1999). Water accounting to assess uses and productivity of water. *Water Resources Development*, 155 (1 & 2): 55–71.

Molden, D., Murray-Rust, D., Sakthivadevel, R. & Makin, I. (2007). A water productivity framework for understanding and action. In: Molden, D. (ed.), *Water for Food, Water for Life: A Comprehensive Assessment of Water Management in Agriculture*. London, Earthscan. pp. 279–310.

Molle, F. (2008). Nirvana concepts, narratives and policy models: Insight from the water sector. *Water Alternatives*, 1(1): 131–156.

OECD (2011). Organization for Economic Co-operation and Development. *Water Governance in OECD Countries A Multi-level Approach*, OECD Studies on Water, Paris, OECD Publishing.

OECD (2012). Organization for Economic Co-operation and Development. *Water Governance in Latin America and the Caribbean: A Multi-level Approach*, OECD Studies on Water, Paris, OECD Publishing. [Online] Available from: dx.doi.org/10.1787/9789264174542-en. [Accessed August, 2013].

Perry, C. (2012). Accounting for water: stocks, flows, and values. In: UNU-IHDP and UNEP, *Inclusive Wealth Report 2012. Measuring progress toward sustainability*, Cambridge, Cambridge University Press.

Pochat, V. (2008). *Principios de gestión integrada de los recursos hídricos: Bases para el desarrollo de planes nacionales*, Global Water Parstnership.

Regional Process of the America (2012). *Americas' water agenda: targets, solutions and the paths to improving water resources management*. February 2012. Available at: www.gwp.org/Global/GWP-CAm_Files/Americas'%20Water%20Agenda.pdf

Sachs, J.D. (2012). From Millennium Development Goals to Sustainable Development Goals. *Lancet* 2012, 379: 2206–11.

Saleth, M. & Dinar, D. (2004). *The Institutional Economics of Water: A Cross-country Analysis of Institutions and Performance*, World Bank Publications. pp. 398.

Scott, C.A., Varady, R.G., Meza, F., Montaña, E., de Raga, G.B., Luckman, B. & Martius, C. (2012). Science-policy dialogues for water security: addressing vulnerability and adaptation to global change in the arid Americas. *Environment* 54 (3): 30–42. [Online] Available from: doi: 10.1080/00139157.2012.673454. [Accessed August, 2013].

Scott, C.A. (2013). Electricity for groundwater use: constraints and opportunities for adaptive response to climate change. *Environmental Research Letters*, 8 (2013). [Online] Available from: doi: 10.1088/1748-9326/8/3/035005. [Accessed August, 2013].

Scott, C.A., Meza, F., Varady, R.G., Tiessen, H., McEvoy, J., Garfin, G.M., Wilder, M., Farfán, L.M., Pineda Pablos, N. & Montaña, E. (2013). Water security and adaptive management in the arid Americas. *Annals of the Association of American Geographers*, 103(2): 280–289. [Online] Available from: doi: 10.1080/00045608.2013.754660. [Accessed August, 2013].

Scott, C.A. & de Gouvello, B. (2013). Public and private governance of water: outlook and lessons learned. In: C.A. Scott, B. de Gouvello (eds). *The Future of Public Water Governance: Has Water Privatization Peaked?*, London, Routledge.

SELPER (2012). Sociedad Latinoaméricana de Percepción Remota. Proceedings from the XV Symposium, *Earth Observation for a green co-developed world*, Cayenne.

Steffen, W., Grinevald, J., Crutzen, P. & Mcneill, J. (2011). The Anthropocene: conceptual and historical perspectives *Phil. Trans. R. Soc.* A, 369: 842–867. [Online] Available from: doi:10.1098/rsta.2010.0327. [Accessed August, 2013].

Tuinhof, A., Dumars, C., Foster, S., Kemper, K., Garduño, H. & Nanni, M. (2006). *Groundwater Resource Management: an introduction to its scope and practice.* Washington, DC, World Bank. Briefing Note Series 1.

UNDP (2005). United Nations Development Programme. *Human Development Report 2005. International Cooperation at crossroads. Aid, trade and security in an unequal world.* Available from: http://hdr.undp.org/sites/default/files/reports/266/hdr05_complete.pdf [Accessed March, 2013].

UNEP (2009). United Nations Environment Programme. *Water Security and Ecosystem Services: the critical connection.* [Online] Available from: www.unep.org/themes/freshwater/pdf/. [Accessed July, 2013].

UNEP (2011). United Nations Environment Programme. *Towards a Green Economy: Pathways to Sustainable Development and Poverty Eradication.* [Online] Available from: www.unep.org/greeneconomy. [Accessed July, 2013].

UN-Water (2006). *Implementing Integrated Water Resources Management; The inclusion of IWRM in National Plans.* Report from the 4th World Water Forum Theme, UN.

UN-Water (2008). *Status Report on Integrated Water Resources Management and Water Efficiency Plans.* Prepared for CSD16.

UN-Water (2012a). *Water and a Green Economy in Latin America and the Caribbean* (LAC), UNECLAC, Natural Resources and Infrastructure Division, UN-Water Decade Programme on Advocacy and Communication (UNW-DPAC), Santiago, Chile, June 2012.

UN-Water (2012b). *Managing Water under Uncertainty and Risk.* [Online] Available from: www.unesco.org/new/en/natural-sciences/environment/water/wwap/wwdr/wwdr4-2012/ [Accessed August, 2013]. The United Nations World Water Development Report No4.

UN-Water (2013). *Water Security & the Global Water Agenda.* A UN-Water Analytical Brief, United Nations University.

UN Global Compact CEO Water Mandate (2007). [Online] Available from: www.ceowatermandate.org [Accessed August, 2013].

WEC (2010). World Energy Council. *Water for Energy,* London.

Yang, H., Zhou, Y. & Liu, J. (2013). Blue water transfer versus virtual water transfer in China – with a focus on the South-North Water Transfer Project. In: Martínez-Santos, P., Aldaya, M.M. and Llamas, R. (eds). *Integrated Water Resources Management in the 21st Century: Revisiting the paradigm,* Abingdon, Taylor and Francis.

Index

Note: Page numbers in **bold** type refer to **figures**
Page numbers in *italic* type refer to *tables*
Page numbers followed by 'n' refer to notes

accountability 231, 291, 320, 332-7, 380
activism: networks 324; social 21, 318, 321-5
activist institutions 288
activity-specific participation 327
acute myocardial infarction (AMI) 105
adaptive management 385-417
ADERASA 227
advocacy 329; networks 20, 318
affordable tariffs 106, 206, 298, 305, 351
afforestation 60
agricultural commodities 179; markets 120, 178
agricultural economic productivity 203, **204**, **205**
agricultural production 57, 126, 129, 264;
    blue water footprint 182-6, **183**, **184**, **185**;
    expansion 11, 151; green water footprint 182-
    6, **183**, **184**, **185**; improvements 57; increase
    134; land use 56; rain-fed 14
agricultural products 178; consumption 186-9,
    187; demand 14; exports 14
agricultural sustainability: threats 36
agricultural trade openness 135
agriculture 4, 12, 42, 50, 65, 120-1; annual
    gross revenues 57; employment 57, 179;
    expansion 16, 58, 61, 64, 134; export growth
    129; exports 14; and food security 179; goods
    126; green water 262; growth 15, 56, **136**;
    industrialization 60, 64; intensification 74;
    intensification trap 16; intensive 84; irrigated
    196, 200, 348-9; land 57, **75**; modernization
    74; organic 75-6; prices 135; productivity 74;
    rain-fed 8, 15, 35, 120, 178, 179, 193, 200,
    274, 279; technical efficiency 270; trade 121,
    125, 135; trends 200-7; and water 177-212;
    water consumption 5; water quality 42; water
    withdrawal 181-2; world trade 6; yields 131
agro-chemicals 84, 206; Costa Rica 191
agro-diversity 71
agro-exports 322
agro-industry: growth 126
Aguirre Diaz, R. 221
Aldaya, M.M. *177-212*, 261-83
Allan, J.A. 5, 145, 146
Allan, T. 410
allocation efficiency 264, 265
aluminium 43
Amazon basin 34, 321; deforestation 197
Amazon rainforest 50
Amazon River 29
America's Water Agenda (Inter-American

Development Bank) 215-16
Amsterdam 233
Andean Community 101
Andes 50, 391, *392*, **392**
animal feed 126, 131
anthropo-hydrogeodrological cycle 387, 388-91,
    389, 410
Antilles 208
Aqua Rating 225
aquifers 388; depletion 49, 297; sustainability
    359
arable land 60, 179, 200, *201*
arbitration courts 325
Argentina: agricultural productivity 203, **204**, **206**;
    Aguas Argentinas Consortium 351; arable land
    201; automobile exports 14; bio-ethanol 248,
    248; biodiesel 249; blue water exports **194**,
    197, **199**; blue water imports **196**, **199**; blue
    water scarcity 153, **153**; Buenos Aires 220-1,
    222, 233; Buenos Aires water concession 351-
    2; cadastre survey **378**; deforestation 59, 64,
    197; Director Plan Framework 221; education
    enrolment **109**; exports 101; extreme rainfall 48;
    Food Code 220; food security indicators **161**;
    GDP growth **91**, 92, **98**; green water exports
    193; green water footprint 36, **183**, **185**; green
    water imports **195**; Greenhouse Gas Emissions
    (GHG) **70**; grey water footprint **190**, 191,
    274; groundwater mining 39; groundwater use
    38; harvested area **206**; Human Development
    Index (HDI) *402*; hunger 162; immigrants 89;
    inequality **99**; irrigation **156**, 179; Mar del Plata
    217-18; Millennium Development Goals (MDGs)
    171; mining 251-2, *252*, **323**; National
    Communication reports (NCs) 47n; navigation
    40; nuclear plants 255; obesity 105, **164**, 165;
    Payment for Ecosystem Services (PES) *369*, *376*;
    Planning Agency (APla) 221; population 86,
    89; poverty reduction 96; Regulatory Framework
    220; river basin committees 46; trade 102, **130**,
    131, **136**, **137**; urban water 220, 222; virtual
    water flows *394*; water footprint **187**, 191, **274**,
    275, 394; water legislation 295; Water Poverty
    Index (WPI) 97, **98**; water responsibilities 47,
    290, 406; Water and Sanitation Regulatory
    Entity (ERAS) 221; water security indicators *155*;
    water utilities 351; water withdrawals **38**, **93**,
    182; yield growth *202*
arsenic 42, 43, 222, 230, 244

asparagus **184**, **194**
avocados **184**, 186, **194**

Bahamas: arable land *201*; child mortality **226**;
crop yields *202*; food security progress *161*;
grey water footprint 191; imports and exports
127; natural hazards 112; sanitation **226**; virtual
water exports **199**; water access **226**; water
footprint 186, **187**
Bali Action Plan (2007) 68
Ballestero, M. 412
bananas **68**, 165, 182, **185**, **193**, **194**, **195**,
196, **205**, **206**, 271
banks: multilateral development 307-8; private
307-8
Barbados: arable land *201*; crop yields *202*; food
security progress *161*; imports and exports **127**;
pollution 42, 189; virtual water exports **199**;
water footprint **187**
barley 76
basins *see* river basins
beans 71
Belize: agricultural yields 154; animal protein 159;
floods 157; grey water footprint 191; obesity
164; pollution 191; rainfall **31**; renewable water
resources 30; storms 48; water hazards 18
benchmarking 336, *336*
bilateral agreements 369, 371
bio-alcohol 240; production costs 248, *248*
bio-energy 242
biochemical oxygen demand (BOD) 45
biodiesel 240, 247, 249
biodiversity 7, 15, 369; conservation 56, 57,
71, 240, 372; data 69; loss 84; markets 373;
payments 373; services 366
bioethanol 247, 248
biofuels 64, 84, 127, 208, 240, 247, 255,
395; demand 57, 64; EU 132-3; factories 249;
production 241; production plants 249; USA
132-3; water footprint 396; water and land
needs 247-9
biomass 247
biophysical context 11-12
biophysical sphere 8, **8**, 9
Biosphere Reserves (BR) 69, 73
Birmingham (UK) 233
block tariffs 351
blue virtual water 386, 393
blue water 5, 35-6, 70, 178, 241, 265;
availability 152; Brazil 153; Chile 153;
depletion rates 197; economic productivity 203;
efficiency 154; exports 197; Mexico 153; Peru
153; productivity 268; scarcity 153, **153**
blue water footprint 36, 180, 203; agricultural
production 182-6, **183**, **184**, **185**; domestic
water supply 273; and economic development
263, **263**
Bocanegra, E. 217-18

Bolivia: animal protein 159; carbon stocks 67;
Cochabamba conflict 324; crop yields 188;
deforestation 59, 64, 197; droughts 391;
eco-tourism 73; economic water productivity
203; emigration 89; exports 127; food security
indicators 144; grey water footprint 191;
groundwater mining 39; hunger 104; informal
employment 100; Millennium Development Goals
(MDGs) 16; mining 251, 252, 256; payments
for ecosystem services (PES) 381; population
growth 49; rainfall 48; Titicaca lake 29, 41;
water access 154; water footprint 186; water
quality 228
bottled water 230
Boyd, D.R. 374
Brazil: agricultural productivity 203, **204**, **206**;
arable land *201*, 397; bio-ethanol 248,
248; biodiesel 249; blue water exports **194**,
197; blue water footprint **185**, 186, **193**,
263, **273**; blue water imports **194**, **196**; blue
water scarcity 153, **153**; cadastre survey **378**;
decentralization 291; deforestation 58-9, 64;
education enrolment **109**; energy-food nexus
397; ethanol 396; exports 101; fees 347; food
security indicators *161*; Forestry Code 372-
3; GDP growth **91**; green water exports **193**;
green water footprint **185**; green water imports
195; Greenhouse Gas Emissions (GHG) **70**;
grey water footprint **190**, 274; groundwater
mining 39; Human Development Index (HDI)
402; hunger *162*; hydropower 40; immigrants
89; industry 331; inequality **99**; irrigation 271,
349; land footprint 193; land use planning
410; Manaus 352; Millennium Development
Goals (MDGs) *17*; mining **323**; Movement
People Affected by Dams 324; National
Communication reports (NCs) 47n; National
Management System of Water Resources 291;
National Water Agency (ANA) 349; National
Water Council 327; National Water Plan 292;
National Water Resource Strategy (1997) 405;
nutritional transition 165-6; obesity **164**; Paraíba
do Sul Basin 347-8; Payment for Ecosystem
Services (PES) *369*, *376*; poultry production
187-8; poverty reduction 96; Produtor Agua
22; reforestation 61-3; Rio de Janeiro 219-20;
river basin committees 405; sanitation 411; Sao
Paulo 62-3, **63**, 218-19, 389; State Water
Councils 327; swine production 187-8; tariffs
352; trade **130**, **131**, **136**, **137**; virtual water
flows *394*; Water for All 303; water footprint
187, 191, **274**, **275**, *394*; Water Law (1997)
291, 330, 347; Water Poverty Index (WPI)
97, **98**; water quality 411; water reforms 289;
Water Resources National Policy 291; water
security indicators *155*; yield growth *202*
Buenos Aires 222; AySA 233; participatory models
of governance 233; Sanitation+Work Plan 233;

urban water 220-1; Water+Work Plan 233; water concession 351-2

Build-Operate-Transfer projects 308, 311

bundle payments 373, *373*

Cabrera, E.: and Custodio, E. 213-17

cadastre survey **378**

Caffera, M. 354

calorie consumption 131, 151

Campuzano, C.: *et al* 27-53

cap and trade 375

capacity: building 412; gap 404, 407

capital: foreign 380; markets 308; natural 15-16, *56, 57,* 366, 367, 374, 382; social 356

carbon 369, 371; payments 372; sequestration *56, 57,* 67, 69; services 366; stocks 367

Carbon Disclosure Project's Water Initiative 331

carbon footprint: and Human Development Index (HDI) **401**, *402*

Caribbean: global context 4-5

Carter, T.R.: and Mäkinen, K. 110

cassava *72*, **184**, **185**, 203, **204**, **205**, **206**

castor oil 249

CAWMA 200, 206

Central American Water Tribunal (CAWT) 303

cereals 64, 203

charges 292, 307, 308, 345, 350; agriculture 344; Costa Rica 346; Mexico 346; urban consumers 349-52

chemical oxygen demand (COD) 45

chemical plants 249

chemical substances 14, 41, 42, 223, 246, 254

child mortality 4; and sanitation 226, **226**

children: obesity 105, 165; overweight 164-5, **164**; stunting 166, **167**, **168**; undernourishment 164, **164**

Chile 64; agricultural productivity 203, **204**, **206**; arable land *201*; blue water 153; blue water exports **194**; blue water footprints **183**, **184**, 186, **263**, **273**, *274*; blue water imports **196**; blue water scarcity 153, **153**; cadastre survey **378**; decentralization 291; desalination 34; discharge fee programmes 353-5; droughts 48, 147-8; Echaurren Norte 110; economic productivity of water 242, *243*; education enrolment **109**; fees 347; food security indicators *161*; GDP growth **91**, 92; green water exports **193**; green water imports **195**; Greenhouse Gas Emissions (GHG) **70**; grey water footprint **190**, 274; groundwater use 38; Human Development Index (HDI) *402*; hunger *162*; hydropower 40; immigrants 89; inequality **99**; irrigation 271; Limari Valley 358, 360; Millennium Development Goals (MDGs) *17*; mining 251-2, *252*, **323**; National Communication reports (NCs) 47n; National Water Code 289; obesity **164**, 165; Payment for Ecosystem Services (PES) *369*; pollution 42-3; population growth 359; poverty

reduction 96; privatization 350; Programa Nacional de Agua Potable Rural 303; rainfall variability 32-3; sanitation 35; Santiago 217; trade **130**, **131**, **136**, **137**; virtual water flows *394*; water availability 33; Water Code 289, 290-1, 344; water footprint **187**, 191, **274**, **275**, *394*; water management 327; water markets 13, 356-60; Water Poverty Index (WPI) 97, **98**; water reforms 293; water rights 357, 358, 359, *359*; water security indicators *155*; water trade mechanisms 344; water user associations (WUA) 357, 358; water withdrawals **93**; yield growth *202*

China 120, 126, 132, 197, 263; hydropower 275

chlorination 151, 214-15, 218, 227

cholera 226

cities: Latin America *215*; water security 213-37

civil society 20-1, 288, 291, 305, 410-11; organizations 318, 319, 320, 325, 327, 328, 329, 331, 333, 334, 407; role 10

Clean Development Mechanism (CDM) 372

climate *7*

climate change 28, 31, 49-50, 82, 83, 114, 396; Andean region 47; and migration 112; vulnerability to 110, *111*; and water resources 47-8, **49**

climate-change, impacts, adaptation and vulnerability (CCIAV) assessments 110

climatic variability 28

closed water cycle 246

coal 255

coca crops: eradication programmes 65

Cochabamba conflict (Water War) 324

coffee 191, 203, 209

Colombia 222; agricultural productivity 203, **204**, **205**; arable land *201*; biodiesel 249; blue water exports **194**, **197**; blue water footprint **183**, **263**, **273**; blue water imports **196**; blue water scarcity 153, **153**; cadastre survey **378**; discharge fee programmes 354; education enrolment **109**; energy 40; food security indicators *161*; GDP growth **91**; green water exports **193**; Greenhouse Gas Emissions (GHG) **70**; grey water footprint **190**; Human Development Index (HDI) *402*; hunger *162*; inequality **99**; irrigation **156**; Millennium Development Goals (MDGs) *17*; mining 251-2, **323**; National Communication reports (NCs) 47n; national environmental law 381; obesity **164**; Payment for Ecosystem Services (PES) *369*, *376*; population **89**; Porce River Basin 209; poverty reduction 97; reforestation 65; trade **130**, **131**; virtual water flows *394*; water footprint **187**, 191, **274**, **275**, *394*; water legislation 289-90; water security indicators *155*; water withdrawals **93**; watershed payments 381; yield growth *202*

command-and-control (CAC) instruments 353
commercial demand 220
commercial forestry 71
commodities: agricultural 120, 178, 179; markets 122; price indices **123**; prices 400; value 125
companies: state-owned 305
competition 131, 294
Comunidad Andina (CAN) 90
CONAGUA 290
concessions 304, 305, **306**, 322, 346, 349, 350, 305
conciliation 329
conflicts: water 321-5, *321*
conservation 56, 57, 71, 179, 240, 372, 389; policies 61; subsidies 266; technologies 265-6
Constitutions 301, **301**, 324, 333, 373-4, **375**
consumers 4, 50, 150, 307, 330, 331
consumption: agricultural products 186-9, **187**; changes 102; food 102-3, **103**, 165; patterns 12, 82, 83, **103**, 165; water **92**
consumptive uses 35
contamination 262
cooling water cycle 245
copper 20, 126, 242, 253
corruption 234, 304
Corzo-Juárez, C.: and Hansen, A.M. 45
Costa Rica: agricultural productivity **205**; agrochemicals 191; arable land *201*; Association Law 328; blue water footprint **183**, **263**, **273**; blue water scarcity 153, **153**; charges 346; discharge fee programmes 354; droughts 48; extreme rainfall 48; FONAFIFO 374; food security indicators *161*; Forestry Law 374; GDP growth **91**; Greenhouse Gas Emissions (GHG) **70**; groundwater 38, 391; Human Development Index (HDI) *402*; hunger *162*; hydropower 40; immigrants 89; inequality **99**; information flow 407-8; Integrated Water Resources Management (IWRM) 412; irrigation **156**; Millennium Development Goals (MDGs) *17*; mining **323**; National Communication reports (NCs) 47n; National Plan of Integrated Water Resources Management 294; Payment for Ecosystem Services (PES) 369, *376*; payments for ecosystem services (PES) 369, 373; poverty reduction 97; virtual water flows *394*; wastewater treatment 411; water footprint **187**, 191, *394*; water information system 391; Water Law 328; water policy 291-2, 294, 327, 407-8, **408**; Water Poverty Index (WPI) 97; water resources management 294; water security 158; water security indicators *155*; water withdrawals **93**; yield growth *202*
cotton 71
Court of the People 325
courts 324-5
coverage 19, 20, 34, 35, 209, 221, 300, 303, 352, 360

crop production 247, 274-5; water footprint **274**
crop water consumption 182, 272
crops 69, 71, 134, 189, 203, **206**, 207, 361, 389; for biofuels 240; diversification 181; green water 192-3; trends *72*; yields 74-6, 200, *202*, 203, **204**, **205**, 206, 207, 249, 268
Cuba: agricultural yields 154; arable land *201*; child mortality **226**; crop yields *202*; eco-cultural sites 73; fertility 85; floods 157; food security indicators 159, **167**; food security progress *161*; governance 290, 403; groundwater 182; hunger 104, *162*; irrigation efficiency **156**; Millennium Development Goals (MDGs) *17*; salinity 42, 189; sanitation **226**; sprinkler irrigation 182; virtual water imports **192**, **199**; water access **226**; Water Act **295**; water extractions 70; water footprint 36, **183**, **187**; water hazards 18; water ownership 296; water security progress *155*; water withdrawals **38**
Cuevas, A.: et al 164-5
cultivated area 60, 203, 206, **206**
cultural convergence 11
cultural services 73, 74
Custodio, E. 39-40, 413; and Cabrera, E. 213-37; and Garrido, A. 239-58
customary tribunals 21
cyanide 253, 256

Dagnino, M.: and Ward, F.A. 266
dairy farming: decline 62-3
dams 245, 395
Dávara, J. 220
de Schutter, O. 163
De Stefano, L. 285-315, 317-42; Garrido, A. and Willaarts, B. 3-24
debt crisis 306-7
decentralization 291, 325
decision-making processes 318, 319, 326, 329
deforestation 15-16, 56, 57, *59*, 61, 68, 74, 76, 178, 197, 372-3; Amazon basin 197; Brazil 58-9, 64; drivers 63-5, **66**; and energy 65; Guatemala 65; Honduras 65; and livestock 65; and mining 65; Nicaragua 65
demand: agricultural products 14; food 57, 64, 75, 134, 208; water 12, 18, 28
democracy 22
democratization 289, 291
demographic dividend 22
demographic transition 82, 83, 84, 104
dengue fever 110, 114
desalination 34, 358, 388, 389
developing countries 45, 57, 67, 68, 88, 110, 131, 151, 207, 325, 352
development 7; agencies 318, 320; economic 90, 154, 166, 169, 263, 288-9, 307; human 207; regional 7, 8; sustainable 62
development banks: multilateral 307-8
diarrhoea 226

diet *57, 71, 170*; patterns *102, 113*
direct investment 12
disasters: natural 148, 409
discharge fee programmes 353
diseases 110, 114, 226; chronic 104; infectious
    104, 105; malaria 110, 114, 151, 226
disinfection 226, 229
domestic consumption 8, 35, **149**, 230, 248
domestic storage tanks 228-9
domestic water supply 50, 289; blue water
    footprint 273; private 303-5; public 303-5;
    water footprint *273*, **273**
Dominica: water footprint 186
Dominican Republic 152; agricultural yields 154;
    economic development 90; grey water footprint
    191; pollution 42; stunting children 160; total
    actual renewable water resources (TARWR) 152;
    water footprint 186; water qaulity 189
Donoso, G. 33; *et al* 261-83
drinking water 16, 18, 19, 34-5, 214, 226, 232-
    3, 360; access 110, 114, 144, **156**, 166,
    167, 168, 169, 180; chlorine 151; Mexico
    411; prices 217, 218; standards 218; status
    231
drinks: soft 230
drought 6, 18-19, 31, 47, 50, 83, 110, 112,
    154, 157, 221, 358, 391, *392*, **392**; Chile
    48, 147-8
Dublin Statement on Water and Sustainable
    Development (1992) 345, 345n
dysentry 226

e-agriculture 108
eco-cultural sites *73*
eco-tourism 73-4
ecological footprint 400, **401**, *402*
economic centre of gravity 121, 121n
Economic Commission for Latin America and the
    Caribbean (ECLAC) 85
economic crisis 403
economic development 6-7, 16, 22, 62, 90, 154,
    166, 288-9, 307; and blue water footprint 263,
    263; regional *57*; sustainable 169; and water
    6-7
economic efficiency 264, 265, 268-9; agriculture
    277, **277**; industry 277, **277**; production 276-7
economic globalization 8, *57*, 64, 165
economic growth 84, 90, 101, 113, 262, 279,
    319, 400; and trade openness 82; and water
    use 263-4
economic instruments 13, 21, 265, 343-64, 411,
    412; water quality management 353-5
economic integration 4
economic liberalization 289
economic productivity: agricultural 203, **204**, **205**
economic water efficiency 148, *277*, **277**
economic water productivity **149**, 203, **205**, **206**,
    276, **206**; agriculture and industry 276, **276**;

blue 203; Chile 242, *243*
economic-environmental zoning 397
economy: green 386, 399-400, 411, 412;
    informal 83, 100-1, **100**, 114; world 121-2
ecosystem productivity *57*, *111*
ecosystem services 11, 13, 15, 287; changes
    73; economic incentives 365-84; and land use
    changes 67-74; laws 373-5, *376*, *377*; legal
    framework 365-84; payments 13; trends 55-80,
    69; watershed payments 370, **370**
ecosystems 28, 40, 41, 50, 73, 278, 412;
    aquatic 154; destruction 322
ecotourism 56
Ecuador: animal protein 159; biodiversity
    conservation 71; constitution 374, *374*;
    deforestation 59; economic water productivity
    203; emigration 89; exports 127; FONAG 22;
    food security indicators 144; forest degradation
    60; glacier melting 48; hunger 104; information
    334; Millennium Development Goals (MDGs)
    16; mining 65; Payment for Ecosystem services
    (PES) 369, 381; payments for ecosystem services
    (PES) 369, 381; Water Poverty Index (WPI) 97;
    watershed management 46
education 22, 108, **109**; primary *17*, 94, *95*,
    108, 410
efficiency 13, 50, 144, 320; definition 264;
    environmental impacts 278; formula 267; status
    and trends 261-83; technical 264, 265, 269-
    72, 280
El Niño 30-1, 48, 50, 110, 147, 154
El Salvador: animal protein 159; food security
    indicators 144, 162; geothermal electricity 246,
    398; tariffs 135; Water Poverty Index (WPI) 97;
    water quality 38
electricity **398**; geothermal generation 246-7,
    247; thermoelectricity 245-6, *246*, 255; water
    footprint 396, **396**, 397-9, **399**
emerging economies 6, 113, 126, 165
Empinotti, V.: *et al* 317-42
employment 94; agriculture 179; non-agricultural
    100; and trade 136
energy 12, 20, 240, 242, 395, 396; Colombia
    40; and deforestation 65; nuclear 240;
    production 255, 257, 275-6; renewable 247,
    247; supply 159; and water 243-9, *244*; for
    water 243-4
energy production: water for 244
energy-food nexus: Brazil 397
engineers: training 232
entrepreneurship **107**
environment: conservation 21; damage 318;
    degradation 20, 37, 61, 148, 154; irrigation
    modernization 278-9; laws 297, *374*; policy
    399
environmental knowledge: citizens 232
environmental migrants 112
environmental rights 374

environmental security (ES) 145, 395
environmental taxation 21, 345
EPANET programme 226-7
equity: social 353
ethanol 132, 247; Brazil 396
European Single Market 138
European Union (EU) 197; biofuels 132-3; Water
    Framework Directive (WFD) 25, 148, 234
eutrophication 41
evaporation 246; dams 245; hydroelectricity 245
evapotranspiration 41, 182, 267, 269
exports 15, 120, 126, **128**, 130; agricultural 14,
    129; Argentina 101; biofuels 127; blue water
    197; Brazil 101; manufacturing 125; Mexico
    14, 101, 125; primary goods 126; value **127**

factories 249-50
farming 322; water rights 360
Federal Water Control Act 215
fees: Brazil 347; Chile 347; Peru 347
FEMSA Foundation 372
Ferro, G.: and Lentini, E. 353
fertility 104, 113; total fertility rate (TFR) 85
fertilizers 16, 42, 134, 151, 181, 189, 191
financialization 122, 122n
financing 306-8
Fioriti, J. 220-1
fishing 40
floods 6, 18, 83, 110, 112, 147, 154, 157,
    391, 392, **392**, 409
fluoride 42, 222, 244
fluorine 230
fodder crops 191
food: access 144, 160; availability 144, 159-60;
    crisis (2007-9) 144; demand 57, 64, 75, 134,
    208; grain outputs 207; imports 6, 162; price
    crisis (2007-9) 160, 162; prices 151, 160-1;
    processing 241, 250; production 179, 200,
    240, 395; pyramid 165, **166**; quality 150; right
    to 163; sovereignty 6; stability **9**, 145, 151,
    160; supply 102, 151, 160; supply chains
    138; utilization 160; and water security 7-9
food consumption 165; Mexico 102-3, **103**;
    patterns 12, **103**
food insecurity 164; definition 149-50
food and nutrition indicators **152**
food and nutritional security (FNS) 144, 145, 146,
    163, 164; concepts and dimensions 149-52;
    indicators 159
food security 264; and agriculture 179;
governance 287; hunger 103-4; improvements 18,
360-1;
    indicators 12, 150, 151, 159-65, 166, 169;
    macro 162; micro 162; progress and links 143-
73,
    161; and water 5-6, 166-9
food systems 120; management 6
food-water 5, 8, 10, 14, 146

food-water-energy nexus 121, 395-6, **395**
food-water-environmental security 16
footprint family 400
footprints: carbon **401**, 402; human 395, see also
    blue water footprint; green water footprint; grey
    water footprint; water footprint
foreign capital 380
foreign direct investment (FDI) 134; inward 120,
    124, **124**
Forest Carbon Partnership Facility (FCPF) 68
Forest Investment Program (FIP) 68
forest transitions 60; Sao Paulo!! 62-3, **63**
forestry: commercial 71
forests 38, 40, 57; conservation 63, 67-8;
    degradation 60, 76; policy 62, 63; products
    71-3; recovery 60, 62, 63; reduction 58;
    secondary 60
formal participation: and negotiation 326-30, 326
fracking 254
free trade agreements 9, 129-30
freshwater 262, 279; abstractions 70; biodiversity
    155; use 69, 70; withdrawals **93**
fruits 196
fuel production 240, 241; water needs 250, see
    also biofuels
funding **309**; international financing 308-11;
    national financing 307-8; water 287

G20 nations 5, 160-1
Galli, A. et al 400
Garrido, A.: and Custodio, E. 239-58; et al 119-
    42, 343-64; Willaarts, B. and De Stefano, L.
    3-24
General Treaty of Central American Integration 101
geopolitical gravity shifts 4
geopolitical power 10
geothermal electricity generation 246-7, 247
geothermal energy 240
GINI index 94
glacier melts 48; Andes 50
Global Environment Facility (GEF) 372
Global Environmental Fund (GEF) 355
Global Food Security Index (GFSI) 151
Global Forest Resource Assessment (FRA) 69
global governance architecture: role 15
Global Hunger Index (GHI) 151
Global Water Partnership (GWP) 87, 387
globalization 7, 14-15, 22, 82, 113, 197, 288;
    economic 8, 57, 64, 165; and trade 10, 119-
    42
gold 20, 240, 253, 256, 297
Gómez Reyes, E. 278-9
good living (buen vivir) 374
governability 412
governance 13, 22, 138, 180, 218, 225, 265,
    386, 402-3; blue 287; definition 287; multi-
    level 403-5, **405**; roles 290; shared basins 50;
    stakeholders 13; weak 208, 231

governance structures: reforming 285-315
governance systems 8-9, **9**; global 9
government: role 290, *290*
grains 132, 247
grants: public 309
grapes **184**, **185**, 186, **190**, **193**, 203, **206**
grassroots movements 324
grazing 178, 186
green economy 50, 386, 399-400, 411, 412
green growth 399-400
green virtual water 386
green water 5, 15, 35, 153-4, 178, 265;
    agriculture 262; crops 192-3; dependency 15;
    exports 178; productivity 268
green water footprint *36*, 178, 180, 203;
    agricultural production 182-6, **183**, **184**, **185**
greenhouse gas emissions (GHG) 67, **70**
Grenada: grey water footprint 191; imports and
    exports **127**
grey water 265
grey water footprint 180, 181, 189-91, **190**,
    274; agricultural product consumption 191
gross domestic product (GDP) 84, 122, *122*, 276;
    evolution *92*; growth **91**, 92, 135, 270, 400;
    per capita **91**, 92, **93**; per capita growth **98**
groundwater 12, 29, 203, 216, 217, 219-20,
    222, 297, 388, 389, 393; abstraction 39,
    219, 298; Costa Rica 391; quality degradation
    49; rights 297; role 218-19; storage 29
groundwater mining 254; Argentina 39; Brazil 39;
    Latin America 39-40; Mexico 39; Peru 39
groundwater use 28, 37-8; Argentina 38; Chile
    38; Costa Rica 38; Mexico 38
Guanajuato State (Mexico) 297, **298**
Guatemala: arable land 60; crop yields 74;
    deforestation 65; droughts 110; fertility 85; food
    security indicators 144; geothermal energy 246;
    undernourishment 160, *160*
Guyana: agricultural water use 179; floods 157;
    food security indicators 144; forests 38; social
    exposure risk 18; stunting children 160, *169*

Haiti 152; animal protein 159; earthquake
    90; economic development 90; food security
    indicators 144; sanitation 154; stunting children
    160; total actual renewable water resources
    (TARWR) 152; Water Poverty Index (WPI) 97
Hansen, A.M.: and Corzo-Juárez, C. 45
hard-path solutions 147
health 82, 94, 144, 170; improvements 151;
    transition 104-5
hepatitis 226
high-value water users 358
Hirata, R. 218-19
Hoekstra, A.Y.: *et al* 43, 180; and Mekonnen,
    M.M. 189, 192, 245, 275, 276
Honduras: animal protein 159; biofuels 127; crop
    yields *74*; deforestation 59, 60, 65, *65*; food

security indicators 144; informal employment
    100; Millennium Development Goals (MDGs)
    16; storms 48, *48*; virtual water exports 192
human development 74, 207
Human Development Index (HDI) 5, 93, 400; and
    carbon footprint **401**, *402*
Human Development Report (1994) 145
human footprint 395
Human Right to Water and Sanitation (HRWS)
    298, 299, 301, **301**, 303, **304**
human rights 299; constitution 302
Human Rights Council (UN, HRC) 299
human well-being 6-7, 12, 74, 84, 145, 146,
    147; progress 93-5; trends *95*
hunger 4, 103-4, 150, *162*, 164, 165
Hunger Reduction Commitment Index (HRCI) 151
hurricanes 31, 47, 48
hydro-climatic variability 18-19, *147*, 157
hydrocarbons 84
hydroelectricity 241, 243, 244-5; evaporation
    245; water use 242
hydrological accounting 269
hydrological cycle 387
hydrological status 148
hydropower 7, 50, 208, 240, 275, 395; Brazil
    40; Chile 40; Costa Rica 40; reservoirs 275-6
hypercholesterolemia 105
hypertension 105

illness: causes 226, *see also* diseases
immigrants 88, 89
imports 120, 126; food *162*; value **127**
in-kind payments 370, 381
income: distribution 82, **99**; growth and poverty
    reduction 95-9
income share (held by highest 10 *95*
increasing block tariffs (IBTO) 351
India 120, 263
indigence line 96, **96**
industrial production: water footprint **273**, 274
industrialization 62; agriculture 60, 64
industry 7, 12, 19-20, 50, 240, 241, 242, 255;
    Brazil 331; and water 249-51; water intensive
    37, 250
inequality 18, 82, 87, 89, 101, 135, 145, 224,
    303, 319
informal economy 83, 100-1, **100**, 114
information: access 333-5; gap 404; water 409
Information and Communication Technology (ICT)
    83, 106, **107**, 108, 113-14; revolution 121
information technology: and integrated
    management 409
information transparency 332-7; laws 21, 318,
    333-5, **334**, 337
infrastructure 227, 306, 329, 412; ageing 224,
    225; private participation 124, **125**
insecurity: food 149-50, 164
institutions 287, 291, 292, 294; coordination

402-9; reforms 21, *287, 288*, 304; setup 288-94; state 288
intangible stratus 8
Integrated Water Resources Management (IWRM) 13, 50, 147, 286, 291, 306; Costa Rica 412; rethinking 385-417
intensification trap 16, 75
intensive goods exports 125
Intensive Trade Margin 130, 130n, **131**
Inter-American Development Bank (IDB) 223, 225, 231, 289, 309, 345, 372
interdependencies 120, 388
Intergovernmental Panel on Climate Change (IPCC) 48, 110
intermittent service 229
international agreements 322
International Covenant on Economic, Social and Cultural Rights (ICESCR) 299n
international food trade flows 10
International Monetary Fund (IMF) 64, 304
international trade 14-15, 22, 120, 126, 160; policy 9
international treaties 46, 101, 322, 325
International Union for Nature Conservation (IUNC) 69
International Water Association (IWA) 225
Internet 106, 333, 337, *337*
investment 305; agreements 325; direct 12; private 233, 287, 310, **310**, 311, **311**; public 63, 154, **157**, 307, 309, **310**; R&D 108
inward foreign direct investment 120, 124, **124**
irrigated agriculture 196, 200, 348-9
irrigation 19, 120, 134, 181-2, 247, 267, 270-1, 395; Brazil 271, 349; charges 348-9; Chile 271; development 70; efficiency 36, **156**, 262, **271**, 279-80, 389; fees 348-9; groundwater 182; impact *207*; improvements 206; investment 361; methods 244; modernization 278-9; potential 179-80; prices 361; subsidies 138; surface water 182; tariffs 345, 348; techniques 182, *183*; technology 266; wastewater 16; water use 182

Jacobi, P.R. 412-13; *et al* 285-315
Jamaica: sprinkler irrigation 182
Joint Monitoring Programme (WHO-UNICEF) 300, 300n
Josefa Fioriti, M. 413
Jouravlev, A. 231

Kuhn, T.S. 387
Kuroiwa, J. 413; *et al* 34

La Niña 30, 31, 48, 50, 110, 147, 154
lakes 29, 388; pollution 41
Lambin, E.F.: and Meyfroidt, P. 62, 63
land: accounting 200; expansion 134; grabbing 133-4; resources 129; sharing 56; sparing 56,

75; stress 134; subsidence 40, 49; tenure 377-9; titling 378, 381
land use **58**, 410; patterns 49; trends 55-80
land use changes 11, **61**, 193; and ecosystem services 67-74
Latin America: global context 4-5; study (2030) 106, **107**
Latin America and the Caribbean (LAC) 5; population structure 22
Latin American Integration Association (ALADI) 101
Latin American Water Funds Partnership 372
Latin American Water Tribunal (LAWT) 303, 325
laws 21, 286, 294; implementation 297; reforms 330
legal frameworks 33, 147, 158, 163, 366, *376*
legal nature of water 294
legislation 344; implementation 336-7
Lentini, E.: and Ferro, G. 353
levies 345, 346; Mexico *346*
liberalization 126, 129, 135, 136
life expectancy 94, 104, 219
Lima 229; SEDAPAL 220; urban water 220; water prices 220
literacy 5
livestock production 186, 196, 209; and deforestation 65; expansion 11; water footprint **275**
Llamas, R. 413
loans 309
lobbying 318, 320, 330
local economies 37, 217
López-Gunn, E. 81-118; *et al* 385-417

macro food security 162
maize 71, 74, 76, 103, 163, *163*, 178, 186, 188, 191, 200, 203, 247
Mäkinen, K.: and Carter, T.R. 110
malaria 110, 114, 151, 226
malnourishment 165
malnutrition 4, 18, 82, 102, 104, 144, 180
managers: training 232
Mar del Plata: water security 217-18
market allocation systems 356
marketable permit programmes 353
markets: agricultural commodities 120, 178; biodiversity 373; capital 308; Chile water 356-60; institutions 288; water 356-60
Marti, J.: *et al* 217
Martinez-Santos, P. 27-53
material stratus 8
Mather, A.S. 60
Mazahua women 324
meadows 58
meat: demand 132; emerging countries 131-2; sales 126
megacities 6, 87, 113, 158, 214
megatrends 4, 10, 121; socio-economic 12, 81-118

Mekonnen, M.M.: and Hoekstra, A.Y. 189, 192, 245, 275, 276
MERCOSUR 90, 101
mercury 253, 256
Mesoamerica 16, **49**, 56, 65, *111*, 112, 154, 202, 398
metal sector 249
meters 221
Mexico: agricultural productivity 203, **204**, **205**; arable land *201*; blue water 153; blue water exports **194**; blue water footprint **183**, **185**; blue water imports **196**; blue water scarcity 153, **153**; Build-Operate-Transfer projects 308; charges 346; decentralization 291; drinking water 411; education enrolment **109**; exports 14, 101, 125; food consumption 102-3, **103**; food security indicators *161*; GDP growth **91**; green water exports **193**; green water footprint **185**; green water imports **195**; Greenhouse Gas Emissions (GHG) **70**; grey water footprint **190**, 274; groundwater mining 39; groundwater use 38; Guanajuato State 297, **298**; Human Development Index (HDI) *402*; hunger *162*; industrial water use *250*; inequality **99**; irrigated agriculture 348-9; irrigation **156**; levies *346*; management units 46; Mazahua women 324; Millennium Development Goals (MDGs) *17*; mining 251-2, *252*, **323**; National Communication reports (NCs) 47n; National Water Commission 348; National Water Law (1992) 45, 290, 348; obesity **164**; Payment for Ecosystem Services (PES) *369*, *376*; population **89**; Santiago River 324; trade **130**, **131**, **136**, **137**; urban water 221; users associations (WUAs) 349; virtual water flows *394*; water footprint **187**, 191, **274**, **275**, *394*; water management 45; water quality policies 45; water reforms 292; water register 391; water security indicators *155*; water use productivity 243; water withdrawals **38**, **93**; yield growth *202*
Mexico City 229; Integrated Water Resources Management Program (PGIRH) 221; urban water 221
Meyfroidt, P.: and Lambin, E.F. 62, 63
micro food security 162
middle-class: definition 132
migrants: environmental 112; Latin American 88; rights 90
migration 6, 82, 83, 84, 88-90, 110, 391; and climate change 112; internal 89; international 85, 88-9; rural-urban 61, 64
Millennium Development Goals (MDGs) 4, 16, *17*, 113, 145, 223, 224, 264, 300, 345, 360; policies 301
Mind the Gaps: Bridge the Gaps (OECD) 403
minerals: prices 242
mining 7, 10, 12, 19-20, 34, 37, 42, 64, 126, 208, 241, 242, 255, 322; conflict **323**; and deforestation 65; dust control 252; exports 14; gold 20; groundwater 39, 254; permits 253; pollution 240, 318; and water 251-4; water consumption 252-3, *252*; water quality 256; and water security 256
mobile phones 106
monopolies 294, 305, 349
Montserrat: crop yields *202*; imports and exports **127**
mortality 104; child 4, 105, 226, **226**; mother-child 105
multi-level governance 403-5, **405**
multilateral development banks 307-8
Muradian, R.: *et al* 368

Naim, M. 4, 4n, 5
National Communications (NCs) 110
National Institute for Space Research (INPE) 58
National Intelligence Council (NIC) 4, 121
natural capital 15-16, 56, 57, 366, 367, 374, 382
natural disasters 148, 409
natural dividend 5, 7
natural hazards 112, **112**, 157, **157**
natural resources 126, 208, 264, 379; management 279
Nature Conservancy 372
navigation 40
net virtual water import: definition 191-2
network water losses 218, **228**
networks: activism 324; advocacy 20, 318; water *332*
New Alliance for Food Security and Nutrition 150
Nicaragua: animal protein 159; crop yields 74; deforestation 59, 65, 65; economic water productivity 203; geothermal electricity 246; groundwater 182; information 334; pollution 42; renewable energy 398; sanitation 154; tariffs 129, 135; undernourishment 160; water footprint 186; Water Poverty Index (WPI) 97; water quality 38; water withdrawals **38**
nitrates 222, 244
nitrogen **44**, 191; pollution 189, 272, 276
non-food water 5, 6, 7, 8, 20, 146; cities 22
non-governmental organizations (NGOs) 288, 303, 318, 320, 323, 327, 329, 330, 331, 380
non-timber forest products (NTFP) 69, 71
North America Free Trade Agreement (NFTA) 101
nuclear energy 240
nuclear plants: Argentina 255
nutrition 151
nutrition transition 104-5, 170; Brazil 165-6

obesity 4, 18, 82, 102, 104, 105, 144, 145, 164-5, 170; children 105, 165
Official Development Aid (ODA) 124, **124**, 154,

309, 309n, 310
oil 255; extraction 253-4; prices 133; refineries 249
oilseeds 64, 126
open defecation 145, 154
operational framework 148, **149**
operational principles 293
oranges **185**, **190**, *202*, **204**, **205**, **206**
organic agriculture 75-6
Organization of American States (OAS) 129
Organization for Economic Cooperation and Development (OECD) 231, 308, 308n, 402, 406; *Mind the Gaps: Bridge the Gaps* 403
Organization of the Petroleum Exporting Countries (OPEC) 132-3
Other Official Flows (OOF) 154, 309
outputs 15, 129, 179, 206, 207, 269, 278
overweight children 164-5, **164**
OXFAM 150

Pacific Alliance 5
Pacific basin 34
palm oil 249
Pan-American Health Organization (PAHO) 102
Panama: arable land 16; Millennium Development Goals (MDGs) 16; sprinkler irrigation 182; Water Poverty Index (WPI) 97
paper industry 71-2
Paraguay: arable land 60; biofuels 127; Conservation Trust 373; Constitution 373; deforestation 64, 197; emigration 89; exports 163; green water use 154; grey water footprint 191; groundwater 29; hunger 104, 162; hydroelectricity 255; informal employment 100; Maize 71; navigation 40; protein 159; soya exports 163; Water Poverty Index (WPI) 97
participation 327; social 292; societal 292
participatory forums 328-9
pastures 58, 60, 65, 131, 197
payments: stacked 373
payments for ecosystem services (PES) 21-2, 344, 345, 360, 366, 367, 368-73, *369*, *371*, *377*, *378*, *379*; Costa Rica 369, 373; definition 368; Ecuador 369; government coordination 381; implementation 380-2; institutional arrangements 379-80; laws 373-5, *376*, *377*; legislation 375; private sector 380; public funds 381
payments for environmental services 355-6, *356*, *357*
payments for watershed services (PWS) 344, 356, 370
per capita GDP 113
per capita growth 82, 90
per capita income 144, 169, 170
per capita water consumption: urban 49
Perez-Espejo, R.: *et al* 81-118
Perry, C.: *et al* 266

Peru: agricultural productivity 203, **204**, **205**; arable land *201*; blue water 153; blue water exports **194**; blue water footprint **183**, **185**; blue water imports **185**; blue water scarcity 153, **153**; Chancay-Lambayeque irrigation system 349; decentralization 291; desalination 34; education enrolment **109**; fees 347; food security indicators *161*; GDP growth **91**; General Water Law 289; green water exports **193**; green water footprint **185**; green water imports **195**; Greenhouse Gas Emissions (GHG) **70**; grey water footprint **190**, *274*; groundwater mining 39; Human Development Index (HDI) *402*; hunger *162*; inequality **99**; irrigation **156**; Juntas de Usuarios y Comités 327; Law of Water Resources (2009) 347; Lima 220, 352; Madre de Dios river 297; Millennium Development Goals (MDGs) *17*; mining 251-2, *252*, **323**; National Communication reports (NCs) 47n; National Mining, Energy and Petroleum Association 330; National Water Authority 291; obesity **164**; Payment for Ecosystem Services (PES) *369*, *376*; population **89**; potatoes 186, 188-9; poverty reduction 97; rainfall 34; river basin councils 405-6; tariffs 352; trade **130**, **131**, **136**, **137**; urban water 220; virtual water flows *394*; Water Act (2009) 289; water availability *34*; water councils 410; water footprint **187**, 191, **274**, **275**, *394*; water policy reform 289; Water Poverty Index (WPI) 97; Water Resources Law (2009) 330; water scarcity 411; water security indicators *155*; water withdrawals **93**; yield growth *202*
pesticides 16, 42, 189
petrochemicals 249
phosphorous **44**
piped water: slums **87**
plantain 209
policy tools 126, 130, 147
political decision makers 232
pollination 57
pollutants 12, 41, 45, 180, 230
polluter-pays-principle 21, 344, 345, 360
pollution 7, 16, 19, 40-1, 158, 189, 219, 251, 262, 268, 297, 319; charges 13, 360; Chile 42-3; control 28, 43, 412; diffuse source 41; indexes 209; and irrigation 42; lakes 41; management 20; metals 42-3; mining 240, 318; nitrogen 189, 272, 276; non-point source 43; point source 41, 43; reduction 354
population: ageing 104; urban 86, **86**; urban-rural change **89**
population distribution 34, **153**
population growth 19, 28, 49, 57, 70, 82, 83-4, 85, **92**, 165, 262, 391; Chile 359
Porce River Basin: water footprint 209
ports 122-3, *123*
positive environmental externalities 368

potatoes 200, 203, 209; Peru 186, 188-9
poultry production 187-8
poverty 4, 18, 83, **96**, 136, 145, 179; alleviation 16; and trade **137**; and trade liberalization 135
Poverty and Hunger Index (PHI) 151
poverty line **96**, 97
poverty reduction 101, 120, 135, 206, 207; Argentina 96; Brazil 96; Chile 96-7; Colombia 97; Costa Rica 97; and income growth 95-9; Peru 97
power: decay 5; definition 4n; diffusion 4
Power Purchasing Parity 132
Preferential Trade Agreements (PTA) 129
prices: crisis (2007-9) 160, 162; food 151, 160-1; oil 133; volatility 6, 113, 133, 160, 161
pricing: policies 13, 351, *351*; water 138, 220, 234, 351, *351*
primary commodities 92
primary education *17*, 94, 95, 108, 410
primary goods 113
primary-based economy 65
private investment 233, 287, 310, **310**, 311, **311**
private sector 330, 331
privatization 20, 303n, 304, 305, 318, 322, 324, 349-50; Chile 350
pro-poor policies 353
PROAL-COOL 248
production: biofuels 241, 249; economic efficiency 276-7; food 179, 200, 240, 395, *see also* agricultural production
productivity: agricultural economic 203, **204**, **205**; agriculture 74; blue water 203, 268; total factor productivity (TFP) 92, *see also* water productivity
Programme for Sustainable Agriculture on the Hillsides of Central America (PASOLAC) 355
property rights 65, 277n, 377-9, 380
protein: animal 159
public authority 21, 293, 333
public good 232, 234
public health 50, 215
public institutions 231
public investment 63, 154, **157**, 307, 309, **310**
public management 233
public ownership 295
public participation 319, 325, 380, 404, 406, 411
public policies 83, 114, 209, 330, 331
public sector 407
public-private cooperation strategies 50
public-private partnerships 292
public-private urban water management 233
Puerto Rico 241; irrigation efficiency 271; urbanization 86
pulp 71-2, 73, 74

rain harvesting 388, 389
rain-fed agriculture 8, 15, 35, 120, 178, 179, 193, 200, 274, 279
rainfall 30, 39, 47, 207; annual **31**, **37**; Chile 32-3; extreme 48; Peru 34
rainwater 35
Reducing Emissions from Deforestation and Forest Degradation (REDD+) 56, 64, 67-8, 372
reforestation 60, 63; Brazil 61-3; Colombia 65
regional development 7, 8
regional opportunities 3-24
Regmi, A.: *et al* 102
regulation 305
regulatory frameworks 220, 231, 304, 349, 367, 375, 380, 381
renewable energy 247
renewable resources per capita **32**
Report on Migrations in the World (IOM) 88
research and development (R&D) 83; investment 108
reservoirs 40, 217, 245, 388
resource portfolios 390
resources: equitable access 138-9; natural 126, 208, 264, 279, 379
reusability 267
rice 186, 188, 189, 191, 200
rights: basic 7; environmental 374; groundwater 297; human 299; property 65, 277n, 377-9, 380, *see also* water rights
Rio de Janeiro: water security 219-20
Rio Grande river basin 197
river basin committees 318, 320; Argentina 46
River Basin Organizations 46, 322, 405
river basins 197, 209, 327, 410; Amazon 34, 197, 321; authorities 410; councils 327; management 306; management plans 406; plans 393; transboundary 46
rivers 29, 40, 154, 256, 322, 388; Amazon 29; pollution 251; quality 412; sustainability 359
runoff 29, *29*, 30, 33, 41, 49, 152, 306
rural development 114
rural population 62, 64, **95**, 112
rural-urban migration 61, 64

Saint Lucia **127**, *161*; virtual water exports **199**; water footprint **187**
Saint Vincent and the Grenadines: grey water footprint 191; imports and exports **127**
salaries 39, 251, 254
salinity 16, 20, 42, 219, 244, 246
sanitation 12-13, 16, 18-20, 34, 94, 144-5, 154, **156**, 166, **167**, **168**, 289; access 169; Brazil 411; and child mortality 226, **226**; Chile 35; networks 214, 217; right to 286, 298-305, **299**, **300**, **302**; rural 216; status 231
Santiago 217
Santos-Baca, A. 102-3
Sao Paulo!! 218-19
savannahs 57
savings: water 388

Scaling Up Nutrition movement 150
scenarios 83, 85, 106, **107**, 121
Schneider, F.G. 100
schooling rates 94
scientific progress 387
Scott, C.A. 387; *et al* 410
seasonal variability 30
seasonality 30, 197, 255, 389
seawater: desalination 241; intrusion 40
seed funding 380
Seoul: Multi-year Action Plan on Development 160
service seller 366, 368, 371, *377*
services: biodiversity 366; carbon 366; cultural
    73, 74; payments for watershed services (PWS)
    344, 356, 370; urban water 223-5; water
    214, 223, 224, 225-9, 230-4, *see also*
    payments for ecosystem services (PES); payments
    for environmental services
sewage 35, 43, 180; effluent 189; networks 220;
    treatment 215, 219; untreated 37
shale gas expansion 132
Shannon-Wiener index 69
shared waters 322
shrub areas: development 60
Silva, G.C. da 219-20
silver 20
slums: piped water **87**
Snow, J. 214
social activism 21, 318, 321-5
social actors 319n, 320, 328
social capital 356
social equity 353
social losses 219
social movements 324-5, 327, 329
social participation 327
social tariffs 352, 353
social welfare 349, 387
societal participation 292
socio-economic context 11-12
socio-economic development 165
socio-economic megatrends 12, 81-118
soda 103
soft drinks 230
soft-path solutions 147
soil: degradation 75; fertility 63; water 388, 393
soil moisture 5, 153, 179, 180, 262, 264, 265,
    269
Sommerville, M.M.: *et al* 368
South Cone 179, 395, **398**, **399**
Southern Common Market 101, 127
soya 126, 132, 163, 395
soybeans 71, 74, 76, 197, 247, 249
spaces of negotiation 318-20, 325, 330, 331
Spain 88
Spanish Fund for Water and Sanitation 303
stacked payments 373
stakeholders 290, 292, 303, 317-42,
    319n, 380, 409; engagement 370, 380;

organizations 319-21
state demand 220
state forest policies 63
state institutions 288
STEEP approach 93, 94
storage tanks: domestic 228-9; water 228-9
storms 48, 112
stunting children 166, **167**, **168**
sugar cane 71, 186, 191, 203, 209, 247, 248,
    248
sugar plants 255-6
sulphates 42
sunflowers 71, 247
supply networks 214, 215
surface water 37, 203, 388, 393; storage 37
Suriname: agricultural water use 179; animal
    protein 159; irrigation 348
sustainability 264; agricultural 36; aquifers 359;
    goals 7
sustainable consumption 262, 279
sustainable development 62
sustainable intensification 16
Swemmer, F.F. 233
swine production 187-8
Swiss Agency for Development and Cooperation
    (SDC) 355
systemic perspective 389

tanneries 249-50
tariffs 135, 287, 292, 306, 307, 345, 346-52;
    Brazil 352; levels and structures 350-2; Peru
    352; social 221, 352, 353; urban 360
taxation: environmental 21, 345
technical efficiency 264, 265, 269-72; agriculture
    270; basin scale 267; urban 280; urban and
    industrial 269-70
technological change 106; socio-economic impacts
    106-8
technology: irrigation 266; role 82; water
    conservation 265-6
tenure security: definition 379n
Thackray, J.E. 233
thermo-power plants 249
thermodynamics 245
thermoelectricity 245-6, *246*, 255
timber 69
Timmer, P. 161-2
Tittonell, P.A. 75
total actual renewable water resources (TARWR)
    152
total factor productivity (TFP) 92
total fertility rate (TFR) 85
trade 82; agreements 101, 126, **130**; and
    agricultural growth **136**; agriculture 121, *125*,
    135; and employment 136; flows 12; and GDP
    123; and globalization 10, 119-42; growth
    122; and income **137**; international 9, 14-15,
    22, 120, 126, 160; openness 135; partners

130; patterns 197; and poverty **137**; and poverty linkages 134-7; regulations 138; trends 125-7; and wages 136; and water access 133; and water use 133

trade liberalization 83, 101, *102*, 113; agriculture 136; policies 64; and poverty 135

trade-growth nexus 135

trade-offs 148, 154, 208, 397

traded goods: value 14

transboundary basins 46

transboundary resources 46, *47*

transboundary water resources management 28

transnational corporations (TNCs) 331

transparency 231, 291, 380, 411; laws 321, 333-5, *335*; portals 334; urban water management 234

transportation: expansion 122-3; water 222, 243

treaties 101, 325; international 46, 101, 322, 325

treatment plants 220, 249

tree felling 64

trihalomethane 227

Trinidad and Tobago: grey water footprint 191; imports and exports **127**; virtual water dependency 186; virtual water imports 192

triple-M revolution 4

trust funds 371-2

UN-Water 146

UNASUR (Union of South American Nations) 90, 101

undernourishment 4-5, 16, 18, 166, **167**, **168**, 169, 170; children 164, **164**; Guatemala 160

Union of South American Nations 101

United Nations Children's Fund (UNICEF): -WHO Joint Monitoring Programme 300, 300n

United Nations Educational, Scientific and Cultural Organization (UNESCO) 73

United Nations Framework Convention on Climate Change (UNFCCC) 67, 68, 110, 372; Conference of Parties (COP) 68; Subsidiary Body for Scientific and Technological Advice (SBSTA) 68

United Nations (UN): Department of Economic and Social Affairs (DESA) 88; Development Program (DP) 93; Food and Agriculture Organization (FAO) 149-50, *377*, 378n; General Assembly (GA) 298, 299; Human Rights Council (HRC) 299; International Covenant on Economic, Social and Cultural Rights (ICESCR) 299n

United States of America (USA): biofuels 132-3; Environmental Protection Agency (EPA) 181, 224, 227; immigrants 88

unsustainability 224, 225

urban growth 215

urban population 86, **86**

urban supply 19, 34, 158, 214, 216, 220, 222, 276

urban water 262; cycle 19-20, 214; services 223-5; sources 216, 222

urban water management: problems 270; transparency 234

urban-rural population change **89**

urbanization 6, 13, 49, 62, 70, 83, 84, 86-7, 165, 367; patterns 18, 28; rapid 22, 262

Uruguay 64, 263, 303; agricultural water use 179; animal protein 159; blue water 186; Constitution 301; economic water productivity 203; forestry 64; groundwater 29; hunger 104; land grabbing 133; malnutrition 18; obesity 165; paper industry 71; pulp 73, 74; urbanization 86; virtual water exports 192; water access 303; Water Poverty Index (WPI) 97

vegetables 196

Venezuela 89, 222; Constitution 301; deforestation 58, 59; federalism 46; grey water footprint 273; hunger 104; hydroelectricity 397; maize 71; migration 89; mining 256; obesity 105, 164; pollution 42, 189; population growth 49; sprinkler irrigation 182; urbanization 86; virtual water imports 192; water asymmetry 34

virtual water 386, 387, 388, 393; exports 15; import dependency definition 186; trade 12, 15, 138, 178, 208

virtual water flows *394*; agricultural products 191-9, **192**, **193**, **194**, **195**, **196**, **197**, **198**, **199**

voluntary transactions 355, 368

vulnerability: definition 110

vulnerability assessment 110, *111*

Ward, F.A.: and Dagnino, M. 266

Washington Consensus 126

wastewater: discharges 12; and irrigation 16; management policies 7

wastewater treatment 19, 43, 46, 180, 215; Costa Rica 411; industrial 223; investment 154; lack of 229; urban 223

water: abundant 129; and food security 5-6, 166-9; frameworks 381; legal nature 294; quantity 181-9, 222

water access 154, 206, 287, 319; and trade 133

water accounting 181-99, 208, 280, 388, 389; techniques 10; tools 332

Water Acts 289, 294, **295**, 297

water allocation 329; and water markets 356-60

water availability 18, 28, 29-35, 278; Chile 33; per capita **32**, 70, 129, 31; Peru *34*; and population density 32; and population distribution 34

Water Basin Committees 292

water certification 320, 330-1

water conflicts *321*

water conveyance systems 255

water councils 21, 318, 320

water cycle: urban 19-20, 214
water flow: regulation 57
water footprint 35, 178, 180, 330-1, *394*;
agricultural consumption 180, 181; agricultural
production 180-1; by continent 29; crop
production **274**; domestic water 273, **273**;
external 181; indicators 12, 393; industrial
production **273**, 274; internal 181; livestock
production **275**; poultry production 187-8;
production **272**; swine production 187-8, *see
also* blue water footprint; green water footprint;
grey water footprint
Water Footprint Assessment (WFA) 209
Water Framework Directive (EU, WFD) 25, 148,
234
water governance *see* governance
water hazards 6, 18, **112**, 144, 145, 146, *147*,
148, 157
water losses 219, 227, **228**
water management 12, 147, 288, 289;
decentralization 289-90, 291; institutions *406*;
ministries 405, *406*; privatization 290; rural
113; stakeholders 317-42; sustainable 178,
208, 223; urban 216
water ownership 13, 294-6, *296*, 305, 335
water policy 21, 186, 286, 291; challenges
293; Costa Rica 407-8, **408**; decentralization
403; enforcement 404; financing 306, 307;
instruments 280; integration 405; models 406-7,
407; national 231
Water Pollution Level (WPL) 43; nitrogen **44**;
phosphorous **44**
water poverty index (WPI) 97, **98**
water pricing 138; Lima 220; policies 351, *351*;
political 234
water productivity 264, 265; evaluation 268;
improvement 272; and production 272-6
water quality 28, 38, 41-6, 50, 180, 189-91,
222; agriculture 42; Brazil 411; control 226-7;
management 43, 46; Mexico 45; mining 256;
standards 146
water quality trading (WQT) 367[0]
water reforms 320, 326, 326n; implementation
292-4; Mexico 292; triggers and trends 288-92
water resources: assessment 27-53; commissions
46; management 306, **413**; ownership 294-5,
296, *296*; renewable *30*
water rights 21, 158, 206, 216, 242, 286,
289, 294-6, 298-305, **299**, **300**, **302**, 324,
356, 381, 411; Chile 357, 358, 359, *359*;
conditions 296-7
water risk 20, 148, 409
water saving technologies 265, 266
water scarcity 29, 31, 393; Peru 411, *411*
water sector: financing 306-11, **306**
water security 108, 143-73, 412-13; cities 213-
37; concepts and metrics 146-8; Costa Rica
158; definition 146-7, 410; goals 148, 158;

government priorities 158; hard-path solutions
147; improving 360-1; indicators 12, 166,
169; indicators and operational frameworks
148, **149**; and mining 256; performance 152-
8, **158**; progress *155*; soft-path solutions 147
water services 214; cover 223; economic
unsustainability 224; poor quality 225-9;
problems 230-4; standards 233
water supply: low pressure 228-9
water trust funds *369*, 371
water use 5, 35-41, **37**, 241; agriculture 270-2;
consumed fraction 267; efficiency evaluation
266-9; non-consumed fraction 267; productivity
(Mexico) 243; and trade 133
water user associations *47*, *326*
Water War 324
water withdrawals 35, **38**
water-electricity nexus 397-9, **399**
water-food-energy nexus 4
waterfalls 395
watershed: committees 21; conservation 360;
contamination 75; initiatives 355; management
292
watershed payments: Colombia 381; ecosystem
services 370, **370**
waterways 40, 360
wealth: distribution 97
well-being 6-7, 12, 74, **84**, 93-5, 145, 146, 147
wells 217, 218, 219; drilling bans 298
wheat *76*, 103, 163, *163*, 188, 200, 203
Wilbanks, T.J.: *et al* 112
Willaarts, B.: De Stefano, L. and Garrido, A. 3-24;
et al 55-80, 143-73, 365-84
wind 16, *47*, **396**, 398, **398**
wood 74
World Bank 64, 68, 160, 289, 304, 308, 308n,
355
World Health Organization (WHO) 226; -UNICEF;
Joint Monitoring Programme 300, 300n
World Heritage Sites (WHS) 69, 73, 367
World Summit for Sustainable Development (2002)
223
world trade 6
World Trade Organization (WTO) 15, 138, 139,
160
World Water Council 387
Worldwide Trade Partners 129
Wunder, S. 355, 368

yield gaps 200
yields 74-6, 134, 154, 200, *202*, 203, **204**,
205, 206, 207, 249, 268; growth 207

Zarate, E.: *et al* 177-212

Printed and bound by CPI Group (UK) Ltd, Croydon, CR0 4YY

22/10/2024

01777639-0004